Molecular Targets and Therapeutic Interventions against Neurodegenerative Diseases

This book comprehensively explores the latest advancements in the understanding, diagnosis, and treatment of neurodegenerative diseases. The chapters provide an in-depth review of current approaches and treatment strategies for Alzheimer's disease, offering insights into the latest developments and breakthroughs. It also reviews the cutting-edge research on potential novel targets for Alzheimer's pharmacotherapy, with a focus on JNK3, GSK3β, and Fyn kinase inhibitors, providing an update on related approaches and their implications. It discusses the potential of ethnomedicines as a promising tool for mitigating Alzheimer's disease, offering insights into traditional remedies and their modern applications. Additional chapters explore the influence of microglia, the neuropharmacological mechanisms associated with SARS-CoV-2, and the molecular intricacies of Parkinson's disease. The book further covers the evolving role of artificial intelligence and machine learning in the management of neurodegenerative disorders. The chapters also examine the role of nanotechnology in addressing the challenges of diseases like multiple sclerosis. Toward the end, the book examines the role of oxidative damage in neurodegeneration and its management in related disorders. This book is an important source for neuroscientists, neurologists, and students of neuroscience.

Molecular Targets and Therapeutic Interventions against Neurodegenerative Diseases

Edited by
Vaishali Manikrao Patil, Dileep Kumar,
and Neeraj Masand

CRC Press
Taylor & Francis Group
Boca Raton London New York

CRC Press is an imprint of the
Taylor & Francis Group, an **informa** business

Designed cover image: Shutterstock

First edition published 2025
by CRC Press
2385 NW Executive Center Drive, Suite 320, Boca Raton FL 33431

and by CRC Press
4 Park Square, Milton Park, Abingdon, Oxon, OX14 4RN

CRC Press is an imprint of Taylor & Francis Group, LLC

ISBN: 9781032828046 (hbk)
ISBN: 9781032828060 (pbk)
ISBN: 9781003506317 (ebk)

DOI: 10.1201/9781003506317

Typeset in Times
by Deanta Global Publishing Services, Chennai, India

Contents

Preface

Neurodegenerative diseases, namely, Alzheimer's, Parkinson's, Huntington's and multiple sclerosis, are multifactorial debilitating diseases associated with the nervous system. Neurodegenerative diseases affect millions of individuals worldwide, and with the increasing population of older individuals, these diseases are becoming ubiquitous. The multifactorial nature of these neurodegenerative diseases is supported by various evidence such as abnormal protein dynamics causing defective protein degradation and aggregation, oxidative stress, compromised bioenergetics, mitochondrial dysfunction, impaired phosphorylation, and toxicity due to, for example, metals/pesticides. Studies have extensively focused on understanding the pathophysiology of the associated proteinopathies, but still clarity is elusive for their use as potential targets for therapeutic interventions. Despite technological developments in the field of drug discovery, there is a substantial gap between design and implementation of potential therapeutics for neurodegenerative diseases. The primary objective of this book, titled *Molecular Targets and Therapeutic Interventions against Neurodegenerative Diseases*, is to present a comprehensive approach based on molecular, preclinical, clinical and translational research, and simultaneously address the associated limitations along with recent advances. The area of neurodegenerative diseases is broad and multidisciplinary, and providing full coverage would be beyond the book's purview. However, the book sheds light on the novel mechanistic pathways, which will undoubtedly be thoroughly studied and established in the years to come. It is gratifying to receive 15 exceptional contributions authored by a team of prominent experts focusing on various therapeutic targets and approaches related to neurodegenerative diseases that skillfully compose the book. Each chapter attempts to address thought-provoking and challenging ideas that could interest researchers from multidisciplinary areas of drug discovery.

In the first chapter, titled "Comprehensive Review on Alzheimer's Disease: Recent Approaches and Treatment Strategies," Nandini et al. elaborate on the current and emerging pharmacological targets, along with coverage of various reported inhibitors for the treatment of Alzheimer's, which is a progressive neurodegenerative disease. Chapter 2 by Singh and Krishnamurthy elaborates on molecular targets reported for Alzheimer's Disease, Parkinson's disease, Huntington's disease and amyotrophic lateral sclerosis, and associated challenges in translational research. Chapter 3 by Jahangir and Akhter, titled "Potential Novel Targets for Alzheimer's Pharmacotherapy: Update on JNK3, GSK3β and Fyn Kinase Inhibitors and Related Approaches," provides insights into the role of kinases in the development of anti-Alzheimer's therapeutics.

Chapter 4 by Shareena et al. highlights the role of the triggering receptor expressed on myeloid cells 2 (TREM2) in inflammatory diseases driven by amyloid plaque. It covers the structure, function and supporting studies describing the supportive role of its genetic variants toward an elevated risk of Alzheimer's. TREM2 protein has the potential for use as a biomarker for early identification and diagnosis among Alzheimer's patients. Among the various reported therapeutic targets for Alzheimer's, the phosphodiesterase enzyme has been identified as a potential target. Chapter 5 elaborates the role of phosphodiesterase (PDE) and its isoforms in the development of safe and potential therapeutics. A large number of heterocyclic nuclei have been assessed for PDE inhibitory potential using preclinical models. Some of them have entered the clinical phase of development, such as BI-409306, EHT0202 and Roflumilast. The anti-Alzheimer's mechanism based on the NMDA receptor hypothesis is discussed by Shareena et al. in Chapter 6.

Chapter 7 by Wal et al. deals with the advances based on artificial intelligence and machine learning approaches for the discovery of therapeutics for the management of various neurodegenerative diseases. In Chapter 8, Karati and Kumar discuss the role of microglia in neurodevelopmental and neurodegenerative disorders along with a brief overview of related drug targets. Chapter 9 by Kaushik et al. discusses the neuropathological aspects of acute and chronic SARS-CoV-2 infection among different age groups and concludes its impact on associated post-COVID consequences.

Chapter 10 by Kumar et al. focuses on the target prioritizing approach along with the significance of the pyrazole scaffold, its derivatives and structure-activity relationship in Alzheimer's therapy. Chapter 11 by Mishra et al. elaborates on the role and therapeutic perspectives of oxidative stress in neurodegenerative diseases. Pal et al. describe the role of various herbs and natural products in neurodegenerative pathologies, mainly Alzheimer's and Parkinson's disease, in Chapter 12.

Further, Chapter 13 by Sharma and Sharma focuses on the role of nanotechnology in multiple sclerosis. Applications of nanocarriers as therapeutic and diagnostic tools are elaborated in this chapter. Along the same line, Chapter 14 by Gaur et al. aims to provide therapeutic advances for the treatment of neurodegenerative diseases with a major focus on nanotechnology-based delivery systems for the central nervous system. Additionally, Yadav et al. elaborate on nanotechnology-based therapeutics, particularly for the treatment of Alzheimer's disease, in Chapter 15.

In summary, this book provides broader coverage of various molecular targets related to neurodegenerative diseases and interprets their potential for the development of therapeutics. The chapters are interdisciplinary and have attempted to cover results derived from preclinical and clinical phases.

We would like to acknowledge all authors for their valuable contributions and cooperation during the editorial process. The editors and publication teams appreciate the researchers who collaborated on peer assessment and editing.

About the Editors

Vaishali Manikaro Patil holds the position of Principal/Head at Charak School of Pharmacy, Chaudhary Charan Singh University in Meerut, India, and also serves as a Visiting Professor at the Faculty of Pharmaceutical Sciences, UCSI University in Kuala Lumpur, Malaysia. With over 17 years of academic and 3 years of industrial experience, her primary research lies in medicinal chemistry and computer-aided drug design, focusing on quantitative structure–activity relationship studies, drug–receptor interactions, and the synthesis of small heterocyclic molecules. Dr Patil, a recipient of prestigious awards, has authored numerous research papers, books, and book chapters, contributing globally through lectures and participation in scientific communities including the World Health Organization, American Chemical Society, and APTI-India.

Dileep Kumar, holding degrees from Manipal University and S.G.S.I.T.S Indore, serves as an Assistant Professor at Poona College of Pharmacy and also has a visiting faculty position at the University of California, Davis. With research interests in the design of GluN2B selective NMDA receptor antagonists and amyloid-β protein aggregation inhibitors for Alzheimer's disease treatment, Dr Kumar actively contributes to cancer research at UC Davis Comprehensive Cancer Center.

Neeraj Masand, Assistant Professor at Lala Lajpat Rai Memorial Medical College in Meerut, earned degrees from North Maharashtra University and Dr A. P. J. Abdul Kalam Technical University, Lucknow. With over 16 years of academic and research experience, his current research focuses on designing gamma-secretase inhibitors for Alzheimer's disease treatment and synthesizing novel heterocyclic analogues. Dr Masand has authored numerous research papers, books, and book chapters.

Contributors

Akshay Kumar
Department of Pharmacy
Guru Ghasidas Vishwavidyalaya (A Central University)
Bilaspur, Chhattisgarh, India

Anand Maurya
Department of Medicinal Chemistry, Faculty of Ayurveda
Institute of Medical Sciences, Banaras Hindu University
Varanasi, Uttar Pradesh, India

Anjali Murmu
Department of Pharmacy
Guru Ghasidas Vishwavidyalaya (A Central University)
Bilaspur, Chhattisgarh, India

Anjana Sharma
Lloyd Institute of Management and Technology (Pharmacy)
Greater Noida, Uttar Pradesh, India

Ankita Wal
PSIT–Pranaveer Singh Institute of Technology–Pharmacy
Kanpur, Uttar Pradesh, India

Anupam K. Sachan
Dayanand Dinanath College of Pharmacy
Kanpur, Uttar Pradesh, India

Balaji Wamanrao Matore
Department of Pharmacy
Guru Ghasidas Vishwavidyalaya (A Central University)
Bilaspur, Chhattisgarh, India

Bommaiah Prakash Kumar
Sri Adichunchagiri College of Pharmacy, University
Madhya Pradesh, India

Deepti Katiyar
KIET School of Pharmacy
KIET Group of Institutions
Ghaziabad, Uttar Pradesh, India

Dileep Kumar
Department of Pharm. Chemistry
Manipal College of Pharmaceutical Sciences
Manipal Academy of Higher Education
Manipal, Karnataka, India
and
UC Davis Comprehensive Cancer Center
Department of Entomology and Nematology, University of California
Davis, CA, USA

Dipanjan Karati
Department of Pharmaceutical Technology, School of Pharmacy
Techno India University
Kolkata, West Bengal, India

Erwin Faller
Graduate School, Centro Escolar University
Manila, Philippines
and
School of Allied Health Sciences
Pharmacy Department
Davao City, Philippines

Gadde Shareena
Poona College of Pharmacy, Department of Pharmaceutical Chemistry
Bharati Vidyapeeth (Deemed to be University)
Pune, Maharashtra, India

Harsh Yadav
Department of Pharmacy
Indira Gandhi National Tribal University
Amarkantak, Madhya Pradesh, India

Himangi Vig
PSIT–Pranveer Singh Institute of Technology–Pharmacy
Kanpur, Uttar Pradesh, India

Jagdish Singh
Department of Pharmacy
Guru Ghasidas Vishwavidyalaya (A Central University)
Bilaspur, Chhattisgarh, India

Kanak Lata
Metro College of Health Science & Research
Greater Noida, Uttar Pradesh, India

Madhu Tanya Singh
Department of Pharmacology, JSS College of Pharmacy
JSS Academy of Higher Education and Research
The Nilgiris, Tamil Nadu, India

Manisha Singh
Department of Pharmacy, Physiopharmacotherapia Centre of Excellence
School of Medical and Allied Sciences
K.R. Mangalam University
Gurugram, Haryana, India

Mangla Nand Singh
Dr B. R. Ambedkar Pooja College of Pharmacy (Affiliated to Dr APJAK Technical University, Lucknow)
Gorakhpur, Uttar Pradesh, India

Manoj Kumar Katul
University College of Pharmacy
Guru Kashi University
Bhatinda, Punjab, India

Manukonda Upender Kumar
University College of Pharmaceutical Sciences, Kakatiya University
Warangal, Telangana, India

Mariya Jahangir
Department of Pharmaceutical Chemistry, SPER
Jamia Hamdard, New Delhi, India

Moumita Barman
I.T.S College of Pharmacy
Ghaziabad, Uttar Pradesh, India
and
MVNL Chaitanya, School of Pharmaceutical Sciences
Lovely Professional University
Phagwara, Punjab, India

Mymoona Akhter
Drug Design and Medicinal Chemistry Lab
School of Pharmaceutical Education and Research
Jamia Hamdard, New Delhi, India

Nandini H. S.
NPD-Pet Division, Natural Remedies Pvt. Ltd.
Bengaluru, Karnataka, India

Neeraj Masand
Department of Pharmacy
Lala Lajpat Rai Memorial Medical College
Meerut, Uttar Pradesh, India

Nitin Sharma
Department of Pharmaceutics, Amity Institute of Pharmacy
Amity University
Noida, Uttar Pradesh, India

Partha Pratim Roy
Department of Pharmacy
Guru Ghasidas Vishwavidyalaya (A Central University)
Bilaspur, Chhattisgarh, India

Pragya Gawande
Department of Pharmacy
Guru Ghasidas Vishwavidyalaya (A Central University)
Bilaspur, Chhattisgarh, India

Prashant Tiwari
Department of Pharmacology and Toxicology, College of Pharmaceutical Sciences
Dayananda Sagar University
Bangalore, Karnataka, India

Praveen Kumar Gaur
Metro College of Health Science & Research
Greater Noida, Uttar Pradesh

Praveen Thaggikuppe Krishnamurthy
Department of Pharmacology, JSS College of Pharmacy
JSS Academy of Higher Education and Research
The Nilgiris, Tamil Nadu, India

Preeti Kumari
Noida Institute of Engineering and Technology (Pharmacy Institute)
Greater Noida, Uttar Pradesh, India

Prem Shankar Mishra
Department of Pharmacy
Galgotias University
Greater Noida, Uttar Pradesh, India

Priyanka,
CT College of Pharmacy
Jalandhar, Punjab, India

Raju Singh
Metro College of Health Science & Research
Greater Noida, Uttar Pradesh, India

Rakhi Mishra
Noida Institute of Engineering and Technology (Pharmacy Institute)
Greater Noida, Uttar Pradesh, India

Ramu Singh
Department of Pharmacy
Indira Gandhi National Tribal University
Amarkantak, Madhya Pradesh, India

Rashmi Saxena Pal
School of Pharmaceutical Sciences
Lovely Professional University
Phagwara, Punjab, India

Rosaline Mishra
Metro College of Health Science & Research
Greater Noida, Uttar Pradesh, India

Sabyasachi Maiti
Department of Pharmacy
Indira Gandhi National Tribal University
Amarkantak, Madhya Pradesh, India

Saroj Verma
Department of Pharmacy
School of Medical and Allied Sciences, K.R. Mangalam University
Gurugram, Haryana, India

Sarvesh Paliwal
Department of Pharmacy, Banasthali Vidyapith
Jaipur, Rajasthan, India

Satheesh Kumar Sellamuthu
Karpagam College of Pharmacy
Coimbatore, Tamil Nadu, India

Satish Dubey
Laboratory of Molecular Taxonomy & Medicinal Plant Biology
Department of Botany, Guru Ghasidas Vishwavidyalaya
Bilaspur, Chhattisgarh, India

Sharma Vedika Vijaypal
CT College of Pharmacy
Jalandhar, Punjab, India

Shikha Kaushik
KIET School of Pharmacy
KIET Group of Institutions
Ghaziabad, Uttar Pradesh, India

Shubhas P
Department of Pharmacy, Indira Gandhi National Tribal University
Amarkantak, Madhya Pradesh, India

Sonika Srivastava
Department of Pharmacology, SGT College of Pharmacy
SGT University
Gurugram, Haryana, India

Souranava Jana
Department of Pharmacy, Guru Ghasidas Vishwavidyalaya (A Central University)
Bilaspur, Chhattisgarh, India
and
Haldia Institute of Pharmacy
Haldia, West Bengal, India

Swati Kaushik
Department of Pharmacy, Physiopharmacotherapia Centre of Excellence
School of Medical and Allied Sciences, K.R. Mangalam University
Gurugram, Haryana, India

Vaishali Manikrao Patil
Department of Pharmaceutical Chemistry, Charak School of Pharmacy
Chaudhary Charan Singh University
Meerut, Uttar Pradesh, India
and
Faculty of Pharmaceutical Sciences
UCSI University
Kuala Lumpur, Malaysia

Yogendra Pal
School of Pharmaceutical Sciences
RIMT University, Mandi Gobindgarh
Punjab, India

1 Comprehensive Review on Alzheimer's Disease

Recent Approaches and Treatment Strategies

Nandini H. S.[1], *Prashant Tiwari*[2*], *and Erwin Faller*[3]

[1]NPD-Pet Division, Natural Remedies Pvt. Ltd.,
Electric City, Bengaluru, Karnataka, India
[2]Department of Pharmacology and Toxicology, College of Pharmaceutical Sciences, Dayananda Sagar University, Bengaluru, India
[3]School of Allied Health, Pharmacy Department, San Pedro College, Davao City, Davao del Sur, Philippines
*Corresponding author

INTRODUCTION

The world's population is aging quickly, and dementia diagnoses are rising. Globally, 35 million individuals (about twice the population of New York) are estimated to have Alzheimer's disease (AD) or another type of dementia, and 65 million people are predicted to have dementia-related issues by 2030 (Prince et al., 2013). Dementia is a clinical illness associated with a steady decline in brain function that impairs the ability of the affected person to carry out daily tasks adequately (Gilman, 2010). Memory lapses or difficulty finding the correct phrases are frequently the first indicators of Alzheimer's disease. Language, thinking, decision-making, visuospatial function, attention, and orientation memory loss are some of the symptoms that progressively worsen over time. One of the most prevalent forms of dementia is AD (Plassman et al., 2007).

There is no known cause for AD, a progressive multifactorial neurodegenerative brain condition that has both modifiable and immutable risk factors. The most significant non-genetic risk factor is age (Ott et al., 1995; Querfurth and LaFerla, 2010). The nerve cells in the brain are harmed structurally and functionally. When a disease first manifests, it also affects the connection between nerve cells in brain circuits, which is crucial for memory and other cognitive activities (Selkoe, 2002). The dominant autosomal mutation in either the presenilin gene on chromosomes 1 and 14 or the amyloid precursor protein (APP) gene on chromosome 21 appears to be the cause of the genetic component of AD.

Additionally, there is a higher risk of early-onset AD among those with Down syndrome (trisomy 21). Despite the fact that the genetics of AD are more complicated and poorly understood, one known risk factor for the onset of sporadic AD is the epsilon four allele of the apolipoprotein E (APOE) gene, which is located on chromosome 19 (Reiman et al., 2005). Lower levels of vitamin D (1,25-dihydroxyvitamin D3) were found to be linked to all forms of dementia and AD, according to studies (Gezen-Ak, Yilmazer, and Dursun, 2014; Sommer et al., 2017). 1,25-D3, the active form of vitamin D, controls the expression of neurotrophins such as nerve growth factor, neurotrophin 3, and glial-derived neurotrophic factor as well as the survival, growth, and functionality of neural cells (Fernandes de Abreu, Eyles, and Féron, 2009; Annweiler et al., 2013). Vitamin D stimulates

DOI: 10.1201/9781003506317-1

macrophages in vitro, increasing the phagocytic clearance of amyloid plaques (Masoumi et al., 2009; Mizwicki et al., 2012). Numerous studies have shown that vitamin D may act as a neuroprotectant, and that 50 nmol/L is adequate in cases of dementia risk. The cost and design of randomized controlled trials to determine if vitamin D supplements can help older persons postpone or prevent the onset of dementia and AD could be improved with the use of this knowledge (Littlejohns et al., 2014).

CAUSES OF AD

Most dementia cases (between 60% and 70%) are caused by AD. It is a long-term neurodegenerative condition that often begins mildly and progresses over time. According to one idea, plaques prevent nerve cells in the brain from properly communicating. The process through which the cells obtain the nutrients they need may be hampered by tangles. It makes sense that as Alzheimer's progresses, more and more nerve cells, also known as neurons, are lost as the disease progresses.

Age: The single most important risk factor for the onset of AD is old age. Once you reach age 65, your chance of getting the ailment doubles every five years.

Down syndrome: Those who have Down syndrome have an increased risk of getting Alzheimer's. Because some people may get AD as a result of the genetic flaw that causes Down syndrome, amyloid plaques can accumulate in the brain over time.

Genetics: According to analyses of twin and family studies, the genetic heritability of AD (and its memory-related symptoms) ranges from 49% to 79%. With an onset before the age of 65, familial types of autosomal (not sex-linked) dominant inheritance make up about 0.1% of the cases. Early-onset familial AD is the term used to describe this type of illness. Even though it's uncommon, only a small fraction of people have AD before they are 65 years of age. APP, PSEN1, and PSEN2 are three genes that have been related to the development of AD as a result of mutations in them (Holtzman, Herz, and Bu, 2012).

ALZHEIMER'S DISEASE WITH LATE ONSET

Apolipoprotein E (APOE) is a gene related to AD, which often manifests after age 65. APOE has three different variants, among which the e4 variant raises the risk of AD. Other genes linked to AD include SORL1, CLU, PICALM, and CR1 (Holtzman, Herz, and Bu, 2012). Three genes that encode—PSEN1, PSEN2, and APP—can be mutated, and these mutations account for the majority of autosomal dominant familial AD cases.

APP (Amyloid Beta Precursor Protein)

The APP gene produces a cell surface receptor and a transmembrane precursor protein that secretases cleave into a number of peptides that, when released, bind to the acetyltransferase complex APBB1/TIP60 to promote transcriptional activation and create the protein structure of the amyloid plaques found in the brains of Alzheimer's patients. Chromosome 21 contains this gene. The etiology of AD is linked to the constitutive overexpression of soluble amyloids, which subsequently leads to the development of amyloid plaques. Due to its limitations and side effects, anti-amyloid therapy, which consists of the monoclonal antibodies solanezumab, crenezumab, and gantenerumab that target soluble and insoluble amyloid beta (Aβ) aggregates, has not been able to improve the clinical results of AD (Rygiel, 2016; Hardy and De Strooper, 2017).

APOE

The APOE gene produces a protein that binds to a particular receptor in the liver and other peripheral cells and is necessary for the proper breakdown of triglyceride-rich lipoprotein components. This gene, as well as the apolipoprotein C1 and C2 genes, is located on chromosome 19. Type III hyperlipoproteinemia (HLP III), caused by mutations in this gene, is characterized by elevated plasma cholesterol and triglycerides as a result of poor clearance of chylomicron and VLDL remnants.

Most mutations in the APP and presenilin genes result in an increase in the amount of Aβ-42, a tiny protein that makes up the majority of senile plaques. Without increasing the amounts of Aβ-42, some mutations change the ratio between Aβ-42 and the other major forms, particularly Aβ-40. This shows that presenilin mutations can cause disease even if they reduce the overall amount of Aβ produced and may indicate alternative functions for presenilin or a role for changes in the function of APP and/or its fragments aside from Aβ. The APP gene can be found in protective versions. These versions produce too much amyloid-beta peptide, a hazardous protein fragment, as a result of mutation. The tau protein malfunctions as these pieces aggregate into amyloid plaques in the brain. The tau protein fragments clump together to create neurofibrillary tangles, which lead to the death of brain cells and the onset of AD symptoms.

AD PATHOPHYSIOLOGY

The transentorhinal cortex is where tangles first begin to accumulate, followed by the CA1 region of the hippocampus, entorhinal cortex, and finally the cortical association areas, where the temporal, frontal, and parietal lobes are most severely impacted. Much more so than the quantity of amyloid-β plaques, the degree and location of tangle development correlate well with the severity of dementia (Kumar, Singh, and Ekavali, 2015).

The accumulation of tau proteins is inversely correlated with hippocampal atrophy, cognitive decline, and brain atrophy. AD causes atrophy of the frontotemporal cortex and loss of neurons, which results in inflammation and the deposition of amyloid plaques, abnormal clusters of protein fragments, and tangled bundles of fibers. As a result, the presence of monocytes and macrophages in the cerebral cortex increases, and the microglial cells in the parenchyma are activated (Mishra and Palanivelu, 2008). A summary of AD pathophysiology is shown in Figure 1.1.

Amyloid Beta and Hyperphosphorylated Tau Protein Hypothesis

The development of senile plaques (SPs), which are brought on by the buildup of Aβ, is one of the primary pathogenic characteristics of AD. The precursor protein of amyloid (APP) is broken into soluble tiny peptides called Aβ by the action of the enzymes β-secretase, α-secretase, and γ-secretase. Depending on the degree of oligomerization, the imbalance between Aβ synthesis and clearance results in protofibrils, fibrils, and plaques, three different forms of hazardous oligomers. Although the cause of the production of Aβ is yet unknown, the sequence, concentration, and conditions of Aβ's stability are crucial elements (Liu et al., 2017). Numerous variables, including mitochondrial dysfunctions/oxidative stress, amyloid/tau toxicity, and cholinergic dysfunction, are thought to contribute to the pathogenesis of AD (Mohamed, Shakeri, and Rao, 2016).

Oxidative Stress Hypothesis

Humans produce reactive nitrogen and oxygen species through a variety of healthy and unhealthy activities. These two species play a dual role because they both have advantageous roles in cellular signaling pathways and toxic actions that can destroy cellular structures (including the cell membrane, lipids, proteins, and DNA). The brain is especially susceptible to oxidative stress due to its high oxygen consumption, which uses 20% more oxygen than other mitochondrial respiratory tissues. The brain's fundamental functional unit, the neuron, is composed largely of polyunsaturated fatty acids. For oxidative stress injury, a decrease in glutathione in neurons is one of the reasons, as it reacts with reactive oxygen species and causes lipid peroxidation and molecular cell damage (Liu et al., 2017).

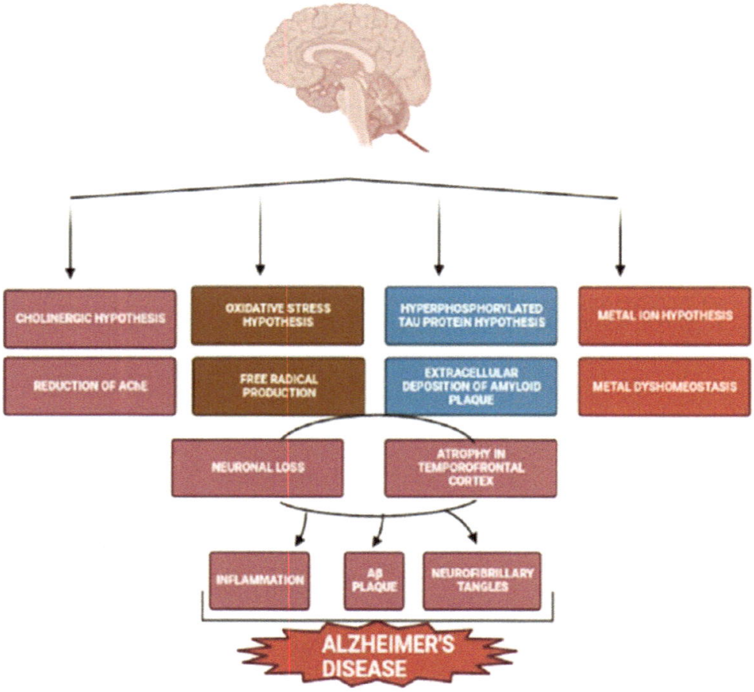

FIGURE 1.1 Pathophysiology of Alzheimer's disease.

Metal Ion Hypothesis

Metal dyshomeostasis has a role in the development and pathophysiology of illnesses, including cancer and neurodegenerative disorders. A number of these compounds are employed in clinical studies. Metal chelators and ionophores are the known modulators of metal homeostasis transition. Other medications besides those that target transition metal homeostasis are also effective (Weekley and He, 2017). According to current findings, the balance of redox transition metals, primarily copper, iron, and other trace metals, is changing. In AD and other neurological diseases, the brain levels of copper, manganese, aluminum, and zinc are altered (Prakash et al., 2017).

Cholinergic Hypothesis

Acetylcholinesterase inhibitors (AChEIs) are commonly used to alleviate symptoms of AD, and the apolipoprotein E (APOE) genotype significantly impacts their effectiveness, particularly influencing patient responsiveness and treatment outcomes. The mainstay of treating AD is AChEIs, and the most significant risk factor for AD is the APOE genotype. Given that cholinergic neurons are not the primary target of AD, this lack of significant effect of APOE is examined in relation to the 1976 "cholinergic hypothesis" of AD (Ashford, 2015).

With mild to moderate AD, cholinergic receptor binding is decreased in particular brain areas, which is associated with neuropsychiatric symptoms. Lower receptor binding may result in reduced processing speed in older, healthy persons. Cholinergic receptor binding *in vivo* may provide a possible molecular therapy target and disclose connections to other significant alterations in the brain related to aging and AD (Sultzer and Marder, 2017). Drugs that tend to normalize acetylcholine transmitter levels, such as cholinesterase inhibitors (ChEIs) and donepezil, have been the cornerstone of symptomatic therapy for AD for more than 20 years. Acetylcholine-mediated neurotransmission decline is due to an excessive loss of cholinergic neurons (Chase, Farlow, and Clarence-Smith, 2017).

AD BIOMARKERS

In the process of developing drugs for AD, biomarkers can be extremely helpful in identifying the best therapeutic candidates for costly phase 3 clinical trials. Along with a positive impact on the clinical course, biomarkers will be crucial for demonstrating how a treatment impacts the disease's underlying pathophysiology, which will be necessary for the medicine to be classified as having a disease-modifying effect, as shown in Figure 1.2A (Blennow, 2010).

GALANTAMINE

In recent years, the extraction of galantamine from a variety of Amaryllidaceae plants, including the genera *Leucojum*, *Narcissus*, and *Galanthus*, has proven to be a crucial treatment strategy for delaying the degradation of neurons in AD (Heinrich and Teoh, 2004; Annweiler et al., 2013). Long-term therapy with galantamine is well tolerated. Galantamine, marketed as Nivalin in Austria and Germany, has already received approval for AD and is used for other conditions such as facial neuralgia (Shu, 1998).

CANNABINOIDS

The cannabinoid system has emerged as a possible therapeutic strategy for treating AD. Numerous cannabinoid receptors (CBs), including the well-known CB1 and CB2 receptors, make up the cannabinoid system. Results show that activating CB1 and CB2 receptors with non-psychoactive doses of natural or synthetic agonists has positive effects in Alzheimer's experimental models by reducing the harmful effects of beta amyloid peptide action and tau phosphorylation, as well as by promoting the brain's natural repair mechanisms (Aso and Ferrer, 2014).

ANTIBODIES

Because it alters Aβ and Aβ-1 levels, intravenous immunoglobulin or pure Aβ antibodies may be recommended as a therapeutic treatment for AD (Dodel et al., 2002).

MTDL

In the multifaceted etiology of the disease, multi-target-directed ligand (MTDL) approaches have the potential approach to find novel AD medications. The majority of the new molecules have heterodimeric structures that enable them to interact with a variety of targets when combined with pharmacophores; they may be medicines or molecules with natural origins (e.g., tacrine, donepezil, galantamine, and memantine). Some of the medications discussed here seem to be potential treatments, while others could serve as valuable inspiration for the hunt for novel and potent treatments for AD (Guzior et al., 2015).

METAL ASSEMBLAGES

Long-lived fluorescent ruthenium and iridium metal complexes that significantly alter their electronic spectra when bound to Aβ fibrils may serve as effective probes to learn more about the molecular processes behind amyloid production and aggregation. Aβ peptides of different lengths interact with complexes with kinetically inert ions to produce adducts with different toxicity and aggregation propensities. Metal complexes must meet the demanding standards of therapeutic medicines in order to have a chance of being effective therapeutic inhibitors of amyloid production and Aβ toxicity (Hayne, Lim, and Donnelly, 2014).

Polyphenolic Substances

Due to their potent antimicrobial, antioxidant, anti-inflammatory, and anticancer properties, polyphenolic compounds are increasingly being considered as natural sources for treating a wide range of ailments, including neurological problems. There is growing data that suggest different methods by which polyphenolic chemicals from a range of sources can enhance brain neuropathy and cognitive function in animal models of AD (Wang et al., 2014).

Metals

In some neurodegenerative illnesses, such as Alzheimer's disease, lithium has also been considered a neuroprotective agent and a potential therapeutic candidate. Lithium regulates a number of homeostatic systems involved in mitochondrial function, inflammation, oxidative stress, neurotrophic response, and autophagy, which underlie its purported neuroprotective benefits. Such a broad spectrum of intracellular reactions may be a result of lithium's two main effects, which include its suppression of the enzymes inositol monophosphatase (IMP) and glycogen synthase kinase-3 beta (GSK-3) (Forlenza, De-Paula, and Diniz, 2014).

Stem Cells

Stem cells may be used to treat AD, according to recent preclinical data. The mechanisms of stem cell-based therapy for AD include neuroprotection and trophic effects, anti-amyloidogenesis, helpful modulation, and neuron replacement. Modeling AD using induced pluripotent stem cells (iPSCs), researching familial and sporadic AD pathophysiology, and searching for anti-AD medications are all possible (Forlenza, De-Paula, and Diniz, 2014).

THERAPEUTIC TARGETS FOR AD

Extracellular beta-amyloid protein and intracellular neurofibrillary tangle deposits that contribute to senile plaques are two pathological features of AD. The majority of tissue samples that are currently available come from advanced stages of the disease, making it challenging to investigate disease-initiating pathways and AD progression in humans. As a result, in vitro research and the use of animal models for AD are the main methods used by researchers studying AD. P1 adenosine and P2 nucleotide receptors for UTP, ATP, and their metabolites make up the extensive family of proteins known as purinergic receptors. This family of receptors may play a role in the etiology of neurodegenerative illnesses like AD as it has been demonstrated to regulate a wide range of physiological and pathological processes, including neuroinflammation (Woods et al., 2016).

There have been reports of many illness indicators, including amyloid deposits surrounding neurons, hyperphosphorylated tau protein, oxidative stress, abnormal bio-metal balance, and low levels of acetylcholine (ACh) (Kumar et al., 2016). Figure 1.2 presents an overview of the many treatment targets for AD. The majority of receptors in the central nervous system are G-protein coupled receptors, which are also connected to intricate downstream pathways and may be therapeutically useful when altered (Guerram, Zhang, and Jiang 2016). According to the amyloid hypothesis, the Aβ peptide is the primary cause of the pathogenic cascade that results in dementia and other symptoms of AD (Adeniji, Adams, and Mody, 2017).

Studies have demonstrated that oxidative stress-induced respiratory chain malfunction, loss of mitochondrial biogenesis, abnormalities of mitochondrial dynamics, and mtDNA mutations are all

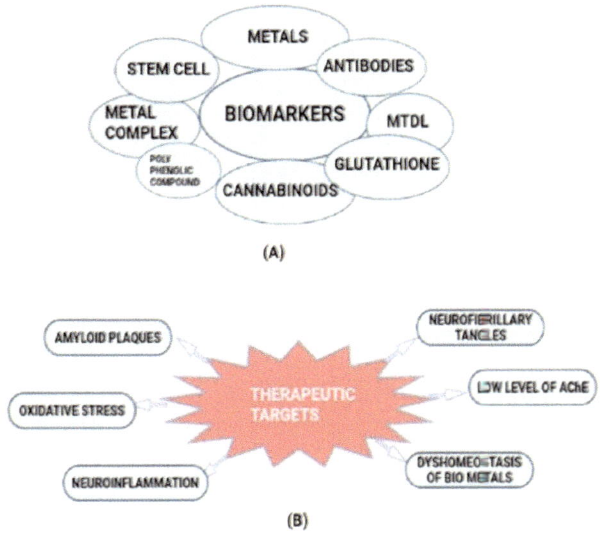

FIGURE 1.2 Biomarkers and therapeutic targets for AD.

signs of mitochondrial dysfunction, which have been linked to AD. The mitochondria may be a target for drug treatment for AD. Currently, mitochondria are not specifically targeted in AD therapy. The only two types of medications recognized by the Food and Drug Administration (FDA) for the treatment of AD are cholinesterase inhibitors (rivastigmine, galantamine, and donepezil) and the N-methyl-D-aspartate (NMDA) receptor antagonist memantine (Moreira et al., 2009).

AD has no known cure, and medication treatment for the condition is still in its early stages. The symptoms of AD can be controlled with drugs that have been approved for the treatment of probable AD, but the disease itself is not slowed or reversed. Drug and non-drug therapies are both necessary for the whole management of AD. Non-pharmacological therapies primarily target behavioral abnormalities and other causes of cognitive impairment (e.g., task simplification, environment adjustment, and minimal excess stimulation). Pharmacological approaches have been centered on controlling neurotransmitter changes brought on by the disease; these approaches can be classified as symptomatic or neuroprotective.

Although neuroprotective therapy will have a cumulative benefit that lasts even after the medication is stopped, symptomatic and neuroprotective pharmacologic treatments may have identical clinical trial result features. The effectiveness of current treatments, such as ChEIs and NMDA receptor antagonists, is evaluated by how well they can prevent the onset of symptoms in the cognitive, behavioral, and functional domains. Based on the cholinergic theory of memory failure, early pharmacologic treatments for AD aimed to improve cholinergic transmission in the brain. Acetylcholinesterase (AChE) inhibition, one of the various methods used to boost synaptic levels of ACh, has so far proven to be the most effective. It may help improve cholinergic transmission to inhibit the butyrylcholinesterase (BuChE) enzyme, which is a small component in healthy brains but is amplified in AD patients' brains in connection with plaques and tangles (Thancharoen et al., 2019).

Current pharmacological treatments for AD lessen symptoms without changing the disease's primary pathological features. Therefore, it is crucial to create innovative and successful medicines. There are numerous possible anti-AD therapies. Research is being done on novel therapy strategies, such as those that are more closely related to the etiology of the disease. Amyloid-peptide vaccines, secretase inhibitors, cholesterol-lowering medications, metal chelators, and anti-inflammatory pharmaceuticals are some of these potentially disease-modifying therapies (Scarpini, Scheltens, and Feldman, 2003). Several potential treatments are now being researched, including monoamine oxidase-B (MAO-B) inhibitors, hormone replacement therapy, anti-inflammatory drugs, free radical scavengers, and antioxidants. It is debatable and mostly dependent on retrospective research that estrogen or non-steroidal anti-inflammatory medicines have a protective effect. To conclusively prove the advantages of long-term estrogen or NSAID treatment in AD prevention, more controlled prospective research is required. Although selective prevention MAO-B inhibitors like lazabemide and free radical scavengers or antioxidants like idebenone are well tolerated, further research is needed to show that they have preventive effects. Additionally, strategies that include anti-amyloid treatments that target beta amylase secretion, aggregation, and toxicity, seem promising; additionally, researchers are in the very early stages of developing therapies that prevent the formation of neurofibrillary tangles and induce nerve growth factor (NGF) (Sramek and Cutler, 1999).

Many natural products and dietary components, including antioxidants and anti-inflammatory compounds, have been studied for their therapeutic potential in the treatment of AD. A small pharmacological pipeline is available, and many therapeutic targets have been researched for treating AD. Despite these efforts, developing drugs for AD has proven to be very challenging, and the majority of clinical trials have produced unimpressive outcomes (Geldenhuys and Darvesh, 2015). A potential focus for the development of an effective treatment for AD is anti-inflammatory medicines. Since immune proteins and immune-competent microglial cells are associated with senile plaques in both Alzheimer's and healthy aging, it is possible that these medications can alter the course of the disease by either preventing the development of senile plaques or by reducing the inflammation that these plaques are associated with (Geldenhuys and Darvesh, 2015). Non-genetic alterations that affect learning and memory, including DNA methylation and histone modifications, have lately been identified as promising prospective targets for AD treatments. The pathogenic pathways for AD include dynamic and latent epigenetic changes, which offer useful reversible targets for AD and other neurological illnesses (Adwan and Zawia, 2013).

One pathogenic mechanism behind this neurodegenerative condition has been hypothesized to be a change in the metabolism of insulin in the brain. According to this hypothesis, Alzheimer's patients exhibit lowered brain insulin receptor sensitivity. The fact that insulin reduces the vulnerability of hippocampal synapses to amyloid β, a peptide thought to be responsible for the onset of AD, provides a strong justification for the hypothesis that pharmacological methods that increase brain insulin signaling, such as giving insulin (intranasally), may have significant potential in the prevention and treatment of AD (Freiherr et al., 2013).

AD PHARMACOTHERAPEUTICS

Only four medications, all of which have limited efficacy, are now approved and sold for the treatment of AD-associated dementia. Donepezil, galantamine, and rivastigmine are three of these medications that operate on the cholinergic pathways in the central nervous system (CNS). The alkaloid galantamine, a natural product, is active as an allosteric modulator at nicotinic acetylcholine receptors, and all three of the medicines show anticholinesterase activity. According to the results of cognitive testing, each of these medications is now available in generic formulations and is approved for treating mild to severe dementia, but they are frequently used by patients with earlier stages of predementia that are associated with significant progressive memory impairment (Graham, Bonito-Oliva, and Sakmar 2017). The medication most recently approved for AD in the US is memantine, which is notable for being the first AD medication to target both glutaminergic and NMDA receptor pathways. Glutamatergic regulation impacts dendritic spine clustering in a mouse model of AD, and excess glutamate at excitatory synapses with concomitant cytotoxicity, presumably caused by reduced glutamate absorption from microglia, has recently been proposed as a pathophysiological mechanism in AD. Riluzole, an inhibitor of glutamate release and postsynaptic glutamate receptor signaling, is consequently being tested in a phase II study of moderate AD patients. Memantine disrupts the endocytosis pathway of APP, which is necessary for secretase-mediated cleavage, and lowers levels of both in AD models and in elderly animals. This causes reduced production. Along with the recognized indications for the drugs memantine and donepezil, both have been approved for use in monotherapy to treat the symptoms of AD (Ito et al., 2017). Together, the medications memantine and donepezil exhibit additional and advantageous effects for the patient due to their complementary and distinct modes of action. Initial evidence showed that the combination of memantine and donepezil can be used safely based on clinical research results in healthy volunteers. Memantine exhibits a favorable safety profile in AD patients when given in conjunction with steady ChEI treatment (Ito et al., 2017).

NON-PHARMACOLOGICAL THERAPY OF AD

People with AD can have a higher quality of life thanks to non-pharmacological treatments (Olazarán et al., 2010). There is evidence from a small number of high-quality randomized controlled trials that different non-pharmacological approaches, such as cognitive rehabilitation, cognitive stimulation therapy, and cognitive training, confer modest but significant benefits in the treatment of AD patients' cognitive symptoms, and that there may be additive benefits when combined with cholinesterase inhibitor therapy (Ballard et al., 2011). Numerous sleep and wakefulness issues associated with aging may be caused by a reduction in the amplitudes of circadian rhythms. Older adults and Alzheimer's patients may benefit from non-pharmacological modification of circadian rhythms using a variety of environmental stimuli to enhance sleep and cognitive functioning (van Someren, Mirmiran, and Swaab, 1993).

Sleep disruption appears to be a contributing factor to AD and its pathogenesis. The results of numerous studies have shown that, directly or indirectly, sleeping patterns affect AD. When sleep is inadequate, Aβ begins to build up, which may early on cause memory loss and later progress to AD. Non-rapid-eye-movement (NREM) sleep disruption, Aβ, and AD are linked by connections and a tenable mechanism. NREM sleep disruption is a unique mechanism that links poor hippocampus-dependent memory consolidation to cortical dysfunction. Proper sleep seems to be a new therapy focusing on elderly patients, providing therapeutic and anticipatory benefits. NREM sleep and the pathophysiology of AD have a causal relationship that is bidirectional and may influence the risk and course of the disease. NREM sleep disruption serves as a novel biomarker for AD. Cortical Aβ's interference with hippocampus-dependent memory consolidation may take a new channel through the disturbance of NREM sleep. It is possible to use the disturbance of NREM sleep physiology as a non-invasive biomarker of a disease, AD risk, and pathophysiological progression. Evidence links sleep disturbance to both the cause and progression of AD; it is changeable and has the potential to be treated both preventatively and therapeutically (Lautenschlager, Anstey, and Kurz, 2014).

Brain health is influenced by both indirect and direct biological processes, which have been related to physical and cognitive activity. Future studies must examine specifics on the kind, degree, length, and combination of interventions (Lautenschlager, Anstey, and Kurz, 2014). Numerous studies show that physical activity (PA) has a preventive effect on brain health, especially by lowering the chance of AD, the neurodegenerative dementia that it causes. The emergence of neurological illnesses can be influenced by a number of other factors, including heredity. But for the most part, moderate PA helps older persons' physical and mental health. Older people may benefit cognitively from moderate-intensity aerobic exercise, weight training, and stretching, toning, and range-of-motion activities. PA should be encouraged since it enhances the quality of life for all older persons, even though the precise processes by which it lowers the risk for dementia are not entirely understood. According to the research now available, dementia rates could be lowered if more people engaged in physical activity. It's possible that PA will overtake other behavioral factors to promote healthy mental and physical aging. Regardless of age, the most recent evidence supports PA's short- and long-term cognitive benefits (Gallaway et al., 2017).

MUSIC THERAPY

Although some studies have shown that music therapy is effective in treating the behavioral and psychological symptoms of dementia, the majority are small and uncontrolled. For the treatment of agitation and anxiety in moderately severe and severe AD, music therapy is a secure and successful strategy (Svansdottir and Snaedal, 2006). Failures and successes in our understanding of the pathophysiology of AD have raised concerns about potential mistakes in clinical trial design, drug development, therapeutic target selection, and diagnosis. Numerous in vitro and in vivo studies are still being conducted, but the priority should be finding a cure for Alzheimer's disease. Thus, this step forward in the creation of medications for the condition needs to be reevaluated. Studying clinical trials helps us analyze patterns of pharmacological interactions as opposed to single potential medication targets, and it frequently gives us new knowledge for treating Alzheimer's disease. We are getting closer to developing the ideal pharmaceutical strategy for treating Alzheimer's with the help of a number of ongoing randomized controlled trials that show positive results and provide a way to increase cooperation between basic researchers, clinical researchers, and pharmaceutical companies (Mangialasche et al., 2010).

POTENTIAL HERBAL DRUGS FOR AD

There are several popular Indian traditional plants that can be utilized to treat neurological illnesses like Alzheimer's and dementia. Ayurvedic Rasayana medications are abundant in antioxidants and immunomodulatory substances. It has already been established that some of these medications have high antioxidant properties. The ability to scavenge free radicals or to activate the oxidant defense of cells can be considered their principal mode of action because the bulk of diseases are associated with disruption of the balance between antioxidants and oxidants. Many plants, including brahmi, shankhapushpi, jatamansi, jyotshmati, ashwagandha, and vacha fall under the category of Rasayana plants. These herbs are categorized as brain tonics or rejuvenators since they are specialized in brain tissues (Farooqui and Farooqui, 2017). In addition to this, various Rasayana medicines have been confirmed in our laboratory to have a positive effect on memory deficiencies, warranting their prospective function in dementia and other neurological disorders, as well as their potential benefit in AD (Thakur, Chatterjee, and Kumar, 2013; Thakur et al., 2016; Thakur and Raj, 2017).

AD AND DIABETES

Researchers have discovered a strong link between the food we eat and Alzheimer's disease and dementia via a similar pathway that causes type 2 diabetes. As a result, they refer to Alzheimer's as type 3 diabetes. We already know that some diseases, like diabetes, are connected to our diet, but they are not the only ones. AD worsens the pace of glucose metabolism in the brain. Of the three kinds of diabetes, type 2 diabetes mellitus has been linked to an increased risk for AD, although the exact mechanisms are still unclear (Abbas, Faivre, and Hölscher, 2009).

It has been demonstrated in earlier research that the rise in glucagon-like peptide 1 (GLP-1) contributes to the normalization of insulin signaling in type 2 diabetes. GLP-1 plays a significant role in both neuronal activity and brain processes. A GLP-1 receptor knockout mouse model was used to test the precise role of GLP-1 receptors in synaptic plasticity and cognitive processes. It was discovered that since the brain's lack of GLP-1 receptor function affects synaptic plasticity and cognitive processes, GLP-1 receptors play a significant role in brain function. Researchers mimicked the symptoms of AD in an animal study by interfering with the brain's insulin signaling. Our brains manufacture insulin, which is crucial for healthy brain signaling and whose disturbance can cause dementia (Agrawal and Gomez-Pinilla, 2012).

Diabetes and AD are related because both conditions promote the formation of brain plaques (Agrawal and Gomez-Pinilla, 2012). Our brain gradually shuts down its insulin signaling as a result of the constant high levels of insulin that result from overindulging in sweets and wheat. According to a study, eating a diet high in fructose and low in omega-3 fatty acids decreases insulin's affinity for its receptor, resulting in chronic insulin resistance as shown by a decline in the phosphorylation of the insulin receptor and its downstream effector Akt in rats after just six weeks (Biessels and Despa, 2018).

Both maternal and child malnutrition may have an effect on a kid's physical and mental health. Type 2 diabetes mellitus, hypertension, insulin resistance, and cardiovascular problems can all develop in adulthood as a result. The growth and function of the brain, as well as the CNS, are both impacted by protein energy deficiency. Early episodes of starvation may cause long-lasting alterations in the function of brain neural receptors, according to neuropharmacological research (Nyaradi et al., 2013).

In a different investigation, it was discovered that Alzheimer's results in a decrease in the levels and activity of insulin–PI3K–AKT signaling pathway components. It was proposed that this reduction in insulin–PI3K–AKT signaling could cause the primary tau kinase, glycogen synthase kinase-3 beta, to become active. This could result in inappropriate tau hyperphosphorylation and abnormal neurofibrillary tangle development inside the cell (Liu et al., 2011).

STRUCTURAL CHANGES IN AD BRAIN

AD causes tissue loss and the death of nerve cells, and over time, the brain's size decreases, impairing all its activities. Loss of brain cells in the cortical area harms the brain's capacity for planning, remembering, and thinking. The hippocampus, a region of the brain crucial for the creation of new memories, is where shrinkage is most severe. The ventricles, the fluid-filled spaces in the brain, are growing in addition to the region of the brain shrinking. An AD patient's brain has fewer neuron cells and synapses than a healthy individual's, but it has a higher buildup of tangles and plaques, which may be the cause of this cell loss. These plaques obstruct cell-to-cell communication and stir up immune cells that attack damaged cells and create inflammation. Tau proteins break down into tangled strands known as tangles, which prevent nutrition and other vital supplies from moving through the cells, causing the cells to die.

CONCLUSION

We have discussed several justifications and potential treatment plans for AD in this review. According to numerous studies, the causal metabolic pathways for AD's neurodegenerative nature include extracellular amyloid plaques, intracellular neurofibrillary tangles, synaptic degeneration, and neuronal death. At any given age, genetics is responsible for around 70% of the risk of AD. The epsilon 4 allele of the gene for apolipoprotein E is the most prevalent genetic risk factor for AD. A vitamin D-deficient diet, whose active form appears to influence nerve growth factor in addition to the genetic and molecular aspects, appears to be another cause of AD. Moreover, diabetes is caused by a reduction in brain glucose metabolism in AD for reasons that are still unclear. Therefore, we draw the conclusion that stem cell therapy and biomarkers may be cutting-edge approaches to treating and detecting AD at an early stage.

ACKNOWLEDGMENT

The authors are grateful to the College of Pharmaceutical Sciences, Dayananda Sagar University, Bengaluru, India, for providing the library facility for scientific literature support to present this chapter in an organized manner.

REFERENCES

Abbas T, Faivre E, Hölscher C. Impairment of synaptic plasticity and memory formation in GLP-1 receptor KO mice: interaction between type 2 diabetes and Alzheimer's disease. *Behavioral Brain Res* (2009) 205(1):265–271.

Adeniji AO, Adams PW, Mody VV. Amyloid – hypothesis in the development of therapeutic agents for Alzheimer's disease, drug discovery approaches for the treatment of neurodegenerative disorders: Alzheimer's disease. In: Adeboye A, editor. *Drug discovery approaches for the treatment of neurodegenerative disorders*. Academic Press (2017) 109–143.

Adwan L, Zawia NH. Epigenetics: a novel therapeutic approach for the treatment of Alzheimer's disease. *Pharmacol Therap* (2013) 139(1):41–50.
Agrawal R, Gomez-Pinilla F. "Metabolic syndrome" in the brain: deficiency in omega-3 fatty acid exacerbates dysfunctions in insulin receptor signalling and cognition. *J Physiol* (2012) 590(10):2485–2499.
Annweiler C, Montero-Odasso M, Hachinski V, Seshadri S, Bartha R, Beauchet O. Vitamin D concentration and lateral cerebral ventricle volume in older adults. *Mol Nutrition Food Res* (2013) 57(2):267–276.
Ashford JW. Treatment of Alzheimer's disease: the legacy of the cholinergic hypothesis, neuroplasticity, and future directions. *J Alzheimer's Dis: JAD* (2015) 47(1):149–156.
Aso E, Ferrer I. Cannabinoids for treatment of Alzheimer's disease: moving toward the clinic. *Front Pharmacol* (2014) 5: 1–11.
Ballard C, Khan Z, Clack H, Corbett A. Nonpharmacological treatment of Alzheimer disease. *Canad J Psychiatry. Revue Canadienne de Psychiatrie* (2011) 56(10):589–595.
Biessels GJ, Despa F. Cognitive decline and dementia in diabetes mellitus: mechanisms and clinical implications. *Nature Rev Endocrinol* (2018) 14(10):591–604.
Blennow K. Biomarkers in Alzheimer's disease drug development. *Nature Medi* (2010) 16(11):1218–1222.
Chase TN, Farlow MR, Clarence-Smith K. Donepezil Plus Solifenacin (CPC-201) treatment for Alzheimer's disease. Neurotherapeutics: *J Am Soc Exp NeuroTherapeutics* (2017) 14(2):405–416.
Dodel R, Hampel H, Depboylu C, Lin S, Gao F, Schock S, Jäckel S, Wei X, Buerger K, Höft C, Hemmer B, Möller HJ, Farlow M, Oertel WH, Sommer N, Du Y. Human antibodies against amyloid beta peptide: a potential treatment for Alzheimer's disease. *Annals Neurol* (2002) 52(2):253–256.
Farooqui T, Farooqui AA. *Neuroprotective effects of phytochemicals in neurological disorders.* Wiley 2017.
Fernandes de Abreu DA, Eyles D, Féron F. Vitamin D, a neuro-immunomodulator: implications for neurodegenerative and autoimmune diseases. *Psychoneuroendocrinol* (2009) 34(Suppl 1):S265–277.
Forlenza OV, De-Paula VJR, Diniz BSO. Neuroprotective effects of Lithium: implications for the treatment of Alzheimer's disease and related neurodegenerative disorders. *ACS Chem Neurosci* (2014) 5(6):443.
Freiherr J, Hallschmid M, Frey WH, Brünner YF, Chapman CD, Hölscher C, Craft S, Felice FG De, Benedict C. Intranasal insulin as a treatment for Alzheimer's disease: a review of basic research and clinical evidence. *CNS Drugs* (2013) 27(7):505–514.
Gallaway PJ, Miyake H, Buchowski MS, Shimada M, Yoshitake Y, Kim AS, Hongu N. Physical activity: a viable way to reduce the risks of mild cognitive impairment, Alzheimer's disease, and Vascular Dementia in older adults. *Brain Sci* (2017) 7(2):22.
Geldenhuys WJ, Darvesh AS. Pharmacotherapy of Alzheimer's disease: current and future trends'. *Exp Rev Neurotherapeutics* (2015) 15(1):3–5.
Gezen-Ak D, Yilmazer S, Dursun E. Why vitamin D in Alzheimer's disease? The hypothesis'. *J Alzheimer's Disease: JAD* (2014) 40(2):257–269.
Gilman S. *Oxford American handbook of neurology.* Oxford University Press (2010) 1–481.
Graham WV, Bonito-Oliva A, Sakmar TP. Update on Alzheimer's disease therapy and prevention strategies. *Annual Rev Medi* (2017) 68:413–430.
Guerram M, Zhang LY, Jiang ZZ. G-protein coupled receptors as therapeutic targets for neurodegenerative and cerebrovascular diseases. *Neurochem Int* (2016) 101:1–14.
Guzior N, Wieckowska A, Panek D, Malawska B. Recent development of multifunctional agents as potential drug candidates for the treatment of Alzheimer's disease. *Curr Medi Chem* (2015) 22(3):373–404.
Hardy J, Strooper B De. Alzheimer's disease: where next for anti-amyloid therapies?. *Brain J Neurol* (2017) 140(4):853–855.
Hayne DJ, Lim S, Donnelly PS. Metal complexes designed to bind to amyloid-β for the diagnosis and treatment of Alzheimer's disease. *Chem Soc Rev* (2014) 43(19):6701–6715.
Heinrich M, Teoh HL. Galantamine from snowdrop – the development of a modern drug against Alzheimer's disease from local Caucasian knowledge. *J Ethnopharmacol* (2004) 92(2–3):147–162.
Holtzman DM, Herz J, Bu G. Apolipoprotein E and apolipoprotein E receptors: Normal biology and roles in Alzheimer disease. *Cold Spring Harb Perspect Med* (2012) 2(3):a006312.
Ito K, Tatebe T, Suzuki K, Hirayama T, Hayakawa M, Kubo H, Tomita T, Makino M. Memantine reduces the production of amyloid-β peptides through modulation of amyloid precursor protein trafficking. *Eur J Pharmacol* (2017) 798:16–25.
Kumar A, Nisha CM, Silakari C, Sharma I, Anusha K, Gupta N, Nair P, Tripathi T, Kumar A, Current and novel therapeutic molecules and targets in Alzheimer's disease. *J Formosan Medical Association = Taiwan yi zhi* (2016) 115(1):3–10.
Kumar A, Singh A. Ekavali. A review on Alzheimer's disease pathophysiology and its management: an update. *Pharmacol Rep* (2015) 67(2):195–203.

Lautenschlager NT, Anstey KJ, Kurz AF. Non-pharmacological strategies to delay cognitive decline. *Maturitas* (2014) 79(2):170–173.

Littlejohns TJ, Henley WE, Lang IA, Annweiler C, Beauchet O, Chaves PHM, Fried L, Kestenbaum BR, Kuller LH, Langa KM, Lopez OL, Kos K, Soni M, Llewellyn DJ. Vitamin D and the risk of dementia and Alzheimer disease. *Neurology* (2014) 83(10):920.

Liu Y, Liu F, Grundke-Iqbal I, Iqbal K, Gong CX. Deficient brain insulin signalling pathway in Alzheimer's disease and diabetes. *J Pathology* (2011) 225(1):54–62.

Liu Z, Zhang A, Sun H, Han Y, Kong L, Wang X. Two decades of new drug discovery and development for Alzheimer's disease. *RSC Adv* (2017) 7(10):6046–6058.

Mangialasche F, Solomon A, Winblad B, Mecocci P, Kivipelto M. Alzheimer's disease: clinical trials and drug development. *Lancet Neurology* (2010) 9(7):702–716.

Masoumi A, Goldenson B, Ghirmai S, Avagyan H, Zaghi J, Abel K, Zheng X, Espinosa-Jeffrey A, Mahanian M, Liu PT, Hewison M, Mizwicki M, Cashman J, Fiala M. 1 Alpha,25-dihydroxyvitamin D3 interacts with curcuminoids to stimulate amyloid-beta clearance by macrophages of Alzheimer's disease patients. *J Alzheimer's Dis: JAD* (2009) 17(3):703–717.

Mishra S, Palanivelu K. The effect of curcumin (turmeric) on Alzheimer's disease: an overview. *Annals Indian Acad Neurology* (2008) 11(1):13–19.

Mizwicki MT, Menegaz D, Zhang J, Barrientos-Durán A, Tse S, Cashman JR, Griffin PR, Fiala M. Genomic and nongenomic signaling induced by 1α,25(OH)2-vitamin D3 promotes the recovery of amyloid-β phagocytosis by Alzheimer's disease macrophages. *JAD* (2012) 29(1):51–62.

Mohamed T, Shakeri A, Rao PPN. Amyloid cascade in Alzheimer's disease: recent advances in medicinal chemistry. *Eur J Med Chem* (2016) 113:258–272.

Moreira P, Cardoso S, Pereira C, Santos M, Oliveira C. Mitochondria as a therapeutic target in Alzheimer's disease and diabetes. *CNS Neurol Disord Drug Targets* (2009) 8(6):492–511.

Nyaradi A, Li J, Hickling S, Foster J, Oddy WH. The role of nutrition in children's neurocognitive development, from pregnancy through childhood. *Front Human Neurosci* (2013) 7:35–77.

Olazarán J, Reisberg B, Clare L, Cruz I, Peña-Casanova J, Ser T Del, Woods B, Beck C, Auer S, Lai C, Spector A, Fazio S, Bond J, Kivipelto M, Brodaty H, Rojo JM, Collins H, Teri L, Mittelman M, Orrell M, Feldman HH, Muñiz R. Nonpharmacological therapies in Alzheimer's disease: a systematic review of efficacy. *Dement Geriatr Cogn Disord* (2010) 30(2):161–178.

Ott A, Breteler MB, Van HF, Claus J, Van Der CTJM, Grobbee DE, Hofman A. Prevalence of Alzheimer's disease and vascular dementia: association with education. The Rotterdam study. *BMJ (Clinical Research Ed.).* (1995) 310(6985):970.

Plassman BL, Langa KM, Fisher GG, Heeringa SG, Weir DR, Ofstedal MB, Burke JR, Hurd MD, Potter GG, Rodgers WL, Steffens DC, Willis RJ, Wallace RB. Prevalence of dementia in the United States: the aging, demographics, and memory study. *Neuroepidemiology* (2007) 29(1–2):125–132.

Prakash A, Dhaliwal GK, Kumar P, Majeed ABA. Brain biometals and Alzheimer's disease – boon or bane? *Int J Neurosci* (2017) 127(2):99–108.

Prince M, Bryce R, Albanese E, Wimo A, Ribeiro W, Ferri CP. The global prevalence of dementia: a systematic review and metaanalysis. *Alzheimer's Dementia J Alzheimer's Ass* (2013) 9(1):63–75.

Querfurth HW, LaFerla FM. Alzheimer's disease. *New Eng J Med* (2010) 362(4):329–344.

Reiman EM, Chen K, Alexander GE, Caselli RJ, Bandy D, Osborne D, Saunders AM, Hardy J. Correlations between apolipoprotein E epsilon4 gene dose and brain-imaging measurements of regional hypometabolism. *Proc Natl Acad Sci U S A (PNAS)* (2005) 102(23):8299–8302.

Rygiel K. Novel strategies for Alzheimer's disease treatment: an overview of anti-amyloid beta monoclonal antibodies. *Ind J Pharmacol* (2016) 48(6):629.

Scarpini E, Scheltens P, Feldman H. Treatment of Alzheimer's disease: current status and new perspectives. *Lancet Neurol* (2003) 2(9):539–547.

Selkoe DJ. Alzheimer's disease is a synaptic failure. *Science (New York, N.Y.)* (2002) 298(5594):789–791.

Shu YZ. Recent natural products-based drug development: a pharmaceutical industry perspective. *J Nat Prod* (1998) 61(8):1053–1071.

Someren EJW, van Mirmiran M, Swaab DF. Non-pharmacological treatment of sleep and wake disturbances in aging and Alzheimer's disease: chronobiological perspectives. *Behav Brain Res* (1993) 57(2):235–253.

Sommer I, Griebler U, Kien C, Auer S, Klerings I, Hammer R, Holzer P Gartlehner G. Vitamin D deficiency as a risk factor for dementia: a systematic review and meta-analysis. *BMC Geriatr* (2017) 17(1):1–13.

Sramek JJ, Cutler NR. Recent developments in the drug treatment of Alzheimer's disease. *Drugs Aging* (1999) 14(5):359–373.

Sultzer DL, Marder SR. Older brains are different: brain-behavior studies and their clinical utility. *Am J Geriatr Psy Off J Am Ass Geriatr Psy* (2017) 25(1):11–12.

Svansdottir HB, Snaedal J. Music therapy in moderate and severe dementia of Alzheimer's type: a case-control study. *Int Psychogeriatr* (2006) 18(4):613–621.

Thakur AJ, Raj P. Pharmacological perspective of *Glycyrrhiza glabra* Linn.: a mini-review. *J Analyt Pharm Res* (2017) 5(5):00156.

Thakur AK, Chatterjee SS, Kumar V. Beneficial effects of Brassica juncea on cognitive functions in rats. *Pharm Bio* (2013) 51(10):1304–1310.

Thakur AK, Rai G, Chatterjee SS, Kumar V. Beneficial effects of an Andrographis paniculata extract and andrographolide on cognitive functions in streptozotocin-induced diabetic rats. *Pharm Bio* (2016) 54(9):1528–1538.

Thancharoen O, Limwattananon C, Waleekhachonloet O, Rattanachotphanit T, Limwattananon P, Limpawattana P. Ginkgo biloba extract (EGb761), Cholinesterase inhibitors, and Memantine for the treatment of mild-to-moderate Alzheimer's disease: a network meta-analysis. *Drugs Aging* (2019) 36(5):435–452.

Wang J, Bi W, Cheng A, Freire D, Vempati P, Zhao W, Gong B, Janle EM, Chen TY, Ferruzzi MG, Schmeidler J, Ho L, Pasinetti GM. Targeting multiple pathogenic mechanisms with polyphenols for the treatment of Alzheimer's disease-experimental approach and therapeutic implications. *Front Aging Neurosci* (2014) 6(MAR):81318.

Weekley CM, He C. Developing drugs targeting transition metal homeostasis. *Curr Opin Chem Bio* (2017) 37:26–32.

Woods LT, Ajit D, Camden JM, Erb L, Weisman GA. Purinergic receptors as potential therapeutic targets in Alzheimer's disease. *Neuropharmacol* (2016) 104:169–179.

2 Novel Molecular Targets at the Forefront of Neurodegenerative Research

*Madhu Tanya Singh and Praveen Thaggikuppe Krishnamurthy**

Department of Pharmacology, JSS College of Pharmacy, JSS Academy of Higher Education and Research, Ooty, The Nilgiris, Tamil Nadu, India

*Corresponding author

INTRODUCTION

Neurodegenerative diseases are characterized by the loss of neurons, resulting in significant physical and cognitive disabilities worldwide. Over the past 30 years, neurodegenerative disorders have impacted nearly 15% of the global population, and this number continues to rise (Feigin et al., 2020). Moreover, the risk of chronic neurodegenerative diseases is anticipated to double in the next two decades (Schependom and D'haeseleer, 2023).

The term "neurodegenerative diseases" encompasses a broad range of diseases distinguished by the specific central nervous system (CNS) region involved. The most prevalent among them include Alzheimer's, Parkinson's, Huntington's and amyotrophic lateral sclerosis. Alzheimer's disease (AD) is characterized by the progressive loss of neurons in the hippocampus, a region linked to episodic memory, resulting in a gradual decline in both memory and cognitive function. In Parkinson's disease (PD), there is a loss of neurons in the substantia nigra, leading to a myriad of symptoms such as tremors, bradykinesia and postural instability (Mendez and McMurtray, 2007). Huntington's disease (HD) is an autosomal dominant neurodegenerative disorder recognized by involuntary choreatic movements and cognitive-behavioral disturbances. This condition arises from the expansion of cytosine, adenine and guanine (CAG) trinucleotide repeats on the short arm of chromosome 4p16.3 within the huntingtin (HTT) gene. A mutation in the HTT gene results in abnormal expansion of polyglutamine in the HTT protein, leading to aggregation, accumulation and subsequent neurodegeneration (Ajitkumar and De Jesus, 2024). Amyotrophic lateral sclerosis (ALS), previously known as Lou Gehrig's disease, is a neurodegenerative disorder affecting the brain and spinal cord. This condition results in the loss of voluntary muscle control, impacting essential movements such as talking, swallowing and walking (Kiernan et al., 2011). There are currently numerous approved medications for treating neurodegenerative diseases, but a substantial portion of them mainly address symptoms without offering disease-modifying therapies. Developing a comprehensive, systems-level understanding of the pathophysiological mechanisms and targets associated with the etiology and progression of neurodegenerative diseases holds the potential to enhance preventive measures and refine pharmacological treatment strategies (Lampтey et al., 2022).

Numerous novel molecular targets are being explored to identify disease-modifying therapeutics such as beta-secretase 1 (BACE1), triggering receptor expressed on myeloid cells 2 (TREM2), sirtuin, caspases, glycogen synthase kinase-3 (GSK3), monoamine oxidase-B (MOA-B), serotonin, N-methyl-D-aspartate receptors (NMDARs) and nuclear factor erythroid 2-related factor 2 (Nrf2) in Alzheimer's disease (Blaikie et al., 2019; Ray and Buggia-Prevot, 2021); glucocerebrosidase, leucine-rich repeat kinase 2 (LRRK2), adenosine A2A receptor, beta 2 adrenergic receptors, histamine 4

DOI: 10.1201/9781003506317-2

receptors, phosphodiesterase 4, angiotensin receptor, sphingosine-1-phosphate receptor (S1PR), purinergic 2Y type 12 receptor (P2Y12R), high-mobility group box-1, rabphilin-3A (Rph3A), colony stimulating factor-1 receptor, vanilloid 4 receptor, alanine-serine-cysteine transporter 2, GPCR-coupled estrogen receptor and mitochondrial antiviral signaling protein in Parkinson's disease (Pingale and Gupta, 2021); huntingtin, mammalian target of rapamycin complex 1 (mTORC1), dopamine receptors, brain-derived neurotrophic factor (BDNF), glutamate receptors and histone deacetylases (HDACs) in Huntington's disease (Williams et al., 2008; Mason and Barker, 2016); and TAR DNA binding protein-43 (TDP-43), superoxide dismutase (SOD1), chromosome 9 open reading frame (C9ORF72), and glutamate transport and excitotoxicity in ALS. Understanding these targets and their molecular mechanisms will pave the way for developing innovative therapeutic strategies for better management of neurodegenerative diseases. This chapter focuses on exploring novel molecular targets implicated in neurodegenerative diseases, including Alzheimer's, Parkinson's, Huntington's and ALS.

COMMON PATHOLOGICAL HALLMARKS AND MOLECULAR PATHWAYS IN NEURODEGENERATIVE DISEASES

Neurodegenerative diseases such as Alzheimer's, Parkinson's, Huntington's and ALS are characterized by five common hallmarks: pathological protein aggregation, oxidative stress, synaptic dysfunction, neuroinflammation and neurodegeneration (Figure 2.1). This section briefs these five most common hallmarks of neurodegenerative diseases.

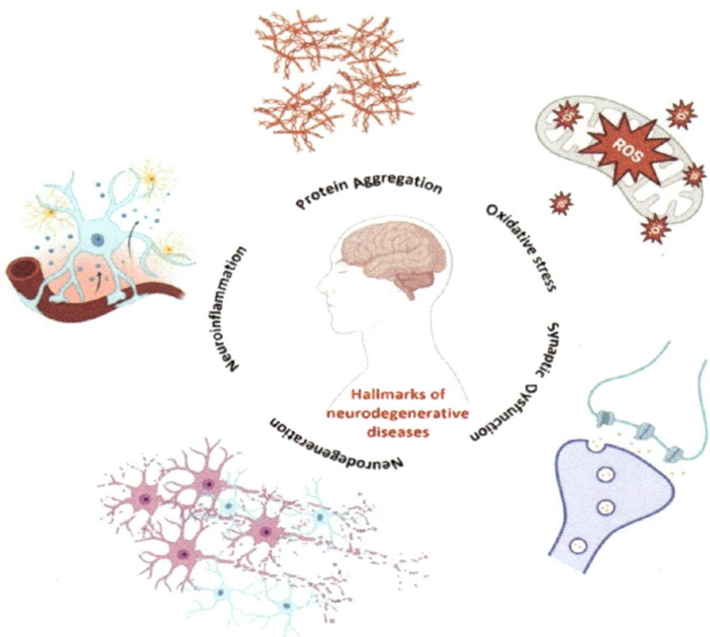

FIGURE 2.1 The five common pathological hallmarks of neurodegenerative diseases: protein aggregation, oxidative stress, synaptic dysfunction, neuroinflammation and neurodegeneration.

Protein Misfolding and Aggregation

The formation of characteristic protein aggregates is a prominent pathological feature shared among various neurodegenerative diseases, serving as a crucial factor in diagnosis and disease classification. These diseases, collectively termed proteinopathies, encompass a spectrum of conditions including AD, PD, primary tauopathies (such as progressive supranuclear palsy [PSP], corticobasal degeneration [CBD] and tau-linked frontotemporal dementia [FTD-tau]), frontotemporal dementia (FTD), ALS, synucleinopathies (including Lewy body dementia [LBD] and multiple system atrophy [MSA]), Huntington's disease, related polyglutamine diseases (including spinocerebellar ataxia [SCA]) and prion diseases (PrD).

The identification of causal mutations in rare inherited forms of neurodegenerative diseases has significantly contributed to understanding the underlying pathogenic mechanisms. A common feature among various neurodegenerative diseases is the causal link between disease-causing mutations and the heightened aggregation of the encoded protein, indicating a toxic gain-of-function mechanism. This connection has been demonstrated in genes associated with specific neurodegenerative diseases, including amyloid precursor protein (APP) in AD; tau in AD and tauopathies; alpha-synuclein in PD and synucleinopathies; PrPC in PrD; SOD1, TDP-43 and FUS in ALS/FTD; and huntingtin (Htt) in HD.

The fact that mutations in different neurodegenerative diseases enhance the aggregation of characteristic proteins suggests a central pathogenic role for protein aggregation. Notably, mutations in some genes, such as presenilin 1 and 2 (PSEN1, PSEN2), may not directly lead to the aggregation of the encoded protein but instead increase the aggregation of key neurodegenerative proteins. Another gain-of-function mechanism involves the generation of aggregating dipeptide repeats (DPRs) through hexanucleotide repeat sequences, produced by repeat-associated non-ATG (RAN) translation. The close correlation between the aggregation process and symptom progression in most neurodegenerative diseases, along with the impact of aggregating proteins on crucial neuronal functions and neurodegenerative hallmarks, further supports the concept of a toxic gain of function. However, it is acknowledged that alternative mechanisms may contribute to neurotoxicity in certain cases (Mason and Barker, 2016).

Oxidative Stress

Oxidative stress arising from an imbalance between pro-oxidants and antioxidants is a crucial factor in neurodegenerative diseases. This imbalance results in the excessive accumulation of reactive oxygen species (ROS) and decreased antioxidant defense activity, leading to potential cellular damage. Under physiological conditions, the concentration of ROS is maintained at a low level through endogenous antioxidant mechanisms. However, with age and various risk factors, the redox system becomes unbalanced, resulting in the overproduction of ROS.

While ROS in moderate concentrations play essential roles in physiological processes, their overproduction and imbalance in the endogenous antioxidant defense system leads to oxidative damage. This damage includes post-translational modifications and the oxidation of proteins, lipids and DNA/RNA, shared features in many neurodegenerative diseases. Patients with neurological disorders experience ROS overproduction in the brain, resulting in increased peroxidation of membrane lipids, elevated malondialdehyde levels, excessive accumulation of metals and mitochondrial dysfunction. These events contribute to neuronal apoptosis, highlighting the significant role of oxidative stress in the progression of neurodegenerative diseases (Chen et al., 2012).

Neuroinflammation

Neuroinflammation, characterized by microgliosis and astrogliosis, stands as a pathological hallmark in various neurodegenerative diseases including AD, PD, ALS and HD. In postmortem brain samples of neurodegenerative patients, the presence of inflammation is consistent, and its definitive role in neurodegeneration is demonstrated in prototypic neuroinflammatory diseases.

Microgliosis is consistently observed in all neurodegenerative diseases, playing essential roles in sensor functions, housekeeping and defense mechanisms in the brain. Activated microglia respond to pathogens or signs of injury by producing cytokines/chemokines, engaging in phagocytosis and generating ROS. In acute stress situations, the inflammatory process is typically resolved, but in neurodegenerative diseases, failure to resolve leads to chronic inflammation, contributing to neurodegeneration. Microglial activation in neurodegenerative diseases can be triggered by danger signals such as protein aggregates, misfolded proteins, damaged synapses, Ca2+ influx or mitochondrial ROS mechanisms associated with other neurodegenerative disease hallmarks.

Genetic studies support the active role of microglia in neurodegenerative diseases, exemplified by the identification of TREM2 loss-of-function mutations. TREM2, expressed exclusively in microglia, plays a crucial role in CNS degeneration. TREM2 genetic variants are associated with AD risk, and both TREM2 and apolipoprotein E (ApoE) contribute to the transition of homeostatic microglia to disease-associated phenotypes. ApoE and TREM2 modify protein aggregation and neurodegeneration in preclinical neurodegenerative disease models, with their functions being protective or detrimental depending on disease phases. Furthermore, genetic links between microglia and other neurodegenerative disease-associated genes, such as GRN and Htt, highlight the intricate involvement of microglia in neurodegenerative processes. Understanding the roles of different microglial populations is crucial for effective therapeutic targeting in neurodegenerative diseases (Wilson et al., 2023).

Synaptic Dysfunction

In neurodegenerative diseases, symptoms often arise from disturbances in specific neuronal networks, with synaptic failure and toxicity being early events preceding neuron degeneration. The precise functioning of these networks relies on regulated synaptic stability, neurotransmitter and calcium modulation, and various cellular processes like axonal transport, cytoskeletal dynamics, proteostasis and energy supply. Synaptic function is intricately linked to mitochondrial function, energy supply and coordinated regulation of cellular processes essential for synaptic maintenance.

Evidence from genetic, preclinical and patient-derived studies underscores the crucial role of synaptic failure in neurodegenerative diseases. Dysfunctional synaptic processes, including hyperexcitation and excitotoxicity, are implicated in various neurodegenerative diseases such as stroke, AD, PD, HD and ALS. In conditions like stroke, excess glutamate leads to excitotoxicity, causing neuronal death through Ca^{2+} influx, mitochondrial dysfunction and energy depletion. Similarly, hyperexcitability and excitotoxicity are considered significant mechanisms in the pathogenesis of ALS and may contribute to neurodegeneration in HD and AD.

Proteins that aggregate in neurodegenerative diseases, such as α-synuclein in PD, Htt in HD, and APP/Ab and tau in AD, often play physiological roles at synapses, and their pathological forms contribute to synaptic failure. Mutations in genes associated with proper synaptic function are also identified in certain neurodegenerative diseases, reinforcing the link between synaptic dysfunction and neurodegeneration.

Synaptic failure and dysfunction are recognized as early events preceding neurodegeneration in various neurodegenerative diseases, including AD, PD, HD, FTD, ALS and stroke. In AD, synaptic failure occurs before neuronal loss, and similar patterns are observed in PD, HD and ALS, where synaptic defects precede neurodegeneration. The complex interplay of hypo- and hyperactive synapses contributes to the deregulation of neurotransmission in the brain during the progression of neurodegenerative diseases (Wilson et al., 2023).

Neurodegeneration

Neurons exhibit inherent properties that render them particularly vulnerable to cell death in neurodegenerative diseases. These vulnerabilities include their post-mitotic nature, leading to the gradual accumulation of age-associated damage and an inability to replicate their high energy requirements, extended axons and dendrites, and dependence on glial cells for maintenance. These factors, compounded by an age-related decline in resilience mechanisms, make neurons more susceptible to cell death. Various neurodegenerative disease hallmarks, such as protein aggregation, oxidative stress and neuroinflammation collectively contribute to neuronal cell loss, overriding intrinsic neuronal resilience to internal and external insults.

The cell death mechanisms in neurons are diverse, encompassing intrinsic and extrinsic apoptosis, necrosis, necroptosis, ferroptosis, phagoptosis, autophagic cell death, pyroptosis and mitochondrial permeability transition. The specific classification of neuronal death depends on the mechanism or inducer, with crosstalk between these mechanisms. Neuronal death can result in brain volume loss, measurable through volumetric MRI, and the release of intraneuronal proteins into biofluids, such as tau and neurofilaments. Various triggers for neuronal cell death include axotomy, aberrant cell-cycle re-entry, excitotoxicity, loss of connected neurons, protein aggregation, unfolded protein response, lysosomal rupture, autophagy, oxidative stress and DNA damage. Multiple neurodegenerative disease hallmarks act as inducers of neuronal cell death, working synergistically to drive the neurodegenerative process (Wilson et al., 2023).

NOVEL MOLECULAR TARGETS IN ALZHEIMER'S DISEASE

Beta-Site Amyloid Precursor Protein Cleaving Enzyme 1

Beta-site amyloid precursor protein cleaving enzyme 1 (BACE1) is a crucial enzyme involved in the generation of various forms of amyloid-β (Aβ), including $A\beta_{42}$, a key contributor to AD pathology (Figure 2.2). Increased BACE1 concentrations and activity are observed in AD brains, supporting its significant role in the disease. Despite past clinical trials with BACE1 inhibitors being discontinued, likely due to futility or safety concerns, BACE1 still remains a validated therapeutic target for AD. Targeting BACE1 is aimed at slowing down Aβ production in early AD stages. BACE1 also has substrates beyond the amyloidogenic pathway, suggesting additional roles in synaptic plasticity

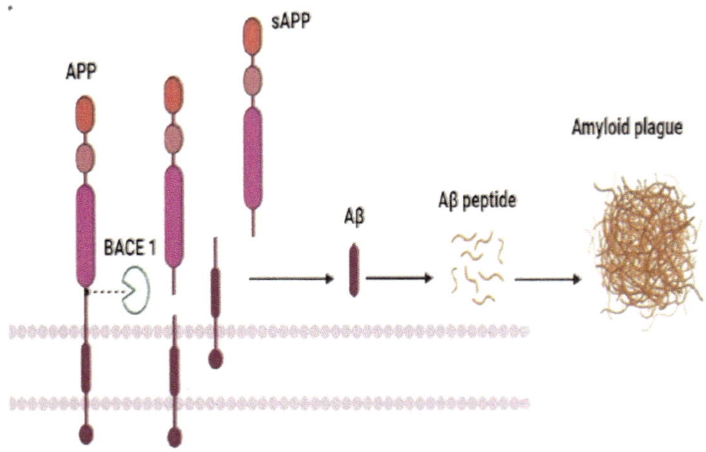

FIGURE 2.2 Role of BACE1 in amyloid plaque formation and Alzheimer's progression.

and homeostasis. Preliminary evidence indicates that a sustained reduction of BACE1 activity by 20% to 30% over an extended period might be effective in preventing AD. This suggests that even a moderate and long-term decrease in BACE1 activity could have a preventive impact on the development of AD. Future research should explore the physiological and pathophysiological aspects of Aβ and BACE1, with the goal of identifying safe and effective compounds for AD treatment (Blaikie et al., 2019; Hampel et al., 2021).

Triggering Receptor Expressed on Myeloid Cells 2

Triggering receptor expressed on myeloid cells 2 (TREM2) is an immunomodulatory receptor discovered in 2000. It belongs to the immunoglobulin superfamily. It is a transmembrane protein encoded by the TREM2 gene containing an extracellular ectodomain that can bind to substances like Aβ, lipopolysaccharide (LPS) and apolipoproteins (Lemarié and Gibot, 2020). The TREM2 extracellular domain can be cleaved to release soluble TREM2 (sTREM2). TREM2 is primarily expressed in peripheral macrophages, dendritic cells, myeloid cells and microglia in the CNS. TREM2 is implicated in inflammatory responses, phagocytosis and cell proliferation, emphasizing its crucial roles in immune regulation (Li et al., 2023).

Microglia, the immune cells of the CNS, play a dual role in neuroprotection and exacerbation of nerve injury in the brain. TREM2 has been identified as a key regulator of these microglial functions (Figure 2.3). Overexpression of TREM2 *in vitro* has demonstrated anti-inflammatory effects and enhanced microglial phagocytosis. *In vivo* studies, involving lentivirus transfection to increase TREM2 expression, have shown improved cognitive function and reduced neuropathological changes in AD mice. However, the specific role of TREM2 in the human brain, especially in AD patients, is yet to be fully explored. It is noteworthy that lentivirus transfection strategies pose risks, limiting their application in human *in vivo* studies. TREM2 missense mutants have been associated with neuroimmune defects, suggesting potential impacts on nervous system immune

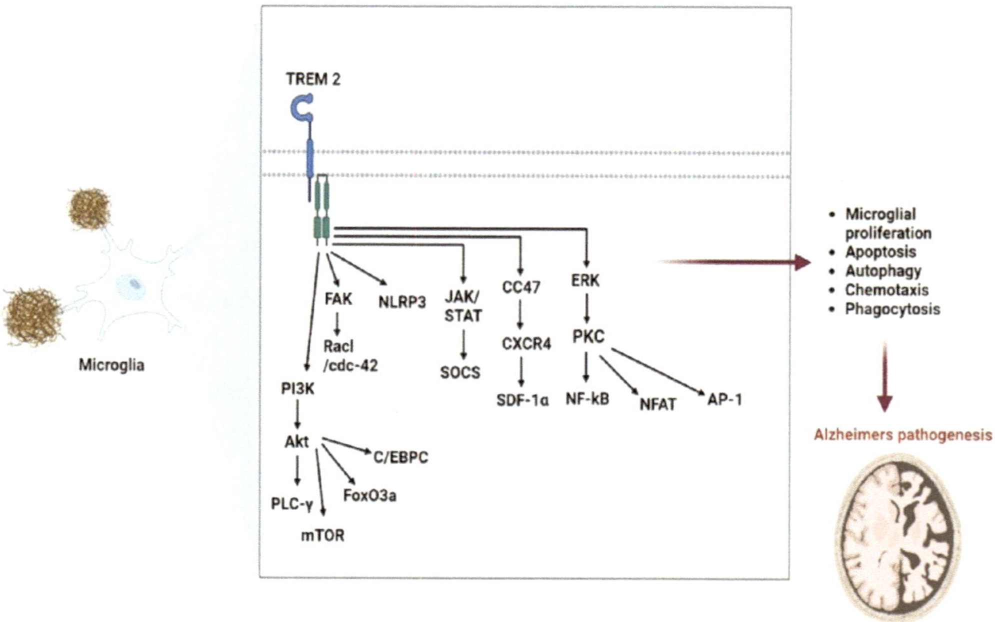

FIGURE 2.3 Mechanism of TREM2 in the microglial.

function. Future investigations should focus on unraveling how TREM2 variants affect AD, particularly whether altered phagocytosis or increased pro-inflammatory molecules contribute to disease progression. While TREM2 has shown promise as a target for neuroprotective drugs in AD treatment, questions remain about its consistent role across disease stages and its potential role in disease induction. Understanding the precise involvement of TREM2 and microglia in AD will be crucial in determining the feasibility of targeted therapies. Moreover, given TREM2's ability to inhibit inflammatory responses, targeting TREM2 may offer an intriguing strategy for treating inflammation-related diseases, including AD. Ongoing research on TREM2 and microglia holds the promise of discovering safer and more effective drugs for AD treatment in the future (Li et al., 2023).

Sirtuin

Sirtuins, a class of enzymes with deacetylase and ADP-ribosyl transferase activity, play a crucial role in processes linked to AD, such as oxidative stress, inflammation and glucose homeostasis (Qin et al., 2006; Lalla and Donmez, 2013; Silva et al., 2014). SIRT1, the extensively studied sirtuin, has been associated with a positive impact on AD models, reducing amyloid aggregation in both *in vitro* and *in vivo* settings (Lalla and Donmez, 2013). However, genetic studies present conflicting evidence, with some suggesting unchanged SIRT1 levels in AD models. Conversely, another hypothesis proposes downregulation of SIRT1 in AD patients, advocating for its neuroprotective potential and a reduction in amyloid deposition through the promotion of non-amyloidogenic processing of APP (Cacabelos et al., 2019). Further research is needed to reconcile these contradictory findings and clarify the role of SIRT1 in AD pathophysiology.

Resveratrol, a natural polyphenolic phytochemical, is known for activating sirtuin activity and has demonstrated benefits in AD models. Its antioxidant properties and protective effects against Aβ toxicity have been established (Sawda and Moussa, 2017). While resveratrol exhibits neuroprotective effects through the activation of SIRT1, other polyphenolic compounds like curcumin, benzoic acid and quercetin have shown similar benefits. Studies indicate that SIRT1 overexpression and resveratrol treatment in cellular models share inhibitory effects on Aβ fibril formation. Based on the above research, activating sirtuin using specific ligands can have beneficial effects in AD treatment (Blaikie et al., 2019).

Glycogen Synthase Kinase-3

Glycogen synthase kinase-3 (GSK3), a serine/threonine kinase, plays a regulatory role in crucial cellular processes such as glycogen synthesis and glucose metabolism. Its dysregulation is implicated in the pathogenesis of AD. Elevated levels of GSK3 have been associated with neurodegenerative changes, contributing to hyperphosphorylation of tau, formation of toxic Aβ, loss of cholinergic neurons, inflammation, apoptosis and synaptic loss (Blaikie et al., 2019). Studies indicate an upregulation of GSK3 in peripheral lymphocytes and the hippocampus of AD patients. The role of GSK3β in AD pathogenesis has led to the exploration of GSK-3 inhibition as a potential therapeutic strategy. Inhibition of GSK3β has shown promise in preclinical studies, with compounds like lithium and tideglusib demonstrating efficacy in reducing tau phosphorylation and Aβ levels in AD mouse models (Machado and Manji, 2009). However, the transition to clinical trials has presented challenges due to the ubiquity of GSK3β in cellular processes, raising concerns about potential off-target effects and toxicity. While some inhibitors, like AZD2558 and AZD1080, demonstrated promising preclinical results, they faced significant challenges in clinical trials, leading to discontinuation (Bhat et al., 2018). Tideglusib reached phase II trials but did not show significant clinical improvements (Congdon et al., 2023; del Ser et al., 2013; Lovestone et al., 2015). Lithium, a less specific GSK3β inhibitor, demonstrated positive outcomes in small clinical studies suggesting its potential as a disease-modifying therapy for AD (Machado and Manji, 2009). However, challenges in achieving therapeutic efficacy and addressing safety concerns underscore the complexities of targeting GSK3β in AD treatment. Exploring more specific and safe GSK3β inhibitors can provide a better therapeutic approach as a disease-modifying therapy for AD.

Nuclear Factor Erythroid 2-Related Factor-2 (Nrf2)

Nrf2 is a transcriptional pathway crucial for cellular defense. Nrf2 exhibits varying expression levels across different organs, with higher levels in metabolic organs (Riedl et al., 2009). In the CNS, Nrf2 is expressed in neurons, astrocytes and glial cells, with higher levels in astrocytes than in neurons (Cuadrado et al., 2018). Activation of Nrf2 in astrocytes induces neuroprotective effects (Lee et al., 2003). A decline in Nrf2 expression is observed in aged and AD brains (Silva et al., 2018). Nrf2 has been found to assist the clearance of β-amyloid precursor protein (APP) and tau through autophagy mechanisms (Pajares et al., 2016). A lack of Nrf2 and dysfunction is implicated in cognitive impairment and AD pathogenesis. Moreover, GSK3β negatively regulates Nrf2, and elevated levels of GSK3β observed in AD patients contribute to a substantial reduction in Nrf2, enhancing neuronal susceptibility to oxidative stress (Engel et al., 2006). Consequently, the development of activators targeting the Nrf2 transcriptional pathway is an attractive approach to inhibit the progression of AD. The induction of Nrf2 has shown promise in ameliorating cognitive impairment and terminating key AD pathogenic factors, making Nrf2 a potential therapeutic target for AD (Osama et al., 2020).

NOVEL MOLECULAR TARGETS IN PARKINSON'S DISEASE

Glucocerebrosidase

Glucocerebrosidase (GCase) is a 497 aa lysosomal enzyme that breaks down glucocerebroside into ceramide and glucose (Boer et al., 2020). Homozygous individuals with pathogenic GBA variants develop Gaucher's disease due to the excessive storage of glucocerebroside in various organs (Bohra and Nair, 2011). Heterozygous carriers of GBA variants have an increased risk of parkinsonism and dementia, with GBA pathogenic variants being a common genetic cause of PD. PD patients with GBA variants exhibit earlier onset, faster progression and reduced survival compared to those without GBA variants (Tayebi et al., 2003). GCase deficiency results in glucocerebroside accumulation in neurons, leading to toxic oligomer formation and lysosomal dysfunction, primarily affecting α-synuclein (α-syn). Conversely, elevated α-syn inhibits normal GCase function, creating a bidirectional pathological loop in PD and other synucleinopathies (Choi et al., 2011; Mazzulli et al., 2011; Aflaki et al., 2017). This understanding has led to the development of various therapeutic strategies targeting GCase to address these conditions.

Therapeutic strategies targeting GCase include small molecular chaperones that enhance lysosomal GCase activity in neurons. Ambroxol hydrochloride, a molecular chaperone, showed promising results in improving lysosomal activity and reducing PD symptoms in clinical trials (McNeill et al., 2014). Other small-molecule GCase activators, including LTI-291 and AT3375, are also under investigation (Khanna, 2012).

Glucosylceramide synthase inhibitors, such as venglustat, have demonstrated efficacy in reducing glucosylceramide levels and improving outcomes in GBA-associated PD mouse models. The ongoing MOVES-PD global study is evaluating the safety and efficacy of venglustat in GBA mutation carriers with PD.

AAV-mediated gene therapies targeting GBA, such as AAV5-GBA and AAV9-GBA, show promise in preventing dopamine neuron loss and reducing α-syn deposits (Rocha et al., 2015). An ongoing clinical trial is assessing the safety of intracisternal PR001A (AAV9-GBA1) administration in patients with moderate to severe PD carrying GBA pathogenic variants (Ntetsika et al., 2021).

Leukine-Rich Repeat Kinase-2

Leucine-rich repeat kinase 2 (LRRK2), belonging to the Ras-of-complex (ROC) protein family, is associated with autosomal dominant PD, particularly in specific ethnic groups (West, 2017). Pathogenic variants of the LRRK2 gene, such as Gly2019Ser, are a relatively common cause of both familial and sporadic PD (West, 2017; Tolosa et al., 2020). These variants lead to a toxic gain-of-function mechanism, hyperactivating the LRRK2 kinase. This hyperactivity is implicated in the pathogenesis of PD. Novel therapeutic approaches for LRRK2-associated PD and idiopathic PD involve the development and utilization of LRRK2 kinase inhibitors to target and mitigate the harmful effects of these pathogenic variants.

Following successful inhibition of LRRK2 activity with small molecular inhibitors, such as DNL201 and DNL151 in animal models, clinical trials have been initiated to evaluate their safety, tolerability and target engagement. DNL201 has shown positive results in placebo-controlled trials in healthy volunteers, and studies are ongoing to assess its effects in patients with PD (Tolosa et al., 2020). Additionally, DNL151 is undergoing evaluation in clinical trials for both healthy volunteers and PD patients. Other LRRK2 kinase inhibitors in preclinical development, including MLi-2 and PF-06685360, are actively being explored. These advancements reflect a promising direction in therapeutic strategies targeting LRRK2-associated PD and potentially idiopathic PD.

An alternative strategy to reduce LRRK2 activity involves the use of anti-sense oligonucleotides (ASOs) to decrease LRRK2 expression selectively. This approach allows for CNS-specific LRRK2 inhibition via intraventricular injection, avoiding potential adverse effects observed in peripheral LRRK2 knockout mice (Herzig et al., 2011). In preclinical studies using Gly2019Ser mice, intraventricular injection of LRRK2-targeted ASOs led to a reduction in α-synuclein inclusions, preservation of nigral dopaminergic neurons and improvement in motor deficits (Zhao et al., 2017). Building on these findings, an ongoing phase I clinical trial (NCT03976349) is assessing the safety, tolerability and pharmacokinetic profile of intrathecal administration of BIIB094 in patients with PD. The study includes patients both with and without verified pathogenic or likely-pathogenic LRRK2 variants. This approach leverages genetic insights into PD, providing more precise tools for patient selection and pharmacodynamic measurements (Ntetsika et al., 2021).

Adenosine A2A Receptors

In recent years, targeting the adenosine A2A receptor has emerged as a promising approach for treating PD (Cieślak et al., 2008). The A2A receptor, a member of the G-protein-coupled receptor family, is highly expressed in specific brain regions and various cell types, including neurons, astrocytes, oligodendrocytes and microglia. Its involvement in dopaminergic and glutamatergic neurotransmission, neuroinflammation, and neurodegeneration has made it a focal point in understanding PD pathophysiology (Cervetto et al., 2017). Particularly, the A2A receptor's high expression in GABAergic and glutamatergic striatopallidal medium spiny neurons, where it forms functional complexes with dopamine D2 receptors, has gained a lot of attention (Cieślak et al., 2008; Diao et al., 2017). Activation of A2A receptors leads to motor inhibition, and their interaction with glutamate receptors, especially mGlu5, facilitates glutamate release, contributing to NMDA glutamate receptor activation (Beggiato et al., 2016). Consequently, inhibiting A2A receptor activity has become a therapeutic target in both preclinical and clinical studies for PD.

After over two decades of research, istradefylline has become the first non-dopaminergic medication approved by the Food and Drug Administration (FDA) for PD (Chen and Cunha, 2020). It has shown significant benefits, reducing off-time duration, improving motor outcomes measured by the UPDRS motor score and increasing the duration of time without troublesome dyskinesia (Paton, 2020). PET studies have demonstrated increased A2A receptor expression in the putamen of PD patients with dyskinesia, and istradefylline has been proven to bind dose-dependently to A2A receptors in the ventral striatum, caudate and putamen (Paton, 2020).

Other A2A antagonists, such as preladenant, tozadenant and vipadenant, have demonstrated promise in early studies but faced challenges in subsequent phase III trials due to issues like unexpected safety concerns (Hauser et al., 2014). KW-6356, a second-generation A2A antagonist, has exhibited positive results in early-stage PD, both in combination with levodopa and as monotherapy (Chen and Cunha, 2020). The approval of Istradefylline opens new therapeutic possibilities for A2A antagonists, emphasizing the importance of identifying patient subgroups that would benefit from these agents.

Sphingosine-1-Phosphate Receptors

Sphingosine-1-phosphate (S1P) serves a crucial role in the brain as both a neuromodulator and a neuroprotector. The primary risk factor for neurodegenerative diseases, aging, is associated with mitochondrial dysfunction, oxidative stress and cellular apoptosis. Mitochondria are responsible for oxidative energy metabolism and contribute to Ca^{2+} homeostasis and the regulation of ROS generation (Witte et al., 2014). Within the mitochondria, sphingosine kinase 2 (SphK2) produces S1P, which interacts with a key factor prohibitin 2 (PHB2) responsible for mitochondrial development and function. S1P inhibits oxidative stress-activated apoptosis through mitochondrial-dependent pathways (Aoki et al., 2016). Additionally, S1P inhibits ceramide-mediated sphingomyelinase generation. The S1P signal pathway regulates autophagy, which is crucial for cell survival. The dynamic equilibrium of S1P formation and degradation, crucial for autophagy , is catalyzed by SphKs, S1P phosphatases (SGPPs) and sphingosine phosphate lyase 1 (SGPL1). Ceramide promotes autophagy-associated cell death, which can be blocked by S1P through the transactivation of mTOR via S1P receptors (Hao et al., 2020). S1P-induced apoptosis can be prevented by SphK2 inhibitors. This suggests that S1P-induced autophagy acts as a cellular defense mechanism, supporting cell survival during nutrient deficiency and exhibiting apoptosis characteristics through reabsorption function.

In PD, S1P plays a crucial role in α-syn degradation, thereby preventing its accumulation and aggregate formation via mTOR pathway activation. FTY720, a structural sphingosine analogue and S1PR modulator, was found to reverse oxidative stress, mitochondrial dysfunction, inflammation and protein misfolding. Mitochondrial dysfunction in nigral neurons decreased complex I activity, which is crucial for PD progression. The interaction between α-syn and S1P has been demonstrated to influence mitochondrial homeostasis and protect dopaminergic neurons in PD models (Wang et al., 2023).

Purinergic 2Y Type 12 Receptor

The ADP-sensitive P2Y12-receptors (P2Y12R), primarily located on platelets in the periphery, play a crucial role in antithrombotic drug action to prevent stroke and myocardial infarction (Jacobson et al., 2020). In the CNS, these receptors are mainly expressed by microglia, allowing for their distinction from other myeloid lineage cells (Moore et al., 2015). Previous research has linked P2Y12R to the regulation of pain modalities, with receptor inhibition reducing pro-inflammatory cytokine levels and regulating the ROCK pathway (Aflaki et al., 2017). Recent advances in neurodegenerative research have identified novel microglia–neuron interaction sites where P2Y12R regulates neuronal activation and pathology. Microglia lacking P2Y12R display normal baseline motility but impaired responses to injury sites (Haynes et al., 2006).

András Iring et al. have investigated the role of P2Y12R in PD (Iring et al., 2022). The findings of this study indicated that P2Y12R inhibition has the potential to protect against neurodegenerative loss in PD. Neurodegeneration in PD can occur through "cell-autonomous" mechanisms or indirect degeneration involving interactions with resident glial cells, particularly microglia. Microglial activation is implicated in PD pathological mechanisms, but whether microglia contribute to the initiation of dopaminergic cell death or respond to signals from degenerating neurons remains unclear. The distinct functions of microglia compared to other myeloid cells in the CNS suggest their unique contribution to injury and repair processes. Exploring the dual roles of P2Y12R in CNS has unveiled innovative strategies for managing neuroinflammation and altering the course of disease progression in PD (Iring et al., 2022).

Rabphilin-3A

Rabphilin-3A (Rph3A) is a synaptic protein primarily recognized for its role in regulating exo- and endocytosis processes at presynaptic sites, specifically synaptic vesicle-associated functions. At excitatory dendritic spines, Rph3A plays a crucial role in the synaptic retention of N-methyl-D-aspartate receptors (NMDARs) through direct interaction with the GluN2A regulatory subunit. This interaction is essential for the molecular and structural modifications of dendritic spines induced by long-term potentiation (LTP), which contribute to synaptic plasticity. Notably, Rph3A downregulation has been implicated in the pathogenesis of PD, leading to synaptic dysfunction and early neurodegeneration.

Studies indicate that Rph3A is enriched at dendritic spines and its accumulation in potentiated spines is observed during LTP. Conversely, silencing of Rph3A results in spine loss and impedes the activity-dependent formation of new spines. MPTP-treated monkeys and postmortem brain samples from PD patients exhibit decreased Rph3A mRNA levels in specific brain regions. Aberrant Rph3A synaptic localization has been linked to L-DOPA-induced dyskinesias, associated with increased NMDAR activity. The restoration of Rph3A synaptic levels and activity presents a novel therapeutic strategy to counteract postsynaptic dysfunction and early motor defects in PD (Ferrari et al., 2022).

NOVEL MOLECULAR TARGETS IN HUNTINGTON'S DISEASE

Mammalian Target of Rapamycin Complex 1

Mammalian target of rapamycin complex 1 (mTORC1), a serine–threonine kinase, plays a vital role in responding to environmental factors like nutrient and oxygen availability. Activation of mTORC1 stimulates cell growth and metabolism by facilitating enhanced protein synthesis and activating downstream effectors. Conversely, inactivation of the mTORC1 pathway is associated with increased macroautophagy and mechanisms promoting cell survival.

Amino acid availability, particularly leucine, serves as a critical factor influencing protein synthesis and acts as a key upstream signal for activating the mTORC1 pathway. Upstream effectors in this pathway, such as sestrin1 and sestrin2, detect and bind leucine (Ben-Sahra et al., 2013). This binding triggers the dissociation of sestrin from the multiprotein complex GATOR2, initiating a signaling cascade that activates mTORC1 (Saxton and Sabatini, 2017). The activation of mTORC1 results in enhanced protein synthesis and mRNA translation initiation, as evidenced by the phosphorylation of ribosomal protein S6 and the negative regulator eukaryotic translation initiation factor 4E-binding protein 1 (4E-BP1), which are reliable markers of mTORC1 activation.

In the brain, mTORC1 is neuroprotective and supports processes like myelination, axon growth and regeneration (Park et al., 2008). The absence of mTORC1-activating subunits, such as Rheb or Rhes, leads to deficits in cortical thickness and myelination, with implications for conditions like HD (Zou et al., 2011).

In the context of HD, mTORC1 pathway dysregulation is observed in the patient striatum. Conversely, genetic restoration of mTORC1 activity alleviated neuropathology, striatal atrophy and motor deficits in HD mouse models. However, the delicate balance of mTORC1 activity is crucial, as chronic activation is associated with adverse conditions. Fine-tuning the modulation of mTORC1 activity emerges as a potential therapeutic approach for conditions like HD, where restoring or optimizing mTORC1 activity may offer benefits without the drawbacks associated with chronic activation (Ravikumar et al., 2004).

Neurotransmitter Receptors and Ion Channels

In HD, alterations in neurotransmitter receptors within the basal ganglia, including reduced levels of dopamine (Dl and D2), adenosine (A2A) and muscarinic acetylcholine receptors, along with increased GABA A receptors, have been observed in HD mice (Kumar et al., 2014; Glass et al., 2004). Clinical trials using tetrabenazine, a dopamine receptor blocker, showed a reduction in chorea severity in HD patients (Paleacu, 2007). Olanzapine, an antipsychotic affecting various receptors, demonstrated symptom improvement in other trials (Videnovic, 2013). Reduced expression of cannabinoid CB1 receptors in the basal ganglia of HD patients and mice suggests a potential avenue for modulation. Recent drug screening identified cannabinoids as potential neuroprotective agents (Glass et al., 1993; Denovan and Robertson, 2000).

Moreover, ion channel subunits linked to neurotransmitter receptors, such as voltage-sensitive sodium channels and inward-rectifying potassium channels, show altered gene expression in HD (Ariano et al., 2005). Decreased levels of potassium channel subunits Kir2.1, Kir2.3 and Kv2.1 in medium spiny neurons of HD mice, associated with changes in potassium conductance, present potential therapeutic targets. These findings open avenues for developing compounds targeting neurotransmitter receptors and ion channels in the treatment of HD (Hannan, 2005; Mason and Barker, 2016).

Brain-Derived Neutrophic Factor (BDNF)

Brain-derived neurotrophic factor (BDNF) plays a crucial role in neuronal maturation and function, regulating neurotransmission and synaptic plasticity. In HD, BDNF levels are decreased in the striatum, potentially due to disrupted cortico-striatal protein trafficking. Normal huntingtin protein, implicated in vesicle trafficking, may be required for the anterograde transport of BDNF down the cortico-striatal tract (Gauthier et al., 2004). Loss of huntingtin function and gain of function contribute to HD pathology, as wild-type huntingtin upregulates BDNF expression (Ferrer et al., 2000). Therapeutically, peptide mimetics of BDNF, rather than direct protein delivery, are being explored as a potential approach (O'Leary and Hughes, 2003). Targeting TrkB, the primary tyrosine kinase receptor for BDNF, is considered an alternative strategy with promising therapeutic implications. Electrophysiological abnormalities in HD mice's GABAergic function can be ameliorated by BDNF application, highlighting its potential as a therapeutic target in HD (Hannan, 2005).

Glutamate Receptors

There is mounting evidence implicating disrupted NMDA receptors and their postsynaptic signaling complexes in HD, with NR2B-subtype NMDA receptors mediating greater excitotoxicity in HD mice (Luthi-Carter et al., 2000, 2002; Chan et al., 2002). Genetic studies suggest NR2B and NR2A subunits act as genetic modifiers influencing the age of disease onset. Targeting abnormal NMDA receptor signaling and excitotoxicity presents promising therapeutic avenues. Amantadine, an NMDA receptor antagonist, reduced dyskinesias in HD patients. However, clinical trials with remacemide and co-enzyme Q10 demonstrated mixed results.

AMPA receptors, another ionotropic glutamate receptor type, are potential therapeutic targets with altered protein levels observed in R6/1 HD mice (Luthi-Carter et al., 2002). Kainate receptors, particularly the GluR6 subunit, influence HD onset. Group II mGluR agonists are considered potential therapeutics, given evidence of reduced mGluR2/3 levels in HD mouse cortico-striatal synapses. Group I mGluRs and intraneuronal calcium signaling are implicated, with studies supporting the therapeutic potential of mGluR2/3 agonists and mGluR5 antagonists.

Riluzole, an inhibitor of glutamate release, showed therapeutic effects in HD. Lamotrigine, another glutamate release inhibitor, demonstrated limited efficacy. Glutamate transporters, such as GLT-1 on astrocytes, are potential targets for modulating glutamate uptake and synaptic transmission in HD, given their downregulation in R6/2 HD mice associated with disrupted glutamate uptake. These findings highlight the complexity of glutamatergic dysregulation in HD and potential avenues for therapeutic intervention (Hannan, 2005).

Histone Deacetylase (HDACs)

Numerous pieces of evidence supporting the involvement of transcriptional dysregulation in HD pathogenesis are bolstered by recent drug studies focusing on histone acetylation as a potential therapeutic target (Steffan et al., 2001). HDAC inhibitors, such as suberoylanilide hydroxamic acid, sodium butyrate and phenylbutyrate, demonstrated beneficial effects in various HD mouse models, particularly alleviating motor symptoms (Hockly et al., 2003; Gardian et al., 2005). This highlights HDAC inhibitors as a promising class of therapeutics for HD and potentially other neurodegenerative disorders.

Beyond HDACs, several transcription factors and co-factors implicated in HD pathogenesis emerge as potential therapeutic targets. Gene expression profiling in HD striatum and studies on transgenic mice tissues suggest disruptions in cAMP-responsive and retinoid-responsive gene expression, indicating a role for altered CRE-mediated transcription in HD. Specific transcription factors like CREB-binding protein (CBP), Sp1 and TAFII130 are also implicated in HD with mithramycin, a regulator of DNA binding for transcriptional activators, significantly extending survival in HD mice (Obrietan and Hoyt, 2004). These findings underscore the potential of targeting transcriptional dysregulation pathways for developing effective therapies against HD (Hannan, 2005).

NOVEL MOLECULAR TARGETS IN AMYOTROPHIC LATERAL SCLEROSIS

TAR DNA Binding Protein-43 (TDP-43)

TDP-43 is a crucial member of the heterogeneous ribonucleoprotein family and is highly conserved DNA/RNA-binding protein. It exhibits a preference for UG-rich and TG-rich motifs in RNA and DNA, respectively. This protein is ubiquitously expressed in all cell types and is primarily localized to the nucleus, although it can also be found in the cytoplasm and mitochondria. TDP-43 proteinopathies are characterized by pathological features such as the mislocalization of TDP-43

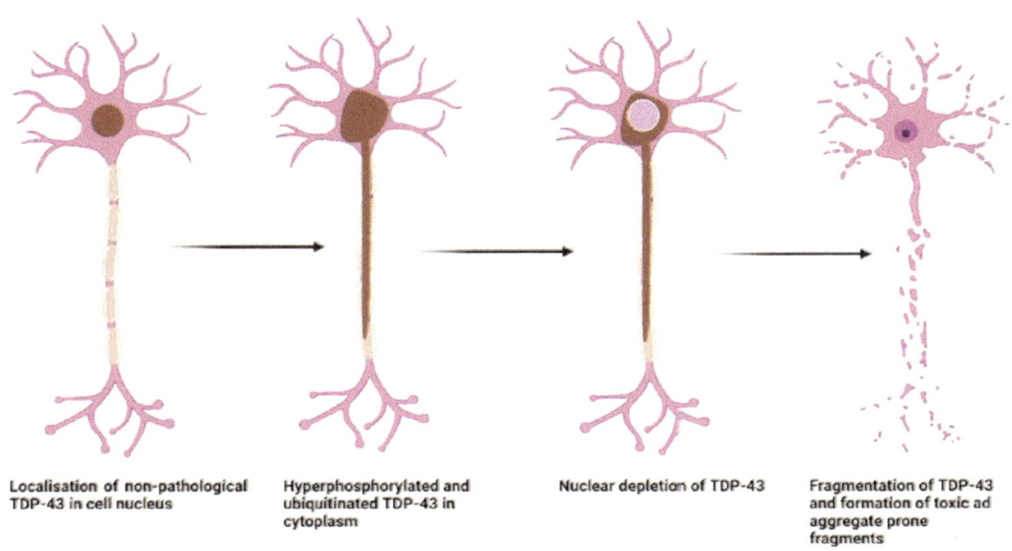

FIGURE 2.4 Illustration demonstrating progressive development of TDP-43 pathology in the brain.

protein from the nucleus to the cytoplasm, formation of inclusion bodies containing ubiquitinated and hyperphosphorylated TDP-43, the truncation of the protein leading to the creation of toxic C-terminal fragments and overall protein aggregation (Figure 2.4). This disrupts normal cellular functions and contributes to the death of motor neurons. The presence of sporadic or familial mutations can exacerbate these processes, contributing to early disease onset. Additionally, abnormal phosphorylation and ubiquitination of TDP-43 further enhance its toxicity. The scope of TDP-43 as a novel target for ALS therapy is promising. Strategies aimed at modulating TDP-43 levels, preventing its mislocalization or promoting its clearance could potentially mitigate the progression of ALS (Suk, 2020).

Superoxide Dismutase 1 (SOD1)

Superoxide dismutase (SOD) plays a crucial role in detoxifying superoxide radicals generated during metabolic processes, with activation occurring upon binding to cofactors like Zn^{2+} and Cu^{+2} ions. Variations in this binding can accelerate protein aggregation and reduce SOD enzyme stability, contributing to the pathology of ALS. Drug design strategies for ALS target this aggregation and stabilization (Sehgal et al., 2018). Through a computer-based approach, lead molecules inhibiting aggregation were sought using mutant SOD and dopamine complex. Docking studies identified 2,3,5,4-tetrahydrostilbene and hesperidin as molecules with higher affinity toward SOD, making them potential agents for designing new drugs in ALS treatment. Inhibiting SOD oligomerization emerged as an effective target for screening and developing novel therapeutics against SOD-associated ALS.

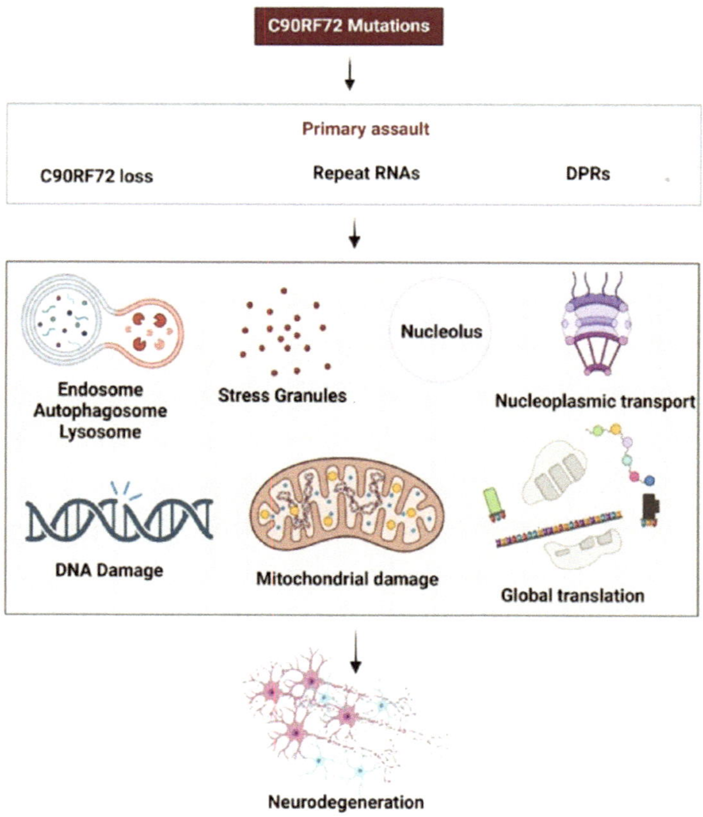

FIGURE 2.5 Role of C90RF72 mutation in neurodegeneration.

Chromosome 9 Open Reading Frame (C9ORF72)

Function mutations in the C9ORF72 gene result in hexanucleotide repeats in ALS patients, impacting multiple downstream pathways. These pathways include DNA damage, nucleolar dysfunction, nucleocytoplasmic transport deficits, endoplasmic reticulum stress, dysfunctional autophagy, inhibition of translation, proteasome inhibition and alterations in stress granule dynamics (Tang et al., 2020). Potential therapeutic targets within these pathways include CRISPR-mediated genome editing to remove C9ORF72 hexanucleotide repeats, inhibition of transcription products by targeting elongation factors SUPT4H and PAF1C, targeting repeat-containing RNA using antisense oligonucleotides or RNA interference, and reducing the export of toxic repeat-containing RNA to the cytoplasm with selective inhibitors like KTP350. These targeted approaches offer avenues for potential interventions in ALS associated with C9ORF72 mutations (Figure 2.5) (Jiang and Ravits, 2019).

Targeting Glutamate Transport and Excitotoxicity

Glutamate, a key excitatory neurotransmitter in the CNS, is crucial for normal brain physiology. Its release and subsequent binding to receptors are regulated by excitatory amino acid transporter 2 (EAAT2) present in astrocytes (Foran and Trotti, 2009). Studies have revealed lower EAAT2 levels in the spinal cord and cortex of ALS patients, leading to increased glutamate levels, which contribute to motor neuron death and degeneration (King et al., 2016).

High glutamate concentrations beyond normal levels can lead to excitotoxicity, a condition where overactivation of glutamate transporters occurs (Dong et al., 2009). This excess glutamate triggers cell death, making cells more susceptible to excitotoxic damage. Another mechanism involves N-methyl D-aspartate (NMDA) and alpha-amino-3-hydroxy-5-methyl-4-isoxazolepropionic acid (AMPA) receptors, which regulate intracellular calcium levels. Elevated calcium concentrations induce death signals, leading to cell death. Imbalances in calcium influx also affect various enzymes, such as endonucleases, proteases, phosphatases and phospholipases contributing to cell death (Dong et al., 2009). Preclinical studies have demonstrated the role of calcium-permeable AMPA receptors in causing motor neuron damage and death *in vitro* and *in vivo* models. Understanding these mechanisms is vital for developing therapeutic strategies against ALS-related excitotoxicity (Bosch et al., 2000; Corona and Tapia, 2004).

CHALLENGES IN TRANSLATIONAL RESEARCH

Current therapies for managing neurodegenerative diseases have focused on providing symptomatic relief rather than addressing the root causes. The challenge in neurodegeneration lies beyond the blood–brain barrier (BBB), where many existing formulations face limitations. The inability to deliver adequate doses to the brain hampers effective intervention against neurodegenerative diseases. The complex nature of the BBB, combined with the limited permeability of most drugs, underscores the deficiency in viable treatment options for neurodegenerative diseases (Lamptey et al., 2022).

Blood–Brain Barrier (BBB)

The BBB serves as a crucial diffusion barrier, maintaining homeostasis and normal brain function by restricting the entry of substances from the blood into the brain (Ballabh et al., 2004). The complex structure of the BBB involves brain microvascular endothelial cells, tight junctions, neurons, astrocytes and basal membranes, forming a physically tight brain capillary (Greene and Campbell, 2016). The absence of fenestrations and the presence of inter-endothelial junctions limit the diffusion of small molecules and proteins, while the basal lamina, astrocytes and pericytes further hinder access to the brain. Efflux transporters in brain capillaries complement this barrier, returning substances that enter the brain back into the bloodstream. Regulation of BBB permeability involves inter-endothelial junctions, including tight, gap and adherens junctions (Begley, 2004; Hawkins and Davis, 2005). Molecules traverse the BBB through paracellular or transcellular pathways, with their physicochemical properties influencing transport, such as size, molecular weight, surface activity, lipid solubility and charge. These intricacies of the BBB present a formidable challenge for novel molecules to diffuse across the BBB and reach the site of action to demonstrate beneficial pharmacological effects.

Pharmacokinetics of the Frugs

The effectiveness of systematically administered drugs is significantly influenced by their pharmacokinetic characteristics (Dong, 2018). The journey from administration to the target site, particularly the brain, poses challenges that often hinder therapeutic efficacy. Plasma proteins, to which some drugs strongly bind, limit the amount of free drug available for transportation to the brain (Ghuman et al., 2005). Moreover, rapid elimination by major clearance organs further diminishes drug availability in the bloodstream. Interaction between drugs and target cells can impede absorption, affecting channels, membrane potential or cell conformation. This transient effect restricts the cell's response to the administered drug. In general, small lipophilic drug molecules are considered more suitable for brain delivery. Recognizing and overcoming these challenges in drug delivery is essential for enhancing the efficacy of novel therapeutic interventions.

CONCLUSION

In conclusion, the exploration of novel molecular targets has become a forefront focus in neurodegenerative research. The growing impact of neurodegenerative diseases, characterized by widespread neuronal loss and significant physical and cognitive impairments, underscores the urgency to advance our understanding and therapeutic approaches. From the progressive loss of neurons in Alzheimer's to the triad of symptoms in Parkinson's and the autosomal dominant nature of Huntington's disease, each condition presents unique challenges. The lack of disease-modifying therapies in currently approved medications emphasizes the imperative to delve deeper into the pathophysiological mechanisms. The ongoing research into novel molecular targets, spanning various receptors, proteins and genetic factors associated with these diseases, holds significant promise. Understanding these molecular targets at a molecular level offers the potential to revolutionize preventive measures and refine pharmacological treatment strategies for the better management of neurodegenerative diseases.

REFERENCES

Aflaki E, Westbroek W, Sidransky E. The complicated relationship between Gaucher disease and Parkinsonism: insights from a rare disease. *Neuron* (2017) 93(4):737–746.

Ajitkumar A, De Jesus O. Huntington disease. In: *StatPearls [Internet]*. Treasure Island, FL: StatPearls Publishing (2024).

Aoki M, Aoki H, Ramanathan R, Hait NC, Takabe K. Sphingosine-1-Phosphate signaling in immune cells and inflammation: roles and therapeutic potential. *Mediators Inflamm* (2016) 2016:8606878.

Ariano MA, Cepeda C, Calvert CR, Flores-Hernández J, Hernández-Echeagaray E, Klapstein GJ, et al. Striatal potassium channel dysfunction in Huntington's disease transgenic mice. *J Neurophysiol* (2005) 93(5):2565–2574.

Ballabh P, Braun A, Nedergaard M. The blood–brain barrier: an overview. *Neurobiol Dis* (2004) 16(1):1–13.

Beggiato S, Tomasini MC, Borelli AC, Borroto-Escuela DO, Fuxe K, Antonelli T, et al. Functional role of striatal A2A, D2, and mGlu5 receptor interactions in regulating striatopallidal GABA neuronal transmission. *J Neurochem* (2016) 138:254–264.

Begley D. ABC transporters and the blood-brain barrier. *Curr Pharm Des* (2004) 10(12):1295–1312.

Ben-Sahra I, Dirat B, Laurent K, Puissant A, Auberger P, Budanov A, et al. Sestrin2 integrates Akt and mTOR signaling to protect cells against energetic stress-induced death. *Cell Death Differ* (2013) 20(4):611–619.

Bhat RV, Andersson U, Andersson S, Knerr L, Bauer U, Sundgren-Andersson AK. The conundrum of GSK3 inhibitors: is it the dawn of a new beginning? *JAD* (2018) 64(s1):S547–S554.

Blaikie L, Kay G, Kong Thoo Lin P. Current and emerging therapeutic targets of Alzheimer's disease for the design of multi-target directed ligands. *Med Chem Comm* (2019) 10(12):2052–2072.

Boer DEC, van Smeden J, Bouwstra JA, Aerts JMFG. Glucocerebrosidase: functions in and beyond the Lysosome. *J Clin Med* (2020) 9(3):736.

Bohra V, Nair V. Gaucher's disease. *Indian J Endocrinol Metab* (2011) 15(3):182–186.

Cacabelos R, Carril JC, Cacabelos N, Kazantsev AG, Vostrov AV, Corzo L, et al. Sirtuins in Alzheimer's disease: SIRT2-related GenoPhenotypes and implications for PharmacoEpiGenetics. *Int J Mol Sci* (2019) 20:1249.

Cervetto C, Venturini A, Passalacqua M, Guidolin D, Genedani S, Fuxe K, et al. A2A-D2 receptor-receptor interaction modulates gliotransmitter release from striatal astrocyte processes. *J Neurochem* (2017) 140(2):268–279.

Chan EYW, Luthi-Carter R, Strand A, Solano SM, Hanson SA, DeJohn MM, et al. Increased huntingtin protein length reduces the number of polyglutamine-induced gene expression changes in mouse models of Huntington's disease. *Hum Mol Genet* (2002) 11(17):1939–1951.

Chen JF, Cunha RA. The belated US FDA approval of the adenosine A2A receptor antagonist istradefylline for treatment of Parkinson's disease. *Purinergic Signal* (2020) 16(2):167–174.

Chen X, Guo C, Kong J. Oxidative stress in neurodegenerative diseases. *Neural Regen Res* (2012) 7(5):376–385.

Choi JH, Stubblefield B, Cookson MR, Goldin E, Velayati A, Tayebi N, et al. Aggregation of α-synuclein in brain samples from subjects with glucocerebrosidase mutations. *Mol Genet Metab* (2011) 104(1–2):185–188.

Cieślak M, Komoszyński M, Wojtczak A. Adenosine A2A receptors in Parkinson's disease treatment. *Purinergic Signal* (2008) 4(4):305–312.

Congdon EE, Ji C, Tetlow AM, Jiang Y, Sigurdsson EM. Tau-targeting therapies for Alzheimer disease: current status and future directions. *Nat Rev Neurol* (2023) 19:715–736.

Corona JC, Tapia R. AMPA receptor activation, but not the accumulation of endogenous extracellular glutamate, induces paralysis and motor neuron death in rat spinal cord in vivo. *J Neurochem* (2004) 89(4):988–997.

Cuadrado A, Kügler S, Lastres-Becker I. Pharmacological targeting of GSK-3 and NRF2 provides neuroprotection in a preclinical model of tauopathy. *Redox Biol* (2018) 14:522–534.

del Ser T, Steinwachs KC, Gertz HJ, Andrés MV, Gómez-Carrillo B, Medina M, et al. Treatment of Alzheimer's disease with the GSK-3 inhibitor tideglusib: a pilot study. *JAD* (2013) 33(1):205–215.

Denovan-Wright EM, Robertson HA. Cannabinoid receptor messenger RNA levels decrease in a subset of neurons of the lateral striatum, cortex and hippocampus of transgenic Huntington's disease mice. *Neuroscience* (2000) 98(4):705–713.

Diao H-L, Xue Y, Han X-H, Wang S-Y, Liu C, Chen W-F, et al. Adenosine A2A receptor modulates the activity of Globus Pallidus Neurons in rats. *Front Physiol* (2017) 8:897.

Dong X. Current strategies for brain drug delivery. *Theranostics* (2018) 8(6):1481–1493.

Dong, X, Wang, Y, Qin, Zh. Molecular mechanisms of excitotoxicity and their relevance to pathogenesis of neurodegenerative diseases. *Acta Pharmacol Sin* (2009) 30:379–387.

Engel T, Hernández F, Avila J, Lucas JJ. Full reversal of Alzheimer's disease-like phenotype in a mouse model with conditional overexpression of glycogen synthase kinase-3. *J Neurosci Off J Soc Neurosci* (2006) 26(19):5083–5090.

Feigin VL, Vos T, Nichols E, Owolabi MO, Carroll WM, Dichgans M, et al. The global burden of neurological disorders: translating evidence into policy. *Lancet Neurol* (2020) 19(3):255–265.

Ferrari E, Scheggia D, Zianni E, Italia M, Brumana M, Palazzolo L, et al. Rabphilin-3A as a novel target to reverse α-synuclein-induced synaptic loss in Parkinson's disease. *Pharmacol Res* (2022) 183:106375.

Ferrer I, Goutan E, Marín C, Rey MJ, Ribalta T. Brain-derived neurotrophic factor in Huntington disease. *Brain Res* (2000) 866(1–2):257–261.

Foran E, Trotti D. Glutamate transporters and the excitotoxic path to motor neuron degeneration in Amyotrophic Lateral Sclerosis. *Antioxid Redox Signal* (2009) 11(7):1587–1602.

Gardian G, Browne SE, Choi DK, Klivenyi P, Gregorio J, Kubilus JK, et al. Neuroprotective effects of phenylbutyrate in the N171-82Q transgenic mouse model of Huntington's disease. *J Biol Chem* (2005) 280(1):556–563.

Gauthier LR, Charrin BC, Borrell-Pagès M, Dompierre JP, Rangone H, Cordelières FP, et al. Huntingtin controls neurotrophic support and survival of neurons by enhancing BDNF vesicular transport along microtubules. *Cell* (2004) 118(1):127–138.

Ghuman J, Zunszain PA, Petitpas I, Bhattacharya AA, Otagiri M, Curry S. Structural basis of the drug-binding specificity of human serum Albumin. *J Mol Biol* (2005) 353(1):38–52.

Glass M, Faull RL, Dragunow M. Loss of cannabinoid receptors in the substantia nigra in Huntington's disease. *Neuroscience* (1993) 56(3):523–527.

Glass M, van Dellen A, Blakemore C, Hannan AJ, Faull RLM. Delayed onset of Huntington's disease in mice in an enriched environment correlates with delayed loss of cannabinoid CB1 receptors. *Neuroscience* (2004) 123(1):207–212.

Greene C, Campbell M. Tight junction modulation of the blood brain barrier: CNS delivery of small molecules. *Tissue Barriers* (2016) 4(1):e1138017.

Hampel H, Vassar R, De Strooper B, Hardy J, Willem M, Singh N, et al. The β-Secretase BACE1 in Alzheimer's disease. *Biol Psychiatry* (2021) 89(8):745–756.

Hannan AJ. Novel therapeutic targets for Huntington's disease. *Expert Opin Ther Targets* (2005) 9(4):639–650.

Hao Y, Guo M, Feng Y, Dong Q, Cui M. Lysophospholipids and their G-Coupled protein signaling in Alzheimer's disease: from physiological performance to pathological impairment. *Front Mol Neurosci* (2020) 13:58.

Hauser RA, Olanow CW, Kieburtz KD, Pourcher E, Docu-Axelerad A, Lew M, et al. Tozadenant (SYN115) in patients with Parkinson's disease who have motor fluctuations on levodopa: a phase 2b, double-blind, randomised trial. *Lancet Neurol* (2014) 13(8):767–776.

Hawkins BT, Davis TP. The blood-brain barrier/neurovascular unit in health and disease. *Pharmacol Rev* (2005) 57(2):173–185.

Haynes, S., Hollopeter, G., Yang, G. Kurpius D, Dailey ME, Gan WB, et al. The P2Y12 receptor regulates microglial activation by extracellular nucleotides. *Nat Neurosci* (2006) 9:1512–1519.

Herzig MC, Kolly C, Persohn E, Theil D, Schweizer T, Hafner T, et al. LRRK2 protein levels are determined by kinase function and are crucial for kidney and lung homeostasis in mice. *Hum Mol Genet* (2011) 20(21):4209–4223.

Hockly E, Richon VM, Woodman B, Smith DL, Zhou X, Rosa E, et al. Suberoylanilide hydroxamic acid, a histone deacetylase inhibitor, ameliorates motor deficits in a mouse model of Huntington's disease. *Proc Natl Acad Sci U S A* (2003) 100(4):2041–2046.

Iring A, Tóth A, Baranyi M, Otrokocsi L, Módis LV, Gölöncsér F, et al. The dualistic role of the purinergic P2Y12-receptor in an in vivo model of Parkinson's disease: signalling pathway and novel therapeutic targets. *Pharmacol Res* (2022) 176:106045.

Jacobson KA, Delicado EG, Gachet C, Kennedy C, von Kügelgen I, Li B, et al. Update of P2Y receptor pharmacology: IUPHAR review. *Br J Pharmacol* (2020) 177(11):2413–2433.

Jiang J, Ravits J. Pathogenic mechanisms and therapy development for C9orf72 Amyotrophic Lateral Sclerosis/Frontotemporal Dementia. *Neurother J Am Soc Exp Neurother* (2019) 16(4):1115–1132.

Khanna R, Pellegrino L, Soska R, Lun Y, Feng J, Frascella M, et al. Exploring the use of pharmacological chaperone AT3375 alone and in combination with recombinant human ß-Glucosidase for Gaucher disease. *Mol Genet Metab* (2012) 105(2):S40.

Kiernan MC, Vucic S, Cheah BC, Turner MR, Eisen A, Hardiman O, et al. Amyotrophic lateral sclerosis. *Lancet* (2011) 377(9769):942–955.

King AE, Woodhouse A, Kirkcaldie MTK, Vickers JC. Excitotoxicity in ALS: overstimulation, or overreaction? *Exp Neurol* (2016) 275 Pt 1:162–171.

Kumar A, Vaish M, Ratan RR. Transcriptional dysregulation in Huntington's disease: a failure of adaptive transcriptional homeostasis. *Drug Discov Today* (2014) 19(7):956–962.

Lalla R, Donmez G. The role of sirtuins in Alzheimer's disease. *Front Aging Neurosci* (2013) 5:16.

Lamptey RNL, Chaulagain B, Trivedi R, Gothwal A, Layek B, Singh J. A review of the common neurodegenerative disorders: current therapeutic approaches and the potential role of nanotherapeutics. *Int J Mol Sci* (2022) 23(3):1851.

Lee JM, Calkins MJ, Chan K, Kan YW, Johnson JA. Identification of the NF-E2-related factor-2-dependent genes conferring protection against oxidative stress in primary cortical astrocytes using oligonucleotide microarray analysis. *J Biol Chem* (2003) 278(14):12029–12038.

Lemarié J, Gibot S. Soluble triggering receptor expressed on Myeloid Cells-1: diagnosis or prognosis? *Crit Care Clin* (2020) 36(1):41–54.

Li Y, Xu H, Wang H, Yang K, Luan J, Wang S. TREM2: Potential therapeutic targeting of microglia for Alzheimer's disease. *Biomed Pharmacother* (2023) 165:115218.

Lovestone S, Boada M, Dubois B, Hüll M, Rinne JO, Huppertz HJ, et al; ARGO investigators: a phase II trial of tideglusib in Alzheimer's disease. *J Alzheimers Dis* (2015) 45(1):75–88.

Luthi-Carter R, Hanson SA, Strand AD, Bergstrom DA, Chun W, Peters NL, et al. Dysregulation of gene expression in the R6/2 model of polyglutamine disease: parallel changes in muscle and brain. *Hum Mol Genet* (2002) 11(17):1911–1926.

Luthi-Carter R, Strand A, Peters NL, Solano SM, Hollingsworth ZR, Menon AS, et al. Decreased expression of striatal signaling genes in a mouse model of Huntington's disease. *Hum Mol Genet* (2000) 9(9):1259–1271.

Luthi-Carter R, Strand AD, Hanson SA, Kooperberg C, Schilling G, La Spada AR, et al. Polyglutamine and transcription: gene expression changes shared by DRPLA and Huntington's disease mouse models reveal context-independent efects. *Hum Mol Genet* (2002) 11(17):1927–1937.

Machado-Vieira R, Manji HK, Zarate CA. The role of lithium in the treatment of bipolar disorder: convergent evidence for neurotrophic effects as a unifying hypothesis. *Bipolar Disord* (2009) 11(Suppl 2):92–109.

Mason SL, Barker RA. Novel targets for Huntington's disease: future prospects. *Degener Neurol Neuromuscul Dis* (2016) 6:25–36.

Mazzulli JR, Xu YH, Sun Y, Knight AL, McLean PJ, Caldwell GA, et al. Gaucher disease glucocerebrosidase and α-synuclein form a bidirectional pathogenic loop in synucleinopathies. *Cell* (2011) 146(1):37–52.

McNeill A, Magalhaes J, Shen C, Chau KY, Hughes D, Mehta A, et al. Ambroxol improves lysosomal biochemistry in glucocerebrosidase mutation-linked Parkinson disease cells. *Brain* (2014) 137(Pt 5):1481–1495.

Mendez MF, McMurtray AM. Neurodegenerative disorders. In: Fink G, editor. *Encyclopedia of stress* (Second Edition). New York: Academic Press (2007) pp. 840–844.

Moore CS, Ase AR, Kinsara A, Rao VT, Michell-Robinson M, Leong SY, et al. P2Y12 expression and function in alternatively activated human microglia. *Neurol Neuroimmunol Neuroinflamm* (2015) 2(2):e80.

Ntetsika T, Papathoma PE, Markaki I. Novel targeted therapies for Parkinson's disease. *Mol Med* (2021) 27(1):17.

Obrietan K, Hoyt KR. CRE-mediated transcription is increased in Huntington's disease transgenic mice. *J Neurosci Off J Soc Neurosci* (2004) 24(4):791–796.

O'Leary PD, Hughes RA. Design of potent peptide mimetics of brain-derived neurotrophic factor. *J Biol Chem* (2003) 278(28):25738–25744.

Osama A, Zhang J, Yao J, Yao X, Fang J. Nrf2: a dark horse in Alzheimer's disease treatment. *Ageing Res Rev* (2020) 64:101206.

Pajares M, Jiménez-Moreno N, García-Yagüe ÁJ, Escoll M, de Ceballos ML, Van Leuven F, et al. Transcription factor NFE2L2/NRF2 is a regulator of macroautophagy genes. *Autophagy* (2016) 12(10):1902–1916.

Paleacu D. Tetrabenazine in the treatment of Huntington's disease. *Neuropsychiatr Dis Treat* (2007) 3(5):545–551.

Park KK, Liu K, Hu Y, Smith PD, Wang C, Cai B, et al. Promoting axon regeneration in the adult CNS by modulation of the PTEN/mTOR pathway. *Science* (2008) 322(5903):963–966.

Paton DM. Istradefylline: adenosine A2A receptor antagonist to reduce "OFF" time in Parkinson's disease. *Drugs Today (Barc)* (2020) 56(2):125–134.

Pingale T, Gupta GL. Current and emerging therapeutic targets for Parkinson's disease. *Metab Brain Dis* (2021) 36(1):13–27.

Qin W, Yang T, Ho L, Zhao Z, Wang J, Chen L, et al. Neuronal SIRT1 activation as a novel mechanism underlying the prevention of Alzheimer disease amyloid neuropathology by calorie restriction. *J Biol Chem* (2006) 281(31):21745–21754.

Ravikumar B, Vacher C, Berger Z, Davies JE, Luo S, Oroz LG, et al. Inhibition of mTOR induces autophagy and reduces toxicity of polyglutamine expansions in fly and mouse models of Huntington disease. *Nat Genet* (2004) 36(6):585–595.

Ray WJ, Buggia-Prevot V. Novel targets for Alzheimer's disease: a view beyond Amyloid. *Annu Rev Med* (2021) 72:15–28.

Riedl MA, Saxon A, Diaz-Sanchez D. Oral sulforaphane increases Phase II antioxidant enzymes in the human upper airway. *Clin Immunol Orlando Fla* (2009) 130(3):244–251.

Rocha EM, Smith GA, Park E, Cao H, Brown E, Hayes MA, et al. Glucocerebrosidase gene therapy prevents α-synucleinopathy of midbrain dopamine neurons. *Neurobiol Dis* (2015) 82:495–503.

Sawda C, Moussa C, Turner RS. Resveratrol for Alzheimer's disease. *Ann N Y Acad Sci* (2017) 1403(1):142–149.

Saxton RA, Sabatini DM. mTOR signaling in growth, metabolism, and disease. *Cell* (2017) 168(6):960–976.

Sehgal SA, Hammad MA, Tahir RA, Akram HN, Ahmad F. Current therapeutic molecules and targets in neurodegenerative diseases based on in silico drug design. *Curr Neuropharmacol* (2018) 16(6):649–663.

Silva-Palacios A, Ostolga-Chavarría M, Zazueta C, Königsberg M. Nrf2: molecular and epigenetic regulation during aging. *Ageing Res Rev* (2018) 47:31–40.
Silva T, Reis J, Teixeira J, Borges F. Alzheimer's disease, enzyme targets and drug discovery struggles: from natural products to drug prototypes. *Ageing Res Rev* (2014) 15:116–45.
Steffan JS, Bodai L, Pallos J, Poelman M, McCampbell A, Apostol BL, et al. Histone deacetylase inhibitors arrest polyglutamine-dependent neurodegeneration in Drosophila. *Nature* (2001) 413(6857):739–743.
Suk TR, Rousseaux MWC. The role of TDP-43 mislocalization in amyotrophic lateral sclerosis. *Mol Neurodegener* (2020) 15(1):45.
Tang X, Toro A, T.G. S, Gao J, Chalk J, Oskarsson BE, et al. Divergence, convergence, and therapeutic implications: a cell biology perspective of C9ORF72-ALS/FTD. *Mol Neurodegener* (2020) 15(1):34.
Tayebi N, Walker J, Stubblefield B, Orvisky E, LaMarca ME, Wong K, et al. Gaucher disease with parkinsonian manifestations: does glucocerebrosidase deficiency contribute to a vulnerability to parkinsonism? *Mol Genet Metab* (2003) 79(2):104–109.
Tolosa E, Vila M, Klein C, Rascol O. LRRK2 in Parkinson disease: challenges of clinical trials. *Nat Rev Neurol* (2020) 16(2):97–107.
Van Den Bosch L, Vandenberghe W, Klaassen H, Van Houtte E, Robberecht W. Ca(2+)-permeable AMPA receptors and selective vulnerability of motor\neurons. *J Neurol Sci* (2000) 180(1–2):29–34.
Van Schependom J, D'haeseleer M. Advances in neurodegenerative diseases. *J Clin Med* (2023) 2(5):1709.
Videnovic A. Treatment of Huntington disease. *Curr Treat Options Neurol* (2013) 15(4):424–438.
Wang W, Zhao Y, Zhu G. The role of sphingosine-1-phosphate in the development and progression of Parkinson's disease. *Front Cell Neurosci* (2023) 17:1288437.
West AB. Achieving neuroprotection with LRRK2 kinase inhibitors in Parkinson disease. *Exp Neurol* (2017) 298(Pt B):236–245.
Williams A, Sarkar S, Cuddon P, Ttofi EK, Saiki S, Siddiqi FH, et al. Novel targets for Huntington's disease in an mTOR-independent autophagy pathway. *Nat Chem Biol* (2008) 4(5):295–305.
Wilson DM, Cookson MR, Van Den Bosch L, Zetterberg H, Holtzman DM, Dewachter I. Hallmarks of neurodegenerative diseases. *Cell* (2023) 186(4):693–714.
Witte ME, Mahad DJ, Lassmann H, van Horssen J. Mitochondrial dysfunction contributes to neurodegeneration in multiple sclerosis. *Trends Mol Med* (2014) 20(3):179–187.
Zhao HT, John N, Delic V, Ikeda-Lee K, Kim A, Weihofen A, et al. LRRK2 Antisense Oligonucleotides Ameliorate α-Synuclein inclusion formation in a Parkinson's disease mouse model. *Mol Ther Nucleic Acids* (2017) 8:508–519.
Zou J, Zhou L, Du XX, Ji Y, Xu J, Tian J, et al. Rhebl is required for mTORC1 and myelination in postnatal brain development. *Dev Cell* (2011) 20(1):97–108.

3 Potential Novel Targets for Alzheimer's Pharmacotherapy

Update on JNK3, GSK3β, and Fyn Kinase Inhibitors and Related Approaches

*Mariya Jahangir and Mymoona Akhter**

Drug Design and Medicinal Chemistry Lab, School of Pharmaceutical Education and Research, Jamia Hamdard, New Delhi, India

*Corresponding author

INTRODUCTION

Alzheimer's disease (AD), the main cause of dementia, is the most common neurodegenerative disease and the seventh leading cause of death worldwide. Like other neurodegenerative diseases, AD is age-related and mainly affects the elderly. Currently, more than 55 million people suffer from dementia, and every year approximately 10 million new cases are reported. The number is expected to rise to 131.5 million by 2050, with a subsequent increase in the social and financial burden. AD, being the most common form of dementia, contributes to 60–70% of cases. The World Health Organization (WHO) recognizes dementia as a public health priority. In May 2017, the World Health Assembly endorsed the "Global action plan on the public health response to dementia 2017–2025." According to the WHO, in 2019, the estimated total global societal cost of dementia was US$1.3 trillion, and these costs are expected to surpass US$2.8 trillion by 2030 as both the number of people living with dementia and care costs increase (WHO Report 2023).

The signs and symptoms linked to dementia/AD can be understood in three stages: early-stage symptoms like forgetfulness and losing track of time, with progression to the middle stage, when signs and symptoms become clearer and may include becoming confused while at home, difficulty with communication, and behavior changes. The last stage involves total dependence and inactivity, difficulty recognizing relatives and friends, and an increasing need for assisted self-care. Numerous reasons have been associated with the development of AD, and it is currently theorized that a combination of factors may be responsible for the progression of the disease. However, the exact and initial cause are still unknown and may be of varied origins. Other factors, including lifestyle and genetic factors influencing the development of the disease, have also been recently identified.

GENETICS OF AD

AD is often divided into two categories: early-onset AD (EOAD) comprising 1–5% of all AD cases, i.e., before the age of 65 only; and the most common type of AD that appears over the age of 65 as late-onset AD (LOAD) (Bekris et al., 2010; Reitz, Rogaeva, and Beecham, 2020; Rabinovici, 2019) Both sporadic and familial forms of EOAD and LOAD are reported and may occur in people with

DOI: 10.1201/9781003506317-3

a positive family history of AD. The familial form of the disease is most often associated with autosomal dominant mutations in genes such as APP (amyloid precursor protein), PSEN1 (presenilin 1), and PSEN2 (presenilin 2) (Bekris et al., 2010; Reitz, Rogaeva, and Beecham, 2020; Cruchaga et al., 2012).

Although the mutations are very rare, accounting for less than 1% of cases, three of these specific genetic mutations (APP on chromosome 21, presenilin-1 [PS-1] on chromosome 14, and presenilin-2 [PS-2] on chromosome 1) were found in patients with familial early-onset autosomal dominant Alzheimer's disease. All these genes appear to increase cellular Aβ42 production by selectively increasing APP cleavage by β- or γ-secretase.

Apolipoprotein E (APOE), the fourth AD gene on chromosome 19's long arm with three allelic forms (APOE-2, -3, and -4), has been established to be a major risk allele, with an approximately eightfold increase in the risk of AD development (Fish et al., 2019). A number of reports suggest that the APOE-4 allele is disproportionately represented in patients with both late-onset and early-onset Alzheimer's disease and that the APOE-4 allele has a dose-dependent relationship with increased risk for AD and decreasing age at onset. On the other hand, inheriting the APOE-2 allele may be protective, as suggested by several studies. While the exact cause of AD is not yet known, it has been reported that several hallmarks are associated with the pathological development of the disease, including acetylcholinesterase, tau protein, reactive oxygen species, metal ions, and β-amyloid. These problems contribute to synapse loss in the brain due to senile plaques and neurofibrillary tangles in the brain of an individual. The areas of the brain associated with cognition are affected first, and the development of senile plaques in this region is followed by a gradual spread to other areas. Senile plaques consist of insoluble deposits of amyloid β peptide (Aβ), a fragment of the amyloid precursor protein (APP). APP is a large membrane protein that has a significant role in neural growth and repair. Aβ is toxic to neurons, causing inflammation or neurodegeneration upon accumulation. The APP fragments into two Aβ types: the shorter Aβ40 and the longer Aβ42. Among these two Aβ proteins, Aβ42 is considered more toxic than Aβ40. This difference in fragmentation occurs due to the difference in the enzyme that cleaves the APP. If cleavage of APP occurs with γ-secretase, then the shorter Aβ40 is obtained, and if cleavage occurs with β-secretase, then the longer Aβ42 is obtained. Another prominent feature of AD is the build-up of neurofibrillary tangles within the neurons. Neurofibrillary tangles are mostly formed by chemically altered (abnormally folded and hyperphosphorylated) tau protein, a protein involved in microtubule formation (Schachter and Davis, 2000). The severity of disease progression is reflected by the tangles: the more the severity, the more the tau tangles in the brain.

Both amyloid plaques and neurofibrillary tangles are thought to cause an excessive release of glutamate in certain cortical and sub-cortical structures (Lewerenz and Maher, 2015) that can lead to neuronal death through N-methyl-D-aspartate (NMDA) receptor-mediated excitotoxicity (Hu, Ondrejcak, and Rowan, 2012).

Recently, with the intervention of *in silico* techniques like network biology and reverse pharmacology, numerous targets have been identified to develop novel small molecules for the treatment of dementia. A recent review by Mangal and Ding on prospective therapeutic targets of Alzheimer's disease presents a collection of therapeutic targets that have yielded positive results in mice models. The targets include calcium homeostasis, high-density lipoprotein, meningeal lymphatics, mitochondrial homeostasis, genomic instability, cold-shock proteins, and antibodies (Mangal and Ding, 2022).

Interventions related to the management of amyloid plaques and neurofibrillary tangle formation are of primary importance. As a result, anti-amyloid strategies and anti-tau strategies have been concentrated upon in particular. The anti-amyloid strategies involve micromanagement of a number of enzymes like secretase (α, β, and γ), prevention of amyloid aggregation, and use of antibodies. The anti-tau strategy includes the management of tau phosphorylation and aggregation inhibition through enzymes like GSK3 or anti-tau immunotherapy. Other approaches include managing cholinesterase enzymes, oxidative stress, metal ion imbalance, sirtuin, caspases, serotonin, monoamine oxidase

B, NMDA receptors, muscarinic and nicotinic acetylcholine receptors, and nuclear factor erythroid 2-related factor 2. Here in this chapter, we will be discussing newer approaches to manage AD.

A hallmark of AD is the cleavage of APP, which leads to the release of a peptide from neurons known as Aβ peptides, the accumulation of which is a hallmark of AD. Soluble oligomeric Aβ peptides, particularly Aβ42, have neurotoxic functions; however, the molecular mechanisms underlying their generation and toxicity are poorly understood. In a paper published in *Neuron*, Yoon and colleagues described the stages of Aβ42 production and toxicity. They demonstrate that AD is a metabolic condition in which Aβ42 peptides result in a translational block that causes neurotoxicity and the cleavage of APP in a convoluted feed-forward loop that is strictly controlled by JUN N-terminal kinase 3 (JNK3). JNK3 deletion has been shown to improve symptoms in an AD animal model (Yoon et al., 2012; Flemming, 2012).

JNK3

JNK3 has been reported to perpetuate metabolic stress induced by Aβ peptides. JNK3 belongs to the JNK family of stress-activated serine threonine protein kinases that, in turn, belongs to the MAPKs. JNKs differ from classical MAPKs in that they are induced in response to cellular stress rather than mitogens. Additionally, the JNK cascade is controlled as the classic module of 3-MAPK. Various JNK splicing variants are encoded by three genes, namely, Mapk8, Mapk9, and Mapk10. JNKs regulate the activity of several transcription factors, including p53, ATF-2, Elk-1, c-Jun, and c-Myc, and other factors in the Bcl-2 family (Bogoyevitch and Kobe, 2006). From proliferation to cell death, JNKs are involved in the regulation of many cellular activities. In contrast to JNK1 and JNK2, JNK3 is expressed in the CNS at very high levels, and in cardiac smooth muscle and testis at low levels. For this reason, JNK3 is mainly involved in neurodegenerative processes like AD (Yoon et al., 2012; Yarza et al., 2015) cerebral ischemia (Pan et al., 2015), Parkinson's disease (PD) (Kuan et al., 2003), and other CNS disorders (Chintala, Putris, and Geno, 2015; de Lemos et al., 2018).

The JNKs, and particularly JNK3, the distinct isoform for neurons are recognized as vital enzymes for the pathology of various forms of dementia, Alzheimer's included (Musi et al., 2020). The JNK3 is distinctively expressed in the CNS and is stimulated by stress stimuli thus making it a worthy potential target for treating AD. A physical pathway for the assembly of signal transduction is provided by the scaffold protein, which brings together a particular triad of MAP3K, MAP2K, and JNK. This organization develops a "functional signaling module" that makes use of a "conveyor belt" mechanism to simplify signal transduction while amplifying various signals (Bonny, Nicod, and Waeber, 1998; Perry et al., 2019). Scaffold proteins can also undergo phosphorylation and dephosphorylation by themselves (Willoughby et al., 2003). Two crucial regulators of JNK signaling are the scaffold proteins JIP and β-arrestin-2 (Morrison and Davis, 2003). In polarized cells like neurons, this signal control process is much more crucial. The response in neurons is extremely varied depending on whether JNK is activated in the nucleus or in postsynaptic components. While JNK3 phosphorylation of PSD-95 by JNK3 results in the downregulation of AMPA and NMDA receptors and the degradation of PSD-95 itself in the postsynaptic element, negatively regulating synaptic plasticity, JNK3 activation in the nucleus induces c-Jun phosphorylation, activating the neuronal-death program.

THE FUNCTION OF JIP SCAFFOLD PROTEIN

The JNK interacting proteins (JIPs), a family of scaffold proteins, are encoded by four genes (JIP1, JIP2, JIP3, and JIP4). The JIPs coordinate the JNK module and bind other proteins thereby impacting the various stimuli in their cellular response. JIPs influence the cellular response to distinct stimuli like NMDA stimulation, oxidative stress, UV, ischemia, and toll-like receptor 4 by coordinating with the JNK module and by binding to other proteins like kinesin light chain (KLC) and APP, as well as regulate distinct and overlapping cellular functions. JIP proteins connect both positive and negative JNK signaling regulators by activating, enhancing, and speeding up JNK phosphorylation/activation, as well as mediating JNK dephosphorylation/inactivation.

All JNK isoforms and other JNK signaling cascade participants (MKK7, phosphatases, MEKK3, MLK3, and DLK) are specifically bound by JIP-1, whereas JIP-2 interacts with JNK1, JNK2, MKK7, MLK2, MLK3, and DLK. Nevertheless, JIP-2 can also interact with p38, but JIP-1 appears to bind solely to JNK family members (Dhanasekaran et al., 2007). Both scaffolds have increased levels of expression in pancreatic beta cells as well as in neurons, where they are localized specifically in the cytoplasm, axons, dendritic growth cones, and synapses (Davis, 2000). JIP-1 has a higher affinity for JNKs than JIP-2, but the overexpression of both JIP-1 and JIP-2 acts as an inhibitor of JNK signaling, most likely because they sequester cascade actors. JIP-1 and JIP-2 can form homodimers as well as heterodimers with each other (Figure 3.1) (Yasuda et al., 1999). JIP-3, like JIP-1 and JIP-2, has a high affinity for JNK3 (Drerup and Nechiporuk, 2013) and acts as a JNK activator, but it can also bind ERK and act as an inhibitor. JIP-3 has the ability to bind MKK7, MKK4, MEKK1, MLK3, and ASK1 (Dhanasekaran et al., 2007). Furthermore, JIP-3 is a JNK target because it can only be phosphorylated by JNK and not by p38 or ERK (Kelkar et al., 2000).

JIPs play critical roles in neuron development and stress responses. JIP-1 was initially identified as a JNK signaling inhibitor because its overexpression inhibits JNK phosphorylation on its preferred target c-Jun, thereby preventing apoptosis (Bonny et al., 2001). In fact, JBD (JNK-binding domain) inhibitors, such as D-JNKI1, mimic JBD overexpression inside cells, preventing neuronal death in cerebral ischemia by avoiding the link between JNK and its physiological targets. JIPs are primarily found in the cytoplasm, but nuclear localization has been reported in neurons under stress conditions; they also localize in axon growth cones and both pre- and postsynaptic compartments. The localization of the JNK module is even more important in neurons, which are highly polarized cells than in other cell types. JNK can perform different functions in different cellular compartments. JIPs, which can move between cellular districts, are thus very intriguing proteins that can be used to modulate JNK action.

FUNCTION OF β-ARRESTIN

Arrestin proteins are well-known for their dual role in G protein-coupled receptor (GPCR) signaling, but they also serve other functions. In fact, β-arrestins serve as scaffold proteins for clathrin and AP2 (adaptin) components of clathrin-coated endocytosis, but they also serve as scaffolds for multiple kinases and phosphatases involved in cell signaling (such as the MAP kinase module).

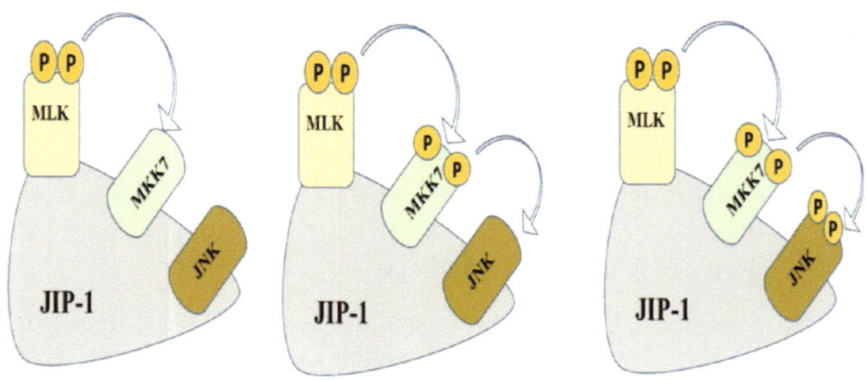

FIGURE 3.1 JNK's scaffold protein JIP-1. JIP-1 is a crucial regulator of the JNK signaling pathway. It links multiple MAPK members, tethering them together and accelerating the reaction cascade. JIP-1 regulates signal transduction and helps localize pathways in components of the cell such as the plasma membrane, the cytoplasm, the nucleus, the Golgi, and in neurons in the pre- and postsynaptic elements (Musi et al., 2020).

Arrestin-1, arrestin-2, arrestin-3, and arrestin-4 are the four members of the arrestin family. Arrestin-1 and arrestin-4 are also known as visual arrestins because they are found in the photoreceptors of the retina, whereas arrestin-2 and arrestin-3 are found everywhere. Arrestin-2 is also known as β-arrestin-1, and arrestin-3 as β-arrestin-2. Both β-arrestin-1 and β-arrestin-2 are expressed by neurons, but adult neurons have 10-/20-fold higher levels of β-arrestin-1 in most brain regions. Among the four arrestin isoforms, arrestin-2 in particular binds to JNK3. By changing conformation, β-arrestin-2 reveals different binding domains that link and recruit specific proteins, specifically MAPK module components. In fact, β-arrestin-2 interacts with the JNK3, MKK4/MKK7, and ASK1 modules, as well as JNK-phosphatase and the MAP kinase phosphatase 7 (MKP7). The β-arrestin-2 scaffold can amplify various signal modules/pathways, and how this module amplifies the signal is explained by the previously described "conveyor belt" model. In the case of JNK3, β-arrestin-2 binds inactive JNK3, bringing it closer to MKK7/MKK4, which can then phosphorylate JNK3. A single phosphorylation of JNK3 decreases its affinity for β-arrestin-2, whereas full phosphorylation decreases its affinity for the scaffold even more, resulting in the dissociation of p-JNK3, allowing another inactive JNK3 molecule to bind β-arrestin-2 (Figure 3.2).

Few JNK small molecule inhibitors have entered clinical trials in recent years for various reasons, but none for the treatment of AD. Bentamapimod (**1**) for treating inflammatory endometriosis; tanzisertib (CC-930) (**2**) for treating inflammatory endometriosis, discoid lupus erythematosus, and idiopathic pulmonary fibrosis; and D-JNKi1 for the treatment of inflammation and stroke are currently being studied (Qin et al., 2022).

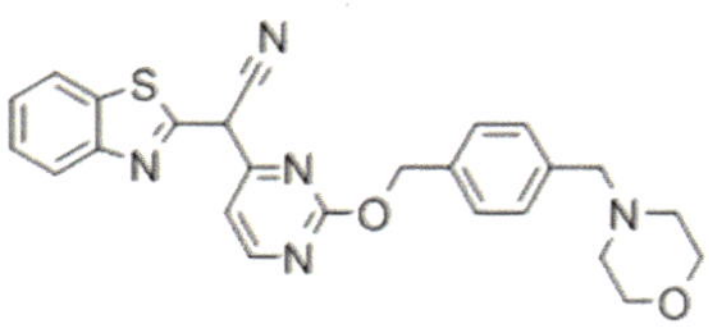

Bentamapimod
(**1**)

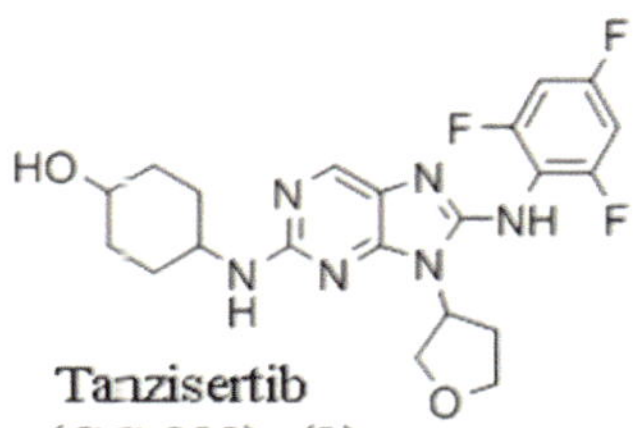

Tanzisertib
(CC-930) (**2**)

Zulfiqar et al. (2020) have identified JNK3 inhibitors for the prevention of neuroinflammation induced by MCAO in rats by repurposing FDA-approved drugs. By performing virtual screening of ~600 FDA-approved drugs against JNK3 using AutoDock Vina 4.2, they identified four drugs—dabigatran (**3**), estazolam (**4**), pitavastatin (**5**), and leucovorin (**6**)—that could be developed for the condition. The assessment of the binding potential of the identified drugs against JNK3 with reference to

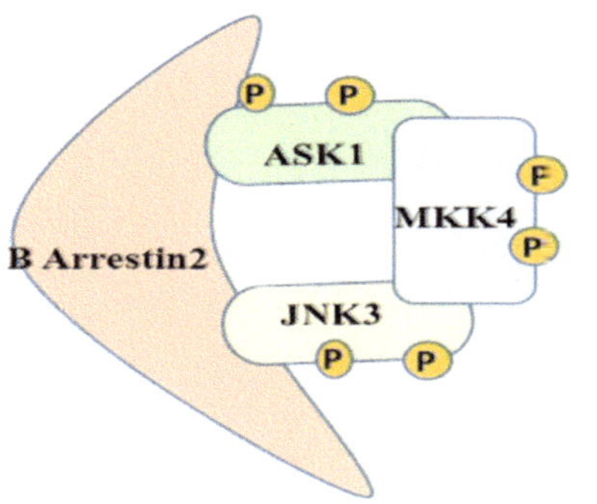

FIGURE 3.2 β-Arrestin-2: JNK3 scaffold. β-Arrestin-2 specifically associates with the JNK3, MKK7/4, and ASK1. Inactive-JNK3 binds β-arrestin-2 close to the MKK7/4 bond-site, facilitating full JNK3 activation. p-JNK3 loses its affinity to the scaffold and, by detaching from it, allows the binding of other inactive JNK3 (Musi et al., 2020).

the standard JNK3 inhibitor SP600125 was reported. To evaluate the interaction profile in terms of dynamics, molecular dynamics (MD) simulations were conducted by utilizing the JNK3-bound inhibitors (dabigatran, estazolam, and pitavastatin) using the Groningen Machine for Chemical Simulations (GROMACS) 5.0.7. As reported, the reprofiled drugs attenuated JNK expression and downstream neuroinflammatory mediators. Additionally, pitavastatin treatment showed a reduction in the infarct area and significant improvement in neurological functions, accompanied by decreased p-JNK expression.

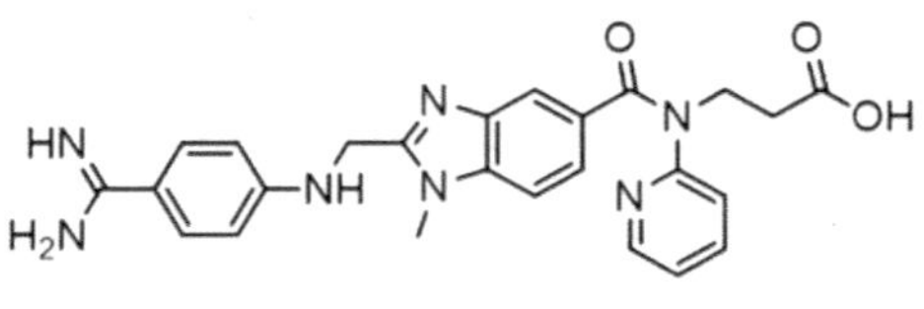

Dabigatran (**3**)

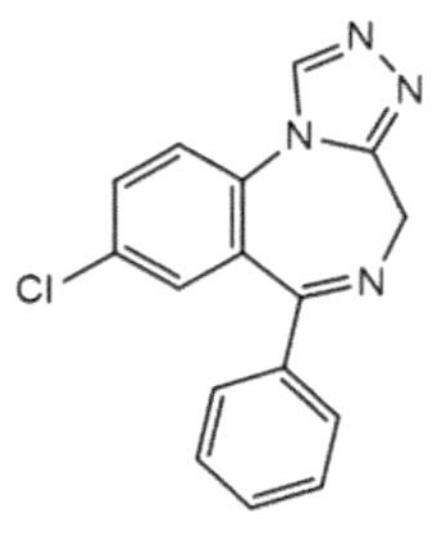

Estazolam (**4**)

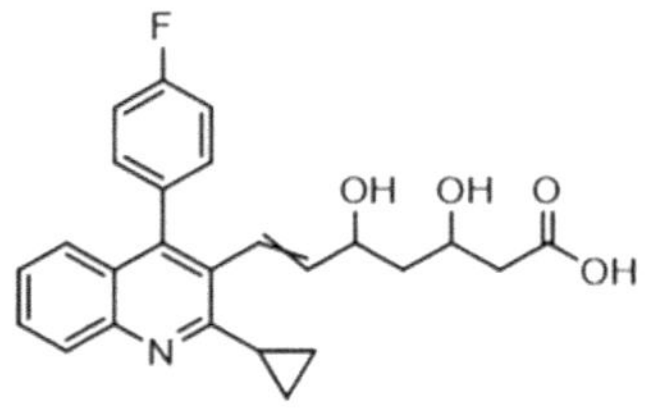

Pitavastatin (**5**)

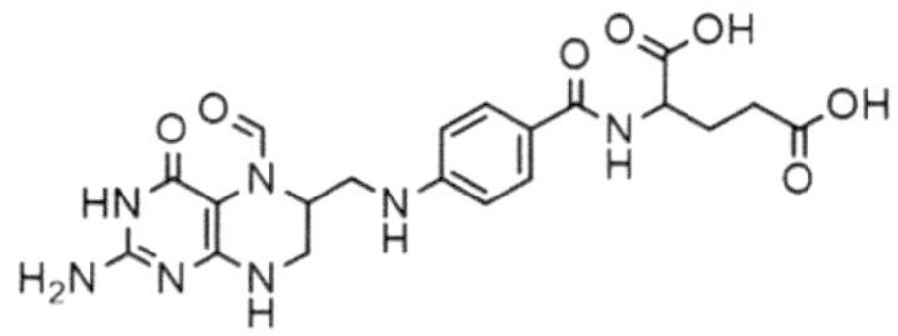

Leucovorin (**6**)

Muth et al. reported tri- and tetra-substituted pyridinylimidazoles as covalent inhibitors of JNK3. The authors synthesized tri- and tetra-substituted pyridinylimidazoles and studied the influence of a 4-fluorophenyl ring at the imidazole-C4 position as well as that of an imidazole N-alkylation on biological activity. The influence of the substitution pattern of both phenyl rings A and B on biological activity has been investigated extensively (Muth et al., 2017).

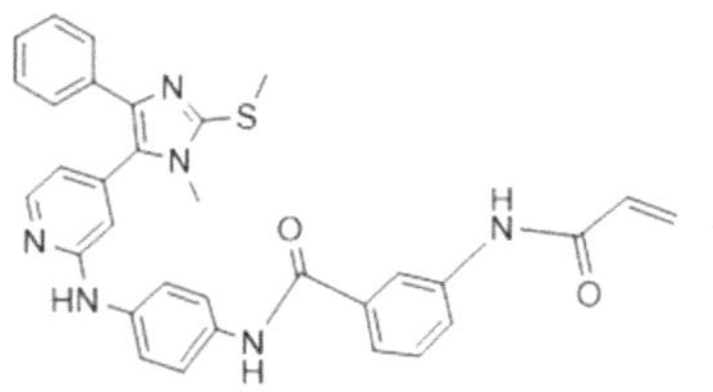

(**7**)

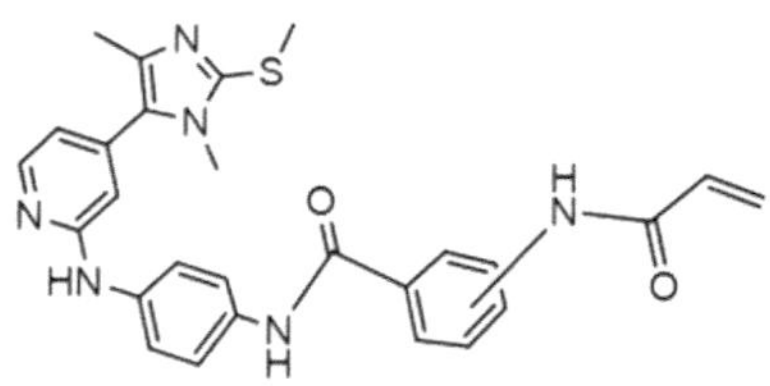

(**8**)

To determine the selectivity of compounds within the kinome, the authors examined compounds **7** and **8** for their metabolic stability and screened them against a panel of 410 kinases. Compound **7** showed inhibition of JNK3 at 0.3 nM and selectivity of 120 toward JNK3 (IC_{50} (p38α)/IC_{50} (JNK3). Compound **7** was optimized by structure modification to yield compound **8** with an IC_{50} of 2 nM and selectivity of 976 over p38α. It is reported that compounds **7** and **8**, with their para/meta substitution pattern, comprise the optimal linker length and angle for the electrophilic warhead to form a covalent bond to the Cys-154. The formation of a covalent bond by compounds **7** and **8** to Cys-154 of the target protein was confirmed by incubation of the compounds with JNK3 followed by mass experiments (Muth et al., 2017).

Compound **7** was further taken up by Rehfeldt et al. (2021) to evaluate the neuroprotective, anti-inflammatory, and antioxidant effects in SH-SY5Y and RAW264.7 cell lines. They reported

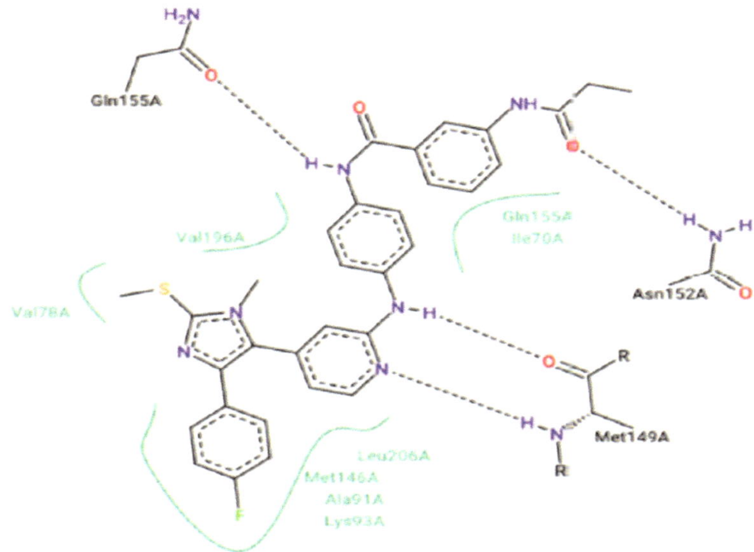

FIGURE 3.3 The 2D protein-ligand interaction plot for 1PMN, emphasizing the interactions between the ligand (compound **7**) and the residues Met149, Asn152, and Gln155.

that the compound decreased neurotoxicity induced by 6-OHDA by approximately 20%, and in RA-differentiated cells, cell death was diminished by approximately 40% and 90% after 24 and 48 hours of treatment, respectively. The authors reported the formation of hydrogen bonds to residues Met149, Asn152, and Gln155, as shown in Figure 3.3, in addition to the interactions reported in the original paper by Muth et al. (2017).

Thiophene-pyrazolourea derivatives as potent, orally bioavailable, and isoform-selective JNK3 inhibitors are reported by Feng et al. (2021). The authors optimized the earlier developed compounds (Zheng et al., 2014), compounds with high general kinase selectivity. A highly potent and selective JNK3 inhibitor (compound **11** with an IC_{50} of 35 nM) has been reported from the optimizations in a panel profiling of 374 wild-type kinases. The authors also reported cocrystal structures of compounds **11** and **12** in human JNK3 with a resolution of 1.84 Å, showing that these JNK3 isoform-selective inhibitors bound to the ATP pocket and had interactions in both hydrophobic pocket I and hydrophobic pocket II.

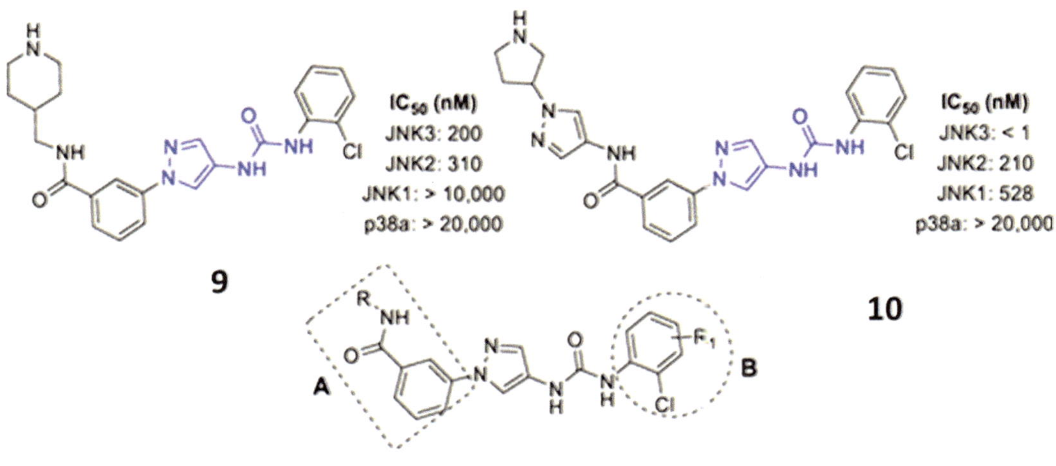

9

10

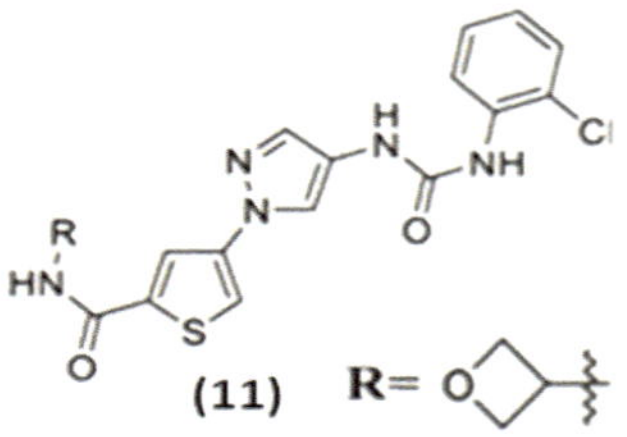

(11)

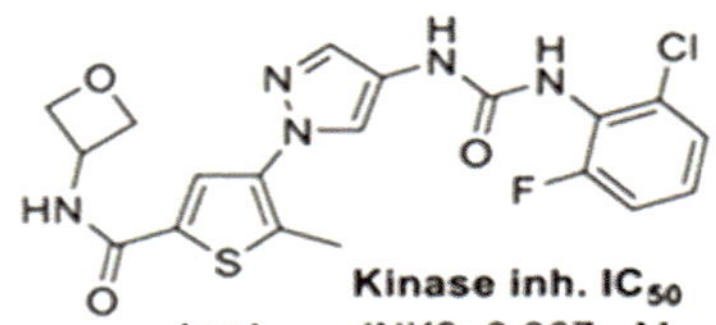

(12)

Feng et al. (2020) also reported N-aromatic-substituted indazole-based, orally bioavailable JNK3 inhibitors. Extensive structure–activity relationship studies of the series, utilizing various *in vitro* and *in vivo* assays, led to development of a highly potent and highly selective JNK3 inhibitor (compound **13**) with good oral bioavailability and high brain penetration. Compound **13** (IC_{50} of 0.005 μM) showed significant inhibition (>80% at 1 μM) of only JNK3 and JNK2 in a panel profiling of 374 wild-type kinases and demonstrated high potency in functional cell-based assays. The cocrystal structure of compound **13** in human JNK3 at a 2.1 Å resolution has also been reported (Figure 3.4) (Feng et al., 2021).

Dou et al. constructed an unbiased benchmarking data set for virtual screening against JNK3 and identified three compounds—**14**, **15**, and **16**—that showed JNK3 IC_{50} values at the nanomolar level. Compound **15** and its derivatives were further developed by optimization based on the structural scaffolds. These identified compounds had a JNK3 IC_{50} value of 40 nM, exhibiting over 2500-fold isoform selectivity against JNK1α1 and JNK2α2 and excellent kinome-wide selectivity against 398 screened kinases. Compound **15**, including its derivatives, is also reported to show comparable neuroprotective activity with LiCl *in vitro* and improved learning and memory abilities in the APPswe/PS1dE9 mice. The compounds act through the reduction of phosphorylation of JNK, APP, and tau protein in the cortex and hippocampus, thereby reducing the diffuse and fibrillar Aβ plaques (Dou et al., 2019).

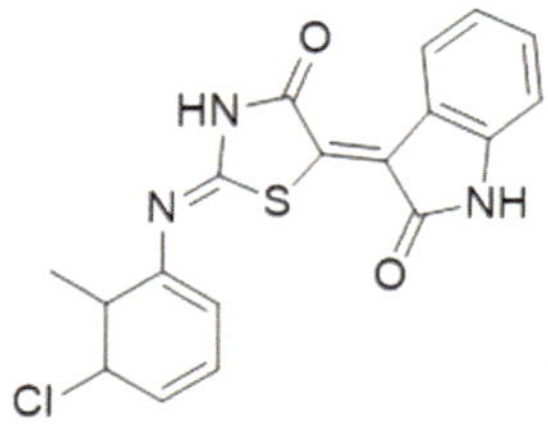

14

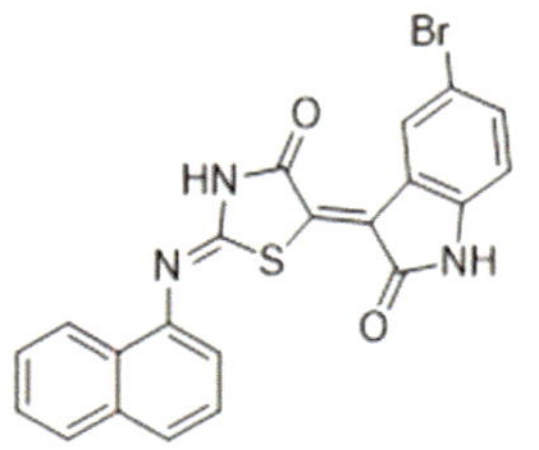

15

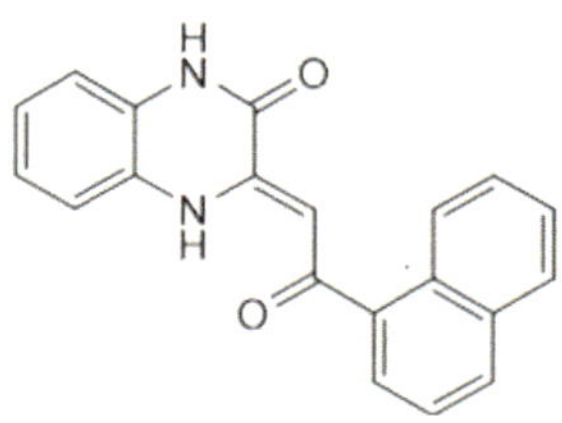

16

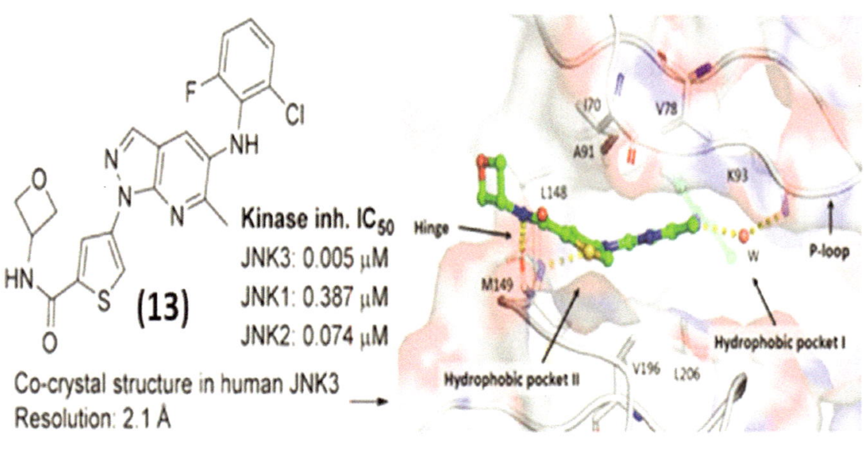

FIGURE 3.4 Structure of compound **13** and its cocrystal structure in human JNK3 (Feng et al., 2021).

Design and synthesis of aminopyrazole derivatives as selective JNK3 inhibitors and their structure–activity relationship studies have been reported, with compounds **17** and **18** as derivatives with good activity (Zheng et al., 2014).

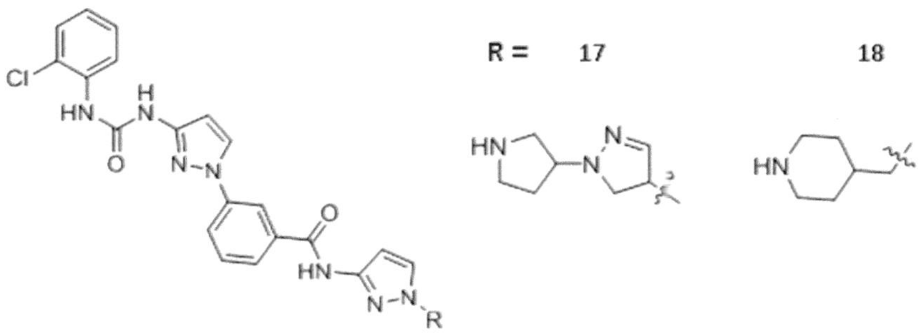

17 and 18

Shuai et al. have reported docking-based virtual screening and structure–activity relationship studies of a series of indazole derivatives and identified **19** with excellent JNK3 inhibitory activity (IC_{50} = 85.21 nM) and 100-fold isoform selectivity for JNK3 over JNK1/2. Compound **19** has shown neuroprotective effects on *in vitro* and *in vivo* Parkinson's disease models by selectively inhibiting JNK3 with ideal blood–brain barrier permeability and low toxicity (Shuai et al., 2023).

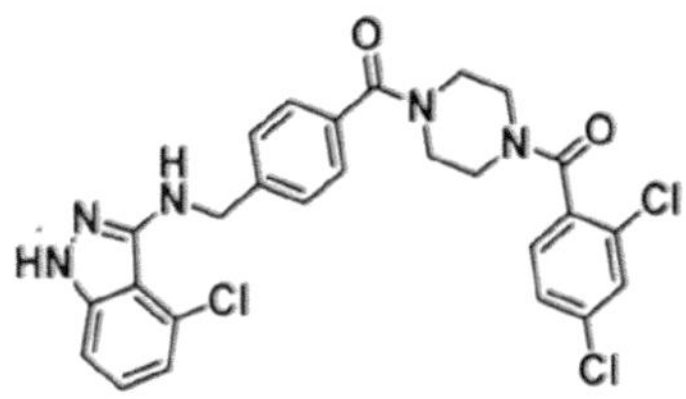

19

Syntheses of 1-(2-aminopyrimidin-4-yl)-2-aryl-1H-benzo[d]imidazole-5,6-diol derivatives as potent inhibitors of the JNK3 isoform have been reported by Jun et al. Thirty-seven derivatives are reported with their enzyme activity against JNK3, where a number of compounds have competitive activities against JNK3 with IC_{50} values of about 20 nM. The novel scaffold, 5, 6-dihydroxy 3-alkyl-2-aryl-1-pyrimidinyl-1H-benzo[*d*]imidazole is presented to be highly useful in the development of JNK3 selective inhibitors as therapeutic agents for neurodegenerative diseases. Compound **20**, with an IC_{50} of 9.7 nM, is presented as the most potent compound against JNK3 with an excellent selectivity profile compared with its activity toward similar protein kinases such as GSK3β, Erk, JNK1, and JNK2. Additionally, compound **20** produced 95.7% neuroprotection in Aβ-induced primary rat cortical neuronal death, as reported by Jun et al. (2021).

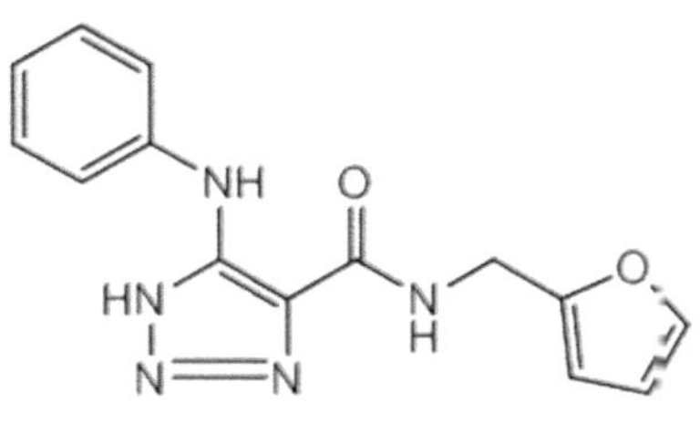

21

Using fluorescence thermal shift assays, screening of a chemically diversified lead-like scaffold library of 2024 pure compounds and native mass spectrometry, Cheng et al. identified novel compounds with JNK3 inhibitory activity. Native mass spectrometry has been used to validate the interaction of hits with JNK3. Binding geometries and interactions of the top two hits were evaluated by docking into a JNK3 crystal structure. Compound **21** with K_d of 21 μM with JNK3 scaffold 5-(phenylamino)-1H-1,2,3-triazole-4-carboxamide is suggested as a novel and selective JNK3 binder (Cheng et al., 2022).

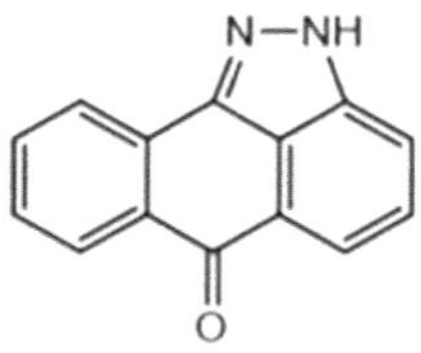

22

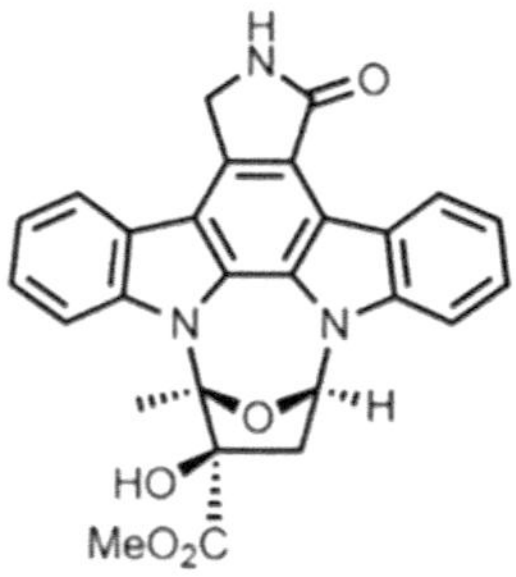

23

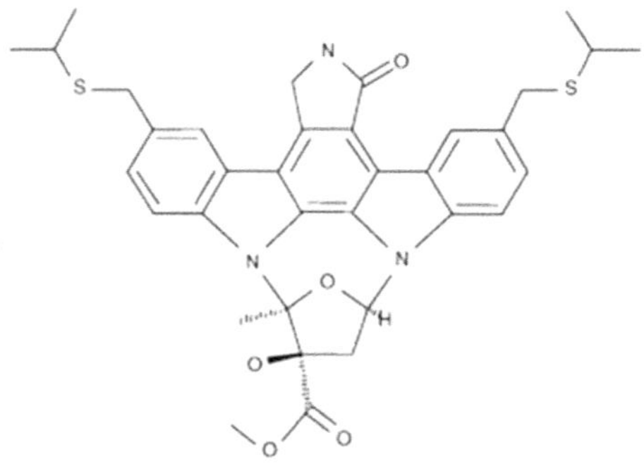

24

Bennett et al. (2001) identified direct ATP-competitive inhibitors of JNK SP600125 (anthra[1,9-*cd*]pyrazol-6-(2H)-one) (**22**) as antitumor agents. Later on, Sharma et al. demonstrated the use of the agent to attenuate JNK in streptozotocin induced neurocognitive deficit and oxidative stress in rats (Sharma, Deshmukh, and Bedi, 2010).

Other direct ATP-competitive inhibitors of JNK3 reported include SP600125 (anthra[1,9-cd]pyrazol-6-(2H)-one) (**22**) (Bennett et al., 2001; Sharma, Deshmukh, and Bedi, 2010) and a few mixed lineage kinase inhibitors such as indolocarbazole K252a (**23**), CEP11004 (**24**), and URMC099 (**25**) BI-78D3 (**26**) (Berndt Plückthun and Waldmann, 2013). These compounds are designed to bind to and inhibit the activity of JNK3, a protein kinase involved in several cellular processes, including inflammation and apoptosis.

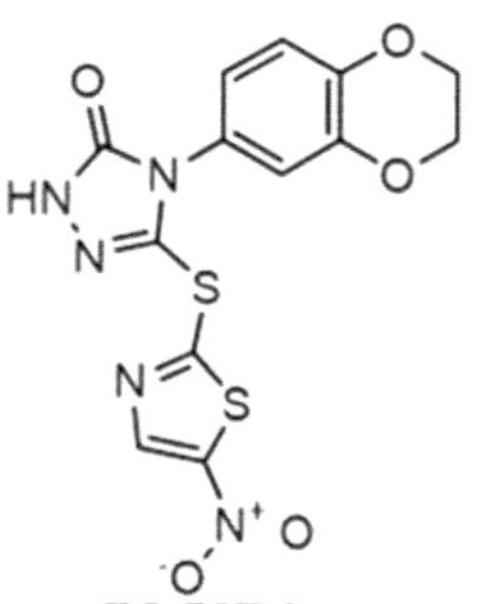

BI-78D3

25

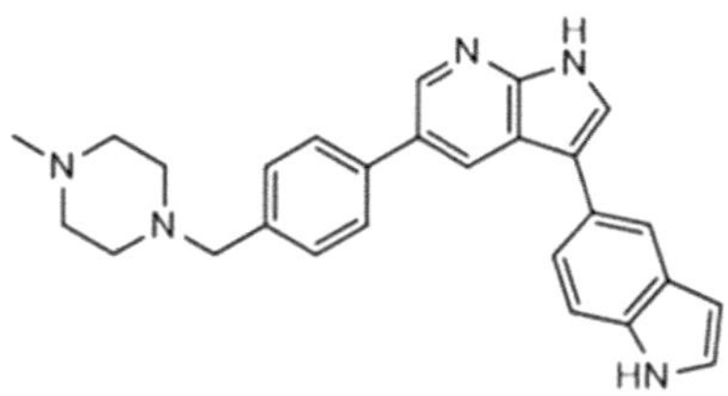

26

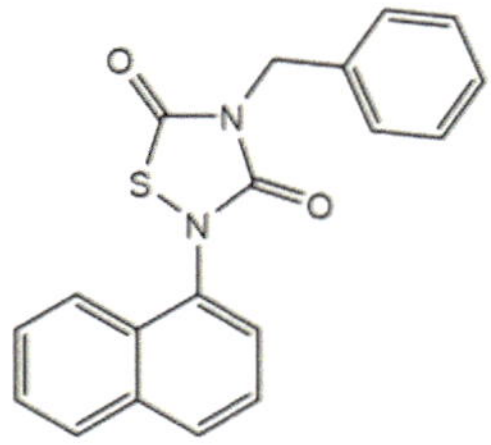

27

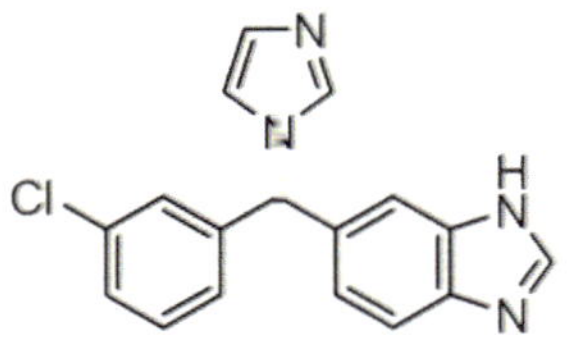

28

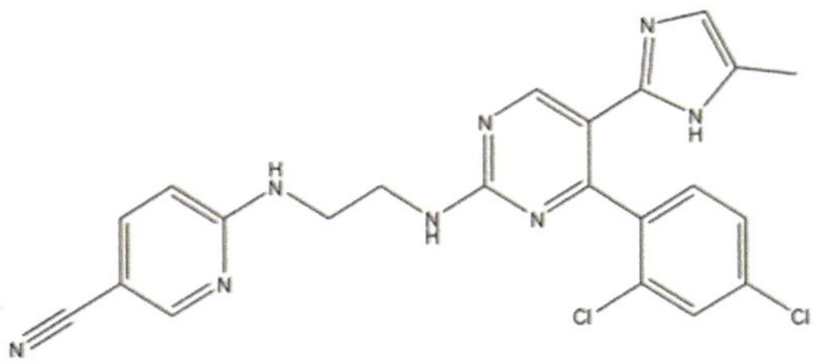

29

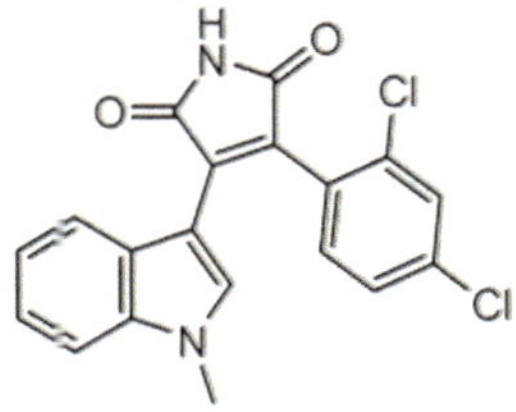

30

However, it is important to note that these compounds are not approved for medical use, and their safety and efficacy have not been established in clinical trials. The structure–activity relationship (SAR) for small molecule inhibitors of JNK3 has been extensively studied, and several key structural features have been identified as important for their activity. Some of the common structural features for JNK3 inhibitors include hydrophobic moieties. Hydrophobic groups are often found in the binding pockets of JNK3 inhibitors and can help increase their binding affinity. Heterocyclic rings, such as pyrazoles and imidazoles, are commonly found in JNK3 inhibitors and can help improve their potency. Substituted amines can also play a role in improving the binding affinity of JNK3 inhibitors. Some JNK3 inhibitors contain chiral centers, which can help optimize their stereochemistry and improve their efficacy. The tight binding with hydrophobic pocket I is essential to improve JNK3 isoform selectivity. The formation of hydrogen bonds with Met149 is key to improving inhibitory activity (Qin et al., 2022). It is important to note that these SAR features are not always essential for JNK3 inhibition, and many other factors, such as conformational flexibility and solvent accessibility, can also play a role in determining the activity of these compounds. Furthermore, the SAR for JNK3 inhibitors can vary depending on the specific JNK3 isoform and the assay used to evaluate their activity.

GLYCOGEN SYNTHASE KINASE-3B

Glycogen synthase kinase-3 (GSK-3) is a highly conserved protein–serine/threonine kinase that is capable of phosphorylating glycogen synthase (Embi, Rylatt, and Cohen, 1980). GSK-3 is encoded by two highly related genes encoding GSK-3α and GSK-3β, respectively. The two enzymes GSK-3α and GSK-3β are ubiquitously expressed, constitutively active kinases involved in a variety of cellular processes in addition to glycogen metabolism (Welsh and Proud et al., 1993), apoptosis (Turenne and Price, 2001), gene transcription (Troussard et al., 1999), and microtubule stability (Anderton et al., 2001; Brion et al., 2001). In the brain, many crucial cellular processes are regulated by GSK-3β, controlling numerous signaling pathways (Doble and Woodgett, 2007; Forde and Dale, 2007).

GSK-3 is intimately involved in the hyperphosphorylation of tau, memory impairment, and the increased production of Aβ and in inflammatory responses (Figure 3.5) (Hooper, Killick, and Lovestone, 2008). It reduces acetylcholine synthesis, which is in accordance with the cholinergic deficit present in AD (Hoshi et al., 1996). In addition to being a key mediator of apoptosis, it might directly contribute to neuronal loss in AD (Turenne and Price, 2001).

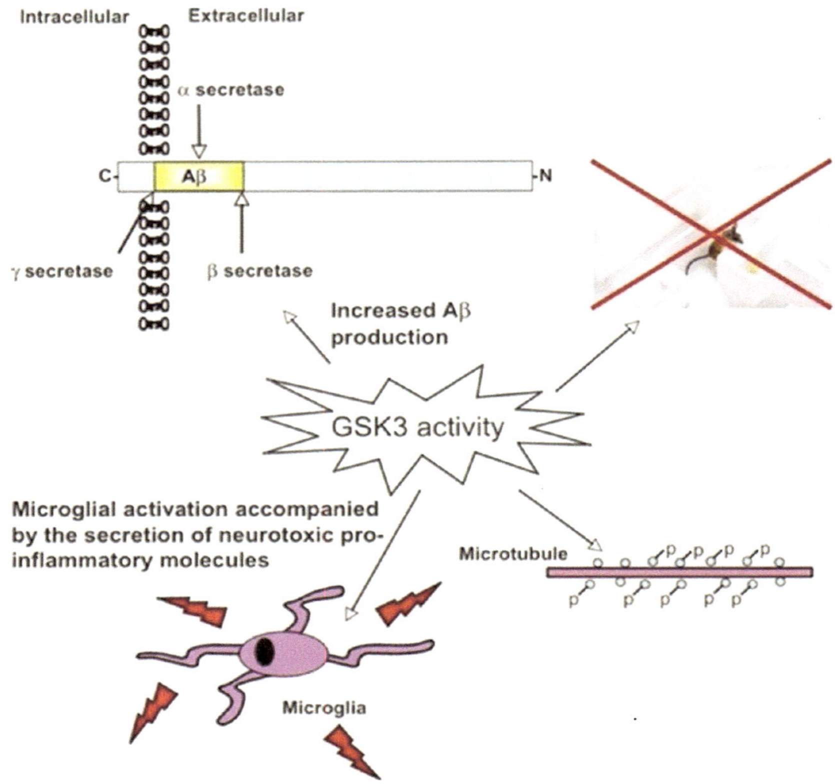

FIGURE 3.5 Possible role of GSK-3 via β- and γ-secretase-mediated increased production of Aβ from APP and impairments in learning and memory (Hooper, Killick, and Lovestone, 2008).

Overactivity of GSK-3 in the brain might play an integral role in disease progression. GSK-3, via β- and γ-secretase-mediated cleavage, leads to increased production of Aβ from APP and impairments in learning and memory. It has been reported to potentially heighten microglial-mediated inflammatory responses in the local vicinity of Aβ plaques (Hooper, Killick, and Lovestone, 2008). *In vitro* and *in vivo* evidence has shown that GSK-3β activation causes Aβ aggregation in the AD brain by incorporating APP cleavage. The cleavage of APP in the body occurs through two known pathways: the amyloidogenic and non-amyloidogenic. In both pathways, the cleavage is done by γ-secretase and α-secretase, respectively. The cleavage performed by α-secretase forms peptides that are easily degraded; on the other hand, in the amyloidogenic pathway, APP, in the beginning, is cleaved by the β-secretase enzyme, and in the next step, it is cleaved by the γ-secretase complex (a component of PS1), leading to the formation of Aβ deposits in the brain. This suggests that APP cleavage and PS-1 activity are disturbed by heightened GSK-3β levels in the brain, resulting in higher Aβ protein production and further deposition in the AD brain. GSK-3β also plays a crucial role in synaptic plasticity and memory formation (Figure 3.6). Therefore, reducing the activity of GSK-3β by using its inhibitors can alter cognitive and memory deficits.

Schematic representation of the potential involvement of GSK-3β in different aspects and pathways relevant to the onset and development of Alzheimer's disease neuropathology is presented in Figure 3.6. It may be summarized as follows: (1) GSK-3β activation promotes tau hyperphosphorylation, leading to dissociation of hyperphosphorylated tau from the microtubules, disrupting other

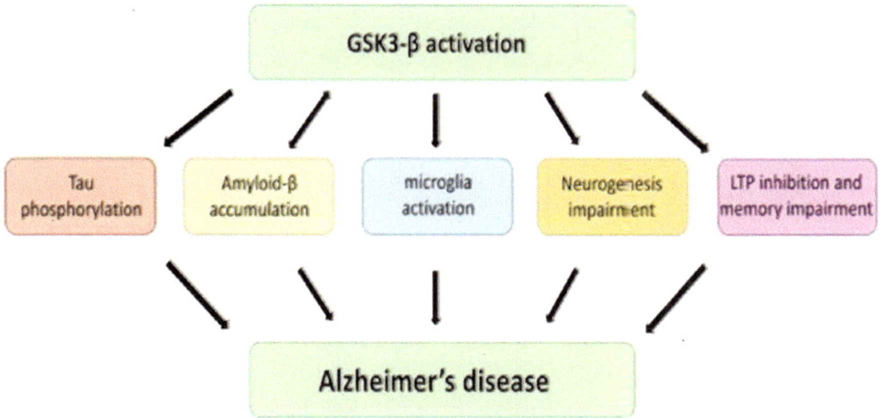

FIGURE 3.6 Role of GSK-3β in various activities leading to Alzheimer's disease (Lauretti, Dincer, and Praticò, 2020).

functions like axonal transport, NFT formation, and neuronal and synaptic dysfunction. (2) GSK-3β induces apoptosis and neuronal damage in AD by promoting amyloid production and its accumulation. (3) GSK-3β also displays pro-inflammatory functions, regulates the biological response of microglia and primary immune cells of the CNS, and promotes production of inflammatory molecules. 4) Additionally, GSK-3β plays a critical role in the regulation of hippocampal neurogenesis, an important process that supports specific forms of learning and memory and is known to be affected in AD pathology. (5) Lastly, GSK-3β is also involved in synaptic plasticity and memory. Overactivation of this kinase has been linked to the inhibition of hippocampal long-term potentiation (LTP), a mechanism required for memory formation (Lauretti, Dincer, and Praticò, 2020).

So far, many GSK-3β inhibitors have been under therapeutic approach, and only a few have reached clinical trials. Tideglusib (**27**), a non-ATP competitive GSK-3β inhibitor, has been successfully shown to lower the levels of tau phosphorylation, decrease Aβ deposition, proliferate astrocytes, and ameliorated memory deficits (Serenó et al., 2009; Wang et al., 2016).

GSK-3β inhibitors are a class of compounds that block the activity of the enzyme GSK-3β. This enzyme has been implicated in a number of biological processes, including glucose metabolism, neuronal function, and the regulation of the cell cycle. GSK-3β inhibitors have been studied for their potential therapeutic applications in a range of diseases, including Alzheimer's disease, depression, bipolar disorder, and type 2 diabetes. Research has shown that GSK-3β inhibitors can have a positive impact on cognition, mood, and insulin sensitivity. However, further clinical trials are needed to determine the safety and efficacy of these compounds in humans. Some common side effects of GSK-3β inhibitors include nausea, dizziness, and headache. Overall, the development of GSK-3β inhibitors is a promising area of research in the pharmaceutical industry, but more studies are needed to fully understand the potential benefits and drawbacks of these compounds. A US patent of indolocarbazole derivatives confirms the therapeutic importance of these derivatives in neurodegenerative diseases (Roder et al., 2000). Several small molecule inhibitors of GSK-3β have been developed and are being studied for their potential therapeutic applications. Some of the well-known GSK-3β inhibitors include (i) liarozole (**28**), a potent and selective inhibitor that has shown beneficial effects in preclinical models of Alzheimer's disease and depression; (ii) CHIR99021 (**29**) has shown neuroprotective effects in preclinical models of neurodegeneration; and (iii) SB216763 (**30**) is a selective inhibitor that has shown positive effects on mood and cognitive function in preclinical models of depression and bipolar disorder.

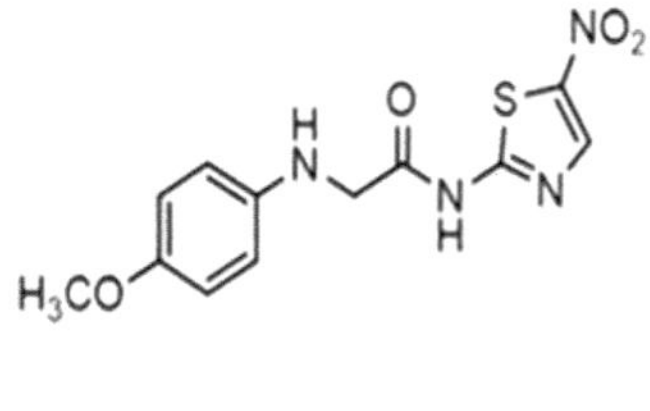

31

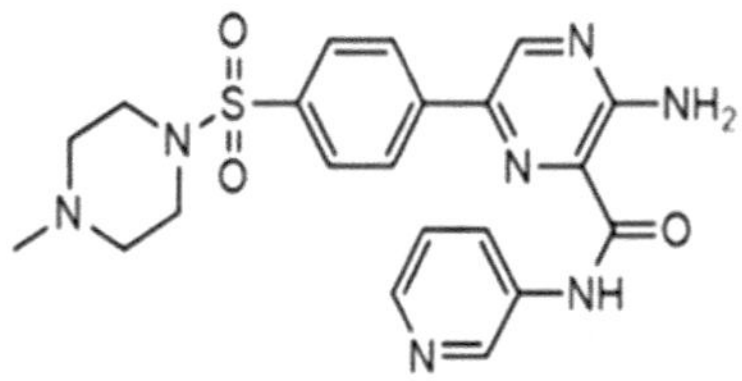

32

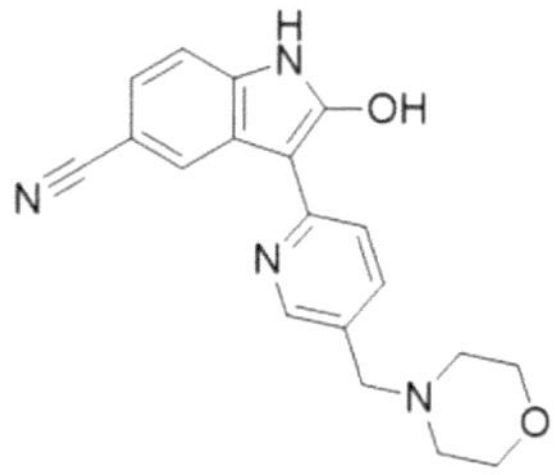

33

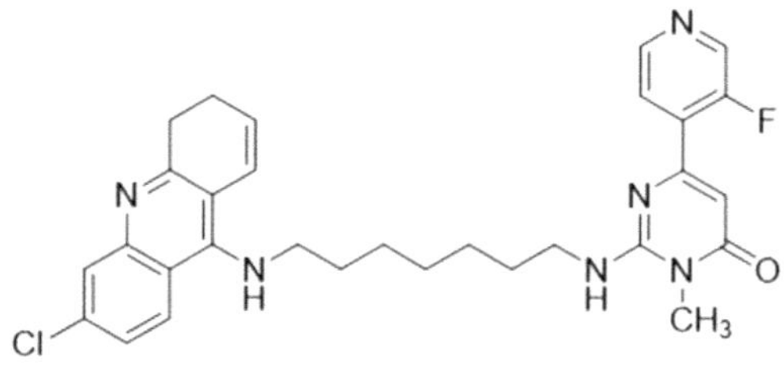

34

These compounds represent just a few of the many GSK-3β inhibitors that have been developed. However, it is important to note that while these inhibitors show promising effects in preclinical studies, further clinical trials are needed to determine their safety and efficacy for use in humans. A few more GSK-3β inhibitors with anti-AD activity *in vivo* are presented here.

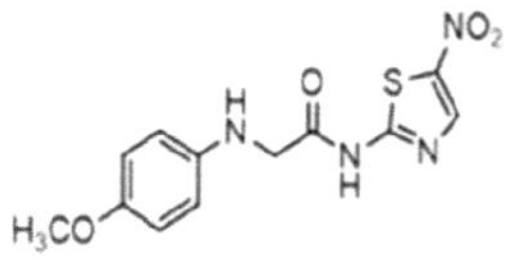

31

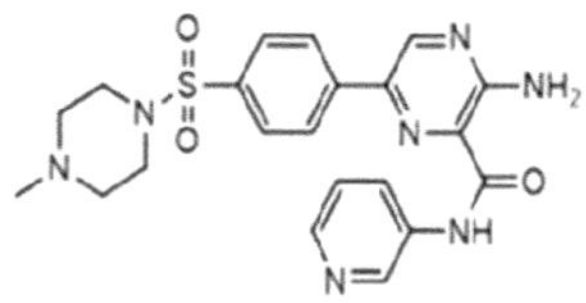

32

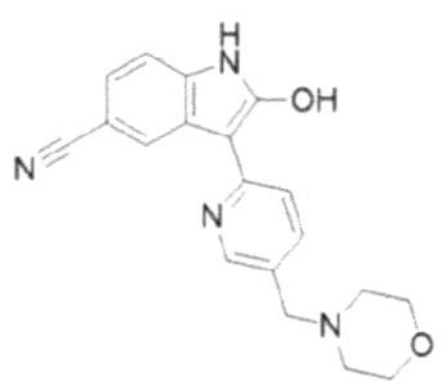

33

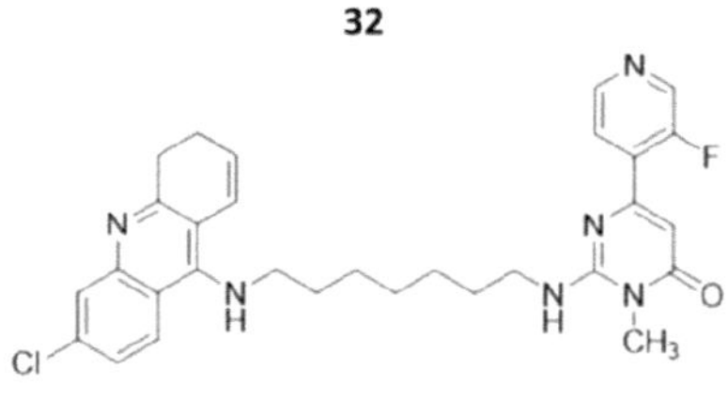

34

The smallest thiadiazolidinone inhibitor that entered the clinical trial for the treatment of AD is tideglusib (**27**) (del Ser, 2010). In addition to GSK-3β inhibitors, AR-A014418 (**31**), AZD2858 (**32**), AZD1080 (**33**), GSK-3β and acetylcholinesterase (AChE) dual inhibitor, i.e., compound **34**, have demonstrated anti-AD effects in animal models (Bhat et al., 2003; Georgievska et al., 2013; Berg et al., 2012; Yao et al., 2021). Most of the reported GSK-3β inhibitors feature the "double-sites occupation" pharmacophore model: a key skeleton interacted with the hinge region by forming two hydrogen bonds with Asp133 and/or Val135, and a moiety connected to the key skeleton as a hydrogen bond acceptor to interact with the Lys85 side chain (Sivaprakasam et al., 2015).

Berg et al. used a high-throughput screening (HTS) approach and cocrystallization of key analogues to optimize and synthesize a novel pyrazine series (Berg et al., 2012). They reported highly potent and selective inhibitors with cellular efficacy and blood–brain barrier penetrance. Three piperazinyl sulfonamide analogues were found potent, and compound **34** with para-substituted sulphonamide had eightfold activity (K_i = 4.9 nM) compared to its precursor compound. The authors also investigated the compounds for selectivity toward GSK-3β versus CDK2, as the two share high sequence similarity in the ATP binding site, and reported the three piperazine sulfonamides as >100-fold selective toward GSK-3β versus CDK2. The binding mode of the lead pyrazine analogues was confirmed from high-resolution X-ray crystal structure of pyrazine bound in the ATP pocket of GSK-3β protein. The interactions revealed include the formation of hydrogen bonds with the backbone NH of Val135 and carbonyl of Asp133 by pyrazine nitrogen and one of the adjacent anilino hydrogens, and formation of a hydrogen bond between the pyridine nitrogen and the terminal amino group of Lys85 due to an internal hydrogen bond of carbonyl in the amide with the aniline group on the pyrazine ring. The sulfonamide group occupies a subpocket that is formed by the salt bridge between Glu137 and Arg141.

Yan et al. designed and synthesized a series of novel thieno[3,2-*c*]pyrazol-3-amine derivatives and reported their GSK-3β inhibitory activity for the treatment of AD. Derivative **35**, with an IC_{50} of 3.1 nM (*in vitro*) and acceptable kinase selectivity is identified as a potent GSK-3β inhibitor. The compound is shown to decrease the phosphorylation of tau at Ser396 in a dose-dependent way as well as increasing the β-catenin, GAP43, N-myc, and MAP-2 expressions, promoting differentiated neuronal neurite outgrowth (Yan et al., 2022).

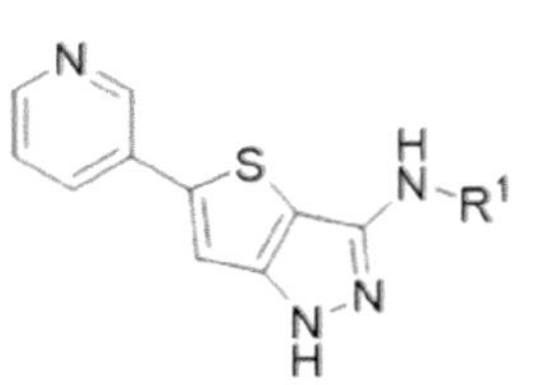

(35) R^1 = isobutyryl

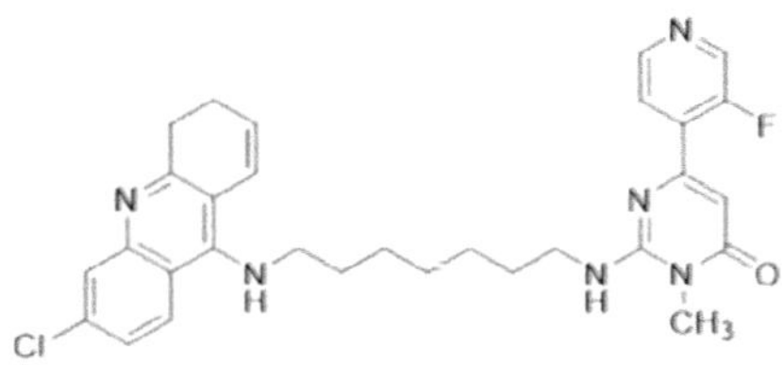

(36)

Novel tacrine-pyrimidone hybrids as potent dual AChE/GSK-3 inhibitors for the treatment of Alzheimer's disease have been developed by Yao et al. Compound **36**, with dual AChE/GSK-3 inhibition (AChE: IC_{50} = 51.1 nM; GSK-3β: IC_{50} = 89.3 nM), has displayed significant amelioration of cognitive deficits and reduction in phosphorylation of tau protein at Ser199 and Ser396 sites in glyceraldehyde (GA)-stimulated differentiated SH-SY5Y cells. Compound **36** is reported to exhibit desirable pharmacokinetic properties, good kinase selectivity, and moderate neuroprotection against GA-induced reduction in cell viability and neurite damage in SH-SY5Y-derived neurons (Yao et al., 2021).

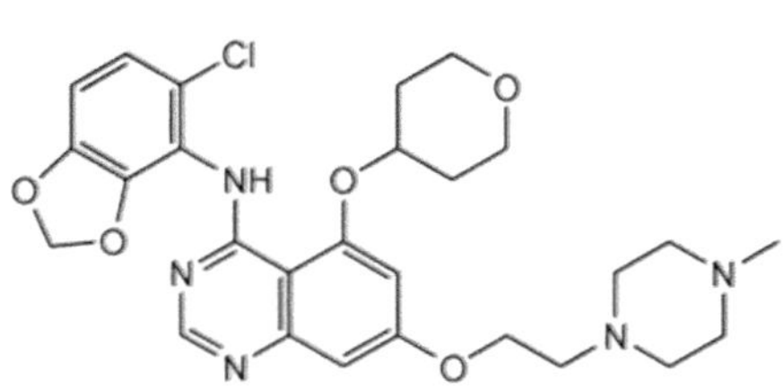

(39)

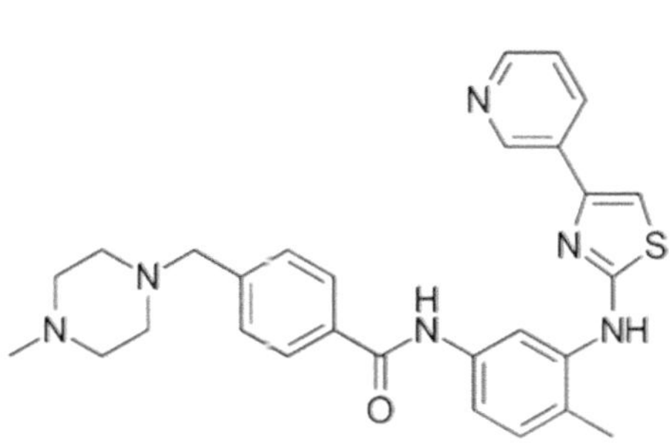

(40)

A new class of GSK-3β inhibitors based on the 6-C-glycosylflavone isoorientin (**37**) for the treatment of AD has been reported by Liang and Li (2018). The new inhibitors are reported to have passive membrane permeability, and constitutively attenuate GSK-3β-mediated tau hyperphosphorylation and amyloid neurotoxicity in an AD cellular model. Compound **37**, with distinct kinase selectivity and a 310-fold increase in activity compared to the precursor compound, is presented as the most active compound among this class.

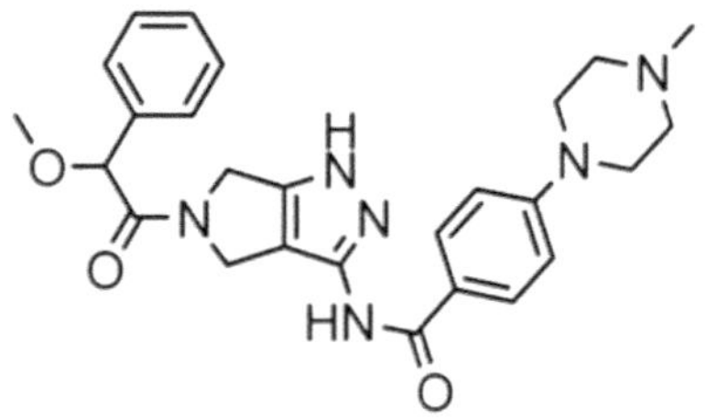

(41)

Another GSK-3β inhibitor, 6-(5-(4-((Pyridin-4-ylamino)methyl)phenyl)-1H-Pyrrolo[2,3-b]pyridin-3-yl)quinolin-8-ol (**38**) discovered with an IC_{50} of 66 nM has a direct effect on Aβ, targets tau and metal dyshomeostasis, and promotes neuronal neurite outgrowth. An increase in concentration in β-catenin abundance, phosphorylated GSK-3β at Ser9, and a decreased phosphorylated tau level in SH-SY5Y cells. In addition to the direct effect on Aβ by inhibiting $A\beta_{1-42}$ aggregation and promoting $A\beta_{1-42}$ aggregate disassociation, selective chelation with Cu^{2+}, Zn^{2+}, Fe^{3+}, and Al^{3+}, and targeted AD metal dyshomeostasis has been reported (Shi et al., 2020).

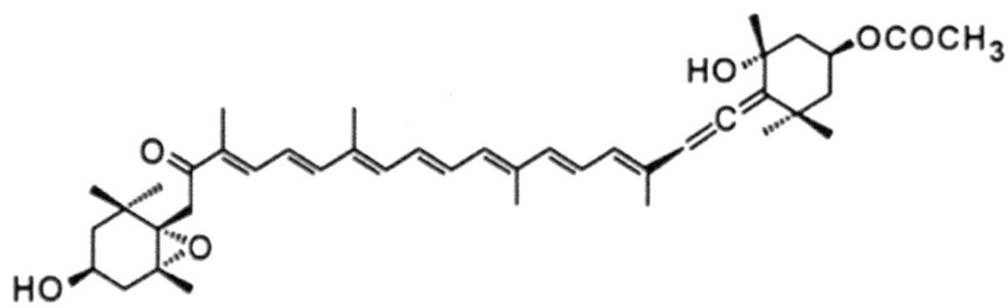

(42)

Drugs from plant sources have always been used to alleviate human sufferings. Polyphenols from *Conyza dioscoridis* (L.) extract have been found to reduce hippocampal GSK-3β and caspase-3 activity and increase the inhibited glutamate receptor expression (AMPA GluR1 subunit and NMDA receptor subunits NR1, NR2A, NR2B). A decrease in hyperphosphorylation of tau protein and $A\beta_{1-42}$ generation in the hippocampus of T2D rats is also reported. In addition, favorable changes in oxidative stress markers and reduction of TNF-α and IL-1β were important findings in this study (Gomaa et al., 2022).

Zha et al. showed that dimers of kojic acid moderately inhibit the deposition of Aβ and selectively chelate Cu^{2+}, Zn^{2+}, Fe^{2+}, Fe^{3+}, and Al^{3+}, which are related to the progress of Alzheimer's disease (Liu et al., 2022). The structure–activity relationship of small molecule inhibitors of GSK-3β is based on the interaction between the inhibitor and the active site of the enzyme. The active site of GSK-3β is composed of several conserved residues, including a catalytic lysine residue, which are essential for the enzyme's activity. In order to inhibit the activity of GSK-3β, small molecule inhibitors must bind to and block access to the catalytic site of the enzyme.

Small molecule inhibitors of GSK-3β are typically classified into two main categories, i.e., ATP-competitive inhibitors and non-ATP-competitive inhibitors. ATP-competitive inhibitors bind to the ATP-binding site of the enzyme and prevent ATP from binding, thereby inhibiting the enzyme's activity. Non-ATP-competitive inhibitors bind to a different site on the enzyme, such as the activation loop, and prevent the enzyme from undergoing the conformational changes necessary for its activity.

Specific structures of small molecule inhibitors of GSK-3β can greatly impact their activity. For example, the length and flexibility of the inhibitor's linker, the presence of polar or hydrophobic residues, and the presence of functional groups such as amides or carboxylic acids can all impact the inhibitor's binding affinity and specificity for GSK-3β. By carefully designing the structure of these inhibitors, researchers can optimize their activity and specificity for the enzyme.

FYN KINASE

Fyn kinase, a non-receptor tyrosine kinase, is one of the 11 members of the Src class of kinases involved in various vital functions relating to the CNS. Due to its role in the signaling pathways that control cell proliferation, migration, invasion, and apoptosis, Fyn received significant attention in cancer research but experienced a setback as a target in oncological applications due to limited use found in clinical trials (Nygaard, van Dyck, and Strittmatter, 2014; Resh, 1998). However, as research developed in the area of Src kinase, so did the understanding of the role of Fyn in other areas, including neurological function. In addition to modulating the neuroinflammatory response in the brain, a key process that is disrupted in neurodegenerative disease, Fyn plays a substantial role in myelination and neurodevelopment (Knox and Jiang, 2015; Panicker et al., 2015; Kramer-Albers and White, 2011; Marotta et al., 2022). Matrone and colleagues' most recent review summarized several of Fyn's most important functions in neurological processes, confirming its central and adaptable position in the brain (Matrone et al., 2020). Here, the role of alterations in Fyn kinase leading to common pathological processes that trigger the neurodegenerative diseases like Alzheimer's and Parkinson's will be briefly discussed, and the potential of Fyn as a therapeutic target to prevent, reduce, or treat disease progression in these degenerative conditions will also be explored. Fyn, a 59 kDa protein, is encoded by the Fyn gene on chromosome 6q21 and exists in three isoforms: Fyn-B, Fyn-T, and Fyn-D7 with intense expression of Fyn-B in the brain. Fyn consists of four domains (SH1–SH4), which are well conserved among SFK members. The very short SH4 domain near the N-terminal is capable of undergoing fat modification and thereby helps Fyn to anchor itself to the cell membrane (Resh, 1993).

The activity of Fyn is regulated by the phosphorylation of both the Y531 motif on the C-terminus and the Y420 motif on the SH1 domain's activation loop (Ponniah et al., 1999; Yang et al., 2011; Nguyen, Liu, and Lombroso, 2002). This results in conformational changes and inactivation of the protein. Similarly, dephosphorylation of the Y531 leads to disruption of the Y531-SH2 intramolecular complex, leading to Fyn activation (Bhandari, Lim, and Fallen, 1998). Fyn kinase and AD was demonstrated by Shirazi and colleagues for the first time (Shirazi and Wood, 1993). This is supported by recent reports of increased levels of FynT in the neocortex of AD patients, a finding associated with neurodegenerative pathology and cognitive impairment (Lee et al., 2016; Low et al., 2021). Successive reports have demonstrated that Fyn kinase plays a key role in the regulation of Aβ, tau, and NMDA signaling, leading to the identification of Fyn as a potential novel therapeutic target for the disease (Yang et al., 2011). In AD, the amyloidogenic pathway is responsible for the cleavage of APP, leading to the formation of Aβ through the sequential action of β- and γ-secretase, result in the aggregation of Aβ monomers into toxic oligomers. These oligomers result in a series of neurodegenerative alterations upon accumulation and subsequent neuronal death throughout the brain (O'Brien and Wong, 2011). Fyn as the residue is crucial for the trafficking of APP in neurons, controlling APP endocytosis and distribution (Poulsen et al., 2017). Increased phosphorylation deteriorates APP endocytosis, leading to the accumulation of APP in acidic neuronal compartments where the β-secretase, BACE-1, is optimally active, promoting the generation of Aβ peptides. Therefore, considering this foreplay of Fyn, it necessitates the inhibition of Fyn kinase to reduce or prevent enhanced tyrosine phosphorylation of APP, thereby decreasing Aβ deposition in the brain.

Fyn kinase interaction is also responsible for tau protein hyperphosphorylation and the formation of neurofibrillary tangles. The amyloid cascade hypothesis implies that abundant extracellular Aβ leads to hyperphosphorylation of tau proteins, which are important in the stabilization of cytoskeletal microtubules in the healthy brain. Further, the hyperphosphorylated tau proteins aggregate, resulting in the formation of neurotoxic intraneuronal inclusions known as neurofibrillary tangles (NFTs) (Hervy and Bicout, 2019; Gotz et al., 2004; Ittner and Gotz, 2011). In short, oligomerization of Aβ increases Fyn phosphorylation and the subsequent Y18 phosphorylation of tau, which abnormally accumulates in dendritic spines, causing NFT formation and eventually leading to neuronal dysfunction and death (Larson et al., 2012; Hoover et al., 2010). Inhibition of Fyn kinase has been explored as a therapeutic strategy for various diseases, including cancer, neurodegenerative disorders, and autoimmune diseases.

Small molecules have been developed as inhibitors of Fyn kinase, including ATP-competitive inhibitors, allosteric inhibitors, and antibodies to treat Alzheimer's. Saracatinib (AZD0530) (**39**), an ATP-competitive inhibitor, is being investigated in clinical trials for the treatment of AD (Nygaard, van Dyck, and Strittmatter, 2014).

Saracatinib was originally developed to target pathogenic signaling pathways in cancer. It is based on quinazoline, an organic compound with a bicyclic base structure consisting of fused benzene and pyrimidine rings, which was successively modified to produce saracatinib (Schenone et al., 2011; Hennequin et al., 2006). Saracatinib acts through reversible ATP inhibition and, in enzymatic assays, inhibits both Fyn isoforms with an IC_{50} value of 8–10 nM (Kaufman et al., 2015; Green et al., 2009). It also inhibits several other SFKs expressed in the brain, including Src (IC_{50} 2.7 nM), Yes (IC_{50} 4 nM), and Lyn (IC_{50} 5 nM), but these kinases are not activated by Aβo *in vitro* (Um et al., 2012). Saracatinib also shows activity against Abl, a kinase related to SFKs (IC_{50} 30 nM) (Hennequin et al., 2006; Green et al., 2009). Using Stat5 phosphorylation as a measure of Abl activity, saracatinib was found to have an IC_{50} of 156 nM against Abl, and plasma from human subjects taking once daily doses up to 125 mg of saracatinib does not alter Abl kinase activity (Nygaard et al., 2015).

Masitinib (**40**), an orally available tyrosine kinase inhibitor, was assessed in a small phase II clinical trial in AD, showing a trend toward improvements across measures of cognitive function and activities of daily living (Piette et al., 2011). Phase III trial (NCT01872598) of masitinib in AD was expected to read out in 2016, but its current status is unclear. Masitinib is selective for colony-stimulating factor 1 receptor (CSF-1R) (IC_{50} 90 nM), c-Kit (IC_{50} 200 nM), platelet-derived growth factor (PDGF; IC_{50} 540 nM), Lyn (IC_{50} 500 nM), Src (IC_{50} 1.87 μM), and Fyn (IC_{50} 240 nM) (Dubreuil et al., 2009; Hermine et al., 2015; Trias et al., 2016). Its promise in AD has been speculated to involve, in part, Fyn inhibition, although this has not been demonstrated experimentally in preclinical AD models or clinical cases (Folch et al., 2015). Publicly available data are currently insufficient to fully determine whether a masitinib dose of 200 mg twice daily, used in the phase II study in AD, will achieve a plasma concentration equal to or greater than the IC_{50} of masitinib against Fyn (Piette et al., 2011; EMA 2014). Moreover, the ability of masitinib to penetrate the blood–brain barrier is not known, and thus the extent to which the drug may engage its intended targets in the brain is unclear.

Danusertib (PHA-739358) (**41**) is an ATP-competitive inhibitor and is in clinical trials for various types of cancers. Glucosylpolyphenols as inhibitors of Aβ-induced Fyn kinase activation and tau phosphorylation have been reported (de Matos et al., 2020).

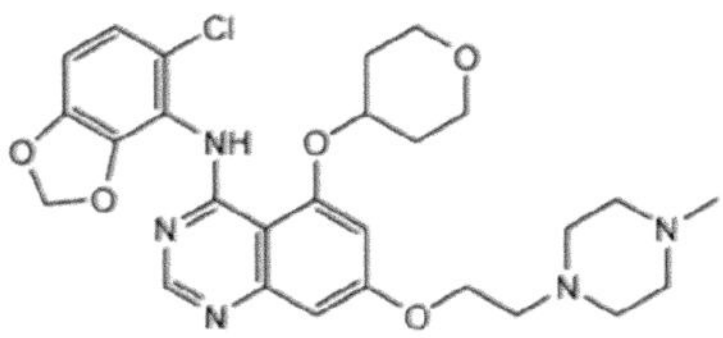

(39)

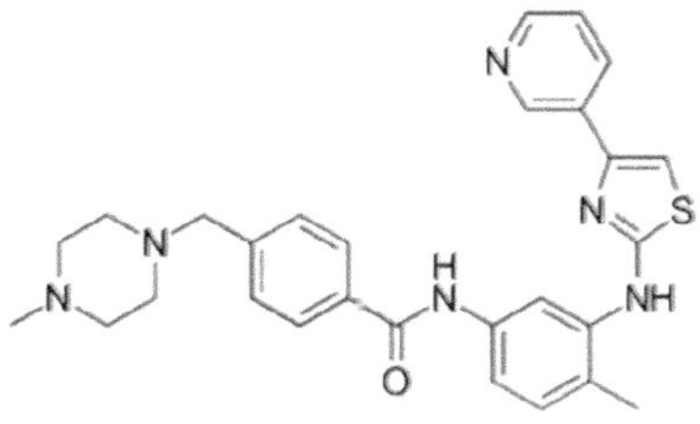

(40)

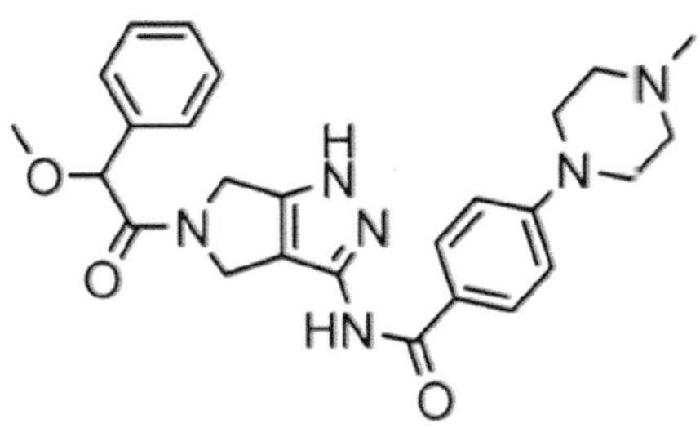

(41)

Fucoxanthin (**42**), an epoxycarotenol that is found in brown seaweed, has been reported to exhibit a neuroprotective effect via activation of Nrf2 through the Akt/GSK-3β/Fyn signaling pathway and upregulation of the antioxidant enzyme expression against Aβ25-35-caused oxidative damage (Lee et al., 2023).

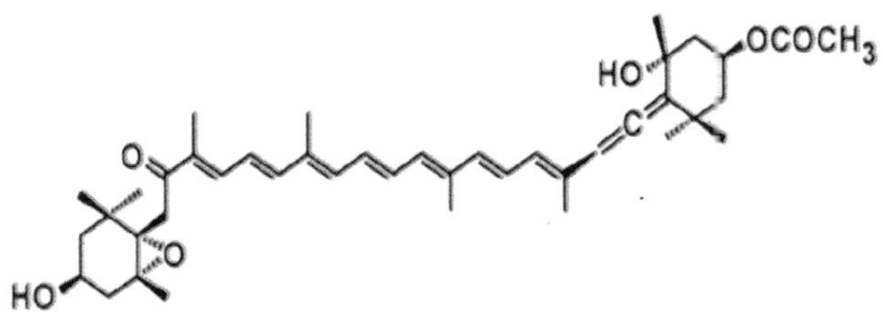

(42)

A number of articles/reviews provide details about the role of Fyn kinase inhibition in neuroinflammatory diseases (Nygaard, 2018; Schenone et al., 2011; Poli et al., 2013). Small molecule inhibitors of Fyn kinase typically target the ATP-binding site or allosteric sites on the kinase domain of Fyn. ATP-competitive inhibitors, i.e., molecules that mimic the structure of ATP and compete with it for binding to the active site of the kinase, thus inhibit its activity. The SAR analysis of ATP-competitive inhibitors has revealed the importance of maintaining a proper orientation of the functional groups in the molecule, as well as correct hydrogen bonding patterns and hydrophobic interactions with the ATP-binding pocket of the kinase. Molecules that bind to allosteric sites bind to a specific site on the kinase that is distinct from the ATP-binding site and alter the conformation of the kinase to prevent its activation. SAR analysis of allosteric inhibitors has revealed the importance of molecular size and shape, hydrophobic interactions, and hydrogen bonding with the allosteric site of the kinase for their potency and specificity. It is important to note that the SAR analysis of Fyn inhibitors is a complex and ongoing process, as the kinase has several regions that can interact with inhibitors, and new insights into the structural basis of kinase inhibition are constantly being discovered.

REFERENCES

Anderton BH, Betts J, Blackstock WP, Brion JP, Chapman S, Connell J, et al. Sites of phosphorylation in tau and factors affecting their regulation. *Biochem Soc Symp* (2001) 67:73–80

Bekris LM, Yu C-E, Bird TD, Tsuang DW. Genetics of Alzheimer disease. *J Geriatr Psychiatry Neurol* (2010) 23: 213–227.

Bennett BL, Sasaki DT, Murray BW, O'Leary EC, Sakata ST, Xu W, Leisten JC, Motiwala A, Pierce S, Satoh Y, Bhagwat SS. SP600125, an anthrapyrazolone inhibitor of Jun N-terminal kinase. *Proc Natl Acad Sci* (2001) 98(24): 13681–13686.

Berg S, Bergh M, Hellberg S, Högdin K, Lo-Alfredsson Y, Söderman P, et al. Discovery of novel potent and highly selective glycogen synthase kinase-3β (GSK-3β) inhibitors for Alzheimer's disease: design, synthesis, and characterization of pyrazines. *J Med Chem* (2012) 55: 9107–9119.

Berndt N, Plückthun A, Waldmann H. Selective inhibition of c-Jun N-Terminal Kinase 3 (JNK3) with bi-specific antibodies. *Angew Chem Int Ed Engl* (2013) 52: 562–566.

Bhandari V, Lim KL, Pallen CJ. Physical and functional interactions between receptor-like protein-tyrosine phosphatase alpha and p59fyn. *J Biol Chem* (1998) 273(15): 8691–8698.

Bhat R, Xue Y, Berg S, Hellberg S, Ormö M, Nilsson Y, et al. Structural insights and biological effects of glycogen synthase kinase 3-specific inhibitor AR-A014418. *J Biol Chem* (2003) 278: 45937–45945.

Bogoyevitch MA, Kobe B. Uses for JNK: the many and varied substrates of the c-Jun N-terminal kinases. *Microbiol Mol Biol Rev* (2006) 70(4): 1061–1095.

Bonny C, Nicod P, Waeber G. IB1, a JIP-1-related nuclear protein present in insulin-secreting cells. *J Biol Chem* (1998) 273(4): 1843–1846.

Bonny C, Oberson A, Negri S, Sauser, C, Schorderet DF. Cell-permeable peptide inhibitors of JNK: novel blockers of beta-cell death. *Diabetes* (2001) 50: 77–82.

Brion JP, Anderton BH, Authelet M, Dayanandan R, Leroy K, Lovestone S, Octave JN, Pradier L, Touchet N, Tremp G. Neurofibrillary tangles and tau phosphorylation. *Biochem Soc Symp* (2001) 67: 81–88.

Cheng C, Liu M, Gao X, Wu D, Pu M, Ma J, Quinn RJ, Xiao Z, Liu Z. Identifying new Ligands for JNK3 by fluorescence thermal shift assays and native mass spectrometry. *ACS Omega* (2022) 7(16): 13925–13931.

Chintala SK, Putris N, Geno M. Activation of TLR3 promotes the degeneration of retinal ganglion cells by upregulating the protein levels of JNK3. *Inv Ophthalmol Vis Sci* (2015) 56(1): 505–514.

Cruchaga C, Chakraverty S, Mayo K, Vallania FLM, Mitra RD, Faber K, et al. Rare variants in APP, PSEN1 and PSEN2 increase risk for AD in late-onset Alzheimer's disease families. *PLoS One* (2012) 7: e31039.

Davis RJ. Signal transduction by the JNK group of MAP kinases. *Cell* (2000) 103: 239–252.

de Lemos L, Junyent F, Camins A, Castro-Torres RD, Folch J, Olloquequi J, Beas-Zarate C, Verdaguer E, Auladell C. Neuroprotective effects of the absence of JNK1 or JNK3 isoforms on kainic acid-induced temporal lobe epilepsy-like symptoms. *Mol Neurobiol* (2018) 55: 4437–4452.

de Matos AM, Blázquez-Sánchez MT, Bento-Oliveira A, de Almeida RFM, Nunes R, Lopes PEM, Machuqueiro M, Cristóvão JS, Gomes CM, Souza CS, El Idrissi IG, Colabufo NA, Diniz A, Marcelo F, Oliveira MC, López Ó, Fernandez-Bolaños JG, Dätwyler P, Ernst B, Ning K, Garwood C, Chen B, Rauter AP. Glucosylpolyphenols as inhibitors of Aβ-induced Fyn kinase activation and Tau phosphorylation: synthesis, membrane permeability, and exploratory target assessment within the scope of type 2 diabetes and Alzheimer's disease. *J Med Chem* (2020) 63 (20): 11663–11690.

del Ser T. Phase IIa clinical trial on Alzheimer's disease with NP12, a GSK3 inhibitor. *Alzheimer's Dement* (2010) 6: S147.

Dhanasekaran DN, Kashef K, Lee CM, Xu H, Reddy EP. Scaffold proteins of MAP-kinase modules. *Oncogene* (2007) 26(22): 3185–3202.

Doble BW, Woodgett JR. Role of glycogen synthase kinase-3 in cell fate and epithelial–mesenchymal transitions. *Cells Tiss Org* (2007) 185: 73–84.

Dou X, Huang H, Li Y, Jiang L, Wang Y, Jin H, Jiao N, Zhang L, Zhang L, Liu Z. Multistage screening reveals 3-substituted indolin-2-one derivatives as novel and isoform-selective c-Jun N-terminal Kinase 3 (JNK3) inhibitors: implications to drug discovery for potential treatment of neurodegenerative diseases. *J Med Chem* (2019) 62(14): 6645–6664.

Drerup CM, Nechiporuk AV. JNK-interacting protein 3 mediates the retrograde transport of activated c-Jun N-terminal kinase and lysosomes. *PLoS Genetics* (2013) 9:e1003303.

Dubreuil P, Letard S, Ciufolini M, Gros L, Humbert M, Casteran N, et al. Masitinib (AB1010), a potent and selective tyrosine kinase inhibitor targeting KIT. *PLoS One* (2009) 4: e7258.

EMA Assessment report, Masiviera (Masitinib). *Eur Medicines Agency* (2014) EMEA/H/C/002659/0000.

Embi N, Rylatt DB, Cohen P. Glycogen synthase kinase-3 from rabbit skeletal muscle: separation from cyclic-AMP-dependent protein kinase and phosphorylase kinase. *Eur J Biochem* (1980) 107(2): 519–527.

Feng Y, Park H, Bauer L, Ryu JC, Yoon SO. Thiophene-pyrazolourea derivatives as potent, orally bioavailable, and isoform-selective JNK3 inhibitors. *ACS Med Chem Lett* (2020) 12(1): 24–29.

Feng Y, Park H, Ryu JC, Yoon SO. N-aromatic-substituted indazole derivatives as brain-penetrant and orally bioavailable JNK3 inhibitors. *ACS Med Chem Lett* (2021) 12(10): 1546–1552.

Fish PV, Steadman D, Bayle ED, Whiting P. New approaches for the treatment of Alzheimer's disease. *Bioorg Med Chem Lett* (2019) 29: 125–133.

Flemming A. JNK3 as new target in AD. *Nat Rev Drug Discov* (2012) 11: 829.

Folch J, Petrov D, Ettcheto M, Pedros I, Abad S, Beas-Zarate C, Lazarowski A, Marin M, Olloquequi J, Auladell C, Camins A. Masitinib for the treatment of mild to moderate Alzheimer's disease. *Expert Rev Neurother* (2015) 15: 587–596.

Forde JE, Dale TC. Glycogen synthase kinase 3: a key regulator of cellular fate. *Cell Mol Life Sci* (2007) 64: 1930–1944.

Georgievska B, Sandin J, Doherty J, Mörtberg A, Neelissen J, Andersson A, et al. AZD1080, a novel GSK3 inhibitor, rescues synaptic plasticity deficits in rodent brain and exhibits peripheral target engagement in humans. *J Neurochem* (2013) 125: 446–456.

Gomaa AA, Farghaly HSM, Makboul RM, Hussien AM, Nicola MA. Polyphenols from *Conyza dioscoridis* (L.) ameliorate Alzheimer's disease- like alterations through multi-targeting activities in two animal models. *BMC Complement Med Ther* (2022) 22(1):288.

Gotz J, Streffer JR, David D, Schild A, Hoerndli F, Pennanen L, Kurosinski P, Chen F. Transgenic animal models of Alzheimer's disease and related disorders: histopathology, behaviour and therapy. *Mol Psychiatry* (2004) 9(7): 664–683.

Green TP, Fennell M, Whittaker R, Curwen J, Jacobs V, Allen J, et al. Preclinical anticancer activity of the potent, oral Src inhibitor AZD0530. *Mol Oncol* (2009) 3: 248–261

Hennequin LF, Allen J, Breed J, Curwen J, Fennell M, Green TP, et al. N-(5- chloro-1,3-benzodioxol-4-yl)-7-[2-(4-methylpiperazin-1-yl)ethoxy]-5-(tetrahydro-2H-pyran-4 yloxy) quinazolin-4-amine, a novel, highly selective, orally available, dual-specific c-Src/Abl kinase inhibitor. *J Med Chem* (2006) 49: 6465–6488.

Hermine O, Moussy A, Mansfield C, Dubreuil P. Masitinib for the treatment of severe systemic mastocytosis: pooled phase 2 study simulation of phase 3 population and response criteria. *7th EMBRN International Mast Cell and Basophil Meeting.* Marseille, France (2015).

Hervy J, Bicout DJ. Dynamical decoration of stabilized microtubules by Tau-proteins. *Sci Rep* (2019) 9(1):12473.

Hooper C, Killick R, Lovestone S. The GSK3 hypothesis of Alzheimer's disease. *J Neurochem* (2008) 104(6): 1433–1439.

Hoover BR, Reed MN, Su J, Penrod RD, Kotilinek LA, Grant MK, et al. Tau mislocalization to dendritic spines mediates synaptic dysfunction independently of neurodegeneration. *Neuron* (2010) 68(6): 1067–1081.

Hoshi M, Takashima A, Noguchi K, Murayama M, Sato M, Kondo S, Saitoh Y, Ishiguro K, Hoshino T, Imahori K. Regulation of mitochondrial pyruvate dehydrogenase activity by tau protein kinase I/glycogen synthase kinase 3beta in brain. *Proc Natl Acad Sci USA* (1996) 93: 2719–2723

Hu NW, Ondrejcak T, Rowan MJ Glutamate receptors in preclinical research on Alzheimer's disease: update on recent advances. *Pharmacol Biochem Behav* (2012) 100: 855–862.

Ittner LM, Gotz J. Amyloid-beta and tau–a toxic pas de deux in Alzheimer's disease. *Nat Rev Neurosci* (2011) 12(2): 65–72.

Jun J, Baek J, Yang S, Moon H, Kim H, Cho H, Hah JM. Discovery of a potent and selective JNK3 inhibitor with neuroprotective effect against amyloid β-induced neurotoxicity in primary rat neurons. *Int J Mol Sci* (2021) 22(20): 11084.

Kaufman AC, Salazar SV, Haas LT, Yang J, Kostylev MA, Jeng AT, et al. Fyn inhibition rescues established memory and synapse loss in Alzheimer mice. *Ann Neurol* (2015) 77(6): 953–971.

Kelkar N, Gupta S, Dickens M, Davis RJ. Interaction of a mitogen-activated protein kinase signaling module with the neuronal protein JIP3. *Mol Cell Biol* (2000) 20: 1030–1043.

Knox R, Jiang X. Fyn in neurodevelopment and ischemic brain injury. *Dev Neurosci* (2015) 37(4–5): 311–320.

Kramer-Albers EM, White R. From axon-glial signalling to myelination: the integrating role of oligodendroglial Fyn kinase. *Cell Mol Life Sci: CMLS* (2011) 68(12): 2003–2012.

Kuan CY, Whitmarsh AJ, Yang DD, Liao G, Schloemer AJ, Dong C, et al.. A critical role of neural-specific JNK3 for ischemic apoptosis. *Proc Natl Acad Sci USA* (2003) 100: 15184–15189.

Larson M, Sherman MA, Amar F, Nuvolone M, Schneider JA, Bennett DA, Aguzzi A, Lesne SE. The complex PrP(c)- Fyn couples human oligomeric Abeta with pathological tau changes in Alzheimer's disease. *J Neurosci* (2012) 32(47): 16857–16871.

Lauretti E, Dincer O, Praticò D. Glycogen synthase kinase-3 signaling in Alzheimer's disease. *BBA-Mol Cell Res* (2020) 1867(5): 118664.

Lee C, Low CY, Francis PT, Attems J, Wong PT, Lai MK, Tan MG. An isoform-specific role of FynT tyrosine kinase in Alzheimer's disease. *J Neurochem* (2016) 136(3): 637–650.

Lee N, Youn K, Yoon JH, Lee B, Kim DH, Jun M. The Role of Fucoxanthin as a Potent Nrf2 Activator via Akt/GSK-3β/Fyn Axis against Amyloid-β Peptide-Induced Oxidative Damage. *Antioxidants (Basel)* (2023) 12(3): 629.

Lewerenz J, Maher P. Chronic glutamate toxicity in neurodegenerative diseases-what is the evidence? *Front Neurosci* (2015) 9: 469.

Liang Z, Li QX. Discovery of selective, substrate-competitive, and passive membrane permeable Glycogen Synthase Kinase-3β inhibitors: synthesis, biological evaluation, and molecular modeling of new C-Glycosylflavones. *ACS Chem Neurosci* (2018) 9(5): 1166–1183.

Liu X, Yu C, Su B, Zha D. Synthesis and properties of the kojic acid dimer and its potential for the treatment of Alzheimer's disease. *RSC Med Chem* (2022) 14(2): 268–276.

Low CYB, Lee JH, Lim FTW, Lee C, Ballard C, Francis PT, Lai MKP, Tan MGK. Isoform-specific upregulation of Fyn kinase expression is associated with tauopathy and glial activation in Alzheimer's disease and Lewy body dementias. *Br Pathol* (2021) 31(2): 253–266.

Mangal R, Ding Y. Prospective therapeutic targets of Alzheimer's disease. *Br Circulat* (2022) 8: 1–5.

Marotta G, Basagni F, Rosini M, Minarini A. Role of Fyn kinase inhibitors in switching neuroinflammatory pathways. *Curr Med Chem* (2022) 29(27): 4738–4755.

Matrone C, Petrillo F, Nasso R, Ferretti G. Fyn tyrosine kinase as harmonizing factor in neuronal functions and dysfunctions. *Int J Mol Sci* (2020) 21(12): 4444.

Morrison DK, Davis RJ. Regulation of MAP kinase signaling modules by scaffold proteins in mammals. *Ann Rev Cell Dev Biol* (2003) 19(1): 91–118.

Musi CA, Agrò G, Santarella F, Iervasi E, Borsello T. JNK3 as therapeutic target and biomarker in neurodegenerative and neurodevelopmental brain diseases. *Cells* (2020) 9(10): 2190.

Muth F, Günther M, Bauer SM, Döring E, Fischer S, Maier J, Drückes P, Köppler J, Trappe J, Rothbauer U, Koch P. Tetra-substituted pyridinylimidazoles as dual inhibitors of p38α mitogen-activated protein kinase and c-Jun N-terminal kinase 3 for potential treatment of neurodegenerative diseases. *J Med Chem* (2017) 58(1): 443–456.

Nguyen TH, Liu J, Lombroso PJ. Striatal enriched phosphatase 61 dephosphorylates Fyn at phosphotyrosine 420. *J Biol Chem* (2002) 277(27): 24274–24279.

Nygaard HB. Targeting Fyn kinase in Alzheimer's disease. *Biol Psychiatry* (2018) 83(4): 369–376.

Nygaard HB, van Dyck CH, Strittmatter SM. Fyn kinase inhibition as a novel therapy for Alzheimer's disease. *Alz Res Ther* (2014) 6(1): 8.

Nygaard HB, Wagner AF, Bowen GS, Good SP, MacAvoy MG, Strittmatter KA, et al. A phase Ib multiple ascending dose study of the safety, tolerability, and central nervous system availability of AZD0530 (saracatinib) in Alzheimer's disease. *Alz Res Ther* (2015) 7: 35.

O'Brien RJ, Wong PC. Amyloid precursor protein processing and Alzheimer's disease. *Annu Rev Neurosci* (2011) 34(1): 185–204.

Pan J, Li H, Zhang B, Xiong R, Zhang Y, Kang WY, Chen W, Zhao ZB, Chen SD. Small peptide inhibitor of JNK3 protects dopaminergic neurons from MPTP induced injury via inhibiting the ASK1-JNK3 signaling pathway. *PLoS One* (2015) 10: e0119204.

Panicker N, Saminathan H, Jin H, Neal M, Harischandra DS, Gordon R, et al. Fyn kinase regulates microglial neuroinflammatory responses in cell culture and animal models of Parkinson's disease. *J Neurosci* (2015) 35(27): 10058.

Perry NA, Kaoud TS, Ortega OO, Kaya AI, Marcus DJ, Pleinis JM, et al. Arrestin-3 scaffolding of the JNK3 cascade suggests a mechanism for signal amplification. *Pro Nat Acad Sci* (2019) 116(3): 810–815.

Piette F, Belmin J, Vincent H, Schmidt N, Pariel S, Verny M, et al. Masitinib as an adjunct therapy for mild-to-moderate Alzheimer's disease: a randomised, placebo-controlled phase 2 trial. *Alz Res Ther* (2011) 3: 1672.

Poli G, Tuccinardi T, Rizzolio F, Caligiuri I, Botta L, Granchi C, Ortore G, Minutolo F, Schenone S, Martinelli A. Identification of new Fyn kinase inhibitors using a FLAP-based approach. *J Chem Inf Model* (2013) 53(10): 2538–2547.

Ponniah S, Wang DZ, Lim KL, Pallen CJ. Targeted disruption of the tyrosine phosphatase PTP alpha leads to constitutive down regulation of the kinases Src and Fyn. *Curr Biol* (1999) 9(10): 535–538.

Poulsen ET, Iannuzzi F, Rasmussen HF, Maier TJ, Enghild JJ, Jorgensen AL, Matrone C. An aberrant phosphorylation of amyloid precursor protein tyrosine regulates its trafficking and the binding to the clathrin endocytic complex in neural stem cells of Alzheimer's disease patients. *Front Mol Neurosci* (2017) 10: 59.

Qin P, Ran Y, Liu Y, Wei C, Luan X, Niu H, Peng J, Sun J, Wu J. Recent advances of small molecule JNK3 inhibitors for Alzheimer's disease. *Bioorg Chem* (2022) 2022: 106090.

Rabinovici GD. Late-onset Alzheimer disease. *Continuum (Minneap Minn)* (2019) 25: 14–33.

Rehfeldt SC, Laufer S, Goettert MI. A highly selective in vitro JNK3 Inhibitor, FMU200, restores mitochondrial membrane potential and reduces oxidative stress and apoptosis in SH-SY5Y cells. *Inter J Mol Sci* (2021) 22(7): 3701.

Reitz C, Rogaeva E, Beecham GW. Late-onset vs nonmendelian early-onset Alzheimer disease. *Neurol Genet* (2020) 6: e512.

Resh MD. Fyn, a Src family tyrosine kinase. *Int J Biochem Cell Biol* (1998) 30(11): 1159–1162.

Resh MD. Interaction of tyrosine kinase oncoproteins with cellular membranes. *Biochim Biophys Acta* (1993) 1155(3): 307–322.

Roder H, Lowinger TB, Britelli DR, VanZandt MC. Indolocarbazole derivatives useful for the treatment of neurodegenerative diseases and cancer. *US Patent 6013646* (2000)

Schachter AS, Davis KL. Alzheimer's disease. *Dial Clin Neurosci* (2000) 2(2): 91–100.

Schenone S, Brullo C, Musumeci F, Biava M, Falchi F, Botta M. Fyn kinase in brain diseases and cancer: the search for inhibitors. *Curr Med Chem* (2011) 18: 2921–2942.

Serenó L, Coma M, Rodriguez M, Sanchez-Ferrer P, Sánchez MB, Gich I, Agulló JM, Pérez M, Avila J, Guardia-Laguarta C, Clarimón J. A novel GSK-3β inhibitor reduces Alzheimer's pathology and rescues neuronal loss in vivo. *Neurobiol Dis* (2009) 35(3): 359–367.

Sharma N, Deshmukh R, Bedi KL. SP600125, a competitive inhibitor of JNK attenuates streptozotocin induced neurocognitive deficit and oxidative stress in rats. *Pharmacol Biochem Behavior* (2010) 96(4): 386–394.

Shi XL, Yan N, Cui YJ, Liu ZP. A unique GSK-3β inhibitor B10 has a direct effect on Aβ, targets Tau and metal Dyshomeostasis, and promotes Neuronal Neurite outgrowth. *Cells* (2020) 9(3): 649.

Shirazi SK, Wood JG. The protein tyrosine kinase, fyn, Alzheimer's disease pathology. *NeuroReport* (1993) 4(4): 435–437.

Shuai W, Bu F, Zhu Y, Wu Y, Xiao H, Pan X, Zhang J, Sun Q, Wang G, Ouyang L. Discovery of novel Indazole chemotypes as Isoform-selective JNK3 inhibitors for the treatment of Parkinson's disease. *J Med Chem* (2023) 66(2): 1273–1300

Sivaprakasam P, Han X, Civiello RL, Jacutin-Porte S, Kish K, Pokross M, et al. Discovery of new acylaminopyridines as GSK-3 inhibitors by a structure guided in-depth exploration of chemical space around a pyrrolopyridinone core. *Bioorg Med Chem Lett* (2015) 25: 1856–1863.

So Y, Park DJ, Ryu JC, Ozer HG, Tep C, Shin YJ, Lim TH, et al. JNK3 perpetuates metabolic stress induced by Aβ peptides. *Neuron* (2012) 75: 824–837.

Trias E, Ibarburu S, Barreto-Nunez R, Babdor J, Maciel TT, Guillo M, et al. Postparalysis tyrosine kinase inhibition with masitinib abrogates neuroinflammation and slows disease progression in inherited amyotrophic lateral sclerosis. *J Neuroinflammation* (2016) 13: 177.

Troussard AA, Tan C, Yoganathan TN, Dedhar S. Cell-extracellular matrix interactions stimulate the AP-1 transcription factor in an integrin-linked kinase- and glycogen synthase kinase 3-dependent manner. *Mol Cell Biol* (1999) 19: 7420–7427.

Turenne GA, Price BD. Glycogen synthase kinase3 beta phosphorylates serine 33 of p53 and activates p53's transcriptional activity. *BMC Cell Biol* (2001) 2: 12.

Um JW, Nygaard HB, Heiss JK, Kostylev MA, Stagi M, Vortmeyer A, et al. Alzheimer amyloid-beta oligomer bound to postsynaptic prion protein activates Fyn to impair neurons. *Nat Neurosci* (2012) 15: 1227–1235.

Wang H, Huang S, Yan K, Fang X, Abussaud A, Martinez A, Sun HS, Feng ZP. Tideglusib, a chemical inhibitor of GSK3β, attenuates hypoxic-ischemic brain injury in neonatal mice. *Biochim Biophys Acta* (2016) 1860(10): 2076–2085.

Welsh GI, Proud CG. Glycogen synthase kinase-3 is rapidly inactivated in response to insulin and phosphorylates eukaryotic initiation factor eIF-2B. *Biochem J* (1993) 294(3): 625–629.

WHO report. Dementia. (2023). https://www.who.int/news-room/fact-sheets/detail/dementia

Willoughby EA, Perkins GR, Collins MK, Whitmarsh AJ. The JNK-interacting protein-1 scaffold protein targets MAPK phosphatase-7 to dephosphorylate JNK. *J Biol Chem* (2003) 278(12): 10731–10736.

Yan N, Shi XL, Tang LQ, Wang DF, Li X, Liu C, Liu ZP. Synthesis and biological evaluation of thieno[3,2-c] pyrazol-3-amine derivatives as potent glycogen synthase kinase 3β inhibitors for Alzheimer's disease. *J Enz Inh Med Chem* (2022) 37(1): 1724–1736.

Yang K, Belrose J, Trepanier CH, Lei G, Jackson MF, Mac Donald JF. Fyn, a potential target for Alzheimer's disease. *JAD* (2011) 27(2): 243–252.

Yao H, Uras G, Zhang P, Xu S, Yin Y, Liu J, et al. Discovery of novel Tacrine-Pyrimidone hybrids as potent dual AChE/GSK-3 inhibitors for the treatment of Alzheimer's disease. *J Med Chem* (2021) 64(11): 7483–7506.

Yarza R, Vela S, Solas M., Ramirez MJ. c-Jun N-terminal Kinase (JNK) signaling as a therapeutic target for Alzheimer 's disease. *Front Pharm* (2015) 6: 321.

Yasuda J, Whitmarsh AJ, Cavanagh J, Sharma M, Davis RJ. The JIP group of mitogen-activated protein kinase scaffold proteins. *Mol Cell Biol* (1999) 19: 7245–7272.

Yoon SO, Park DJ, Ryu JC, Ozer HG, Tep C, Shin YJ., et al. JNK3 perpetuates metabolic stress induced by Aβ peptides. *Neuron* (2012) 75(5): 824–837.

Zheng K, Iqbal S, Hernandez P, Park H, LoGrasso PV, Feng Y. Design and synthesis of highly potent and isoform selective JNK3 inhibitors: SAR studies on aminopyrazole derivatives. *J Med Chem* (2014) 57(23): 10013–10030.

Zulfiqar Z, Shah FA, Shafique S, Alattar A, Ali T, Alvi AM, Rashid S, Li S. Retracted: repurposing FDA approved drugs as JNK3 inhibitor for prevention of neuroinflammation induced by MCAO in rats. *J Infl Res* (2020) 13: 1185–1205.

4 The TREM2 Receptor Puzzle in Alzheimer's Disease

Assembling the Pieces

Gadde Shareena[1], Dileep Kumar[1,2,3], and Satheesh Kumar Sellamuthu[4]*

[1]Poona College of Pharmacy, Department of Pharmaceutical Chemistry, Bharati Vidyapeeth (Deemed to be University), Erandwane, Pune, Maharashtra, India
[2]Department of Pharm. Chemistry, Manipal College of Pharmaceutical Sciences, Manipal Academy of Higher Education, Manipal, Karnataka, India
[3]UC Davis Comprehensive Cancer Center, Department of Entomology and Nematology, University of California Davis, One Shields Avenue, Davis, CA, USA
[4]Karpagam College of Pharmacy, Othakkalmandapam, Coimbatore, Tamil Nadu, India
*Corresponding author

INTRODUCTION

Alzheimer's disease (AD) is a widespread neurodegenerative disorder clinically characterised by a persistent decline in cognitive and mental abilities, along with the pathological accumulation of senile amyloid plaques and tau protein NFTs in the brain. Age influences the AD incidence rates, and beyond the age of 65, the risk doubles every five years. AD patients live 4 to 8 years on average after diagnosis, though they may live close to 20 years (Breijyeh and Karaman, 2020; Tahami et al., 2022). According to the World Health Organization (WHO), 1.6 million mortalities are estimated to be related directly to dementia annually, with AD accounting for the majority of these deaths. Almost 47 million individuals worldwide already suffer from dementia; by 2050, that figure will have risen to more than 131 million. In 2020, the global cost of dementia was estimated to be around \$1 trillion, and this is projected to increase to \$2 trillion by 2030. This cost includes direct medical expenses, such as hospitalisation and medication, and indirect costs, such as lost productivity and caregiver burden (Zhu et al., 2015; Yiannopoulou and Papageorgiou, 2020).

The initial line of immunological defence in the human central nervous system (CNS) is mediated via microglia. It engages with pathogen recognition and triggers the innate immune response, which is pivotal for the brain (Lehnardt, 2010). The elevated risk of developing late-onset advanced AD is attributed to genetic variations in TREM2 that are dominantly expressed in CNS microglia (Jonsson et al., 2013; Ulrich et al., 2014). A rare polymorphism (R47H) in the TREM2 gene modulates inflammatory responses and imposes a risk for AD comparable to that presented by one APOE 4 allele (expanding the risk for AD by nearly three times) and was identified by whole-genome analysis (Hooli et al., 2014). TREM2 decreases the seeding of amyloid plaques and NFTs and regulates the microglial activation to the amyloid plaque vicinity in AD (Parhizkar et al., 2019). TREM2 is also crucial for microglia in the adolescent nervous system to eliminate surplus synapses (Filipello et al., 2018). The implications of TREM2 loss, TREM2 polymorphisms or human TREM2 expression in

DOI: 10.1201/9781003506317-4

amyloid or tau AD models have been explored in other studies. The majority of the studies revealed that TREM2-dependent AD modulation was correlated to disease progression, with findings suggesting TREM2 might up- or downregulate AD-related pathology. These genomic discoveries have drawn attention to the function of TREM2 in controlling microglial activity and the relevance of microglia in neurodegenerative disorders (Calcagno et al., 2018; Nordengen et al., 2019).

Despite substantial studies into AD, no effective treatment is available, and the underlying molecular pathways remain unknown (Livingston et al., 2020). Several signalling pathways connected to AD pathogenesis are being researched to identify possible therapeutic targets, including TREM. Many of the medications used to treat AD have undesirable effects because of their high costs, unclear safety, harmful side effects, poor drug tolerance, and patients' resistance to medication (Mimica and Presecki, 2009). Further knowledge of glutamate neurotransmission alterations associated with AD would be very useful to better analyse and comprehend its mechanisms of action and develop the treatment regimen, as cholinesterase inhibitors and the NMDA antagonist memantine are available medicines for AD (Parsons et al., 2007).

Several challenges, such as disease complexity, complicated diagnoses and limitation on BBB permeability, substantially influence AD patients. Patients nevertheless experience the disease's debilitating signs, such as brain damage, cognitive decline and behavioural issues, in the absence of suitable therapies (Frozza et al., 2018). Although there are presently no viable therapies for AD, new developments in the knowledge of AD pathogenesis have revealed interesting therapeutic targets, such as TREM2 receptors, which are related to microglia, the principal immune cell of the brain. More exploration and technology are needed to create efficient medications and ultimately develop a treatment for this serious disease. It is essential to support current research projects and share knowledge about how AD affects patients and their families (Zheng et al., 2018). This review focusses on TREM2, which is a possible target for eliciting a protective role for microglia and supports the ideal microglial activity necessary to slow the onset of AD, and its relationship with several AD-related indicators.

TREM2 STRUCTURE, FUNCTIONS, VARIANTS AND THEIR ASSOCIATED RISKS IN AD

TREM2 is a receptor protein, which is encoded by the TREM2 gene positioned on human chromosome no. 6p21.1. It is expressed in a wide range of cells, including microglia, macrophages, osteoclasts and monocyte-derived dendritic cells (Yeh et al., 2017). TREM2 receptor is a transmembrane protein comprised of an extracellular domain (ectodomain), intracellular component and membrane-traversing segment. The TREM2 protein's three regions interact synergistically to mediate cell signalling, wherein the ectodomain binds exogenous ligands, and the intracellular domain binds intermediary signalling proteins to facilitate effective cellular activity completion (Jin et al., 2014). In neural tissues, TREM2 is mainly expressed on the microglial cell surface and plays multiple critical roles in cellular survival, proliferation and phagocytosis. The ectodomain of TREM2 can bind different anionic ligand molecules, mainly glycoproteins and lipids, and activate intercellular signalling pathways that modulate innate immune responses. It also includes an Ig-like V-type domain that acts as ligand-receptor binding sites. The ectodomain is altered after protein translation via post-translational modifications, including sulfidation and glycosylation. TREM2's intracellular component does not have any intrinsic signalling ability and signals through DAP10 and DAP12 co-receptor mediators (Feuerbach et al., 2017; Yang et al., 2020). TREM2's ectodomains are processed by metalloprotease and disintegrin, forming and releasing soluble TREM2. sTREM2 has the capability of binding with multiple ligands, stimulating microglia and monitoring immunological responses along AD development. Clinical observations further imply that heightened sTREM2 levels are positively associated with elevated p-tau and t-tau in CSF samples of AD patients, suggestive of sTREM2 as a potent determinant in early-onset AD (Brown and George-Hyslop, 2022).

AD-related risk variants were previously discovered and identified in TREM2 protein's ectodomain. Among many, the p.Arg47His (R47H) TREM2 variant is the most prevalently observed genetic polymorphism in late-onset AD. Genome-wide association studies (GWAS) have indicated that the R47H mutation triples AD-associated risk and is linked to a 23% greater rate of dementia in contrast to non-carriers (Del-Aguila et al., 2018). It modulates TREM2's ligand-binding domain at the cellular level, affecting ligand-receptor interactions. Arg47 is a vital amino acid residue in the functioning of the complementarity-determining region 2 loop that promotes TREM2 interaction with various lipids and other ligands based on an analysis of high-resolution R47H configurations and wild-type TREM2 structures (Ma et al., 2016; Park et al., 2016). The levels of soluble TREM2 and its capacity to bind to cells, APOE and different lipids are reduced significantly when Arg is replaced by His (Park et al., 2016). Research has shown that the R47H variation (heterozygous) may lead to a reduction of TREM2 activity, exacerbate neuritic dystrophy and cause synapse loss around plaques, increasing AD risk. Patients with the R47H mutation displayed onset at a noticeably younger age than those without the variant (Li et al., 2021). Similar findings in the populations of Dutch and Icelandic inhabitants suggest a link between R47H and EOAD susceptibility (Jonsson et al., 2013). In a European population, it was noted that R47H dramatically raised the level of CSF-tau levels but not Aβ42, suggesting that tau dysfunction may be involved in TREM2's function in AD (Zhou et al., 2019). The R47H mutation was not found in the Chinese or Korean populations, implying that TREM2 expression could be related to racial genetic differences and is variably associated with the incidence of AD in different ethnic groups (Bonham et al., 2017). A range of many other rare TREM2 polymorphisms, such as p.Arg62His (R62H, rs143332484), p.His157Tyr (H157Y, rs2234255) and p.Asp87Asn (D87N, rs142232675), has also been reported in AD patients with elevated disease susceptibility. Several investigations have shown that disease-associated variations of TREM2 affect synaptic and neuronal loss, influence neuritic dystrophy, and modify tau hyperphosphorylation and aggregation. Given the lack of a definitive pathogenic mechanism for TREM2 in AD, TREM2 mutations are identifiably great susceptibility markers for the disease (Lu et al., 2015).

TREM2 RECEPTOR AND ITS IMPLICATIONS WITH TAU PATHOLOGY

Tau protein is produced by the MAPT gene through an alternative splicing phenomenon. It is abundant in CNS neurons and plays key functions in preserving the stability of the microtubule in axons (Sjölin et al., 2022). NFTs are made of hyperphosphorylated tau proteins, which mediate synaptic dysregulation and neurodegeneration in brain tissues and are linked to neurological pathologies and dementias, including AD and PD. The instability of the neuronal cytoskeleton is caused by hyperphosphorylated tau peptides, which eventually results in cellular malfunction and neurone and synapse death (Cleveland et al., 1977; Weingarten et al., 1975). Interestingly, tau pathology is another significant pathological marker of AD and has a stronger association with cognitive abnormalities compared to amyloid plaques in AD. There is noteworthy evidence linking functionally rare TREM2 variants, such as p.R47H, with higher amounts of hyperphosphorylated tau in AD patients' brain tissue and CSF. Nevertheless, a recent study by Lue and co-workers revealed that in post-mortem temporal cortex samples from AD cases, TREM2 levels were elevated and positively linked with levels of hyperphosphorylated tau. All these data suggest a possible connection between tau disease and TREM2 (Lue et al., 2015).

Axonal degeneration and synaptic decline are more frequently associated with TREM2 deficiency. As per findings, TREM2 can stimulate microglia to control inflammatory reactions by blocking pro-inflammatory cytokines, including IL-1β, TNF-α and IL-6. This lowers the activity of CDK-5 and GSK-3β kinases, which in response suppresses aberrant tau phosphorylation in P301S tau TG AD mice (Jiang et al., 2016). Others suggest that its loss of function in PS301 TG mice can reduce neuroinflammation and safeguard against neurodegeneration (Vogels et al., 2019). In a humanised tauopathy model, Bemiller and colleagues found that a TREM2 loss accelerates and exacerbates tau hyperphosphorylation and aggregation.

Furthermore, p38, JNK and ERK kinase signalling are drastically downregulated as a result of TREM2 loss in brain regions (Bemiller et al., 2017). Studies found that TREM2 is crucial for controlling tau-related innate immune responses in the CNS and established that TREM2 signalling shields from tauopathies despite indications that it can negatively affect early-onset Aβ-associated pathological events (Vogels et al., 2019). Jiang and colleagues further suggested that TREM2 reduced the deleterious effects of activated microglial cells on neuronal tau hyperphosphorylation by reducing the inflammatory response of activated microglia in neuronal–microglial co-culture AD mice. Together, these results reveal the fundamental processes by which TREM2 guards against tau pathology and propose TREM2 as a possible AD treatment target (Jiang et al., 2018).

MOLECULAR CONNECTIONS BETWEEN AMYLOID PLAQUES AND TREM2 ACTIVITY

The Aβ peptide is composed of nearly 38 to 43 amino acid residues. It is generated following the enzymatic cleavage of APP, a type-1 trans-membrane protein expressed in a multitude of tissues, especially the CNS, by the enzymes β- and γ-secretases (Jakob-Roetne and Jacobson, 2009). APP is crucial for brain development, memory and synaptic plasticity, though its overall physiological significance is yet unknown (Nalivaeva and Turner, 2013). Two distinct mechanisms are possible to disintegrate APP. The principal non-amyloidogenic process involves the APP cleavage by α-secretase, succeeded by γ-secretase, to yield shortened Aβ-17-40/42 peptides or by β-secretase to generate shorter Aβ-1-16 peptides (Chow et al., 2010; Nalivaeva and Turner, 2013). In the amyloidogenic phase, APP is sequentially degraded by β- and γ-secretases, producing entire Aβ peptides Aβ40/42. Numerous sequences of peptides, ranging from Aβ1-38 to Aβ40-43, can be released depending on the precise region of the cleavage by γ-secretase. Nevertheless, Aβ1-40 and, to a minor extent, Aβ1-42 are the most prevalent species released in the cortex (Chow et al., 2010).

Complete research of TREM2 receptors and their relevance to amyloid pathology has been the focus of numerous *in vivo* and *in vitro* investigations in recent decades. Although extensive *in vitro* TREM2 studies strongly support a potential role for TREM2 in the microglial uptake of Aβ-40-42, the findings of *in vivo* TREM2 studies are depicted to be inconsistent. Heterozygous TREM2 point mutations are related to a variety of dysfunctions, including elevated dystrophic neurites and p-tau positive cells. These pathologies additionally involve an overall reduction in the amount of microglia associated with amyloid plaques as well as the inability of the residual microglia to compress beta-amyloid deposits and establish a barrier between Aβ plaques and neurones. Figure 4.1 showcases the implications of TREM2 receptor on amyloid plaque and microglia environment. Aβ deposition in the cortex was not influenced by TREM2 haplo-insufficiency in APPPS1-21 mice. When TREM2 was eliminated in the APPPS1-21 AD model, the two-month-old mice cortex had a decreased Aβ burden, whereas the cortex of eight-month-old mice had a significant Aβ accumulation (Sayed et al., 2018). Wang and colleagues suggested that TREM2 deletion exacerbated Aβ pathology (both overall Aβ load and insoluble A42) in the hippocampal regions but not in the cortical sections of mice, with TREM2-haplo-insufficient mice showing an intermediary pattern in 5XFAD mice. These results indicate that TREM2's impact on Aβ deposition is age- and Aβ burden-dependent (Jay et al., 2015). TREM2 can thus induce more Aβ deposition in the early phases of the disease and less Aβ deposition in the advanced stages. By utilising TREM2-activating antibodies to identify and target TREM2, Price and co-workers performed microglial activation, leading to their recruitment to amyloid plaques, decreasing amyloid production, and potentially improving cognition. TREM2-deficient microglia in amyloid deposit-positive mice failed to adequately transform into phagocytic, disease-associated microglial cells and to express their relevant gene profile. Additionally, myelin debris found cannot be cleared by TREM2-deficient microglia. With heightened microglial clustering around amyloid plaques and reduced amyloid deposition, while utilising a TREM2-activating strategy, their results imply a vital function for TREM2 in microglial phagocytosis (Price et al., 2020).

TREM2 has also been shown to mediate Aβ phagocytosis plaques in the cortex. The V-type immunoglobulin-like sites on the ectodomain of TREM2 serve as a target region for ligand binding

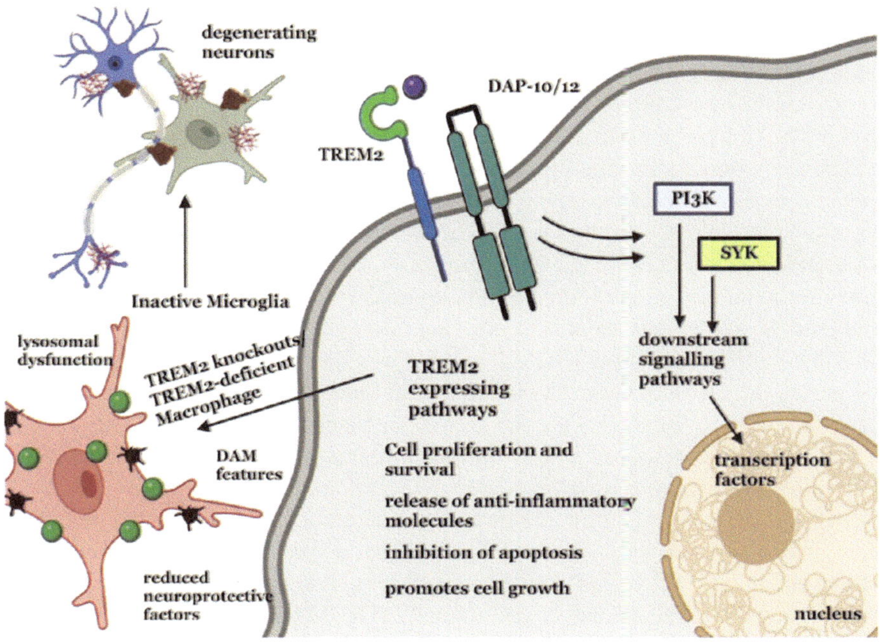

FIGURE 4.1 Alternate roles of TREM2 receptors in physiological neural microglia and TREM2 knockout macrophages.

and mediate Aβ fibril interactions (Wang et al., 2016). By producing several C-terminally shortened synthetic TREM2 proteins and conducting extensive Aβ1-42 binding experiments, Zhong and companions discovered that TREM2's 31–91 amino acid residues are the segment responsible for its interaction with Aβ (Zhong et al., 2018). The interaction involving Aβ oligomers and sTREM2 is characterised by very delayed dissociation that can inhibit Aβ oligomers from engaging with other ligands (Kober et al., 2020). Microglia are triggered to initiate targeted cellular phagocytosis to remove aggregated proteins as soon as the TREM2–Aβ complex is created. Aβ plaques are prevented from migrating into the nearby healthy tissues by TREM2 and sTREM2 as they bind to Aβ plaques (Condello et al., 2015; Yuan et al., 2016). Likewise, in the proximity of amyloid plaques, microglia can be triggered to generate inflammatory mediators to initiate cellular signalling that suppresses Aβ lesion proliferation (Hou et al., 2022). Recombinant sTREM2 was shown to be capable of lowering Aβ deposition and restoring behavioural abnormalities in 5XFAD mouse models (Zhong et al., 2019). Hence, all these findings show that TREM2 and sTREM2 exert protective properties by directly influencing the pathophysiology brought on by Aβ oligomers and plaques.

TREM2-ASSOCIATED MICROGLIAL ACTIVITY IN ALZHEIMER'S DISEASE

Microglia are a group of neuroglial cell subtypes present throughout the cerebral cortex and spinal cord. They represent approximately 10% to 15% of the brain's cells (Dos Santos et al., 2020). They serve as the initial and predominant type of active immune defence in the CNS as they are the resident macrophage cells. They are distributed in large non-overlapping CNS regions throughout the brain (Nayak et al., 2014). They play essential roles in neurodevelopment and synaptic plasticity. At a steady state, microglia engulf surrounding neuronal synapses in the complement pathway, ensuring regulated brain development and synaptic connectivity (Paolicelli et al., 2011). Microglial deficits are highly associated with defective learning, cognitive decline and memory formation abilities. Numerous studies on mice models have revealed the relationship between mutations of the TREM2 gene and its effect on microglial activation, amyloid burden and tau pathology in the brain affected by AD (Malik et al., 2015). Much evidence supports TREM2's involvement in controlling microglia-related neuroinflammatory processes in AD. After the TREM2 ectodomain is cleaved by proteases in response to ligand binding, the intracellular fraction of TREM2 interacts with DAP12 to activate Syk-PI3K/MAPK signalling and release IP3-gated Ca2+ stores for phagocytosis. sTREM2 shedding is hence necessary for microglial activation (Sasaki et al., 2015).

In-depth research has been done on the mechanisms underlying TREM2-related neurodegeneration. Many studies have demonstrated that TREM2 expression is upregulated by microglia in AD mice models and people (Frank et al., 2008; Wang et al., 2015), indicating that the overexpression of TREM2 may be connected to the course of AD. In addition, TREM2 is a significant factor in both AD animal models and human patients, promoting microgliosis. Microglia cluster around plaques in AD mice that express wild-type TREM2, serving as a barrier to the nearby neurones (Ulrich et al., 2014; Yuan et al., 2016). In contrast, the number of microglia associated with plaques is significantly decreased in AD animals lacking the Trem2 gene, and the effect is dose dependent.

Multiple types of microglia can trigger various immune reactions. There have been reports of distinct M1 and M2 phenotypes' involvement in neuroinflammation. The M1 type is cytotoxic and pro-inflammatory, while the M2 type is neuroprotective and anti-inflammatory. According to earlier research, P301S tau transgenic mice with TREM2 overexpression have activated M2 microglia, which reduces the release of pro-inflammatory cytokines (Jiang et al., 2016). In primary microglia from both WT and *Trem2*-knockout mouse models, sTREM2 additionally increases the production of survival by modifying the activity of the Akt/GSK3β/β-catenin signalling network (Zhong et al., 2017; Neumann and Takahashi, 2007). Wang and co-workers demonstrated that TREM2 is a microglial detector that recognises damage-related molecules that have an anionic group and a common lipidic backbone. They discovered that phagocytosis of apoptotic cells is not directly mediated by TREM2 involvement. Yet, by encouraging the survival of active microglia, TREM2 signalling may indirectly enhance phagocytosis. Microglia are activated and proliferate in the presence of Aβ, and TREM2 serves as a co-stimulatory receptor that supports microglia survival. TREM2 also provides a signal that is necessary for microglia survival at low CSF-1 levels (Wang et al., 2015). Additionally, sTREM2 has been demonstrated to improve the elimination of amyloid plaques in an activated microglia-dependent way, which the 5XFAD animal model also verified. This information supports the concept that TREM2, and sTREM2 mediate microglial-associated inflammatory processes and may serve as a protective factor in AD.

Conversely, multiple studies indicated that TREM2 could induce microglia to generate pro-inflammatory cytokines that harm neurons, including TNFα, IL-6 and IL-1β. This creates uncertainty about TREM2's function in regulating immunological responses (Jay et al., 2017). The AD development and the impact of TREM2 on resident microglia must be thoroughly explained to address these contradictory results. TREM2 demonstrates a protective impact in the initial and intermediate stages of AD, but by stimulating the adaptive immune system in the late stages of AD, it switches toward a negative impact (Karanfilian et al., 2020). This perspective aligns with microglia's activation at various illness stages.

CONCLUSION

Research into the connection between the TREM2 protein and AD has grown significantly in recent years. The aetiology of the illness is complicated and includes genetic, environmental and behavioural elements. Recent research has demonstrated that the TREM2 protein is essential for the immune response and neuroinflammation of the brain and that malfunction of this protein contributes to the aetiology of AD. Correlation studies have shown that genetic variants in the TREM2 gene are linked to an elevated risk of Alzheimer's disease development. The specific processes through which the TREM2 protein affects the onset and progression of AD require more study. New treatment approaches may be developed as a result of the link between the TREM2 protein and Alzheimer's disease. By decreasing neuroinflammation and beta-amyloid build-up, targeting the TREM2 protein may offer a potential strategy for treating AD. Moreover, the TREM2 protein has the potential to be used as a biomarker for the early identification and diagnosis of AD, creating new opportunities for deciphering the illness's pathophysiology and creating prospective treatment targets. Future studies should concentrate on determining the specific processes by which the TREM2 protein affects Alzheimer's disease and creating potent treatment approaches to treat this deadly condition.

ACKNOWLEDGEMENTS

All the schematic diagrams were made using Biorender.com software. The authors acknowledge support from Poona College of Pharmacy, Pune.

REFERENCES

Bemiller SM, McCray TJ, Allan K, Formica SV, Xu G, Wilson G, et al. TREM2 deficiency exacerbates tau pathology through dysregulated kinase signaling in a mouse model of tauopathy. *Mol Neurodegener* (2017) 12(1):74.

Bonham LW, Sirkis DW, Fan J, et al. Identification of a rare coding variant in TREM2 in a Chinese individual with Alzheimer's disease. *Neurocase* (2017) 23:65–69.

Breijyeh Z, Karaman R. Comprehensive review on Alzheimer's disease: causes and treatment. *Molecules* (2020) 25(24):5789.

Brown GC, St George-Hyslop P. Does soluble TREM2 protect against Alzheimer's disease? *Front Aging Neurosci* (2022) 13:834697.

Calcagno N, Baufeld C, Madore C, Butovsky O. Tremendous 2 be social. *Immunity* (2018) 48(5):842–843.

Chow VW, Mattson MP, Wong PC, Gleichmann M. An overview of APP processing enzymes and products. *Neuromol Med* (2010) 12(1):1–12.

Cleveland DW, Hwo SY, Kirschner MW. Purification of tau, a microtubule-associated protein that induces assembly of microtubules from purified tubulin. *J Mol Bio* (1977) 1 6 (2):207–25.

Condello C, Yuan P, Schain A, Grutzendler J. Microglia constitute a barrier that prevents neurotoxic protofibrillar Abeta42 hotspots around plaques. *Nat Commun* (2015) 6:6176.

Del-Aguila JL, Fernandez MV, Schindler S, Ibanez L, Deming Y, Ma S, et al. Assessment of the genetic architecture of Alzheimer's disease risk in rate of memory decline. *J Alzheimer's Dis* (2018) 62:745–756

Dos Santos SE, Medeiros M, Porfirio J, Tavares W, Pessôa L, Grinberg L, et al. Similar microglial cell densities across brain structures and mammalian species: implications for brain tissue function. *J Neurosci* (2020) 40(24):4622–4643.

Feuerbach D, Schindler P, Barske C, Joller S, Beng-Louka E, Worringer KA, et al. ADAM17 is the main sheddase for the generation of human triggering receptor expressed in myeloid cells (hTREM2) ectodomain and cleaves TREM2 after Histidine 157. *Neurosci Lett* (2017) 660:109–114.

Filipello F, Morini R, Corradini I, Zerbi V, Canzi A, Michalski B, et al. The Microglial innate immune receptor TREM2 is required for synapse elimination and normal brain connectivity. *Immunity* (2018) 48(5):979–991.

Frank S, Burbach GJ, Bonin M, Walter M, Streit W, Bechmann I, et al. TREM2 is upregulated in amyloid plaque-associated microglia in aged APP23 transgenic mice. *Glia* (2008) 56(13):1438–1447.

Frozza RL, Lourenco MV, De Felice FG. Challenges for Alzheimer's disease therapy: insights from novel mechanisms beyond memory defects. *Front Neurosci* (2018) 12:37.

Hooli BV, Parrado AR, Mullin K, Yip W-K, Liu T, Roehr JT, et al. The rare TREM2 R47H variant exerts only a modest effect on Alzheimer disease risk. *Neurology* (2014) 83(15):1353–1358.

Hou J, Chen Y, Grajales-Reyes G, Colonna M. TREM2 dependent and independent functions of microglia in Alzheimer's disease. *Mol Neurodegener* (2022) 17(1):84.

Jakob-Roetne R, Jacobsen H. Alzheimer's disease: from pathology to therapeutic approaches. *Angew Chem Int Ed Engl* (2009) 48(17):3030–3059.

Jay TR, Hirsch AM, Broihier ML, Miller CM, Neilson LE, Ransohoff RM, Lamb BT, Landreth GE. Disease progression-dependent effects of TREM2 deficiency in a mouse model of Alzheimer's disease. *J Neurosci* (2017) 37:637–647.

Jay TR, Miller CM, Cheng PJ, Graham LC, Bemiller S, Broihier ML, et al. TREM2 deficiency eliminates TREM2+ inflammatory macrophages and ameliorates pathology in Alzheimer's disease mouse models. *J Exp Med* (2015) 212:287–295.

Jiang T, Zhang YD, Chen Q, Gao Q, Zhu XC, Zhou JS, et al. TREM2 modifies microglial phenotype and provides neuroprotection in P301S tau transgenic mice. *Neuropharmacology* (2016) 105:196–206.

Jiang T, Zhang YD, Gao Q, Ou Z, Gong PY, Shi JQ, et al. TREM2 Ameliorates neuronal Tau pathology through suppression of microglial inflammatory response. *Inflammation* (2018) 41(3):811–823.

Jin SC, Benitez BA, Karch CM, Cooper B, Skorupa T, Carrell D, et al. Coding variants in TREM2 increase risk for Alzheimer's disease. *Hum Mol Genet* (2014) 23:5838–5846.

Jonsson T, Stefansson H, Steinberg S, Jonsdottir I, Jonsson PV, Snaedal J, et al. Variant of TREM2 associated with the risk of Alzheimer's disease. *N Engl J Med* (2013) 368:107–116.

Karanfilian L, Tosto MG, Malki K. The role of TREM2 in Alzheimer's disease; evidence from transgenic mouse models. *Neurobiol Aging* (2020) 86:39–53.

Kober DL, Stuchell-Brereton MD, Kluender CE, Dean HB, Strickland MR, Steinberg DF, Nelson SS, Baban B, Holtzman DM, Frieden C, Alexander-Brett J, Roberson ED, Song Y, Brett TJ. Functional insights from biophysical study of TREM2 interactions with apoE and Aβ1-42. *Alzheimers Dement.* (2020) 17(3):475–488.

Lehnardt S. Innate immunity and neuroinflammation in the CNS: the role of microglia in toll-like receptor-mediated neuronal injury. *Glia* (2010) 58(3):253–263.

Li R, Wang X, He P. The most prevalent rare coding variants of TREM2 conferring risk of Alzheimer's disease: a systematic review and meta-analysis. *Exp Ther Med* (2021) 21(4):347.

Livingston G, Huntley J, Sommerlad A, Ames D, Ballard C, Banerjee S, et al. Dementia prevention, intervention, and care: 2020 report of the Lancet Commission. *Lancet* (2020) 396(10248):413–446.

Lu Y, Liu W, Wang X. TREM2 variants and risk of Alzheimer's disease: a meta-analysis. *Neurol Sci* (2015) 36(10):1881–1888.

Lue LF, Schmitz C, Walker DG. What happens to microglial TREM2 in Alzheimer's disease: immunoregulatory turned into immunopathogenic? *Neuroscience* (2015) 302:138–150.

Ma L, Allen M, Sakae N, Ertekin-Taner N, Graff-Radford NR, Dickson DW, et al. Expression and processing analyses of wild type and p.R47H TREM2 variant in Alzheimer's disease brains. *Mol Neurodegener* (2016) 11:72

Malik M, Parikh I, Vasquez JB, Smith C, Tai L, Bu G, et al. Genetics ignite focus on microglial inflammation in Alzheimer's disease. *Mol Neurodegener* (2015) 10:52.

Mimica N, Presecki P. Side effects of approved antidementives. *Psychiatr Danub* (2009) 21(1):108–113.

Nalivaeva NN, Turner AJ. The amyloid precursor protein: a biochemical enigma in brain development, function and disease. *FEBS Lett* (2013) 587(13):2046–2054.

Nayak D, Roth TL, McGavern DB. Microglia development and function. *Annu Rev Immunol* (2014) 32:367–402.

Neumann H, Takahashi K. Essential role of the microglial triggering receptor expressed on myeloid cells-2 (TREM2) for central nervous tissue immune homeostasis. *J Neuroimmunol* (2007) 184:92–99.

Nordengen K, Kirsebom BE, Henjum K, Selnes P, Gísladóttir B, Wettergreen M, Torsetnes SB, Grøntvedt GR, Waterloo KK, Aarsland D, Nilsson LNG, Fladby T. Glial activation and inflammation along the Alzheimer's disease continuum. *J Neuroinflammation* (2019) 16(1):46.

Paolicelli RC, Bolasco G, Pagani F, Maggi L, Scianni M, Panzanelli P, et al. Synaptic pruning by microglia is necessary for normal brain development. *Science* (2011) 333:1456–1458.

Parhizkar S, Arzberger T, Brendel M, Kleinberger G, Deussing M, Focke C, et al. Loss of TREM2 function increases amyloid seeding but reduces plaque-associated ApoE. *Nat Neurosci* (2019) 22(2):191–204.

Park JS, Ji IJ, Kim DH, An HJ, Yoon SY. The Alzheimer's disease-associated R47H variant of TREM2 has an altered glycosylation pattern and protein stability. *Front Neurosci* (2016) 10:618.

Parsons CG, Stöffler A, Danysz W. Memantine: a NMDA receptor antagonist that improves memory by restoration of homeostasis in the glutamatergic system--too little activation is bad, too much is even worse. *Neuropharmacology* (2007) 53(6):699–723.

Price BR, Sudduth TL, Weekman EM, Johnson S, Hawthorne D, Woolums A, et al. Therapeutic Trem2 activation ameliorates amyloid-beta deposition and improves cognition in the 5XFAD model of amyloid deposition. *J Neuroinflammation* (2020) 17(1):238.

Sasaki A, Kakita A, Yoshida K, Konno T, Ikeuchi T, Hayashi S, et al. Variable expression of microglial DAP12 and TREM2 genes in Nasu-Hakola disease. *Neurogenetics* (2015) 16:265–276.

Sayed FA, Telpoukhovskaia M, Kodama L, Li Y, Zhou Y, Le D, et al. Differential effects of partial and complete loss of TREM2 on microglial injury response and tauopathy. *Proc Natl Acad Sci USA* (2018) 115(40):10172–10177.

Sjölin K, Kultima K, Larsson A, Freyhult E, Zjukovskaja C, Alkass K, et al. Distribution of five clinically important neuroglial proteins in the human brain. *Mol Brain* (2022) 15(1):52.

Tahami Monfared AA, Byrnes MJ, White LA, Zhang Q. Alzheimer's disease: epidemiology and clinical progression. *Neurol Ther* (2022) 11(2):553–569.

Ulrich JD, Finn MB, Wang Y, Shen A, Mahan TE, Jiang H, et al. Altered microglial response to Abeta plaques in APPPS1-21 mice heterozygous for TREM2. *Mol Neurodegener* (2014) 9:20.

Vogels T, Murgoci AN, Hromádka T. Intersection of pathological tau and microglia at the synapse. *Acta Neuropathol Commun* 2019 7(1):109.

Wang Y, Cella M, Mallinson K, Ulrich JD, Young KL, Robinette ML, et al. TREM2 lipid sensing sustains the microglial response in an Alzheimer's disease model. *Cell* (2015) 160:1061–1071.

Wang Y, Ulland TK, Ulrich JD, Song W, Tzaferis JA, Hole JT, et al. TREM2-mediated early microglial response limits diffusion and toxicity of amyloid plaques. *J Exp Med* (2016) 213:667–675.

Weingarten MD, Lockwood AH, Hwo SY, Kirschner MW. A protein factor essential for microtubule assembly. *Proc Natl Acad Sci USA* (1975) 72 (5):1858–1862.

Yang J, Fu Z, Zhang X, Xiong M, Meng L, Zhang Z. TREM2 ectodomain and its soluble form in Alzheimer's disease. *J Neuroinflammation* (2020) 17(1):204.

Yeh FL, Hansen DV, Sheng M. TREM2, microglia, and neurodegenerative diseases. *Trends Mol Med* (2017) 23:512–533.

Yiannopoulou KG, Papageorgiou SG. Current and future treatments in Alzheimer disease: an update. *J Cent Nerv Syst Dis* (2020) 12:1179573520907397.

Yuan P, Condello C, Keene CD, Wang Y, Bird TD, Paul SM, et al. TREM2 Haplo deficiency in mice and humans impairs the Microglia barrier function leading to decreased Amyloid Compaction and severe Axonal dystrophy. *Neuron* (2016) 90(4):724–739.

Zheng H, Cheng B, Li Y, Li X, Chen X, Zhang YW. TREM2 in Alzheimer's disease: microglial survival and energy metabolism. *Front Aging Neurosci* (2018) 23(10):395.

Zhong L, Chen XF, Wang T, Wang Z, Liao C, Huang R, et al. Soluble TREM2 induces inflammatory responses and enhances microglial survival. *J Exp Med* (2017) 214:597–607.

Zhong L, Wang Z, Wang D, Martens YA, Wu L, Xu Y, et al. Amyloid-beta modulates microglial responses by binding to the triggering receptor expressed on myeloid cells 2 (TREM2). *Mol Neurodegener* (2018) 13:15.

Zhong L, Xu Y, Zhuo R, Wang T, Wang K, Huang R, et al. Soluble TREM2 ameliorates pathological phenotypes by modulating microglial functions in an Alzheimer's disease model. *Nat Commun* (2019) 10:1365.

Zhou SL, Tan CC, Hou XH, Cao XP, Tan L, Yu JT. TREM2 variants and neurodegenerative diseases: a systematic review and meta-analysis. *J Alzheimers Dis* (2019) 68(3):1171–1184.

Zhu CW, Scarmeas N, Ornstein K, Albert M, Brandt J, et al. Health-care use and cost in dementia caregivers: longitudinal results from the predictors caregiver study. *Alzheimers Dement* (2015) 11(4):444–454.

5 Phosphodiesterase

A Safe and Potential Target for Anti-Alzheimer's Therapeutics

Shikha Kaushik[1], *Sarvesh Paliwal*[2], *Neeraj Masand*[3], *and Vaishali Manikrao Patil*[4,5*]

[1]Department of Pharmaceutical Chemistry, KIET School of Pharmacy, KIET Group of Institutions, Delhi-NCR, Ghaziabad, Uttar Pradesh, Bharat
[2]Department of Pharmacy, Banasthali Vidyapith, Tonk, Jaipur, Rajasthan, India
[3]Department of Pharmacy, Lala Lajpat Rai Memorial Medical College, Meerut, Uttar Pradesh, Bharat
[4]Charak School of Pharmacy, Chaudhary Charan Singh University (CCSU), Meerut Uttar Pradesh, Bharat
[5]Faculty of Pharmaceutical Sciences, UCSI University, Kuala Lumpur, Malaysia
*Corresponding author

INTRODUCTION

Phosphodiesterase (PDE) enzymes play a key role in many cellular processes and control the intracellular levels of cyclic adenosine monophosphate and cyclic guanosine monophosphate. Catalyzing the hydrolysis of 3′-5′-cyclic adenosine monophosphate (cAMP), 3′-5′-cyclic guanosine monophosphate (cGMP), or both is the primary task (Newton et al., 2016). Both cyclic nucleotides are essential as secondary messengers and play an important role in several activities like metabolism and growth, and their levels can be increased only by inhibition of PDEs (Raker et al., 2016; Fajardo et al., 2014; Lorigo et al., 2022; DeNinno, 2012). There are 11 different subfamilies along with over 40 isoforms grouped for different PDEs including (PDE1A, 1B, and 1C) for PDE1, (PDE2A) for PDE2, (PDE3A-3B) for PDE3, (PDE4A, 4B, 4C, 4D) for PDE4, (PDE5A) for PDE5, (PDE6A, 6B, 6C, 6D, 6G, and 6H) for PDE6, (PDE7A-7B) for PDE7, (PDE8A-8B) for PDE8, (PDE9A) for PDE9, (PDE10A) for PDE10, and (PDE11A) for PDE11, depending on whether they are encoded by single or several genes (Bondarev et al., 2022; Francis et al., 2011b) (Figure 5.1).

It has been discovered that the reported PDE subtypes play a significant role in controlling the behavior of cyclic nucleotides in a given cell/region. PDEs 5–11, which belong to some PDE families that are widely expressed in mammalian tissues, are less frequent or expressed in fewer tissues than PDEs 1–4. The PDE6 family has only been discovered in the pineal gland and the outer segments of photoreceptors, where it is concentrated. This suggests that the PDE6 family's distribution is the most restricted (Cote, 2021). Both cAMP and cGMP are essential for neural behavior such as signaling and boosting the power of nerve impulses, resulting in enhanced memory and cognition (Nabavi et al., 2019). Because PDE2 is a highly expressed enzyme in the human brain that is stimulated by cGMP and that degrades both cAMP and cGMP, PDE2 inhibitors work by increasing the levels of cyclic nucleotides to treat cognitive disorders and other disorders of brain function, and they are particularly effective in treating Alzheimer's disease (Morriello et al., 2021). A large number of PDE inhibitors have been developed to control the unique functions of each PDE family and were also used in treating similar symptoms

DOI: 10.1201/9781003506317-5

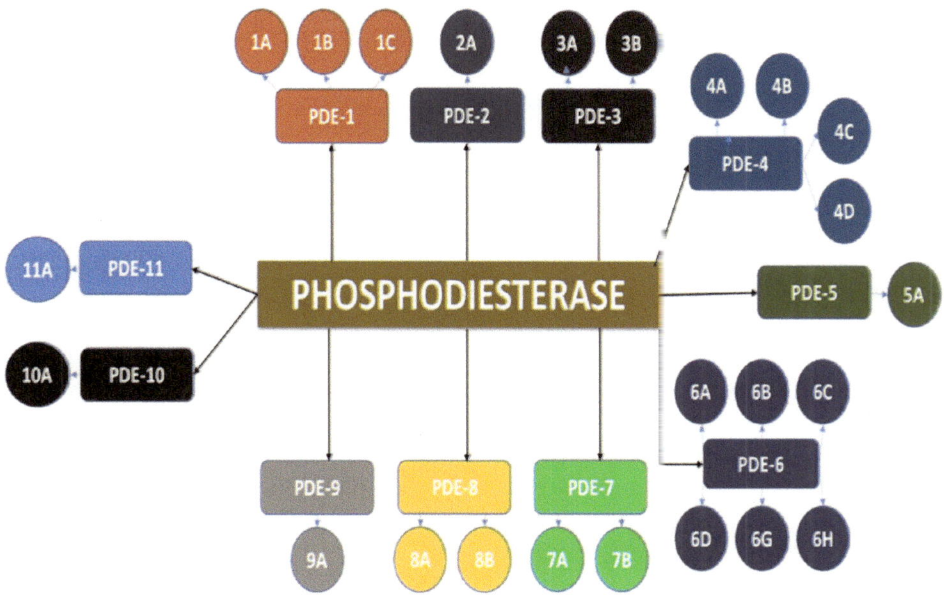

FIGURE 5.1 Various reported types and isoforms for phosphodiesterase enzyme.

caused by severe acute respiratory syndrome coronavirus 2 (SARS-CoV-2) virus infections. This is due to the widespread distribution and moderating role in almost all organs and cellular environments. PDE inhibitors exhibit a broad range of therapeutic effects, including those for asthma, chronic obstructive pulmonary disease (COPD), cardiovascular illness, and neurological disorders (Giorgi et al., 2020). To treat neuro-developmental illnesses that are associated with behavioral and biochemical abnormalities, and memory impairment like dementia, Alzheimer's, Parkinson's, epilepsy, and various learning deficits, many PDE modulators are now being studied. Recent research has indicated that several PDE inhibitors, such as PDE2, PDE3, PDE4, and PDE10, impact several neurological conditions, including autism, Alzheimer's, schizophrenia, and depression (Delhaye et al., 2021). PDE families differ in their specificities toward the cyclic nucleotide substrate: PDE1, PDE2, PDE3, PDE10, and PDE11 hydrolyze both cAMP and cGMP, whereas PDE4, PDE7, and PDE8 have a higher affinity for cAMP; and PDE5, PDE6, and PDE9 have a higher specificity for cGMP. This PDE family exhibits a diverse pattern of expression that is particularly reliant on intracellular localization in the cytoplasm (Francis et al., 2011a). The world's efforts to use PDE enzymes therapeutically have so far resulted in a variety of compounds with different selectivity and therapeutic uses, as well as an increasing number of drugs in development (Baillie et al., 2019). A variety of nitrogen-containing classes make up the majority of the PDE inhibitors (Weinryb et al., 1972; Weishaar et al., 1985). The PDE enzymes have also been shown to be impacted by at least one non-nitrogenous substance, an isoflavone derivative called genistein (Nichols et al., 2000). In early studies, zanthine derivatives were studied as a lead for further development of PDE inhibitors (Francis et al., 2010). Caffeine and theophylline also showed weak inhibitory action, but the substitution of xanthine with a long alkyl chain like the isobutyl group was found to be more potent as compared to theophylline (Choi et al., 1988; Beavo et al., 1970). The Food and Drug Administration (FDA) has approved two categories of compounds: cholinesterase inhibitors and N-methyl-D-aspartate (NMDA) receptor antagonists. It has been found that donepezil, galantamine, rivastigmine, and memantine play an important role in cognitive function but not in curing the disease (Zhou et al., 2021). Numerous neurological conditions limit the brain's ability to discriminate between neural pathways, impairing neuronal plasticity and causing disorders like Alzheimer's and Parkinson's. The brain's capacity to recognize neural circuits is known as neuronal plasticity. Numerous PDE inhibitors have been demonstrated to enhance neural plasticity. The present chapter aims to explore the role of PDE in the discovery and development of safe and effective therapeutics for the treatment of Alzheimer's, a neurodegenerative disorder. It also provides an overview of various PDE inhibitors developed for the different subtypes and their mechanisms of action.

ETIOLOGY OF ALZHEIMER'S DISEASE

Alzheimer's disease (AD) is a progressive neurodegenerative disorder that leads to memory decline and cognitive impairment over time. The cyclic nucleotide phosphodiesterase contains subtypes that control the hydrolysis of cAMP and cGMP, which are involved in signal transduction. The cAMP response element binding protein (CREB), which is abundantly expressed in the hippocampus, is regulated by cAMP/cGMP and promotes neuronal survival through mechanisms associated with the stimulation of synaptic strengthening and memory formation (Teich et al., 2015). Amyloid-β (Aβ) accumulation and phosphorylation of tau protein lead to the downregulation of CREB and brain-derived neurotrophic factor (BDNF), which impair cognitive improvement (Rosa et al., 2015). It has been observed that glycation of proteins with reducing sugars and metal-catalyzed oxidation leads to the formation of advanced end products, which induce tumor necrosis factor alpha (TNF-α) release, causing neuroinflammation. PDE inhibitors (PDEIs) have been reported for their role in improving cognitive ability by inhibiting TNF-α (Meiners et al., 2004). Additionally, acetylcholinesterase (AChE) and butyrylcholinesterase (BuChE) play an important role in signaling. When AChE levels decrease, BuChE levels increase; such a condition may lead to the progression of AD. Some combined therapies, inhibitors of PDE5 along with AChE inhibitors (AChEIs) have shown prominent results in combating various types of dementia disorders (Ni et al., 2018). In one finding, beta-site amyloid precursor protein cleaving enzyme 1 (BACE-1) is required for the generation of all monomeric forms of Aβ. BACE-1 concentrations and its rate of activity are increased in AD (Hampel et al., 2021). Expression of nicotinic acetylcholine receptors partially through β-amyloid leads to the upregulation of inflammatory pathways, leading to cognitive decline in AD (Hoskin et al., 2019). Sometimes overexpression of glutamate receptors, namely, NMDA receptors, causes disturbance in Ca^{2+} handling, causing synaptic dysfunction or neuronal death (Wang et al., 2017). Glycogen synthase kinase (GSK)-3 causes hyperphosphorylation of tau protein, modulates the generation of β-amyloid, and its overexpression causes a decline in cognitive abilities and sometimes neuronal death (Avila et al., 2007). Various mechanisms reported for the development of anti-Alzheimer's disease (anti-AD) therapeutics are depicted in Figure 5.2.

ROLE OF PDES IN AD

PDE1

PDE1 is one of the cyclic GMP phosphodiesterases and is regulated by Ca^{2+}/calmodulin binding. PDE1 isoenzymes are widely expressed in the brain, heart, reproductive system, immune system, and liver. The PDE1 family shows diverse functions like smooth muscle proliferation, learning and cognitive abilities, olfactory sensation, and cardiac hypertrophy (Francis et al., 2001). PDE1 has three isoforms: PDE1A, PDE1B, and PDE1C; out of these, PDE1B was found to be responsible for more than 90% of brain activity, which includes learning and cognizance primarily. Thus, PDE1B was found to be a promising drug target for the treatment of neurodegenerative diseases (Shy et al., 2021). The primary pathology of Alzheimer's disease is linked to the deposition of amyloid β plaques and the aggregation of tau neurofibrillary tangles (Liu et al., 2017). Additionally, decreased concentrations of cAMP/cGMP and BDNF in the brain have been found to be associated with AD progression. The modern lifestyle has led to increased numbers of Alzheimer's patients, which

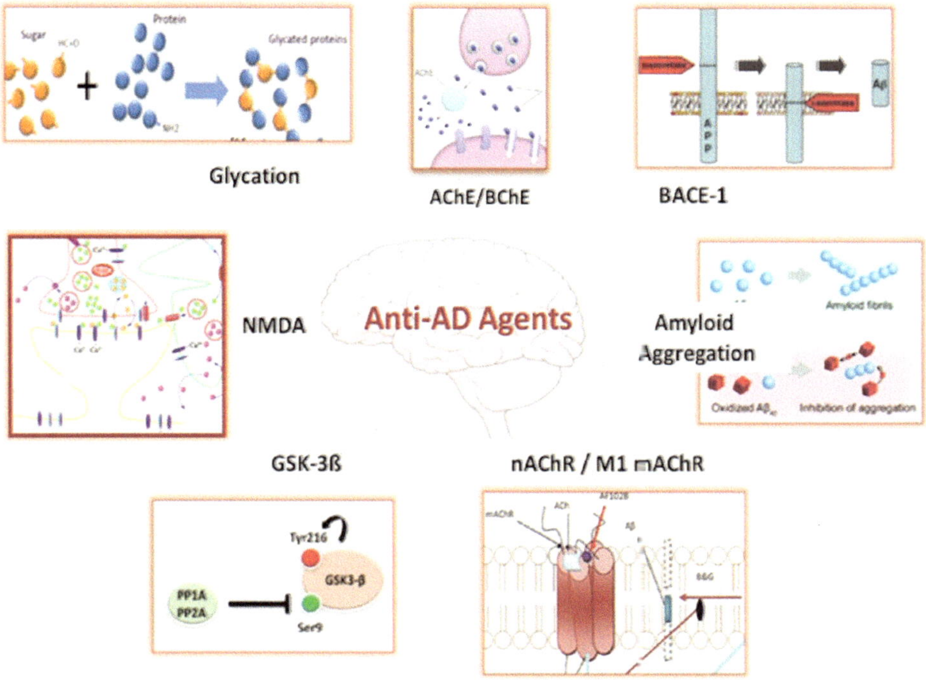

FIGURE 5.2 Various mechanisms for the development of anti-AD therapeutics.

have put a huge burden on the world. Consequently, the investigation of treatments for cognitive decline has increased the interest of researchers. PDE1 inhibitors, such as vinpocetine, have been proven to increase dendritic alignment, which in turn enhances memory and cognition. Vinpocetine has also demonstrated its anti-inflammatory effects, but due to the absence of randomized clinical investigations, this medication remains controversial. Furthermore, vinpocetine showed harmful effects on the fetus, including miscarriages, hypotension, and other side effects (Dubey et al., 2020). Amantadine is a weak NMDA antagonist that has the role of raising dopamine levels. Other PDE1 inhibitors include caffeine, an adenosine receptor antagonist; deprenyl, a monoamine oxidase (MAO) inhibitor that inhibits PDE1A2 but has no effect on PDE1B1; and nimodipine, a Ca^{2+} channel blocker. The reported PDE1 inhibitors have exhibited non-selective outcomes (Medina, 2011) (Table 5.1).

PDE2

PDE2 can hydrolyze both cAMP and cGMP. PDE2A is the only isoform with substantial levels of expression in the brain. The amount of PDE2A in the brains of AD patients was shown to be higher than that of healthy individuals. By blocking PDE2, the inhibitors can aid in raising cAMP and cGMP levels and also aid in activating protein kinase A (PKA) and protein kinase G (PKG). PDE2 is regarded as a potentially effective target for AD therapeutics (Tresadern et al., 2020). In this regard, several compounds have been synthesized and evaluated for biological action and toxicity using *in vivo/in vitro/in silico* methods. Some of them are discussed next.

Triazolopyrazine

A novel series of [1,2,4]triazolo[4,3-α] pyrazine compounds show good inhibitory activities against PDE2 as well as antioxidant activities (Jiang et al., 2021). The cytotoxicity of compound

TABLE 5.1
Details of Some PDE-1 Inhibitors with Their Anti-Neurodegenerative Role/Mechanism

Compound Name	Structure	Anti-Neurodegenerative Role	Reference
EHNA	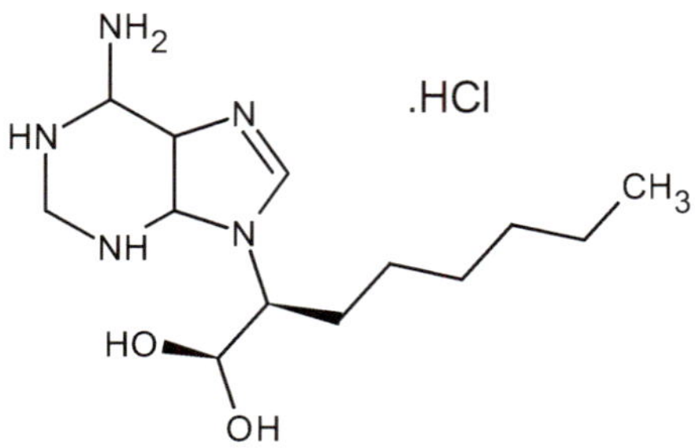	First reported PDE2 inhibitor. Restricted application due to adenosine deaminase and adenosine aminohydrolase inhibitory activity.	Bessodes et al. (1982)
PF-05180999	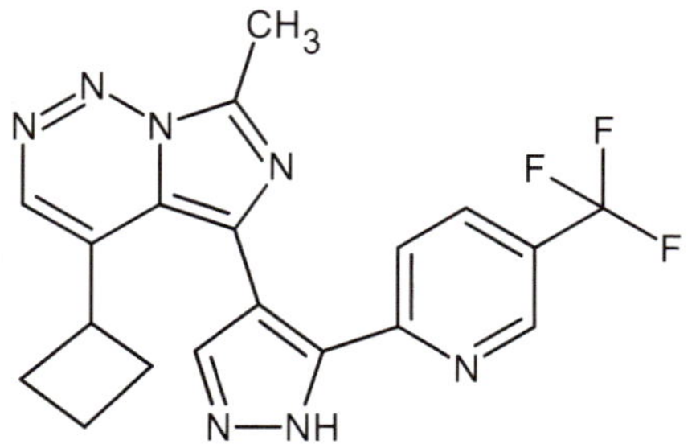	Developed by Pfizer. Entered phase I of clinical trials for the treatment of schizophrenia and migraine.	Wu et al. (2018)
Vinpocetine	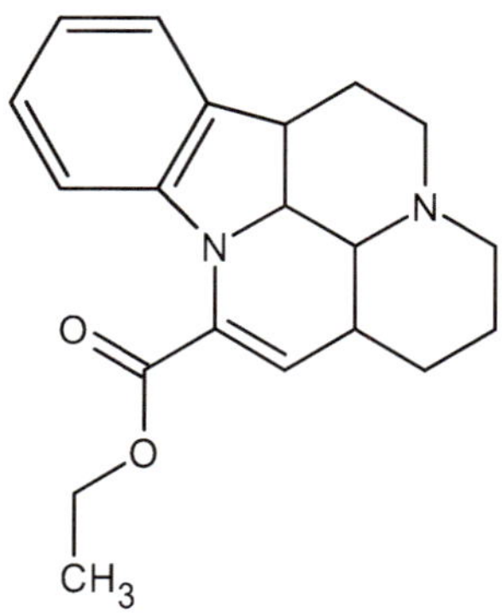	Vinpocetine promotes neuroplasticity. Helps in enhancing memory.	Dubey et al. (2020)
IBMX	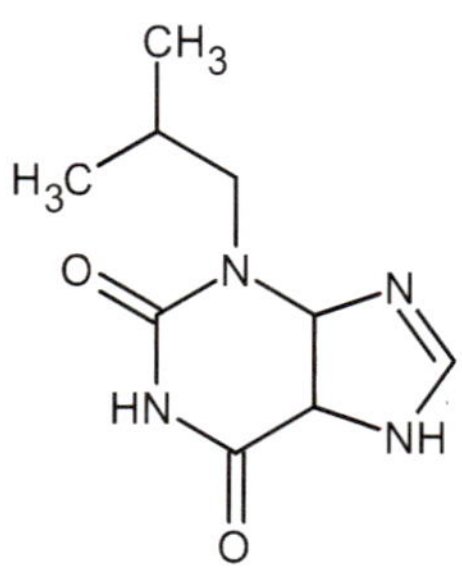	It was not very specific for PDE1 and also inhibits PDE5.	Prickaerts et al. (2017)

Compound Name	Structure	Anti-Neurodegenerative Role	Reference
ITI-214	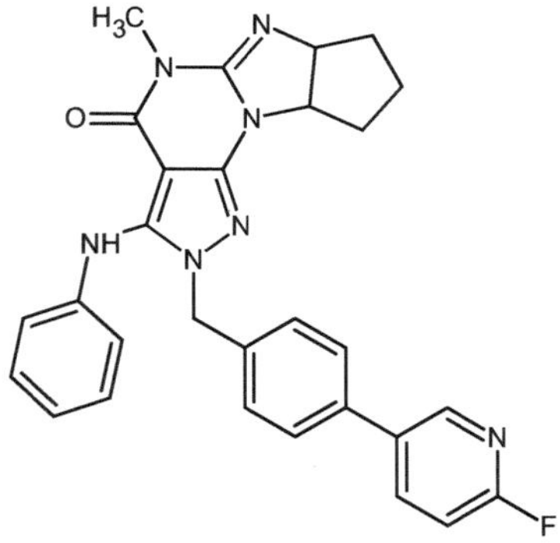	It has a submicromolar binding tendency for all PDE1 isoforms. It can cross the blood–brain barrier in animal models, exerting procognitive effects.	O'Brien et al. (2020)
DNS-0056	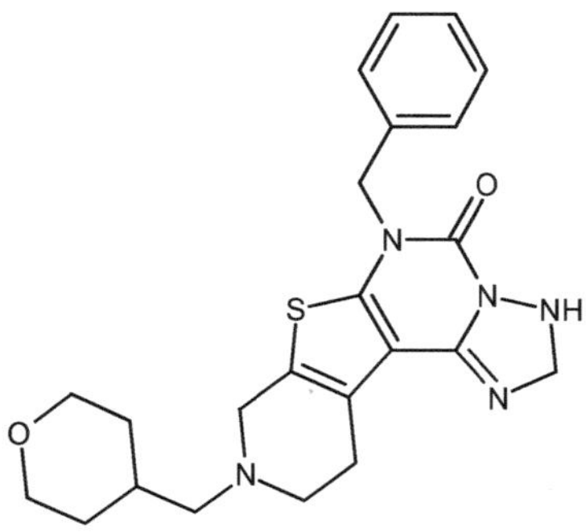	Derivative of thienotriazolopyrimidinone. Exhibited good pharmacokinetic and brain penetrative properties.	Wu et al. (2018)
3M	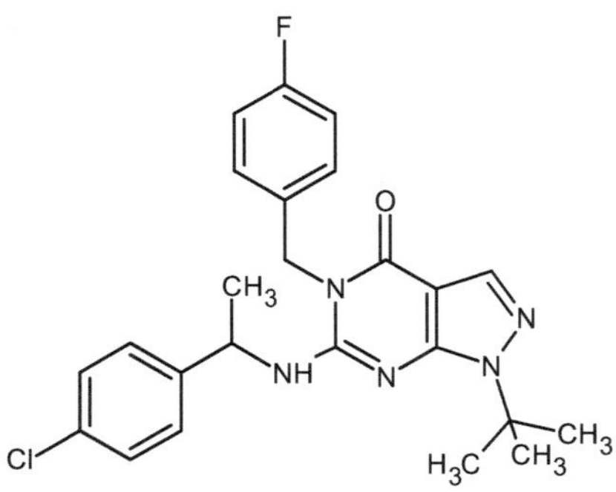	It has a hydrophobic pocket, and the inclusion of a group like a benzyl group can further promote inhibitory activity.	Huang et al. (2022)
PF-04822163	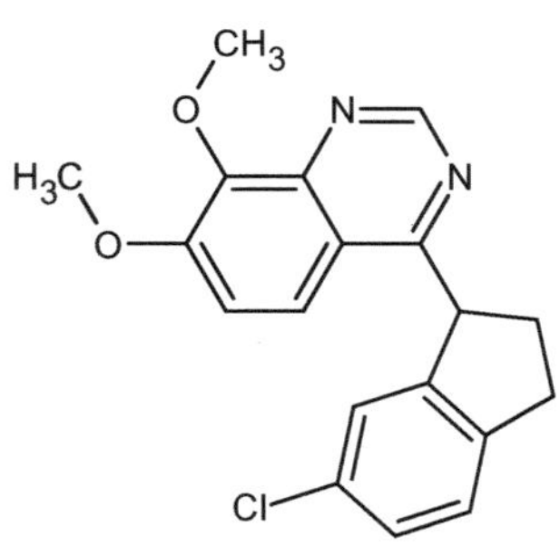	It is a brain penetrating quinazoline-based PDE1 inhibitor	Humphrey et al. (2014)

Compound Name	Structure	Anti-Neurodegenerative Role	Reference
DSR-141562	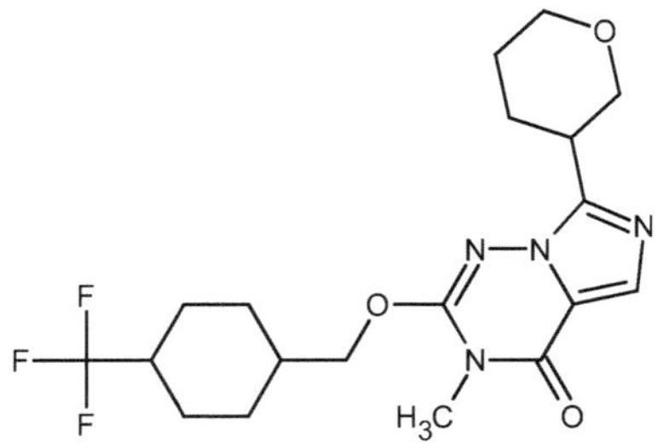	Inhibits PDE1B.	Enomoto et al. (2019)
Deprenil/ selegiline	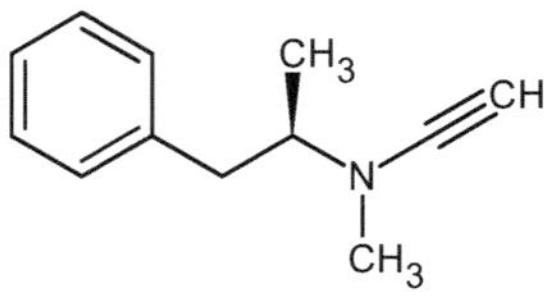	Inhibits PDE1A2 causing short-term improvement of AD.	Sanders and Rajgopal (2020)
Nimodipine	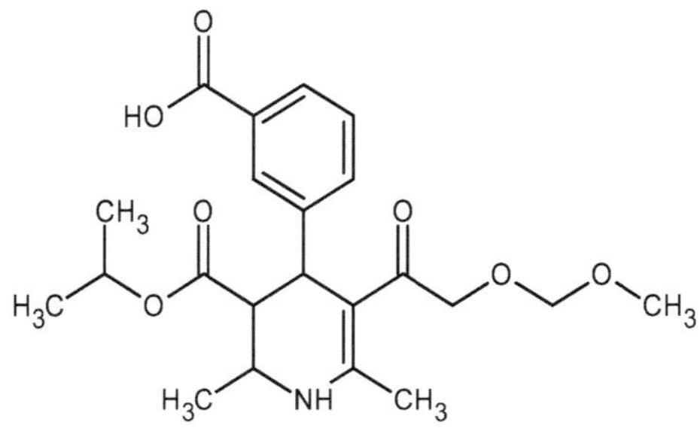	Clinically useful for subarachnoid hemorrhages.	Ansari et al. (2019)
SCH-51866	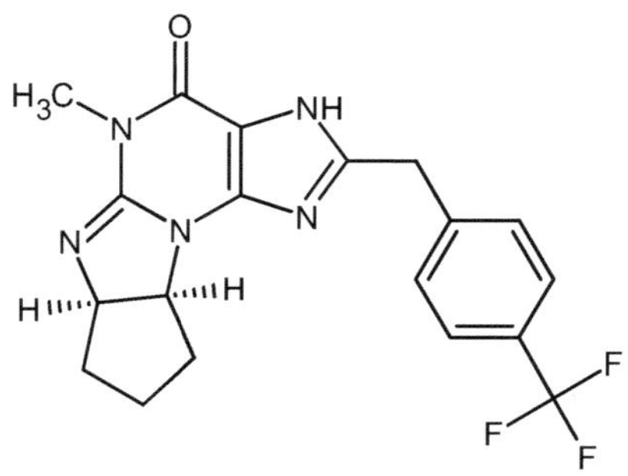	300-fold more selective for PDE1 than PDE5.	Dyck et al. (2017)
PF-04677940	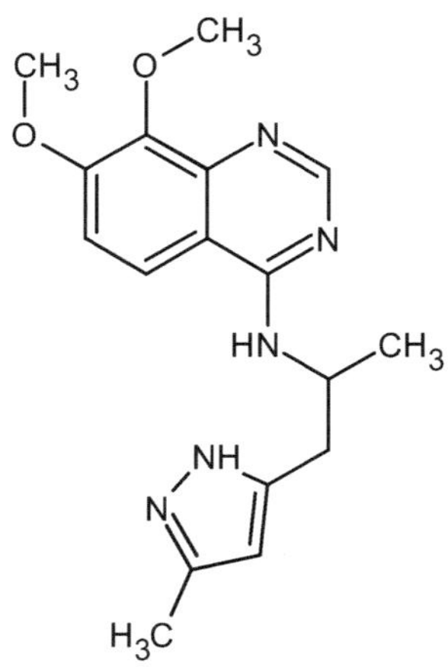	PDE1 inhibitor. Detailed research remains unavailable.	Huang et al. (2022)

1 (Figure 5.3) was tested on the neuroblastoma cell line SH-SY5Y by the CCK8 method and upon incubation for two days, its cellular viability was found to be 53% and 39% at 50 μM and 100 μM, respectively. The IC_{50} of 6.1 nM emphasizes its potential for further investigation for the treatment of AD.

Triazolopyrimidine

Tresadern et al. (2020) reported the synthesis and evaluation of around 100 derivatives of [1,2,4] triazolo[1,5-a]pyrimidine, which are PDE2A inhibitors with IC_{50} in the range of 2000 to 0.5 nM. Among a series of 4-substituted pyridyls, compound **2** (Figure 5.4) with 4-phenyl substituted pyridine was found promising with an IC_{50} of 1.3 ± 0.4 nM. On the basis of free energy perturbation (FEP) calculations, compound **2** was found to have a predicted value of 4.0 kcal/mol. The FEP calculations included multiple structure–activity relationship factors where it showed aromatic interactions with Phe862, and the 4-phenyl substituent showed pi interaction with Ile866. The hydrophobic pocket shows good binding activity and helps in facilitating the FEP binding. Also, compound **2** has shown excellent results for penetration of the blood–brain barrier with a concentration of 596 ± 75 ng/ml.

Pyrazolopyrimidones

In another study of novel N-1-substituted pyrazolopyrimidones, the pharmacophore display was kept similar, but flipping the pyrazole ring gave the lead for synthesizing compounds of another type (Figure 5.5). It was found that compounds **3** and **4** showed excellent PDE2 potency with values of 1.1 and 1.3 nM, respectively. Compound **4** showed good selectivity, i.e., more than 500-fold. It was found that compound **3** possesses a good pharmacokinetic profile in rats, excellent selectivity

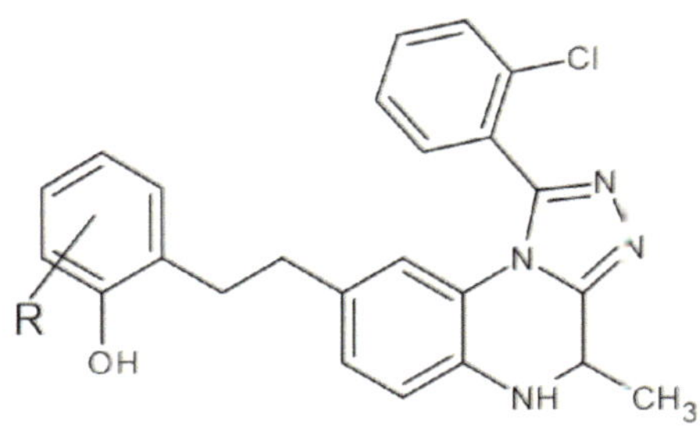

(Compound 1: R = 4-F)

FIGURE 5.3 Chemical structure of triazolopyrazine derivatives.

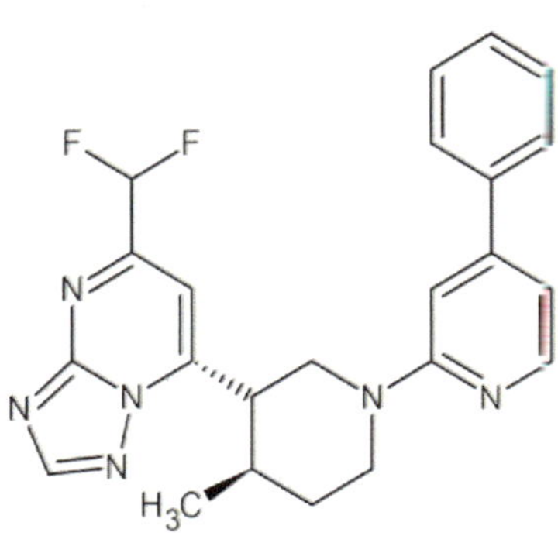

FIGURE 5.4 Chemical structure of compound **2**, triazolopyrimidine derivative.

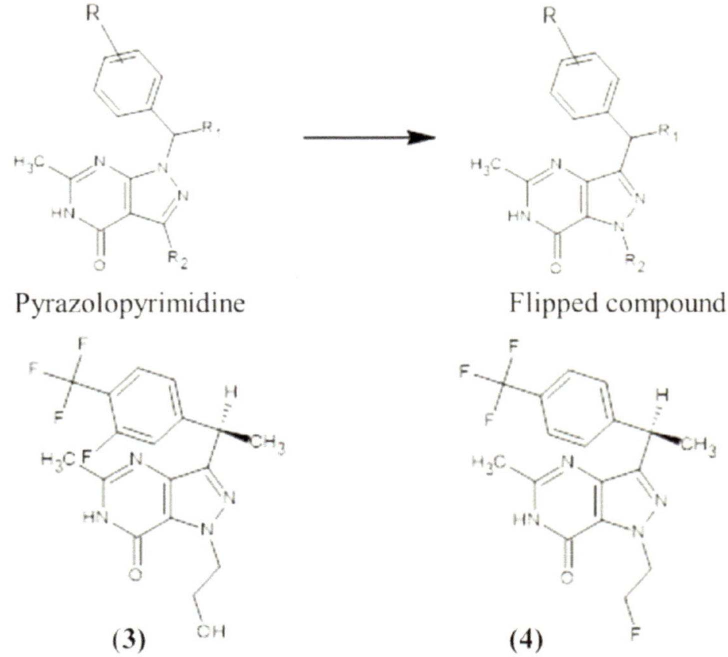

FIGURE 5.5 Design of N-1-substituted pyrazolopyrimidones.

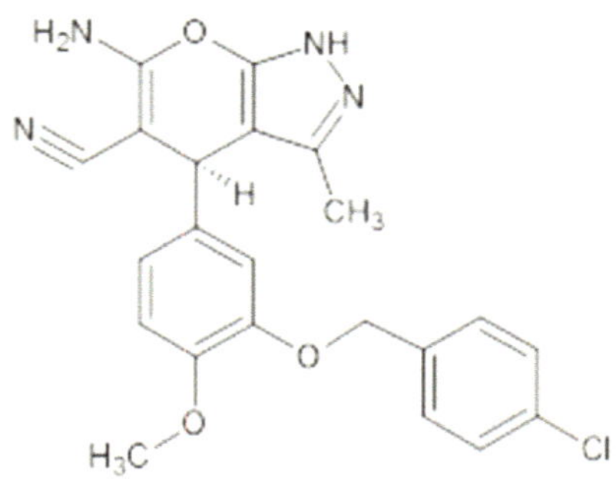

FIGURE 5.6 Chemical structure of LZ-77.

on the cytochrome (CYPs) ion channel and pregnane X receptor (PXR), and good passive permeability (P_{app} = 29.5). Compound **4**, which is a fluoroanalog, was found to have good brain-penetrating properties and a better ancillary profile, which opens the door for further exploration of this series of compounds (Morriello et al., 2021).

Dihydropyranopyrazole

A series of dihydropyranopyrazole derivatives were synthesized, comprising 28 compounds based on the structure of LZ77 (Figure 5.6) with the aim of obtaining inhibitory activity against PDE2. Here, the reference compound taken was erythro-9-(2 hydroxy 3-nonyl)adenine (EHNA) (Figure 5.6) (IC_{50} = 2460 nM), and the inhibitory activity of other targeted compounds was compared. It was found that compound **5** (Figure 5.7) was the most effective PDE2 inhibitor (IC_{50} = 41.5 nM) with an inhibition rate of 93.6% at 0.5 μM, which was even more potent than the hit compound (*R*)-LZ77 (IC_{50} = 261.3 nM). Based on structure–activity relationship studies of compound **5**, it was found that a

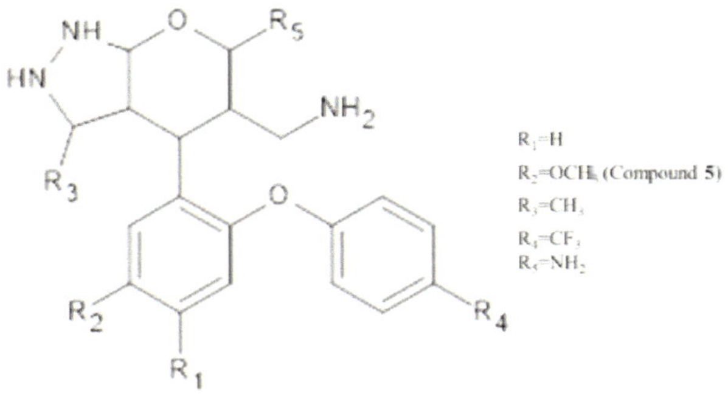

FIGURE 5.7 Chemical structure of dihydropyranopyrazole derivatives.

trifluoromethyl substituent on the benzene ring at R_4 is more active than chlorine, and that at the fifth position, the methoxyl group is more active (IC_{50} = 75.5 nM) than the methyl group (IC_{50} = 153.5 nM). Also, compound **5** has high selectivity against PDE2, and energy decomposition studies show that the side chain at the second position forms strong hydrophobic interactions by entering H-pockets with L770/L809/F862, which further enhances inhibitory activity (Zhou et al., 2021).

PDE3

PDE3 inhibition causes an increased intracellular concentration of cyclic adenosine monophosphate (cAMP). PDE3 enzymes have a unique property wherein cGMP inhibits the catalysis of cAMP at the catalytic position, meaning that cGMP turns off the enzymatic activity for cAMP until it itself gets degraded (Beavo et al., 2007). Various nitro vasodilators activate guanylate cyclase, which inhibits PDE3 by cGMP and increases cAMP. All these studies have been conducted on platelets of rabbits; myocytes of frogs; and cells of rabbits, mice, and humans (Maurice et al., 1990; Fisch et al., 1995). Inhibitor cilostamide is a potent PDE3 inhibitor that has helped in revealing the presence of two PDE3 subfamilies: PDE3A and PDE3B. The structure of PDE3A and PDE3B is identical; the catalytic domain in the C-terminal portion is followed by a hydrophilic C-terminal region (Shakur et al., 2001; Leroy et al., 1996; He et al., 1998). The 44 amino acids in the conserved domain are differently aligned compared to other PDE families. The regulatory N-terminal portions of PDE3A and PDE3B contain large hydrophobic regions. This region shows a helical structure responsible for the association of PDE3B with the intracellular membrane. PDE3A is localized in platelets, vascular smooth muscles, and cardiac tissue, whereas PDE3B is predominantly present in the endoplasmic reticulum and plasma membrane of adipocytes, hepatocytes, and beta cells of the pancreas (Kenan et al., 2000; Shakur et al., 2000). Additionally, it was found that to control the activities and protein interactions, PDE3A and PDE3B get phosphorylated at various positions (Pozuelo et al., 2005). When PDE3B is overexpressed, it inhibits lipolytic activity and raises the desire for high-fat diets in mice, which further promotes the development of diabetes. Given that it is mostly found in the heart and oocytes, the PDE3A isoform exhibits a major impact as a cardiovascular agent and in the development of contraceptive drugs (Wechsler et al., 2002; Ding et al., 2005). Cilostazol (Figure 5.8), which is an important selective PDE3 inhibitor, elevates cAMP concentration (Ikeda, 1999). It was reported in 2006 for the first time that, other than artery obstruction, cilostazol is used to enhance spatial learning memory (Watanabe et al., 2006). The activation of caspase-3 and caspase-9, reduction of oxidative stress, and upregulation of the protein BCl-2 were shown to be some of the processes by which cilostazol exhibits neuroprotective effects. Cilostazol is therefore said to aid in the improvement of cognitive impairment (Oguchi et al., 2017). Studies have also proved that cilostazol reduces amyloid β production and tau phosphorylation, which are responsible for memory dysfunction.

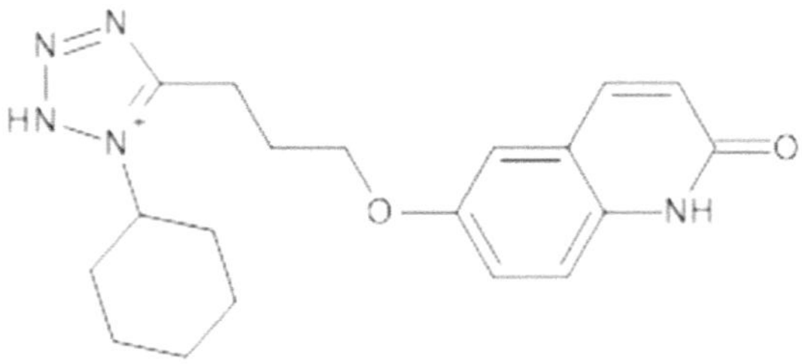

FIGURE 5.8 Chemical structure of cilostazol.

Although PDE3 inhibitors were primarily used for cardiovascular diseases to date, the representative drugs have been repurposed for addressing cognitive dysfunction. It was found that cilostazol, when given in combination therapy with donepezil, showed improvement in patients with cognitive dysfunction (Ihara et al., 2014; Tai et al., 2017). Additionally, this combo therapy demonstrated improved cerebral blood flow in the cortical area, which persisted for six months (Sakurai et al., 2013). Also, combining a PDE3 inhibitor and an AChE inhibitor showed improved cognitive function in animal models and in patients with moderate symptoms of Alzheimer's disease and myocardial infarction (Lee et al., 2007).

PDE4

PDE4 is primarily localized in neurological cells, the cardiovascular system, smooth muscle, and immunity cells. It has shown an effective contribution to improving cognition, treating depression, and controlling neuroinflammation (Houslay et al., 2005; Gurney et al., 2015; Zhang et al., 2002). PDE4 acts specifically on cAMP. PDE4 can significantly reduce inflammation. It participates in the modulation of β2-adrenergic signaling in myocardial cells and pulmonary smooth muscles and subsides inflammation (Fertig et al., 2018; Schafer et al., 2014). PDE4 has 18 different isoforms that are expressed in mammalian cells (Houslay, 2001; Conti et al., 1995). Among PDE4 isoforms, PDE4D1 levels are increased in the hippocampus lobe of Alzheimer's patients (McLachlan et al., 2007). PDE4 selective inhibitors attenuate cognitive disorders, improve neuroplasticity, and enhance memory-related functioning (Zhuo et al., 2016; Wang et al., 2012). Although PDE4 inhibitors are potential medications for memory and cognitive impairment, their main adverse effect is emesis (Miller et al., 1994). The N- and C-terminal domains of the PDE4 protein perform a pivotal role in the conformation and function of the catalytic core. Currently, many PDE4 inhibitors have received approval for clinical practice, including rolipram (Roesler et al., 2014; Kumar et al., 2017; Kaur et al., 2017), roflumilast (Hu et al., 2019), ibudilast (Wang et al., 2014), MK-0952 (Gallant et al., 2010), HT-0712 (Peters et al., 2014), and GEBR-32a (Figures 5.9 and 5.10) (Ricciarelli et al., 2017). Rolipram is an analog of γ-aminobutyric acid (GABA) and significantly improves memory function in animals. Its primary mechanisms involve inhibiting cholinesterase, preventing amyloid build-up, capturing oxidation, and reducing inflammatory reactions. Rolipram also demonstrates a significant impact on dementia associated with AD (Kumar et al., 2017).

A subthreshold dose of donepezil combined with roflumilast, a PDE4 inhibitor, and an anticholinergic medication combination, was found to enhance synaptic plasticity and memory formation in a different investigation (Vanmierlo et al., 2016; Heckman et al., 2020). In a recent study, it was found that when roflumilast is given in combination with clioquinol (Figure 5.10), it forms a hybrid compound **6** with $IC_{50} = 0.399\ \mu M$ and has more penetrating power to the blood–brain barrier (Hu et al., 2019).

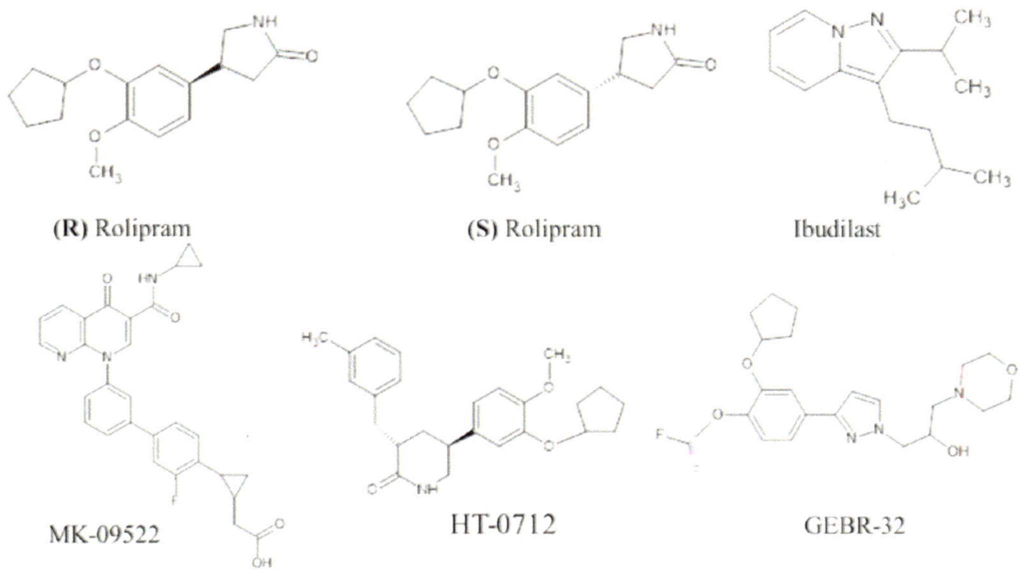

FIGURE 5.9 Chemical structures of some PDE4 inhibitors.

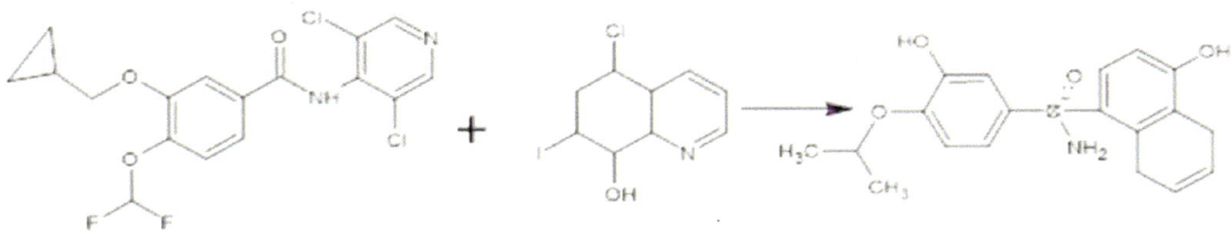

Roflumilast **Ciloquinol** **Compound 6**

FIGURE 5.10 Preparation scheme for compound **6**.

Ibudilast has shown a neuroprotective effect on neural cell death and neuroinflammation in animal models of ischemic brain damage. Ibudilast also improves β-amyloid-induced memory impairments and attenuates neurotoxicity (Wang et al., 2014). Ibudilast shows inhibition in neurodegenerative diseases. The main mechanism of action is by increasing cAMP and cGMP and downregulating inflammatory factors like TNF-α, macrophage inhibitory factor (MIF), NF-kB signaling, and upregulating PKA/CREB pathway (Huang et al., 2006; Oliveros et al., 2023; Angelopoulou et al., 2022; Mizuno et al., 2004; Lee et al., 2012). MK-0952 has also shown its effect in memory impairment and mild cognitive damage. MK-0952 (IC_{50} = 0.53 nM) has been found to be highly penetrating in the brains of rats. It has also completed phase I trials (Gallant et al., 2010). Other inhibitors like HT-0712 ameliorate long-term memory in aged mice (Peters et al., 2014) and GEBR-32a can enhance memory in Alzheimer's transgenic mice (Guo et al., 2017). Another PDE4 inhibitor, piclamilast and PDE5 inhibitor as combination therapy possess antifibrotic and anti-inflammatory effects in case of Duchenne muscular dystrophy (Nio et al., 2017).

PDE5

So far, PDE5 inhibitors have been used for treating erectile dysfunction, pulmonary artery hypertension, and some urological disorders. It has been reported that PDE5I shows its effects not only in the aforementioned disorders but also in cardiac disease, cancers, diabetes, and AD (Samidurai et al., 2023). Alzheimer's severely impairs the nitric oxide or cGMP pathway, which leads to decreased levels of cGMP. It has been observed that PDE5 levels have been upregulated in Alzheimer's patients while depleting cGMP levels (Ugarte et al., 2015). PDE5 inhibitors prevent the phosphorylation of the cGMP pathway, which eventually helps in treating aberrant neuroplasticity and other memory-related issues (Zuccarello et al., 2020). Alzheimer's disease severely impairs the nitric oxide (NO) or cGMP pathway, which leads to decreased levels of cGMP. Important participants in the neuroinflammatory process that may result in brain malfunction, cell death, and subsequent neurodegeneration are cGMP-dependent protein kinase (PKG) and the cGMP controller phosphodiesterase (Boswell-Smith et al., 2006; Domek-Lopacinska and Strosznajder, 2008; Francis et al., 2011b). The major focus in developing PDE5 is to explore the NO–cGMP signaling pathway, but it has been found that many PDE5 inhibitors cause visual abnormalities as they also inhibit PDE6. Therefore, PDE5 inhibitors are being explored based on the approach of multitarget drug ligands in combination with acetylcholine inhibitors and NMDA antagonists (Liu et al., 2019).The only PDE subtype with a single isoform, PDE5A, specifically hydrolyzes cGMP and is highly expressed in peripheral tissues, with limited presence in the brain. It was predominantly used for erectile dysfunction, hypertension, COPD, and cardiac disease (Lakics et al., 2010; Ahmed, 2018). The structures and biological effects of some PDE5 inhibitors are summarized in Figure 5.11 and Table 5.2, respectively.

PDE6

Phosphodiesterase is mainly located in the photoreceptors of the retina. It mainly activates the degradation of cGMP. It has been discovered that many PDE5 inhibitors also inhibit PDE6, causing

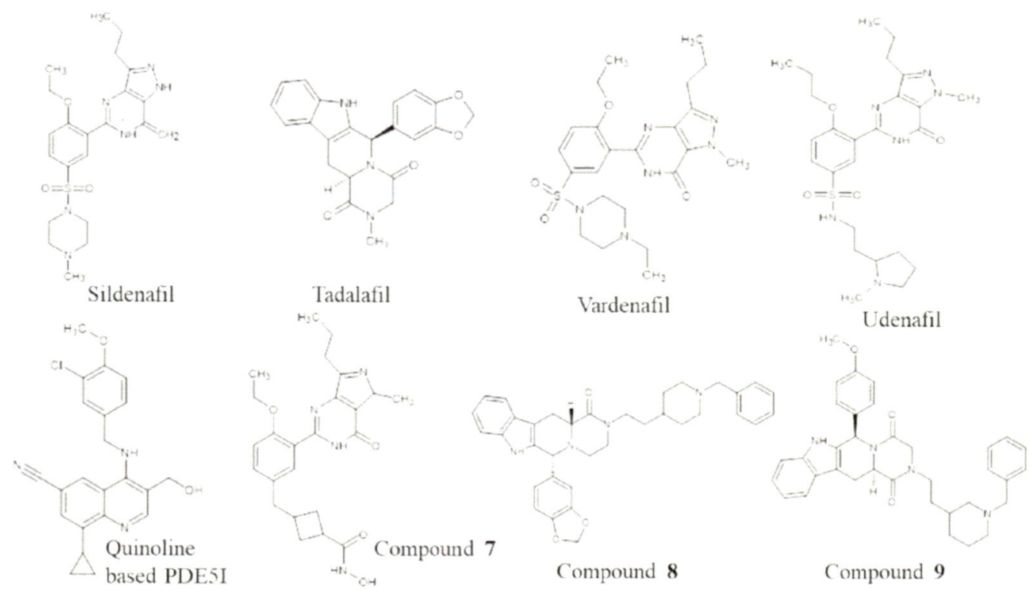

FIGURE 5.11 Chemical structures of some PDE4 inhibitors.

TABLE 5.2
Details of PDE5 Inhibitors and Their Biological Effect/Mechanism of Action

Drug	Biological Effect/Mechanism of Action	Reference
Sildenafil	• It inhibits both PDE5 and PDE6. • It decreases oxidative stress, leading to inhibition in neuroinflammation. • The mechanism of action of sildenafil was found to increase CREB signaling and lower Aβ levels. • IC_{50} = 2.2 and 9.5 nM • Dissociation constant = 0.5 – 6.6 nM • Causes abnormal vision.	Zhang et al. (2013) Tejedor et al. (2011)
Tadalafil	• It was initially used in the treatment of pulmonary arterial hypertension. • It was also found to have inhibition against PDE5 and PDE11 but wasn't explored much because of its poor penetrability of the blood–brain barrier. • It was found to decrease tau protein and oxidative stress. • IC_{50} = 1.2 and 11 nM • Dissociation constant = 0.9 to 6.7 nM • Side effects observed were headache, dyspepsia, and back pain.	Curran et al. (2003) Fiorito et al. (2013) Puzzo et al. (2009)
Vardenafil	• It increased cognition, memory, and attention in a dose of 0.3 to 3 mg/kg, independent of the mechanism of cerebral blood flow. • Dissociation constant = 0.6 to 0.7 nM	Reneerkens et al. (2013) Rutten et al. (2009) Akkerman et al. (2015)
Udenafil	• It showed cognitive improvement in patients with erectile dysfunction.	Shim et al. (2014)
Quinoline-based PDE5I	• It showed selectivity PDE5 (IC_{50} = 0.270 nM) and PDE6 (IC_{50} = 0.339 μM). • It readily crossed the blood–brain barrier and increased the level of cGMP in the hippocampus area of mice and also increased Aβ. • Good water solubility.	Fiorito et al. (2013)
Compound **7**	• Dual target inhibitor for PDE5 and HDACIs. • It is able to cross the blood–brain barrier. • PDE5, IC_{50} = 60 nM • HDAC4, IC_{50} = 91 nM • HDAC1, IC_{50} = 310 nM • HDAC2, IC_{50} = 490 nM • HDAC3, IC_{50} = 322 nM • It was found to work by acetylation of the histones and phosphorylation of CREB in the hippocampus area of mice.	Cuadrado et al. (2017)
Compound **8**	• Dual target inhibitors, i.e., PDE5 and AChE. • Compound was designed using tadalafil parent nuclei. • It showed an excellent effect on PDE5 inhibition by increasing CREB phosphorylation. • The main inhibition that came in its further exploration is its poor water solubility. • AChE inhibition with IC_{50} = 36 nM • Moderate PDE5 inhibition IC_{50} = 153 nM	Ni et al. (2018)
Compound **9**	• Good selectivity for dual target PDE5 and AChE. • PDE5, IC_{50} = 3.23 μM • AChE, IC_{50} = 15 nM • The compound shows good permeability against the blood–brain barrier. • Its citrate enhances water solubility. • It also increases CREB phosphorylation.	Liu et al. (2017)

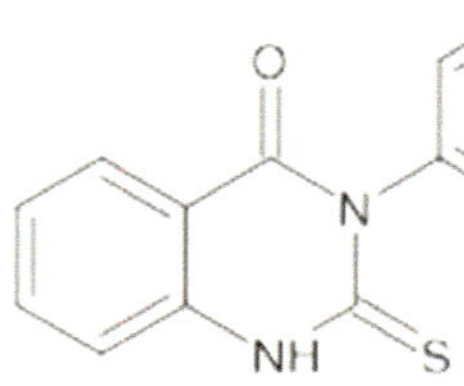

FIGURE 5.12 Chemical structure for S-14.

an effect on visual functioning. This observation can lead to design specific PDE5 medicinal substances (Cote, 2004).

PDE7

PDE7 is majorly expressed in the brain, immune cells, adrenal glands, lungs, pancreas, and heart. It has been found to be a potential drug target for treating various neurodegenerative diseases. PDE7 is massively specific for cAMP (Perez et al., 2003). The main mechanism of PDE7 inhibition involves regulating mitochondrial dysfunction, improving inflammation, and cell loss in the hippocampal region of the brain (Morales-Garcia et al., 2017). S-14 is a potential PDE7 inhibitor, which is a quinazoline derivative. Long-term treatment with S-14 improves spatial learning and alleviates memory impairment in mice by regulating the proliferation of neural cells (Figure 5.12) (Morales-Garcia et al., 2011).

Another PDE7 inhibitor, BRL50481, does not induce emesis because of the low expression of PDE7A and PDE7B in the postrema region of the brain (Torres et al., 2003). The main PDE subtypes found in humans are PDE7A, PDE1A, PDE2A, PDE3A, PDE7B, PDEB1, PDEB2, and PDE5 (Han et al., 1998; Glavas et al., 2001; Sasaki et al., 2000). In Table 5.3 are some PDE7 inhibitors that are effective in AD.

PDE8

PDE8 is mainly expressed in the spleen, heart, ovary, colon, kidney, and testis of humans. The main function of PDE8 is to regulate chemotaxis and calcium channels (Patrucco et al., 2010). Many studies that PDE8 is expressed in the hippocampus, cerebellum, and striatum regions of the brain. Inhibition of PDE8 was found to control fear, improve cognitive function, control anxiety, and improve coordination (Torres et al., 2003; Tsai et al., 2012). It mainly regulates cAMP and also activates dopamine receptor activation, facilitating several motor functions (Sweatt, 2001; Nishi et al., 2008). These findings open up the possibility of the development of PDE8 inhibitors for neurodegenerative diseases.

PDE9

PDE9 expresses itself in the cortex, basal ganglia, hippocampus, and cerebellum areas of the brain. PDE9 inhibitors are more specific toward cGMP and have only one isoform of PDE9A. Inhibition of cGMP activates NO/cGMP/CREB signaling, leading to improvement in memory and other cognitive activities (Yu et al., 2017; Rosenbrock et al. 2019). Some PDE9 inhibitors have been reported in Table 5.4.

PDE10

PDE10 is expressed in the brain and is responsible for eliminating cAMP and cGMP through hydrolysis. The present data does not show much of its involvement in memory and cognition-related disabilities; however, some natural agents and a few drugs show its effectiveness in neurodegenerative

TABLE 5.3
PDE7 Inhibitors, Chemical Structure, and Biological Activity

Drug	Category	Chemical Structure	Biological Activity	Reference
BRL50481	Benzenesulfonamide derivative	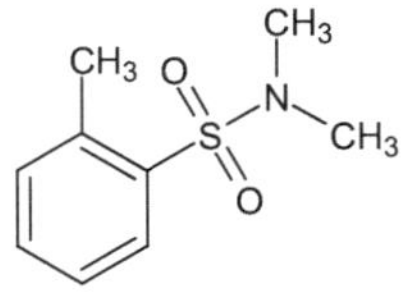	PDE7A IC_{50} = 0.15 µM PDE7B IC_{50} = 12.1 µM	Mestre et al. (2015)
S-14	Thioxoquinazoline derivative	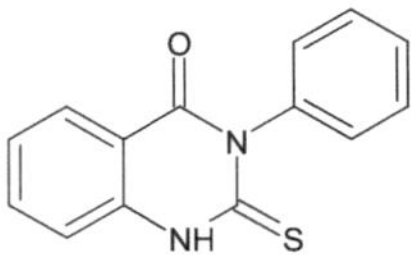	PDE7A IC_{50} = 4.7 µM PDE7B IC_{50} = 8.8 µM	Perez et al. (2012)
TC3.6	Thioxoquinazoline derivative	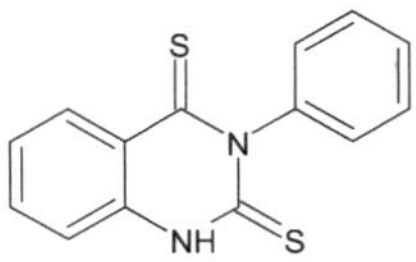	PDE7 IC_{50} = 0.55 µM	Chen et al. (2021)
VP1.15	Iminothiadiazole derivative	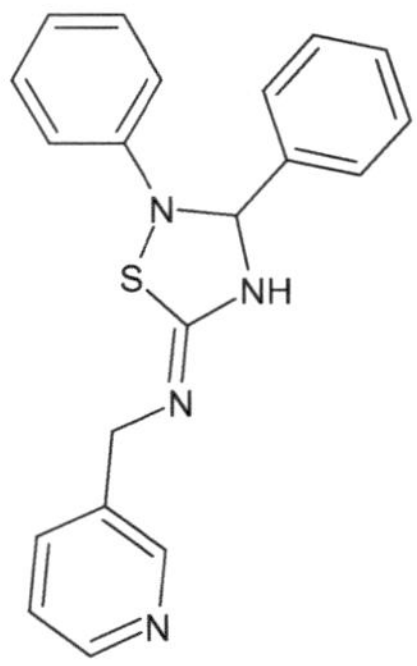	PDE7 IC_{50} = 1.11 µM	Susin et al. (2012)
Compound **10**	Spiroquinazolin derivative	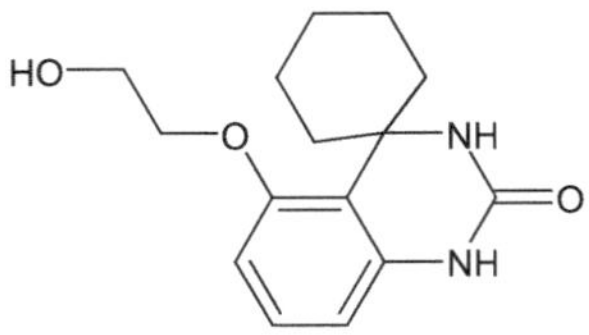	PDE7A IC_{50} = 55 nM PDE7B IC_{50} = 0.203 µM	McQuown et al. (2021)

diseases. Its effectiveness in Alzheimer's disease is still not explored much. PDE10 has an isoform, PDE10A, blockade of which can increase cGMP and CREB signaling (Siuciak et al., 2006). Li et al. (2018) synthesized a series of 13 quinazoline derivatives using papaverine as the lead compound. It was found that most of the drugs showed good lipophilicity, and topological polar surface area showed good results for high penetration of compounds through the blood–brain barrier, although only a few compounds showed good PDE10A inhibitory activities with IC_{50} > 1.0 µM. Compound **11** has shown an IC_{50} of 0.64 µM on PDE10A and a reduction in oxidative stress with an oxygen radical absorbance capacity value of value 2.3 (Figure 5.13) (Li et al., 2018).

Some other drugs like TAK-063 and MP-10 show sensitivity for binding toward cyclic nucleotides, resulting in the activation of striatal pathways both directly and indirectly (Suzuki et al., 2016).

TABLE 5.4
Details of Some PDE9 Inhibitors with Their Anti-Neurodegenerative Role/Mechanism

Drug	Chemical Structure	Biological Effect	Reference
BAY73-6691	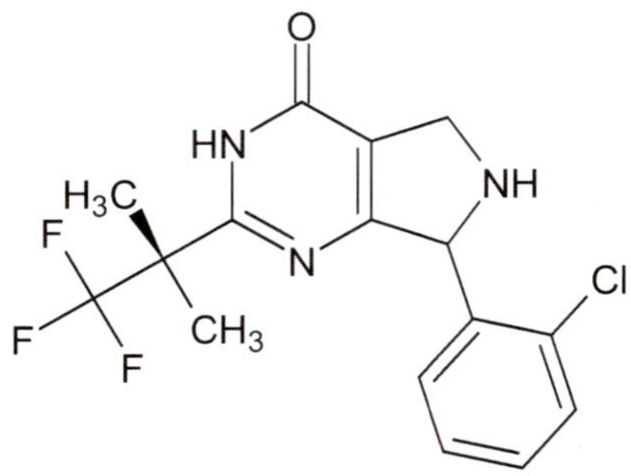	• Restores memory performance. • Protects Aβ-induced oxidative damage. • PDE9 IC_{50} = 55 nM	van der Staay et al. (2008)
PF-04447943	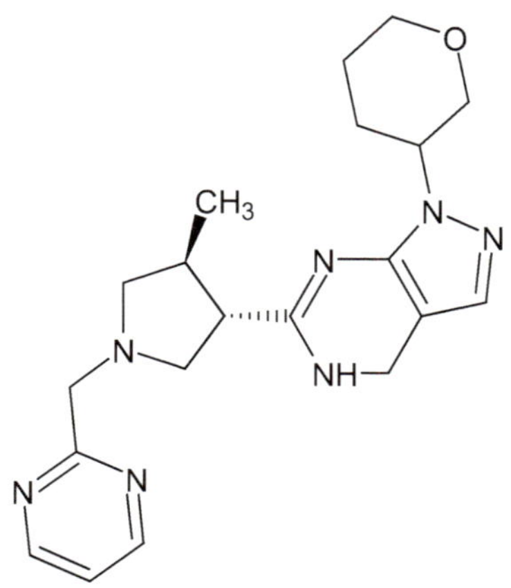	• Pyrazolopyimidinone derivative. • Regulates dendrite spine density in hippocampus region. • PDE9A IC_{50} = 8 nM	Kroker et al. (2014) Kleiman et al. (2010)
BI-409306	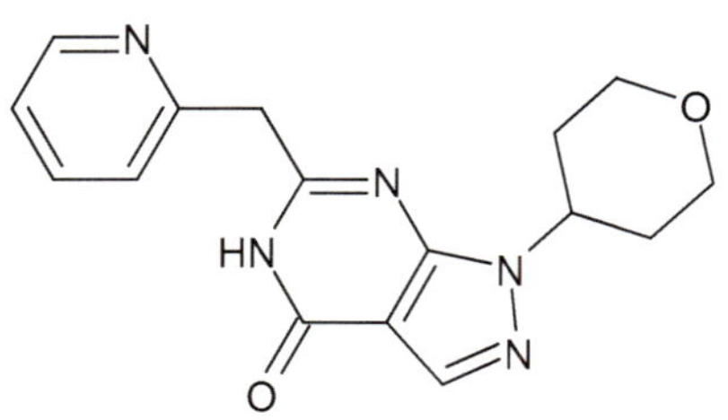	• It increases cerebrospinal fluid. • It activates cAMP response binding protein (CREB) and decreases phosphorylated tau protein levels.	Rosenbrock et al. (2019)

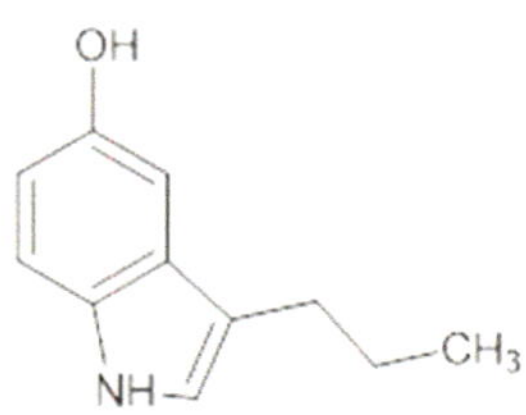

FIGURE 5.13 Chemical structure of compound **11,** a quinazoline derivative.

PDE11

PDE11 regulates intracellular signaling by hydrolyzing both cAMP and cGMP. It has been found to play roles in inflammation and the brain. It is expressed in the hippocampus area of the brain, adrenal glands, prostate, and testis. It has four splice variants: PDE11 A1, A2, A3, and A4 (Kelly, 2015). Many behavioral studies have shown that PDE11 affects various brain functions, such as social memory formation and lithium responsiveness, which opens up new therapeutic avenues (Kelly et al., 2010)

CLINICAL SUCCESS OF PDE INHIBITORS

A more advanced treatment regimen is required due to the worldwide, rapid growth in the number of people with neurodegenerative disorders. Because PDEs are found in practically every part of the body, a wide range of PDE inhibitor designs and developments are possible. Although several PDE inhibitors are now being tested, some of them have already demonstrated their efficacy in treating AD over time. Table 5.5 provides the clinical status of certain PDE inhibitors that are effective against AD.

CONCLUSION

All 11 families of PDEs play a significant role in controlling the behavior of cyclic nucleotides. PDE1 is widely expressed in the brain, heart, and reproductive system, and shows diverse functions like smooth muscle proliferation, learning, and cognitive abilities. PDE1 inhibitors, such as vinpocetine, were found to enhance memory and cognition. IBMX is not very specific to PDE1 and also inhibits PDE5. ITI-214 was found to cross the blood–brain barrier, exerting precognitive effects. Selegiline was found to cause short-term improvement in Alzheimer's disease, while nimodipine was found useful for subarachnoid hemorrhages. PDE2 hydrolyzes both cAMP and cGMP. AD patients were found to have higher PDE2A compared to healthy individuals. Derivatives of triazolopyrazine, triazolopyrimidine, pyrazolopyrimidones, and dihydropyranopyrazole showed good inhibitory activity against PDE2. PDE3 has an affinity for both cAMP and cGMP but more toward cAMP. Cilostamide is a potent PDE3 inhibitor that has helped in revealing the presence of two PDE3 subfamilies (PDE3A and PDE3B). Overexpression of PDE3B causes diabetes, and it is mainly found in the heart and oocytes. Cilostazol is an important selective PDE3 inhibitor that elevates cAMP concentration, activates caspase-3 and caspase-9, reduces oxidative stress and upregulates protein BCl-2, and reduces β-amyloid and tau phosphorylation, which eventually leads to improvement in cognitive impairment. Cilostazol is also given in combination with donepezil, which improves cognitive dysfunction. Also, a PDE3 inhibitor in combination with AChE shows improved cognitive function. PDE4 is primarily localized in neurological cells, cardiovascular tissues, smooth muscle, and immunity cells. PDE4 was found to be very effective in improving cognition and treating depression and neuroinflammation. PDE4 specifically acts on cAMP. PDE4 has 18 different isoforms. PDE4 selective inhibitors attenuate cognitive disorders, improve neuroplasticity, and enhance memory-related functioning. The main side effect of PDE4 is emesis. Many PDE4s have been approved for clinical practice, such as rolipram, roflumilast, ibudilast, HT-0712, and GEBR-32a. A combination of donepezil with roflumilast, along with an anticholinergic medication combination, was found to enhance synaptic plasticity and memory formation. Roflumilast in combination with clioquinol forms a hybrid compound with greater penetrating power to the blood–brain barrier. Ibudilast downregulates inflammatory factors like TNF-α, MIF, and NF-kβ signaling and upregulates the PKA/CREB pathway. MK-9522, HT-0712, and GEBR-32a also showed effects in memory impairment and mild cognitive damage. PDE5 not only showed its effect in treating erectile dysfunction, pulmonary artery hypertension, and urological disorders, but was also

TABLE 5.5
Clinical Status of PDE Inhibitors in AD and Other Neurodegenerative Diseases

Identifier	Year	Drug	Disease	Status
NCT02240693	2018	BI-409306	Alzheimer's disease	Phase 2
NCT05039086	2021	Exposure group: PDE5I (sildenafil, tadafil); Reference group: Endothelin receptor antagonist (bosentan, ambrisentan, mairtentan)	Alzheimer's disease	Active, not recruiting
NCT02392468	2017	Drug: B1409306 (matching placebo) Drug: BI409306	Schizophrenia Alzheimer's disease	Phase 1
NCT01409564	2014	PDE3I Drug: Cilostazol Drug: Placebo	Dementia Mild to moderate Alzheimer	Phase 4
NCT00880412	2009	Drug: EHT0202 or etazolate	Mild to moderate Alzheimer's disease Adjunct therapy of acetylcholinesterase	Phase 2
NCT04570085	2022	Drug: Caffeine Drug: Placebo	Alzheimer's disease	Phase 3
NCT02337907	2018	Drug: Placebo Drug: BI409306 Drug: Donepezil	Alzheimer's disease	Phase 2
NCT00726726	2008	Drug: BMS-708163 interaction with midazolam, warfarin, vitamin K, caffeine, omeprazole, and dextromethorphan	Alzheimer's disease	Phase 1
NCT046586	2023	Drug: Placebo Drug: Roflumilast (50 ug) Drug: Roflumilast (100 ug)	Alzheimer's disease Mild cognitive impairment Mild dementia Amnestic mild cognitive disorder	Phase 2
NCT00692510	2008	Drug: AZD3480 Drug: Placebo Drug: Cocktail mix (caffeine, bupropion, rosiglitazone, omeprazole, midazolam, bilirubin)	Metabolism Alzheimer's disease	Phase 1
NCT02491268	2020	Drug: Cilostazol Drug: Placebo	Mild cognitive impairment dementia	Phase 2
NCT05574400	2023	Drug: Dextrose water Drug: Caffeine citrate	Post-operative delirium Post-operative cognitive dysfunction Mild cognitive impairment	Active, recruiting

reported to have effects in cardiac diseases, cancers, diabetes, and Alzheimer's disease. PDE5 levels have been found to be upregulated in Alzheimer's patients while depleting cGMP levels. PDE5 inhibitors prevent the phosphorylation of cGMP, which improves neuroplasticity and other memory-related issues. Some PDE5 inhibitors, like sildenafil, tadalafil, vardenafil, udenafil, quinolone-based PDE5I, and PDEIs with AChE, have shown some neuroprotective activity. Some PDE5 inhibitors also inhibit PDE6, leading to the design of specific PDE5 medicinal substances. PDE7 is also found in the brain, showing its potential to treat various neurodegenerative diseases. Some PDE7I, like BRL50481, S-14, TC3.6, and VP1.15, also showed their effect in AD. PDE8 has also shown its expression in the hippocampus, cerebellum, and striatum regions of the brain, and is found to control fear and anxiety. PDE9 inhibitors are more specific toward cGMP, and their inhibition leads to improvement in memory and other cognitive activities. PDE9 inhibitors, like BA473-6691, PF-04447943, and BI-409306, have shown neuroprotective effects through different pathways. For PDE10, some natural agents and a few drugs are showing effectiveness in neurodegenerative diseases. PDE11 also expresses itself in the hippocampus area of the brain, adrenal glands, prostate, and testis. Many studies have shown that PDE11 has an effect on social memory formation and lithium responsiveness. Some PDE inhibitors have entered clinical trials and are showing their effectiveness in Alzheimer's and other neurodegenerative diseases. These studies open the fact that PDE inhibition is a promising target for developing drugs for neurodegeneration. Further, proper clinical trials and investigations will open the door to new developments for Alzheimer's patients.

REFERENCES

Ahmed NS. Tadalafil: 15 years' journey in male erectile dysfunction and beyond. *Drug Dev Res* (2018) 80(6):683–701.

Akkerman S, Blokland A, van Goethem NP, Cremers P, Shaffer CL, Osgood SM et al. PDE5 inhibition improves acquisition processes after learning via a central mechanism. *Neuropharmacology* (2015) 97:233–239.

Angelopoulou E, Pyrgelis ES, Piperi C. Emerging potential of the Phosphodiesterase (PDE) inhibitor Ibudilast for neurodegenerative diseases: an update on preclinical and clinical evidence. *Molecules* (2022) 27(23):8448.

Ansari MA, Iqubal A, Ekbbal R, Haque SE. Effects of nimodipine, vinpocetine and their combination on isoproterenol-induced myocardial infarction in rats. *Biomed Pharmacother* (2019) 109:1372–1380.

Avila J, Hernández F. GSK-3 inhibitors for Alzheimer's disease. *Expert Rev Neurother* (2007) 7(11):1527–1533.

Baillie GS, Tejeda GS, Kelly MP. Therapeutic targeting of 3', 5'- cyclic nucleotide phosphodiesterases: inhibition and beyond. *Nat Rev Drug Discov* (2019) 18(10):770–796.

Beavo JA. Phosphodiesterase 3A. *Com Pharmacol Ref* (2007) 9780080552323:1–8.

Beavo JA, Rogers NL, Crofford OB, Hardman JG, Sutherland EW, Newman EV. Effects of xanthine derivatives on lipolysis and on adenosine 3', 5'-monophosphate phosphodiesterase activity. *Mol Pharmacol* (1970) 6(6):597–603.

Bessodes M, Bastian G, Abushanab E, Panzica RP, Berman SF, Marcaccio EJ et al. Effect of chirality in erythro-9-(2-hydroxy-3-nonyl) adenine (EHNA) on adenosine deaminase inhibition. *Biochem Pharmacol* (1982) 31:879–882.

Bondarev AD, Attwood MM, Jonsson J, Chubarev VN, Tarasov VV, Liu W, Schiöth HB. Recent developments of phosphodiesterase inhibitors: clinical trials, emerging indications and novel molecules. *Front Pharmacol* (2022) 13:1057083.

Boswell-Smith V, Spina D, Page CP. Phosphodiesterase inhibitors. *Br J Pharmacol* (2006) 147 (Suppl 1): S252–S257.

Chen Y, Wang H, Wang WZ, Wang D, Skaggs K, Zhang HT. Phosphodiesterase 7(PDE7): a unique drug target for central nervous system diseases. *Neuropharmacology* (2021) 196:108694.

Choi OH, Shamim MT, Padgett WL, Daly JW. Caffeine and theophylline analogues: correlation of behavioral effects with activity as adenosine receptor antagonists and as phosphodiesterase inhibitors. *Life Sci* (1988) 43(5):387–398.

Conti M, Nemoz G, Sette C, Vicini E. Recent progress in understanding the hormonal regulation of phosphodiesterases. *Endocr Rev* (1995) 16(3):370–389.

Cote R. Characteristics of Photoreceptor PDE (PDE6): similarities and differences to PDE5. *Int J Impotence Res* (2004) 16(Suppl 1): S28–S33.

Cote RH. Photoreceptor phosphodiesterase (PDE6): activation and inactivation mechanisms during visual transduction in rods and cones. *Pflugers Arch* (2021) 473(9):1377–1391.

Cuadrado-Tejedor M, Garcia-Barroso C, Sánchez-Arias JA, Rabal O, Pérez-González M, Mederos S et al. A first-in-class small-molecule that acts as a dual inhibitor of HDAC and PDE5 and that rescues Hippocampal Synaptic Impairment in Alzheimer's disease mice. *Neuropsychopharmacology* (2017) 42(2):524–539.

Cuadrado-Tejedor M, Hervias I, Ricobaraza A, Puerta E, Perez-Roldan JM, Franco R et al. Sildenafil restores cognitive function without affecting beta-amyloid burden in a mouse model of Alzheimer's disease. *Br J Pharmacol* (2011) 164(8):2029–2041.

Curran M, Keating G. Tadalafil. *Drugs* (2003) 63(20):2203–2212.

Delhaye S, Bardoni B. Role of phosphodiesterases in the pathophysiology of neurodevelopmental disorders. *Mol Psychiatry* (2021) 26:4570–4582.

DeNinno MP. Future directions in phosphodiesterase drug discovery. *Bioorg Med Chem Lett* (2012) 22(22):6794–800.

Ding B, Abe J, Wei H, Xu H, Che W, Aizawa T et al. A positive feedback loop of phosphodiesterase 3 (PDE3) and inducible cAMP early repressor (ICER) leads to cardiomyocyte apoptosis. *Proc Natl Acad Sci USA* (2005) 102:14771–14776.

Domek-Łopacińska K, Strosznajder JB. The effect of selective inhibition of cyclic GMP hydrolyzing phosphodiesterases 2 and 5 on learning and memory processes and nitric oxide synthase activity in brain during aging. *Brain Res* (2008) 1216:68–77.

Dubey A, Kumar N, Mishra A, Singh Y, Tiwari M. Review on vinpocetine. *Int J Pharm Life Sci* (2020) 11(5):6590–6597.

Dyck B, Branstetter B, Gharbaoui T, Hudson AR, Breitenbucher JG, Gomez L et al. Discovery of selective phosphodiesterase 1 inhibitors with memory enhancing properties. *J Med Chem* (2017) 60:3472–3483.

Enomoto T, Tatara A, Goda M, Nishizato Y, Nishigori K, Kitamura A et al. A novel phosphodiesterase 1 inhibitor DSR-141562 exhibits efficacies in animal models for positive, negative, and cognitive symptoms associated with schizophrenia. *J Pharmacol Exp Ther* (2019) 371(3):692–702.

Fajardo AM, Piazza GA, Tinsley HN. The role of cyclic nucleotide signaling pathways in cancer: targets for prevention and treatment. *Cancers* (2014) 6(1):436–458.

Fertig B, Baillie G. PDE4-mediated cAMP signalling. *J Cardiovasc Dev Dis* (2018) 5(1):8.

Fiorito J, Saeed F, Zhang H, Staniszewski A, Feng Y, Francis YI et al. Synthesis of quinoline derivatives: discovery of a potent and selective phosphodiesterase 5 inhibitor for the treatment of Alzheimer's disease. *Eur J Med Chem* (2013) 60:285–294.

Fisch A, Michael-Hepp J, Meyer J, Darius H. Synergistic interaction of adenylate cyclase activators and nitric oxide donor SIN-1 on platelet cyclic AMP. *Eur J Pharmacol* (1995) 289(3): 455–461.

Francis SH, Blount MA, Corbin JD. Mammalian cyclic nucleotide phosphodiesterases: molecular mechanisms and physiological functions. *Physiol Rev* (2011a) 91(2):651–690.

Francis SH, Houslay MD, Conti M. Phosphodiesterase inhibitors: factors that influence potency, selectivity, and action. *Handb Exp Pharmacol* (2011b) (204):47–84.

Francis SH, Turko IV, Corbin JD. Cyclic nucleotide phosphodiesterases: relating structure and function. *Prog Nucleic Acid Res Mol Biol* (2001) 65:1–52.

Gallant M, Aspiotis R, Day S, Dias R, Dubé D, Dubé L et al. Discovery of MK-0952, a selective PDE4 inhibitor for the treatment of long-term memory loss and mild cognitive impairment. *Bioorg Med Chem Lett* (2010) 20(22):6387–6393.

Giorgi M, Cardarelli S, Ragusa F, Saliola M, Biagioni S, Poiana G et al. Phosphodiesterase inhibitors: could they be beneficial for the treatment of COVID-19? *Int J Mol Sci* (2020) 21(15):5338.

Glavas NA, Ostenson C, Schaefer JB, Vasta V, Beavo JA. T cell activation up-regulates cyclic nucleotide phosphodiesterases 8A1 and 7A3. *Proc Natl Acad Sci U S A* (2001) 98(11):6319–6324.

Guo H, Cheng Y, Wang C, Wu J, Zou Z, Niu B et al. FFPM, a PDE4 inhibitor, reverses learning and memory deficits in APP/PS1 transgenic mice via cAMP/PKA/CREB signalling and anti-inflammatory effects. *Neuropharmacology* (2017) 116:260–269.

Gurney ME, D'Amato EC, Burgin AB. Phosphodiesterase-4 (PDE4) molecular pharmacology and Alzheimer's disease. *Neurotherapeutics* (2015) 12(1):49–56.

Hampel H, Vassar R, De Strooper B, Hardy J, Willem M, Singh N et al. The β-Secretase BACE1 in Alzheimer's disease. *Biol Psychiatry* (2021) 89(8):745–756.

Han P, Fletcher CF, Copeland NG, Jenkins NA, Yaremko LM, Michaeli T. Assignment of the mouse Pde7A gene to the proximal region of chromosome 3 and of the human PDE7A gene to chromosome 8q13. *Genomics* (1998) 48(2):275–276.

He R, Komas N, Ekholm D, Murata T, Taira M, Hockman S et al. Expression and characterization of deletion recombinants of two cGMP-inhibited cyclic nucleotide phosphodiesterases (PDE-3). *Cell Biochem Biophys* (1998) 29:89–111.

Heckman PRA, Roig Kuhn F, Raven F, Bolsius YG, Prickaerts J, Meerlo P, Havekes R. Phosphodiesterase inhibitors roflumilast and vardenafil prevent sleep deprivation-induced deficits in spatial pattern separation. *Synapse* (2020) 74(6):e22150.

Hoskin JL, Al-Hasan Y, Sabbagh MN. Nicotinic Acetylcholine receptor agonists for the treatment of Alzheimer's dementia: an update. *Nicotine Tob Res* (2019) 21(3):370–376.

Houslay MD. PDE4 cAMP-specific phosphodiesterases. *Prog Nucleic Acid Res Mol Biol* (2001) 69:249–315.

Houslay MD, Schafer P, Zhang KY. Keynote review: phosphodiesterase-4 as a therapeutic target. *Drug Discov Today* (2005) 10(22):1503–1519. https://clinicaltrials.gov/ct2/search

Hu J, Pan T, An B, Li Z, Li X, Huang L. Synthesis and evaluation of clioquinol-rolipram/roflumilast hybrids as multitarget-directed ligands for the treatment of Alzheimer's disease. *Eur J Med Chem* (2019) 163:512–526.

Huang MX, Tian YJ, Han C, Liu RD, Xie X, Yuan Y et al. Structural modifications of nimodipine lead to novel PDE1 inhibitors with anti-pulmonary fibrosis effects. *J Med Chem* (2022) 65(12):8444–8455.

Huang Z, Liu S, Zhang L, Salem M, Greig GM, Chan CC et al. Preferential inhibition of human phosphodiesterase 4 by ibudilast. *Life Sci* (2006) 78(23):2663–2668.

Humphrey JM, Yang E, Ende CW, Arnold EP, Head JL, Jenkinson S Small-molecule phosphodiesterase probes: discovery of potent and selective CNS-penetrable quinazoline inhibitors of PDE1. *Med Chem Commun* (2014) 5(9):1290–1296.

Ihara M, Nishino M, Taguchi A, Yamamoto Y, Hattori Y, Saito S, et al. Cilostazol add-on therapy in patients with mild dementia receiving donepezil: a retrospective study. *PLoS One* (2014) 9(2):e89516.

Ikeda Y. Antiplatelet therapy using cilostazol, a specific PDE3 inhibitor. *Thromb Haemost* (1999) 82(2):435–438.

Jiang MY, Han C, Zhang C, Zhou Q, Zhang B, Le ML et al. Discovery of effective phosphodiesterase 2 inhibitors with antioxidant activities for the treatment of Alzheimer's disease. *Bioorg Med Chem Lett* (2021) 41:128016.

Kaur R, Pandey SK. Efficient synthesis of (-)-(R)- and (+)-(S)-rolipram. *Tetrahedron Lett* (2017) 58(46):4333–4335.

Kelly MP. Does phosphodiesterase 11A (PDE11A) hold promise as a future therapeutic target? *Curr Pharm Des* (2015) 21(3):389–416.

Kelly MP, Logue SF, Brennan J, Day JP, Lakkaraju S, Jiang L et al. Phosphodiesterase 11A in brain is enriched in ventral hippocampus and deletion causes psychiatric disease-related phenotypes. *Proc Natl Acad Sci USA (PNAS)* (2010) 107(18):8457–8462.

Kenan Y, Murata T, Shakur Y, Degerman E, Manganiello VC. Functions of the N-terminal region of cyclic nucleotide phosphodiesterase 3 (PDE 3) isoforms. *J Biol Chem* (2000) 275: 12331–12338.

Kleiman RJ, Lanz TA, Finley JE, Bove SE, Majchrzak MJ, Becker SL et al. Dendritic spine density deficits in the hippocampal CA1 region of young Tg2576 mice are ameliorated with the PDE9A inhibitor PF-04447943. *Alzheimer's Dementia* (2010) 6(4S):563–564.

Kroker KS, Mathis C, Marti A, Cassel JC, Rosenbrock H, Dorner-Ciossek C. PDE9A inhibition rescues amyloid beta-induced deficits in synaptic plasticity and cognition. *Neurobiol Aging* (2014) 35(9):2072–2078.

Kumar A, Singh N. Inhibitor of Phosphodiestearse-4 improves memory deficits, oxidative stress, neuroinflammation and neuropathological alterations in mouse models of dementia of Alzheimer's Type. *Biomed Pharmacother* (2017) 88:698–707.

Lakics V, Karran EH, Boess FG. Quantitative comparison of phosphodiesterase mRNA distribution in human brain and peripheral tissues. *Neuropharmacology* (2010) 59(6):367–374.

Lee JH, Park SY, Shin YW, Kim CD, Lee WS, Hong KW. Concurrent administration of cilostazol with donepezil effectively improves cognitive dysfunction with increased neuroprotection after chronic cerebral hypoperfusion in rats. *Brain Res* (2007) 1185:246–255.

Lee JY, Cho E, Ko YE, Kim I, Lee KJ, Kwon SU et al. Ibudilast, a phosphodiesterase inhibitor with anti-inflammatory activity, protects against ischemic brain injury in rats. *Brain Res* (2012) 1431:97–106.

Leroy MJ, Degerman E, Taira M, Murata T, Wang LH, Movsesian MA, Meacci E, Manganiello VC. Characterization of two recombinant PDE3 (cGMP-inhibited cyclic nucleotide phosphodiesterase) isoforms, RcGIP1 and HcGIP2, expressed in NIH 3006 murine fibroblasts and Sf9 insect cells. *Biochemistry* (1996) 35(31):10194–10202.

Li J, Chen JY, Deng YL, Zhou Q, Wu Y, Wu D et al. Structure-based design, synthesis, biological evaluation, and molecular docking of novel PDE10 inhibitors with antioxidant activities. *Front Chem* (2018) 6:167.

Liu L, Xu H, Ding S, Wang D, Song G, Huang X. Phosphodiesterase 5 inhibitors as novel agents for the treatment of Alzheimer's disease. *Brain Res Bull* (2019) 153:223–231.

Liu Z, Zhou T, Ziegler AC, Dimitrion P, Zuo L. Oxidative stress in neurodegenerative diseases: from molecular mechanisms to clinical applications. *Oxid Med Cell Longev* (2017) 2017:2525967.

Lorigo M, Oliveira N, Cairrao E. PDE-mediated cyclic nucleotide compartmentation in vascular smooth muscle cells: From basic to a clinical perspective. *J Cardiovasc Dev Dis* (2022) 9(4):1–32.

Maurice DH, Haslam RJ. Molecular basis of the synergistic inhibition of platelet function by nitrovasodilators and activators of adenylate cyclase: inhibition of cyclic AMP breakdown by cyclic GMP. *Mol Pharmacol* (1990) 37(5):671–681.

McLachlan CS, Chen ML, Lynex CN, Goh DL, Brenner S, Tay SK. Changes in PDE4D isoforms in the hippocampus of a patient with advanced Alzheimer disease. *Arch Neurol* (2007) 64(3):456–457.

McQuown S, Paes D, Baumgärtel K, Prickaerts J, Peters M. Pharmacological inhibition of phosphodiesterase 7 enhances consolidation processes of spatial memory. *Neurobiol Learn Mem* (2021) 177:107357.

Medina AE. Therapeutic utility of phosphodiesterase type I inhibitors in neurological conditions. *Front Neurosci* (2011) 5:1–11.

Meiners I, Hauschildt S, Nieber K, Munch G. Pentoxyphylline and propentophylline are inhibitors of TNF-α release in monocytes activated by advanced glycation end products. *J Neural Transm* (2004) 111:441–447.

Mestre L, Redondo M, Carrillo-Salinas FJ, Morales-García JA, Alonso-Gil S, Pérez Castillo A et al. PDE7 inhibitor TC3.6 ameliorates symptomatology in a model of primary progressive multiple sclerosis. *Br J Pharmacol* (2015) 172(17):4277–4290.

Miller AD, Leslie RA. The area postrema and vomiting. *Front Neuroendocrinol* (1994) 15(4):301–320.

Mizuno T, Kurotani T, Komatsu Y, Kawanokuchi J, Kato H, Mitsuma N et al. Neuroprotective role of phosphodiesterase inhibitor ibudilast on neuronal cell death induced by activated microglia. *Neuropharmacology* (2004) 46(3):404–411.

Morales-Garcia JA, Echeverry-Alzate V, Alonso-Gil S, Sanz-SanCristobal M, Lopez-Moreno JA, Gil C et al. Phosphodiesterase7 inhibition activates adult neurogenesis in Hippocampus and Subventricular zone in vitro and in vivo. *Stem Cells* (2017) 35(2):458–472.

Morales-Garcia JA, Redondo M, Alonso-Gil S, Gil C, Perez C, Martinez A et al. Phosphodiesterase 7 inhibition preserves dopaminergic neurons in cellular and rodent models of Parkinson disease. *PLoS One* (2011) 6(2):e17240.

Morriello GJ, Dwyer MP, Chen Y, Ginetti AT, Xu S, Lu J et al. Discovery of novel N-1 substituted pyrazolopyrimidinones as potent, selective PDE2 inhibitors. *Bioorg Med Chem Lett* (2021) 44:128082.

Nabavi SM, Talarek S, Listos J, Nabavi SF, Devi KP, Roberto de Oliveira M et al. Phosphodiesterase inhibitors say NO to Alzheimer's disease. *Food Chem Toxicol* (2019) 134:110822.

Newton AC, Bootman MD, Scott JD. Second messengers. *Cold Spring Harb Perspect Biol* (2016) 8(8):1–15.

Ni W, Wang H, Li X, Zheng X, Wang M, Zhang J et al. Novel Tadalafil derivatives ameliorates scopolamine-induced cognitive impairment in mice via inhibition of acetylcholinesterase (AChE) and phosphodiesterase 5 (PDE5). *ACS Chem Neurosci* (2018) 9(7):1625–1636.

Nichols MR, Morimoto BH. Differential inhibition of multiple cAMP phosphodiesterase isozymes by isoflavones and tyrphostins. *Mol Pharmacol* (2000) 57(4):738–745.

Nio Y, Tanaka M, Hirozane Y, Muraki Y, Okawara M, Hazama M, Matsuo T. Phosphodiesterase 4 inhibitor and phosphodiesterase 5 inhibitor combination therapy has antifibrotic and anti-inflammatory effects in mdx mice with Duchenne muscular dystrophy. *FASEB J* (2017) 12:5307–5320.

Nishi A, Kuroiwa M, Miller DB, O'Callaghan JP, Bateup HS, Shuto T et al. Distinct roles of PDE4 and PDE10A in the regulation of cAMP/PKA signalling in the striatum. *J Neurosci* (2008) 28(42):10460–10471.

O'Brien JJ, O'Callaghan JP, Miller DB, Chalgeri S, Wennogle LP, Davis RE et al. Inhibition of calcium-calmodulin-dependent phosphodiesterase (PDE1) suppresses inflammatory responses. *Mol Cell Neurosci* (2020) 102:103449.

Oguchi T, Ono R, Tsuji M, Shozawa H, Somei M, Inagaki M et al. Cilostazol suppresses Aβ-induced neurotoxicity in SH-SY5Y cells through inhibition of oxidative stress and MAPK signalling pathway. *Front Aging Neurosci* (2017) 17(9):337.

Oliveros G, Wallace CH, Chaudry O, Liu Q, Qiu Y, Xie L et al. Repurposing ibudilast to mitigate Alzheimer's disease by targeting inflammation. *Brain* (2023) 146(3):898–911.

Patrucco E, Albergine MS, Santana LF, Beavo JA. Phosphodiesterase 8A (PDE8A) regulates excitation-contraction coupling in ventricular myocytes. *J Mol Cell Cardiol* (2010) 49(2):330–333.

Pérez DI, Pistolozzi M, Palomo V, Redondo M, Fortugno C, Gil C et al. 5-Imino-1,2-4-thiadiazoles and quinazolines derivatives as glycogen synthase kinase 3β (GSK-3β) and phosphodiesterase 7 (PDE7) inhibitors: determination of blood-brain barrier penetration and binding to human serum albumin. *Eur J Pharm Sci* (2012) 45(5):677–684.

Pérez-Torres S, Cortés R, Tolnay M, Probst A, Palacios JM, Mengod G. Alterations on phosphodiesterase type 7 and 8 isozyme mRNA expression in Alzheimer's disease brains examined by in situ hybridization. *Exp Neurol* (2003) 182(2):322–334.

Peters M, Bletsch M, Stanley J, Wheeler D, Scott R, Tully T. The PDE4 inhibitor HT-0712 improves hippocampus-dependent memory in aged mice. *Neuropsychopharmacology* (2014) 39(13):2938–2948.

Pozuelo Rubio M, Campbell DG, Morrice NA, Mackintosh C. Phosphodiesterase 3A binds to 14-3-3 proteins in response to PMA-induced phosphorylation of Ser428. *Biochem J* (2005) 392(Pt 1):163–172.

Prickaerts J, Heckman PRA, Blokland A. Investigational phosphodiesterase inhibitors in phase I and phase II clinical trials for Alzheimer's disease. *Expert Opin Investig Drugs* (2017) 26(9):1033–1048.

Puzzo D, Staniszewski A, Deng SX, Privitera L, Leznik E, Liu S. Phosphodiesterase 5 inhibition improves synaptic function, memory, and amyloid-beta load in an Alzheimer's disease mouse model. *J Neurosci* (2009) 29(25):8075–8086.

Raker VK, Becker C, Steinbrink K. The cAMP pathway as therapeutic target in autoimmune and inflammatory diseases. *Front Immunol* (2016) 7(123):1–11.

Reneerkens OA, Sambeth A, Ramaekers JG, Steinbusch HW, Blokland A, Prickaerts J. The effects of the phosphodiesterase type 5 inhibitor vardenafil on cognitive performance in healthy adults: a behavioral-electroencephalography study. *J Psychopharmacol* (2013) 27(7):600–608.

Ricciarelli R, Brullo C, Prickaerts J, Arancio O, Villa C, Rebosio C et al. Memory-enhancing effects of GEBR-32a, a new PDE4D inhibitor holding promise for the treatment of Alzheimer's disease. *Sci Rep* (2017) 7:46320.

Roesler R, Reolon GK, Maurmann N, Schwartsmann G, Schroder N Amaral OB. A phosphodiesterase 4-controlled switch between memory extinction and strengthening in the hippocampus. *Front Behav Neurosci* (2014) 8(91):1–8.

Rosa E, Fahnestock M. CREB expression mediates amyloid-β induced basal BDNF downregulation. *Neurobio Aging* (2015) 36(8):2406–2413.

Rosenbrock H, Giovannini R, Schänzle G, Koros E, Runge F, Fuchs H et al. The novel phosphodiesterase 9A inhibitor BI 409306 increases cyclic guanosine monophosphate levels in the brain, promotes synaptic plasticity, and enhances memory function in rodents. *J Pharmacol Exp Ther* (2019) 371(3):633–641.

Rutten K, Van Donkelaar EL, Ferrington L, Blokland A, Bollen E, Steinbusch HW, Kelly PA, Prickaerts JH. Phosphodiesterase inhibitors enhance object memory independent of cerebral blood flow and glucose utilization in rats. *Neuropsychopharmacol* (2009) 34(7): 1914–1925.

Sakurai H, Hanyu H, Sato T, Kume K, Hirao K, Kanetaka H et al. Effects of cilostazol on cognition and regional cerebral blood flow in patients with Alzheimer's disease and cerebrovascular disease: a pilot study. *Geriatr Gerontol Int* (2013) 13(1):90–97.

Samidurai A, Xi L, Das A, Kukreja RC. Beyond erectile dysfunction: cGMP-specific phosphodiesterase 5 inhibitors for other clinical disorders. *Annu Rev Pharmacol Toxicol* (2023) 63:585–615.

Sanders O, Rajagopal L. Phosphodiesterase inhibitors for Alzheimer's disease: a systematic review of clinical trials and epidemiology with a mechanistic rationale. *J Alzheimers Dis Rep* (2020) 4(1):185–215.

Sasaki T, Kotera J, Yuasa K, Omori K. Identification of human PDE7B, a cAMP-specific phosphodiesterase. *Biochem Biophys Res Commun* (2000) 271(3):575–583.

Schafer PH, Parton A, Capone L, Cedzik D, Brady H, Evans JF. Apremilast is a selective PDE4 inhibitor with regulatory effects on innate immunity. *Cell Signal* (2014) 26(9):2016–2029.

Shakur Y, Holst LS, Landstrom TR, Movsesian M, Degerman E, Manganiello V. Regulation and function of the cyclic nucleotide phosphodiesterase (PDE3) gene family. *Prog Nucleic Acid Res Mol Biol* (2001) 66:241–277.

Shakur Y, Takeda K, Kenan Y, Yu ZX, Rena G, Brandt D. Membrane localization of cyclic nucleotide phosphodiesterase 3 (PDE3). Two N-terminal domains are required for the efficient targeting to, and association of, PDE3 with endoplasmic reticulum. *J Biol Chem* (2000) 275(49): 38749–38761.

Shim YS, Pae CU, Cho KJ, Kim SW, Kim JC, Koh JS. Effects of daily low-dose treatment with phosphodiesterase type 5 inhibitor on cognition, depression, somatization and erectile function in patients with erectile dysfunction: a double-blind, placebo-controlled study. *Int J Impot Res* (2014) 26(2):76–80.

Shy TW, Gaurav A. Pharmacophore modelling and virtual screening studies to discover potential natural products based on PDE1B inhibitor lead compounds. *Cent Nerv Syst Agents Med Chem* (2021) 21:195–204.

Siuciak JA, McCarthy SA, Chapin DS, Fujiwara RA, James LC, Williams RD et al. Genetic deletion of the striatum-enriched phosphodiesterase PDE10A: evidence for altered striatal function. *Neuropharmacol* (2006) 51(2):374–385.

Susín C, Morales-Garcia JA, Aguilar-Morante D, Palomo V, Sanz-Sancristobal M, Alonso-Gil S et al. The new iminothiadiazole derivative VP1.14 ameliorates hippocampal damage after an excitotoxic injury. *J Neurochem* (2012) 122(6):1193–1202.

Suzuki K, Harada A, Suzuki H, Miyamoto M, Kimura H. TAK-063, a PDE10A inhibitor with balanced activation of direct and indirect pathways, provides potent antipsychotic-like effects in multiple paradigms. *Neuropsychopharmacol* (2016) 41(9):2252–2262.

Sweatt JD. The neuronal MAP kinase cascade: a biochemical signal integration system subserving synaptic plasticity and memory. *J Neurochem* (2001) 76(1):1–10.

Tai SY, Chen CH, Chien CY, Yang YH. Cilostazol as an add-on therapy for patients with Alzheimer's disease in Taiwan: a case control study. *BMC Neurol* (2017) 17:40–46.

Teich AF, Nicholls RE, Puzzo D, Fiorito J, Purgatorio R, Fa' M, Arancio O. Synaptic therapy in Alzheimer's disease: a CREB-centric approach. *Neurother* (2015) 12(1):29–41.

Tresadern G, Velter I, Trabanco AA, Van den Keybus F, Macdonald GJ, Somers MVF et al. [1,2,4] Triazolo[1,5-*a*]pyrimidine Phosphodiesterase 2A inhibitors: structure and free-energy perturbation-guided exploration. *J Med Chem* (2020) 63(21):12887–12910.

Tsai LCL, Chan GCK, Nangle SN, Shimizu-Albergine M, Jones GL, Storm DR et al. Inactivation of Pde8b enhances memory, motor performance, and protects against age-induced motor coordination decay. *Genes Brain Behav* (2012) 11(7): 837–847.

Ugarte A, Gil-Bea F, García-Barroso C, Cedazo-Minguez Á, Ramírez MJ, Franco R. Decreased levels of guanosine 3', 5'-monophosphate (cGMP) in cerebrospinal fluid (CSF) are associated with cognitive decline and amyloid pathology in Alzheimer's disease. *Neuropathol Appl Neurobiol* (2015) 41(4):471–482.

van der Staay FJ, Rutten K, Bärfacker L, Devry J, Erb C, Heckroth H et al. The novel selective PDE9 inhibitor BAY 73-6691 improves learning and memory in rodents. *Neuropharmacol* (2008) 55(5):908–918.

Vanmierlo T, Creemers P, Akkerman S, van Duinen M, Sambeth A, De Vry J et al. The PDE4 inhibitor roflumilast improves memory in rodents at non-emetic doses. *Behav Brain Res* (2016) 15(303):26–33.

Wang C, Yang XM, Zhuo YY, Zhou H, Lin HB, Cheng YF et al. The phosphodiesterase-4 inhibitor rolipram reverses Aβ-induced cognitive impairment and neuroinflammatory and apoptotic responses in rats. *Int J Neuropsychopharmacol* (2012) 15(6):749–766.

Wang H, Mei Zl, Zhong KL, Hu M, Long Y, Miao MX. Pretreatment with antiasthmatic drug ibudilast ameliorates Aβ 1-42-induced memory impairment and neurotoxicity in mice. *Pharmacol Biochem Behav* (2014) 124:373–379.

Wang R, Reddy PH. Role of Glutamate and NMDA receptors in Alzheimer's disease. *J Alzheimers Dis* (2017) 57(4):1041–1048.

Watanabe T, Zhang N, Liu M, Tanaka R, Mizuno Y, Urabe T. Cilostazol protects against brain white matter damage and cognitive impairment in a rat model of chronic cerebral hypoperfusion. *Stroke* (2006) 37(6):1539–1545.

Wechsler J, Choi YH, Krall J, Ahmad F, Manganiello VC, Movsesian MA. Isoforms of cyclic nucleotide phosphodiesterase PDE3A in cardiac myocytes. *J Biol Chem* (2002)277(41): 38072–38078.

Weinryb I, Chasin M, Free CA, Harris DN, Goldenberg H, Michel I et al. Effects of therapeutic agents on cyclic AMP metabolism in vitro. *J Pharm Sci* (1972) 61(10):1556–1567.

Weishaar RE, Cain MH, Bristol JA. A new generation of phosphodiesterase inhibitors: multiple molecular forms of phosphodiesterase and the potential for drug selectivity. *J Med Chem* (1985) 28(5):537–545.

Wu Y, Li Z, Huang YY, Wu D, Luo HB. Novel Phosphodiesterase inhibitors for cognitive improvement in Alzheimer's disease. *J Med Chem* (2018) (61):5467–5483.

Yu YF, Huang YD, Zhang C, Wu XN, Zhou Q, Wu D et al. Discovery of novel pyrazolopyrimidinone derivatives as phosphodiesterase 9A inhibitors capable of inhibiting butyrylcholinesterase for treatment of Alzheimer's disease. *ACS Chem Neurosci* (2017) 8(11):2522–2534.

Zhang HT, Huang Y, Jin SL, Frith SA, Suvarna N, Conti M et al. Antidepressant-like profile and reduced sensitivity to rolipram in mice deficient in the PDE4D phosphodiesterase enzyme. *Neuropsychopharmacol* (2002) 27(4):587–595.

Zhang J, Guo J, Zhao X, Chen Z, Wang G, Liu A et al. Phosphodiesterase-5 inhibitor sildenafil prevents neuroinflammation, lowers beta-amyloid levels and improves cognitive performance in APP/PS1 transgenic mice. *Behav Brain Res* (2013) 250:230–237.

Zhou Y, Jinjian Li, Yuan H, Su R, Huang Y, Huang Y et al. Design, synthesis, and evaluation of dihydropyranopyrazole derivatives as novel PDE2 inhibitors for the treatment of Alzheimer's disease. *Molecules* (2021) 26:3034.

Zhuo Y, Guo H, Cheng Y, Wang C, Wang C, Wu J et al. Inhibition of phosphodiesterase-4 reverses the cognitive dysfunction and oxidative stress induced by Aβ25-35 in rats. *Metab Brain Dis* (2016) 31(4):779–791.

Zuccarello E, Acquarone E, Calcagno E, Argyrousi EK, Deng SX, Landry DW et al. Development of novel phosphodiesterase 5 inhibitors for the therapy of Alzheimer's disease. *Biochem Pharmacol* (2020) 176:113818.

6 The NMDA Receptor Hypothesis in Alzheimer's Disease

Unraveling the Mechanisms

Gadde Shareena[1], Dileep Kumar[1,2,3], and Satheesh Kumar Sellamuthu[4]*

[1]Poona College of Pharmacy, Department of Pharmaceutical Chemistry, Bharati Vidyapeeth (Deemed to be University), Erandwane, Pune-, Maharashtra, India
[2]Department of Pharm. Chemistry, Manipal College of Pharmaceutical Sciences, Manipal Academy of Higher Education, Manipal, Karnataka, India
[3]UC Davis Comprehensive Cancer Center, Department of Entomology and Nematology,University of California Davis, One Shields Avenue, Davis, CA, USA
[4]Karpagam College of Pharmacy, Othakkalmandapam, Coimbatore, Tamil Nadu, India
*Corresponding author

INTRODUCTION

Research and technological developments have increased lifespans and life quality. With few effective treatments available, the number of people with neurodegenerative illnesses is steadily rising. Neurodegeneration associated with dementia is the seventh most common cause of death, affecting about 55.2 million individuals globally (Arvanitakis et al., 2019). Almost 6.5 million Americans aged 65 and above are presently affected by Alzheimer's disease (AD). If no medical advances are made to prevent or cure AD, the affected AD population might increase to 13.8 million by 2060 (Breijyeh and Karaman, 2020). AD is a neurological condition that affects older people. Clinical symptoms of AD include behavioral and personality changes, cognitive problems, and memory loss. The neocortical and hippocampal regions are gradually damaged in function and structure as AD progresses (Murphy and LeVine, 2010). Anomalies in both structure and functioning are hallmarks of AD's pathogenesis. As AD worsens, the brain develops significant structural lesions, including amyloid beta (Aβ) senile plaques, phosphorylated tau neurofibrillary tangles (NFTs), and synapse loss. Significant oxidative stress, dysfunctional mitochondria, severe synapse loss, and neuronal death can also be observed as part of AD-related neurodegeneration (Meraz-Ríos et al., 2013; Mota et al., 2014).

Despite substantial studies into AD, no effective treatment is available, and the underlying molecular pathways remain unknown. Several signaling pathways connected to AD pathogenesis are being researched to identify possible therapeutic targets, including N-methyl-D-aspartate (NMDA) and triggering receptor expressed on myeloid cells 2 (TREM2). Many of the medications used to treat AD have undesirable effects because of their high costs, unclear safety, harmful side effects, poor drug tolerance, and patients' resistance to medication (Mimica and Presecki, 2009). Further knowledge of glutamate neurotransmission alterations associated with AD would be very

DOI: 10.1201/9781003506317-6

useful to better analyze and comprehend its mechanisms of action and develop treatment regimens, as cholinesterase inhibitors and the NMDA antagonist memantine are available medicines for AD (Parsons et al., 2007).

Anomalous synaptic Ca^{2+} processing in response to excessive NMDA stimulation leads to synaptic dysfunction (Mota et al., 2014). The majority of excitatory neuronal transmission in brain tissues is mediated by NMDA receptors, a class of L-glutamate receptors and principal iGluRs. These tetrameric ion channels function in tandem with other iGluRs, such as AMPA and kainate receptors, to regulate synaptic plasticity, including LTP (long-term potentiation) and LTD (long-term depression), a crucial component for cognition and learning function (Traynelis et al., 2010; Paoletti et al., 2013). NMDA receptors (NMDARs) demonstrate various features that distinguish them from other ligand-gated iGluRs, including strong Ca^{2+} penetration, voltage-sensitive inhibition by extracellular Mg^{2+}, and relatively slow activation/deactivation rates (Cull-Candy et al., 2001). Furthermore, protons, redox state changes, and endogenous physiological chemicals also influence NMDARs (Hansen et al., 2017). Calcium is highly permeable through NMDARs, and the channel promotes calcium influx by releasing intracellular signals that may induce synaptic currents to LTP and LTD. The NMDAR is activated by the release of Mg^{2+} and agonist binding, which in turn triggers a Ca^{2+}-stimulated signaling cascade that regulates synaptic strength (Chen and Roche, 2007).

NMDARs also have a crucial role in the pathophysiology of various psychiatric, neurodegenerative, and higher cognitive disorders (Akazawa et al., 1994). In pathological conditions, excitotoxicity results from the prolonged activation of glutamatergic signaling. Several studies suggest that glutamate excitotoxicity in neural tissues plays a key role in late-onset neurodegeneration and the development of acute symptoms (Lipton and Rosenberg, 1994).

In order to understand the NMDA receptor's molecular activity and regulation, this chapter aims to provide a detailed understanding of both the receptor's inhibitors and NMDA receptors. In addition, multiple aberrant signaling pathways are also described that intend to trigger AD pathogenicity.

NMDA RECEPTOR AND ROLE OF NMDA SUBUNITS

NMDA receptor is a multi-subunit ligand-gated cellular channel widely expressed in the human and mammalian CNS. There are several different subtypes of the NMDAR, each containing a diverse range of subunits and having a distinct function in the spinal cord and neural tissues (Monyer et al., 1994). Seven distinct NMDAR subunits have been identified across a multitude of human tissues: one GluN1 subunit, four GluN2 subunits (GluN2A, GluN2B, GluN2C, and GluN2D), and two GluN3 subunits (GluN3A and GluN3B). All NMDA subunits are distinctive in that to activate their channels, magnesium ions must be released from the presynaptic synapses at the same time as a substantial postsynaptic membrane depolarization (Paoletti, 2011). Functional NMDARs possess heterotetramers constituting two glutamate-binding GluN2(A-D) units and two D-serine or glycine-binding GluN1 subunits. Four domains constitute the structure of all glutamate receptor subunits: an intracellular carboxyl-terminal domain (CTD), a large extracellular amino-terminal domain (ATD), one pore-forming transmembrane domain (TMD), and a bi-lobed agonist-binding domain (ABD) (Paoletti, 2011). The extracellular ATD is involved in subunit assembling, allosteric regulation, and modulation of ion channel open probabilities and deactivation kinetics. The two discontinuous segments (S1 and S2) that comprise the extracellular ligand-binding domain (LBD) bind agonists and antagonists to regulate ion channel opening. The transmembrane helices (M1, M3, and M4), along with a re-entrant pore loop helix (M2), a component of the channel pore, constitute the TMD (Karakas and Furukawa, 2014). The M2 includes a crucial asparagine residue that controls the channel's calcium permeability and facilitates magnesium blockage. There are several cytosolic proteins with which the intracellular CTD interacts (Hackos and Hanson, 2017). The glutamate binding site is present in GluN2 LBDs, while the glycine (or D-serine) co-agonist binding region is prevalent in GluN1 and GluN3 LBDs. The association of both glycine and glutamate molecules at the extracellular LBDs as well as the release of the magnesium barrier by membrane depolarization at the TMD, are essential for activating GluN1- and N2-associated NMDARs activation (Yu and Lau, 2018). Positively charged residues help glutamate bind through directed molecular diffusion on the GluN2A LBD surface. Inversely, glycine links to the GluN1 LBD by an undirected diffusional technique, wherein glycine detects its binding site mainly through periodic temperature variations (Liu et al., 2019).

ABERRANT EXPRESSIONS OF NMDA RECEPTORS WITH AD HALLMARKS: β-AMYLOID PLAQUES AND NFTS

AD brain tissues are characterized histopathologically by intraneuronal NFTs and extracellular Aβ plaques. NMDAR plays a critical part in synaptic plasticity, which is closely correlated to synaptic impairment in AD. NMDAR-associated excitatory glutamatergic neurotransmission is essential for synaptic plasticity and neuronal survival in the brain (Parameshwaran et al., 2008). Further, increased NMDAR activity promotes excitotoxicity and stimulates cell death showcasing evidence for a potential neurodegenerative pathway in AD. Glutamate concentration, NMDAR channel activity, and functional alteration are the main mechanisms influencing NMDAR signaling in AD (Hoey et al., 2009).

NMDA Association with Aβ Species

According to the Aβ cascade hypothesis, amyloidosis triggers AD, followed by tau deposition before the neuronal loss and cognitive decline (Hardy, 2002; Ittner et al. 2010). According to the amyloid cascade hypothesis, the earliest abnormal Aβ peptide production stages result in the eventual neurodegeneration seen in Alzheimer's disease (Verdile et al., 2005). β-Amyloid peptides constitute the amyloid plaques. The precursor protein, APP, which is ubiquitously expressed in the brain and spinal cord of the CNS, is subdivided to release the peptides. A number of Aβ species are formed as a result of the proteolytic degradation of APP by a range of secretases and proteases.

The Aβ species are formed in the amyloidogenic pathway as a result of APP gene mutations. BACE-1 enzymes endo-proteolytically break down APP, resulting in the formation of sAPP-α and β-CTF. In the process, the β-secretase complex releases AICD and Aβ40-42 after further disintegrating β-CTF. These primarily aggregate, culminating in neuritic plaques and other oligomeric peptide forms that are observed as AD aggregates (Soriano et al., 1999). Much research suggests that Aβ oligomers are more influential than Aβ fibrils in inducing abnormal synaptic functions. Moreover, the most synaptotoxic forms of Aβ1-42 are believed to be the oligomers, which are responsible for the early cognitive abnormalities in AD (Palop and Mucke, 2010). Aβ leads to abnormal excitatory network activation and counter-inhibitory responses involving pathways for learning and memory.

The early cognitive abnormalities seen in AD are attributable to the dysregulation of NMDA receptors elicited by Aβ peptides, oligomers, and the consequent collapse of Ca^{2+} homeostasis. Numerous experiments have discovered that β-amyloid oligomers and NMDAR are fundamental for the synaptic and neuronal deterioration in AD (Roselli et al., 2005; Chang et al., 2016). The influence of Aβ on synapses is to decrease surface glutamate receptors as well as other postsynaptic constituents in cultured hippocampal neurons (Kamenetz et al., 2003).

Research has indicated that permeable and oligomeric β-amyloid plaques have far more severe repercussions than impenetrable β-amyloid aggregates. GluN2A and GluN2B are prominent subunits with essential roles in synapse formation and plasticity in the hippocampus and cerebral cortex regions. As a result of elevated Aβ-peptides' suppression of LTP, NMDA receptor-dependent signaling cascades are transitioned across to LTD (Palop and Mucke, 2010). Hence, accumulating Aβ oligomers interfere with synaptic activity and signal transduction, triggering early cognitive impairments. By preventing glutamate from being assimilated by glial cells, glutamate concentrations in the synaptic cleft increase, leading to Aβ-induced LTD (Zhang et al., 2016). Moreover, astrocytes and presynaptic neurons both discharge more glutamate when associated with Aβ peptides. Thus, an increase in glutamate levels activates extrasynaptic NMDA receptors that incorporate the GluN2B subunit. This yields a gradual increase in postsynaptic Ca^{2+} levels that triggers LTD, which further induces synaptic impairment since the breakdown of dendritic spines negatively impacts synapse progress and plasticity (Wang et al., 2013).

In addition, higher cellular Ca^{2+} levels enhance the expression of GSK-3β, JNK, and MAPKs, which are associated with tau hyperphosphorylation and cell death signaling. Aβ influenced NMDA-facilitated responses and concomitant exposure to Aβ and NMDA-enhanced Ca^{2+} levels mediated by NMDARs that comprise the GluN2B molecule (Li et al., 2011).

To avoid cytosolic Ca^{2+} overloading, mitochondrial cells quickly absorb the increasing cytosolic Ca^{2+}. As a result of the excess Ca^{2+} in mitochondria, nitric oxide and reactive oxygen species (ROS) are generated, ATP synthesis is interrupted, mitochondrial mPTP unfolds, cytochrome-c is liberated, pro-apoptotic caspases are activated, and apoptosis is induced. In addition, Aβ also mediates a wide range of neuroinflammatory processes by activating pro-inflammatory microglia, which have a detrimental effect on the production of pro-inflammatory cytokines, including TNF-α and interleukins (IL-10), which promote neurodegeneration (Calvo-Rodriguez et al., 2020; Esteras and Abramov, 2020).

The finding demonstrates that Aβ oligomers cause glutamate receptors to function irregularly, impairing glutamatergic synaptic transmission and being associated with early cognitive impairments (Esposito et al., 2013). The NMDA receptor might theoretically play a variety of roles in Aβ-associated pathways. Firstly, an NMDA receptor may act as an Aβ receptor or indirectly associate via interactions with Aβ-binding molecules. Secondly, NMDA receptor function might be an important downstream target of Aβ peptides, implying Aβ could induce an alleviated or augmented function of NMDA receptors. Thirdly, NMDA receptors may be crucial for Aβ's effects on synaptic plasticity and transmission. Lastly, the NMDA receptor might regulate Aβ formation (Malinow, 2012). Increased Aβ peptides prevent LTP, which leads to NMDA receptor-dependent signaling cascades to transition to LTD. As a result, the build-up of Aβ oligomers impairs synaptic transmission, which promotes early cognitive deficits. Since glutamate uptake at glial cells is impeded, the concentration of glutamate in the synaptic cleft increases, leading to Aβ-induced LTD. Moreover, astrocytes and presynaptic neuronal cells both release glutamate at higher levels due to Aβ peptides (Olsen and Sheng, 2012). Consequently, an increase in glutamate level activates NMDA receptors that include extrasynaptic GluN2B, which results in a small amount of rise in postsynaptic Ca^{2+}, initiating LTD, and, ultimately, synaptic damage from the collapse of dendritic spines impairing synapse development and plasticity (Vizi et al., 2000).

Kamenetz and colleagues suggested that heightened neuronal activity in organotypic hippocampal AD slices cultured from mice overexpressing APP increased the proportions of Aβ detected in the culture medium. It was discovered that this elevation in Aβ levels was not dependent on resting levels of NMDAR activity; the impact of increased levels of NMDAR stimulation on Aβ secretion was not investigated (Kamenetz et al., 2003). In a recent investigation, *in vivo* microdialysis techniques were used to demonstrate that Aβ secretion is contingent on neuronal activity. These studies contend that the presynaptic segment is where the activity-dependent occurs, implying the absence of NMDAR dependence (Cirrito et al., 2005).

NMDA Receptors and Aberrant NFTs

Tau proteins that stabilize microtubules are typically present in the distal axonal tubules. They are found in low amounts in the nucleus and somatodendritic regions of the CNS but are broadly distributed throughout neural tissues (Tashiro et al., 1997). As microtubule structures are essential for regulating cell shape and axonal stability, variations in their characteristics may contribute to the emergence of dementia-related neurological illnesses (Cleveland et al., 1977). Tau proteins are complex and challenging to define owing to the multiple underlying dynamics. Since tau regulates structural assembly, spatial arrangement, and dynamic properties, it plays a crucial part in microtubule activity. Tau binds to microtubules and either explicitly or inferentially stabilizes them by serving as a link, improving their connectivity to a variety of other cytoskeletal components, such as actin and neurofilaments. Tau and other MAPs bound to microtubules are strongly affected by various factors, such as post-translational modifications (PTMs), to support axonal growth and govern vesicle and organelle trafficking across microtubules (Mandelkow et al., 1995; Gong et al., 2005). It has been discovered that tau is constantly linked to regulatory proteins, controls microtubule-mediated axonal transport, and is essential for axonal differentiation, projection, trafficking, and neuroplasticity (Takei et al., 2009; Naseri et al., 2019).

Tau proteins are regarded as axonal proteins involved in sustaining microtubule stability and form an essential component of NFT (Figure 6.1 depicts the aberrant NMDA-associated transmission propagating both tau NFTs and amyloid plaques). The AD diagnosis is associated with tau protein mislocalization and phosphorylation from the axonal compartment to dendritic spines, resulting in synaptic instability and LTP impairment. Tau accumulates in dendritic spines upon hyperphosphorylation or cleavage, where it interacts with postsynaptic proteins to induce synaptic dysfunction. Augmented tau hyperphosphorylation due to NMDAR stimulation and Aβ deposition may promote NMDAR transmission, which could heighten LTD. Although NMDAR-dependent tau phosphorylation is a reversible process, tau phosphorylation mediated by Aβ is irreversible if it persists for more than five days (Mondragón-Rodríguez et al., 2012; Roberson et al., 2007). In addition, tau protein is also engaged in NMDAR activation and is mediated via Src tyrosine kinase Fyn kinase, which phosphorylates the GluN2B subunit at Tyr1472 and stabilizes its binding with PSD95 in dendritic spines (Nakazawa et al., 2001). Overexpressed Fyn upregulates NMDAR activation, overloading the dendritic spines with detrimental Ca levels, and coincides with excessive tau in AD dendrites. Aβ-induced glutamatergic excitotoxicity is further exacerbated by PSD95 and GluN2B complex stability. Tau phosphorylation is lowered in the GluN2A NMDAR via the PKC/GSK3 pathway (Ittner et al., 2010). AD knockout studies suggest that tau knockouts implicate GluN2B

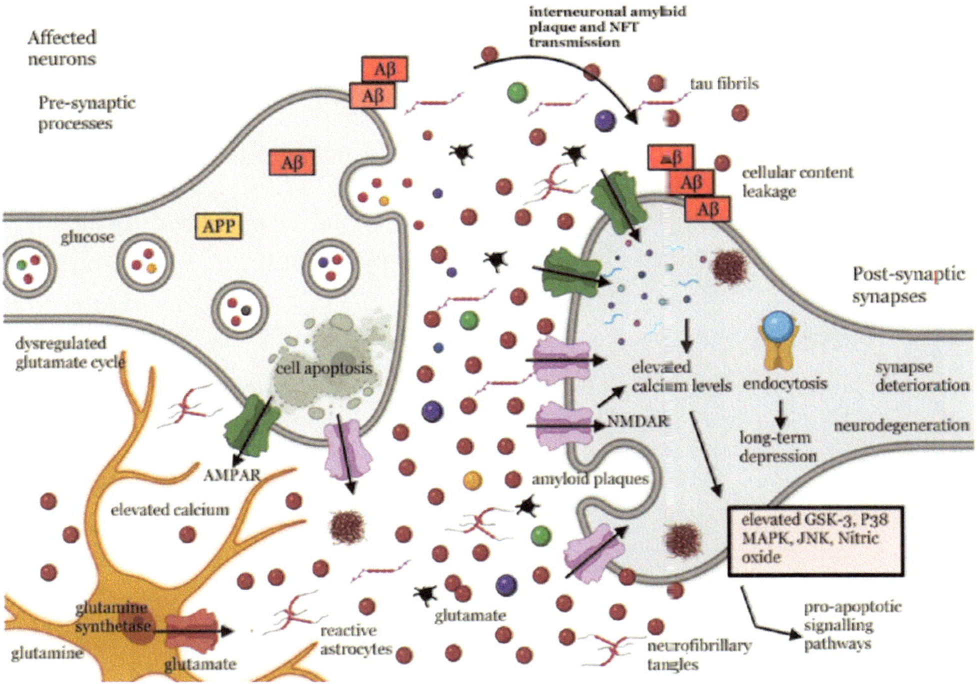

FIGURE 6.1 Aberrant NMDA-linked transmission in Alzheimer's disease. In affected neurons, aberrant glutamate transport occurs along with inter-neuronal migration of NFTs and amyloid plaques. Higher levels of intracellular calcium levels and glutamate dysregulation lead to pro-apoptotic pathway activation, synapse deterioration, and, ultimately, neurodegeneration.

phosphorylation and impair Fyn kinase accumulation at postsynaptic regions, which alleviates Aβ-induced neurotoxicity (Mondragón-Rodríguez et al., 2012). Overall, these findings provide evidence that tau has a pathogenic function in NMDAR interaction and regulation, synaptic compartment, and, more particularly, synaptic impairment.

NMDA INHIBITORS

Memantine

Memantine is a non-competitive, low-affinity, voltage-dependent glutamate receptor NMDA subtype antagonist. It is utilized to decrease neurotoxicity, a key part of AD and other neurodegenerative illnesses. Strong voltage dependence, swift unblocking kinetics, and potent inhibitor of NMDAR-mediated excitotoxicity are all evident in this molecule. Identical to Mg^{2+}, it prevents the NMDAR from activating while facilitating the transmission of temporary physiological impulses that are crucial for learning and memory processes (Danysz and Parsons, 2012). Memantine also enhances astroglial neurotrophic factor release, which results in neuroprotective and survival effects (Wu et al., 2009). Currently, it is the only NMDAR antagonist with clinical approval that operates primarily as an antagonist of extrasynaptic NMDARs. While MEM exhibits a suitable pattern in preclinical AD models, clinical evidence hasn't demonstrated that the therapy of the disease has yielded great outcomes. Even so, drugs that affect these two pathways, including MEM, appear to have no substantial usage in the clinical treatment of this illness. Cholinergic and glutamatergic dysfunctions are currently prominent neurochemical targets for the symptomatology of AD. So, it indicates that adopting combined treatments is the most effective way to treat the present state (Folch et al., 2018).

Ketamine

Ketamine acts as an open-channel blocker and is a non-competitive NMDAR antagonist. In order for ketamine to function, NMDAR channels must be open, as compared to being blocked (MacDonald et al., 1987). Ketamine produces a trapping type of open-channel blockage, identical to its structural analogues MK-801 and phencyclidine, where the medication binds to an electrically deep position within the ion channel, impedes the movement of ions through the open channel, and can remain in the channel after its closure (Zorumski et al., 2016). This latter characteristic explains the long-lasting obstruction that is eliminated following channel opening.

Dextromethorphan

Dextromethorphan (DM) is a prevalent antitussive medication and a non-competitive NMDA receptor antagonist. By suppressing the excitatory amino acids in a non-competitive approach, DM also prevents neuronal damage and modifies pain perception. The principal metabolite of dextromethorphan, dextrorphan, is an NMDA receptor antagonist with properties similar to those of ketamine and phencyclidine (Siu and Drachtman, 2007).

CONCLUSION

Studies on the NMDA receptor have emphasized the pathophysiology of Alzheimer's disease. The NMDA receptor has been recognized as a crucial modulator of synaptic plasticity and cognitive function, and its dysregulation has been linked to the onset and progression of Alzheimer's disease. Numerous studies have shown that NMDA receptor antagonists may help Alzheimer's patients with cognitive function. In addition, mutations in NMDA receptor subunits have been linked to a higher risk of Alzheimer's disease through genetic studies. Many studies have been conducted over the years regarding the role of the neurotransmitter glutamate and its receptors in the activity of synaptic plasticity and the etiology of neurodegenerative disorders. Scientific literature has revealed that outcomes from glutamatergic neurotransmission via NMDARs are multimodal. Neuronal survival is reliant on synaptic NMDAR action. The excess of astrocyte-mediated or presynaptic terminal-mediated glutamate release serves a vital part in antagonizing synaptic pro-survival signaling and shifting the balance in support of excitotoxicity and ultimate neurodegeneration. NMDAR inhibition has been employed to treat AD associated with glutamate-induced excitotoxicity. High-affinity NMDA antagonists have been investigated in clinical trials; however, these tests have so far been ineffectual. Lack of efficacy and unpredictable perplexing adverse effects are accountable for clinical failure. The contrasting functions performed by the multiple NMDAR subtypes, all of which are suppressed by the therapeutics utilized, can be used to explain these unfavorable outcomes. Emerging research attempts to discover efficient disease-modifying treatments that particularly target the pathophysiologic pathways to postpone the development and course of the illness. Future NMDAR-targeted therapeutics will still require persistent and sustained efforts to become therapeutically viable. These results demonstrate the importance of NMDAR research in creating cutting-edge treatment options for Alzheimer's disease. Although more investigation is required to fully comprehend the function of NMDARs in Alzheimer's disease and to create efficient treatments, the knowledge gained from these studies has already opened up significant new research directions and has the potential to enhance the lives of millions of people who are affected by this life-altering condition.

ACKNOWLEDGEMENT

All the schematic diagrams were made using Biorender.com software.

REFERENCES

Akazawa C, Shigemoto R, Bessho Y, Nakanishi S, Mizuno N. Differential expression of five N-methyl-D-aspartate receptor subunit mRNAs in the cerebellum of developing and adult rats. *J Comp Neurol.* (1994) 347(1):150–160.

Arvanitakis Z, Shah RC, Bennett DA. Diagnosis and management of Dementia: review. *JAMA.* (2019) 322(16):1589–1599.

Breijyeh Z, Karaman R. Comprehensive review on Alzheimer's disease: causes and treatment. *Molecules (Basel, Switzerland).* (2020) 25(24):5789.

Calvo-Rodriguez M, Hou SS, Snyder AC, Kharitonova EK, Russ AN, Das S, et al. Increased mitochondrial calcium levels associated with neuronal death in a mouse model of Alzheimer's disease. *Nat Commun.* (2020) 11(1):2146.

Chang L, Zhang Y, Liu J, Song Y, Lv A, Li Y, et al. Differential regulation of N-Methyl-D-Aspartate receptor subunits is an early event in the actions of soluble Amyloid-β(1-40) Oligomers on Hippocampal neurons. *J Alzheimers Dis.* (2016) 51(1):197–212.

Chen BS, Roche KW. Regulation of NMDA receptors by phosphorylation. *Neuropharmacol.* (2007) 53(3):362–368.

Cirrito JR, Yamada KA, Finn MB, Sloviter RS, Bales KR, May PC, et al. Synaptic activity regulates interstitial fluid amyloid-beta levels in vivo. *Neuron.* (2005) 48(6):913–922.

Cleveland DW, Hwo SY, Kirschner MW. Purification of tau, a microtubule-associated protein that induces assembly of microtubules from purified tubulin. *J Mol Biol.* (1977) 116(2):207–225.

Cull-Candy S, Brickley S, Farrant M. NMDA receptor subunits: diversity, development and disease. *Curr Opin Neurobiol.* (2001) 11(3):327–335.

Danysz W, Parsons CG. Alzheimer's disease, β-amyloid, glutamate, NMDA receptors and memantine—searching for the connections. *Br J Pharmacol.* (2012) 167, 324–352.

Esposito Z, Belli L, Toniolo S, Sancesario G, Bianconi C, Martorana A. Amyloid β, glutamate, excitotoxicity in Alzheimer's disease: are we on the right track? *CNS Neurosci Ther.* (2013) 19(8):549–555.

Esteras N, Abramov AY. Mitochondrial calcium deregulation in the mechanism of beta-amyloid and tau pathology. *Cells.* (2020) 9(9):2135.

Folch J, Busquets O, Ettcheto M, Sánchez-López E, Castro-Torres RD, Verdaguer E, et al. Memantine for the treatment of Dementia: a review on its current and future applications. *J Alzheimers Dis.* (2018) 62(3):1223–1240.

Gong CX, Liu F, Grundke-Iqbal I, Iqbal K. Post-translational modifications of tau protein in Alzheimer's disease. *J Neural Transm* (Vienna). (2005) 112(6):813–838.

Hackos DH, Hanson JE. Diverse modes of NMDA receptor positive allosteric modulation: mechanisms and consequences. *Neuropharmacol.* (2017) 112(Pt A):34–45.

Hansen KB, Yi F, Perszyk RE, Menniti FS, Traynelis SF. NMDA receptors in the central nervous system. *Methods Mol Biol.* (2017) 1677:1–80.

Hardy J. The amyloid hypothesis of Alzheimer's disease: progress and problems on the road to therapeutics. *Science.* (2002) 297:353–356.

Hoey SE, Williams RJ, Perkinton MS. Synaptic NMDA receptor activation stimulates alpha-secretase amyloid precursor protein processing and inhibits amyloid-beta production. *J Neurosci.* (2009) 29(14):4442–4460.

Ittner LM, Ke YD, Delerue F, Bi M, Gladbach A, van Eersel J, et al. Dendritic function of tau mediates amyloid-beta toxicity in Alzheimer's disease mouse models. *Cell.* (2010) 142(3):387–397.

Kamenetz F, Tomita T, Hsieh H, Seabrook G, Borchelt D, Iwatsubo T, Sisodia S, Malinow R. APP processing and synaptic function. *Neuron.* (2003) 37(6):925–937.

Karakas E, Furukawa H. Crystal structure of a heterotetrameric NMDA receptor ion channel. *Science.* (2014) 344(6187):992–997.

Li S, Jin M, Koeglsperger T, Shepardson NE, Shankar GM, Selkoe DJ. Soluble A{beta} oligomers inhibit long-term potentiation through a mechanism involving excessive activation of extrasynaptic NR2B-containing NMDA receptors. *J Neurosci.* (2011) 31:6627–6638.

Lipton SA, Rosenberg PA. Excitatory amino acids as a final common pathway for neurologic disorders. *N Engl J Med.* (1994) 330(9):613–622.

Liu J, Chang L, Song Y, Li H, Wu Y. The role of NMDA receptors in Alzheimer's disease. *Front Neurosci.* (2019) 8(13):43.

MacDonald JF, Miljkovic Z, Pennefather P. Use-dependent block of excitatory amino acid currents in cultured neurons by ketamine. *J Neurophysiol.* (1987) 58(2):251–266.

Malinow R. New developments on the role of NMDA receptors in Alzheimer's disease. *Curr Opin Neurobiol.* (2012) 22(3):559–563.

Mandelkow E, Song YH, Schweers O, Marx A, Mandelkow EM. On the structure of microtubules, tau, and paired helical filaments. *Neurobiol Aging.* (1995) 16(3):347–354.

Meraz-Ríos MA, Toral-Rios D, Franco-Bocanegra D, Villeda-Hernández J, Campos-Peña V. Inflammatory process in Alzheimer's disease. *Front Integr Neurosci.* (2013) 7:59.

Mimica N, Presecki P. Side effects of approved antidementives. *Psychiatr Danub.* (2009) 21(1):108–113.

Mondragón-Rodríguez S, Trillaud-Doppia E, Dudilot A, Bourgeois C, Lauzon M, Leclerc N, Boehm J. Interaction of endogenous tau protein with synaptic proteins is regulated by N-methyl-D-aspartate receptor-dependent tau phosphorylation. *J Biol Chem.* (2012) 287(38):32040–32053.

Monyer H, Burnashev N, Laurie DJ, Sakmann B, Seeburg PH. Developmental and regional expression in the rat brain and functional properties of four NMDA receptors. *Neuron.* (1994) 12(3):529–540.

Mota SI, Ferreira IL, Rego AC. Dysfunctional synapse in Alzheimer's disease - a focus on NMDA receptors. *Neuropharmacol.* (2014) 76 Pt A:16–26.

Murphy MP, LeVine H. Alzheimer's disease and the amyloid-beta peptide. *J Alzheimers Dis.* (2010) 19(1):311–323.

Nakazawa T, Komai S, Tezuka T, Hisatsune C, Umemori H, Semba K, Mishina M, Manabe T, Yamamoto T. Characterization of Fyn-mediated tyrosine phosphorylation sites on GluR epsilon 2 (NR2B) subunit of the N-methyl-D-aspartate receptor. *J Biol Chem.* (2001) 276(1):693–699.

Naseri NN, Wang H, Guo J, Sharma M, Luo W. The complexity of tau in Alzheimer's disease. *Neurosci Lett.* (2019) 13(705):183–194.

Olsen KM, Sheng M. NMDA receptors and BAX are essential for Aβ impairment of LTP. *Sci Rep.* (2012) 2:225.

Palop JJ, Mucke L. Amyloid-beta-induced neuronal dysfunction in Alzheimer's disease: from synapses toward neural networks. *Nat Neurosci.* (2010) 13(7):812–818.

Paoletti P, Bellone C, Zhou Q. NMDA receptor subunit diversity: impact on receptor properties, synaptic plasticity and disease. *Nat Rev Neurosci.* (2013) 14:383–400.

Paoletti P. Molecular basis of NMDA receptor functional diversity. *Eur J Neurosci.* (2011) 33:1351–1365.

Parameshwaran K, Dhanasekaran M, Suppiramaniam V. Amyloid beta peptides and glutamatergic synaptic dysregulation. *Exp Neurol.* (2008) 210(1):7–13.

Parsons CG, Stöffler A, Danysz W. Memantine: a NMDA receptor antagonist that improves memory by restoration of homeostasis in the glutamatergic system--too little activation is bad, too much is even worse. *Neuropharmacol.* (2007) 53(6):699–723.

Roberson ED, Scearce-Levie K, Palop JJ, Yan F, Cheng IH, Wu T, Gerstein H, Yu GQ, Mucke L. Reducing endogenous tau ameliorates amyloid beta-induced deficits in an Alzheimer's disease mouse model. *Science.* (2007) 316(5825):750–754.

Roselli F, Tirard M, Lu J, Hutzler P, Lamberti P, Livrea P, Morabito M, Almeida OF. Soluble beta-amyloid1-40 induces NMDA-dependent degradation of postsynaptic density-95 at glutamatergic synapses. *J Neurosci.* (2005) 25(48):11061–11070.

Siu A, Drachtman R. Dextromethorphan: a review of N-methyl-d-aspartate receptor antagonist in the management of pain. *CNS Drug Rev.* (2007) 13(1):96–106.

Soriano S, Chyung AS, Chen X, Stokin GB, Lee VM, Koo EH. Expression of β-amyloid precursor protein-CD3 gamma chimeras to demonstrate the selective generation of amyloid β (1-40) and amyloid β (1-42) peptides within secretory and endocytic compartments. *J Biol Chem.* (1999) 274(45):32295–32300.

Takei N, Miyashita A, Tsukie T, Arai H, Asada T, Imagawa M, et al. Japanese genetic study consortium for Alzheimer disease. Genetic association study on in and around the APOE in late-onset Alzheimer disease in Japanese. *Genomics.* (2009) 93(5):441–448.

Tashiro K, Hasegawa M, Ihara Y, Iwatsubo T. Somatodendritic localization of phosphorylated tau in neonatal and adult rat cerebral cortex. *Neuroreport* (1997) 8(12):2797–2801.

Traynelis SF, Wollmuth LP, McBain CJ, Menniti FS, Vance KM, Ogden KK, Hansen KB, Yuan H, Myers SJ, Dingledine R. Glutamate receptor ion channels: structure, regulation, and function. *Pharmacol Rev.* (2010) 62(3):405–496.

Verdile G, Fuller S, Atwood CS, Laws SM, Gandy SE, Martins RN. The role of β amyloid in Alzheimer's disease: still a cause of everything or the only one who got caught? *Pharmacol Res.* (2005) 50:397–409.

Vizi ES. Role of high-affinity receptors and membrane transporters in nonsynaptic communication and drug action in the central nervous system. *Pharmacol Rev* (2000) 52:63–89.

Wang ZC, Zhao J, Li S. Dysregulation of synaptic and extrasynaptic N-methyl-D-aspartate receptors induced by amyloid-β. *Neurosci Bull.* (2013) 29(6):752–760.

Wu HM, Tzeng NS, Qian L, Wei SJ, Hu X, Chen SH, et al. Novel neuroprotective mechanisms of memantine: Increase in neurotrophic factor release from astroglia and anti-inflammation by preventing microglial activation. *Neuropsychopharmacol.* (2009) 34:2344–2357.

Yu A, Lau AY. Glutamate and Glycine binding to the NMDA receptor. *Structure.* (2018) 3;26(7):1035–1043.

Zhang Y, Li P, Feng J, Wu M. Dysfunction of NMDA receptors in Alzheimer's disease. *Neurol Sci.* (2016) 37(7):1039–1047.

Zorumski CF, Izumi Y, Mennerick S. Ketamine: NMDA receptors and beyond. *J Neurosci.* (2016) 36(44):11158–11164.

7 Artificial Intelligence and Machine Learning-Based Advances in the Management of Neurodegenerative Diseases

Ankita Wal[1], Mangla Nand Singh[2], Himangi Vig[1], Manoj Kumar Katual[3], Bommaiah Prakash Kumar[4], and Manukonda Upender Kumar[5]*

[1]PSIT-Pranaveer Singh Institute of Technology, Pharmacy, Kanpur UP, India
[2]Dr. B. R. Ambedkar Pooja College of Pharmacy (Affiliated to Dr. APJAKTU Lucknow), Gorakhpur, Uttar Pradesh, India
[3]University College of Pharmacy, Guru Kashi University, Bhatinda, Punjab, India
[4]Sri Adichunchangiri College of Pharmacy, University, Madhya Pradesh, India
[5]University College of Pharmaceutical Sciences, Kakatiya University, Telangana, India
*Corresponding author

INTRODUCTION

The word "neurodegeneration" derives its etymology from the term "neuro," which refers to neurons, along with the word "degeneration," which signifies a degradation of function as well as structure. In summary, neurodegeneration is a mechanism by which neurons lose their ability to operate both structurally and functionally. Nevertheless, while being an often-used expression with a simple description, this term has a much deeper meaning along with a specific meaning (Wal et al., 2022). The most prevalent clinical symptoms of neurodegenerative disorder (NDD) include extrapyramidal and pyramidal locomotor problems, as well as cognition or behavioral disorders. Most individuals exhibit multiple clinical characteristics, with a minority possessing pure syndromes (Dugger and Dickson, 2017). A set of maladies known as neurodegenerative illnesses, which include Alzheimer's disease (AD), amyotrophic lateral sclerosis (ALS), Huntington's disease (HD), and Parkinson's disease (PD), are characterized by unique etiologies with unique anatomical and pathological traits. Numerous pieces of research point to a complex set of factors that may contribute to the development of these types of disorders, including aberrant protein dynamics with impeded protein aggregation or degradation, oxidative damage and the production of free radicals, impaired biochemical processes, dysfunctional mitochondria, and contact with chemical pesticides as well as metallic contaminants. While each illness has its unique clinical presentation and molecular process, certain common routes may be identified in various pathologic cascades. They involve simultaneous oxidative stress as well as free radical production, metallic dyshomeostasis, malfunctioning mitochondria, misfolding of proteins and accumulation, and phosphorylation dysfunction (Sheikh et al., 2013). Numerous investigations have revealed that aging, neurological problems associated with aging, and various other ailments associated with mitochondria are caused by abnormalities in the

 DOI: 10.1201/9781003506317-7

mitochondria. The onset and progression of neurological illnesses are thought to be significantly influenced by malfunctioning nonenzymatic complexes, the production of oxygen radicals, irregularities in the permeability of mitochondria and transitional pores, higher arrest of the cell cycle, and anomalies in mitochondrial cellular structure. α-Synucleinopathies, tauopathies, transactivation response DNA binding protein 43 (TDP-43) proteinopathies, and amyloidosis are the most prevalent neurodegenerative illnesses (Wal et al., 2023).

The key histologic characteristics required for a precise neuropathologic identification in these conditions are aberrant conformations of proteins and their cellular and neuroanatomical distributions. Progressive methods to support clinical processes include comprehending similarities from the medical records of the patient and teaching a computer to forecast a determination based on these tendencies. Neurodegenerative conditions are difficult to comprehend and considerably harder to diagnose. With an unidentified root and no recognized therapy neurodegenerative disorders are troublesome and may turn into a source of stress. Early detection of the physiological and biochemical alterations in the nervous system is necessary for the development of appropriate therapeutic interventions, maintenance protocols, and the appropriate treatment of NDD. Brain imaging, biochemical analysis, clinical assessment, and genetic evaluation are the four main methods for early detection of NDD. To identify any changes in the genome, chromosomal, or protein function that may be connected to a specific NDD, genetic analysis entails sequencing genomes. Clinical evaluation determines a disease or injury from the symptoms and signs reported by the individual, medical records, and physical examination in a medical setting (Paulsen et al., 2013). At present, the goal of therapy is to lessen discomfort. Owing to recent developments in artificial intelligence (AI), computational methods aimed at evaluation and monitoring—such as tracking drug side effects, quantifying the progression of illness, improving differential diagnosis, identifying the illness, and detecting disease onset—can be of great assistance. With the use of machine learning techniques, these jobs may either be automated or at the very least enhanced (Tăuţan et al., 2021). Machine learning (ML), one of the latest artificial intelligence ideas, can speed up the development of pharmaceuticals by identifying novel and significant details from the massive amounts of complicated data produced throughout the process of discovering drugs. AI/ML-based techniques have been extensively used in several therapeutic domains in recent years, achieving state-of-the-art efficacy while tackling a variety of drug development issues. Unconventional AI systems possess the capacity to identify acute complicated neurodegenerative emergencies within the optimal time limit. Chronic illnesses can also be identified early, and treatment can be tailored to the patient's needs at much lower prices and with better quality (Moffat et al., 2017).

Massive amounts of data, which can be found in digital databases and assets, have made it possible for methods of AI like machine learning to arise and be very efficient in processing, analyzing, and predicting unknown data. Utilizing these cutting-edge methods has completely changed the paradigm for drug development, leading to an unparalleled surge in the pace of central nervous system (CNS) pharmaceutical discovery initiatives (Vatansever et al., 2021). Furthermore, novel deep learning architectures that have been presented in several recent studies have improved our knowledge of how AI may address large-scale, complicated issues resulting from CNS illnesses. As a result, the current analysis offers thorough and current information on AI/ML-driven work in the field of NDD therapeutics.

ARTIFICIAL INTELLIGENCE/MACHINE LEARNING (AI/ML)

The process of finding and developing new drugs is a drawn-out, multidisciplinary undertaking that is fraught with uncertainty about their potential for success. Information technological advances have established a foundation for the advancement of scientific fields.

Development in contemporary pharmaceutical discovery techniques has resulted in the development of AI and ML, which can analyze complex information generated by biological mechanisms at an unprecedented rate, employing a multitude of increasingly intelligent technologies (Liu et al.,

2018). This is because massive datasets are becoming more prevalent in the current big data age. According to the US Food and Drug Administration (FDA), machine learning is defined as "an artificial intelligence technique that can be used to design and train software algorithms to learn from and act on data," while artificial intelligence is defined as "the science and engineering of making intelligent machines." However, not all AI methods are ML methods. The following major categories apply to AI-related processing methods: unsupervised, semi-supervised, supervised, active, transfer, and multifunction learning. Several learning frameworks employ various algorithms to carry out certain tasks, including grouping or categorization (Liu et al., 2018).

Three distinct phases of the initial drug discovery process—target identification, lead generation and optimization, and preclinical development—have seen the application of AI/ML, as shown in Figure 7.1. AI-based methods have been applied in target development to combine multiple sets of data and find trends to comprehend the molecular pathways behind illnesses and treatment actions. AI/ML algorithms help with the automation and optimization of de novo drug design procedures and enhance scoring algorithms and quantitative structure–activity relationship (QSAR) models utilized in virtual screening pathways for generating leads and optimization. In preclinical studies, absorption, distribution, metabolism, and excretion–toxicity (ADME-T) profiling is progressively optimized by using AI/ML techniques to handle massive amounts of chemical data quickly and provide models for the prediction of biophysical features. Investigators now have a plethora of opportunities to quickly discover several promising medications thanks to the development of AI/ML techniques. The creation of huge data from high throughput screening and computerized evaluation of datasets in both lead and target identification has boosted the accuracy of AI/ML-based systems to a large extent. The widespread availability of contemporary graphics processing units (GPUs) and tensor processing units (TPUs), which can perform trillions of procedures per second with simultaneous processing, has also significantly increased their data processing capacity. The creation of medicinal compounds that will alter a target's activity to modify the progression of a medical condition is a popular strategy in drug development (Lionta et al., 2014). Finding a new target where regulation can produce a therapeutic effect with a manageable risk margin is frequently the first step in the process of developing drugs. After this comes confirming the chosen target's function in the illness in in vivo models and, eventually, clinical trials. Thus, prompt recognition of viable therapeutic targets is critical to the eventual achievement of a drug research endeavor (Moffat et al., 2017). Investigators frequently use high-throughput screening (HTS), a process in vitro that consequently tests huge libraries of chemicals against a particular target to find novel compounds

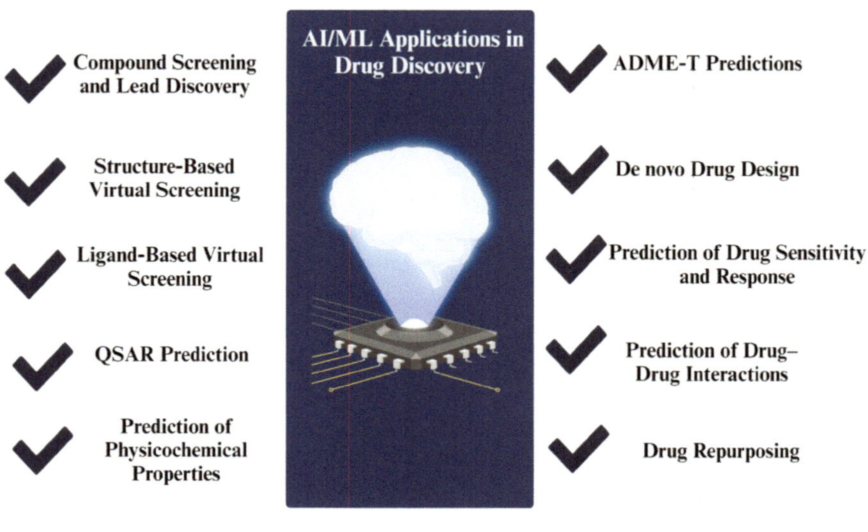

FIGURE 7.1 Overview of important drug discovery steps that are assisted by AI/ML.

that could have potential interactions. Nevertheless, the small hit rate and elevated cost of HTS have accelerated the creation of virtual screening (VS) options, which allow for quicker and cheaper evaluation of bigger compound libraries. VS utilizes a variety of approaches to determine which substances are most probable to attach to a protein of concern; the two main types of VS are ligand-based VS (LBVS) and structure-based VS (SBVS), which both use target protein structures as input. The first type uses knowledge of known inhibitors (Lionta et al., 2014). Because related compounds typically display similar characteristics, LBVS is essentially "analogous" to some level. It also aids in the development of more accurate pharmacophore predictions. LBVS and SBVS are frequently used in tandem: LBVS may be utilized to enhance leads from SBVS, and information from greater hits can be utilized to increase SBVS model accuracy. For both SBVS or LBVS, AI/ML-based techniques have been used to improve performance in VS processes (Zilian and Sotriffer, 2013).

AI's ability to make evidence-based judgments free from cognitive prejudices and psychological effects is only one of its many benefits. This sets AI decision-making procedures apart from human decision-making techniques, which inherently involve intuition and emotion when determining a course of action. Rather, the AI method's ability to operate intuitively comes from analyzing vast amounts of information as well as its user's historical behavior. Healthcare organizations must become data-driven, seeing datasets as a strategic resource, and putting in place procedures and structures to guide decision-making and provide meaningful outcomes if they are to apply AI to the medical industry. Most current methods for incorporating AI platforms in modeling diseases follow the conventional disease-focused methodology. Alternatively, using deep learning procedures to provide a population-based assessment of disease states, AI may hold the key to solving the scientific issue of identifying preclinical neurological illnesses. While it poses a few hurdles to typical ML, automatically detecting preclinical illnesses from a plethora of sensors could be consistent with the intended application of AI in healthcare technology, which is personalized surveillance and therapy (Summers et al., 2017).

In order to improve medical results, data collection for ML and deep learning skills has the opportunity to greatly enhance repeatability in surgical operations, care route design, and diagnostics. Using mathematical models programmed to identify abnormalities in such data, the use of ML concepts can enhance the interpretation of the raw data given by these medical-grade gadgets to clinically meaningful medical data. The interpretation of aberrations by ML decision-support tools could then be comparable to the work of a highly skilled professional physician, sparing a significant amount of valuable time that doctors might otherwise have to spend analyzing unstructured information (Das and Dhillon, 2023).

AI/ML ROLE IN CNS DRUG DISCOVERY

A class of neurological diseases known as CNS illnesses has a major influence on society and the economy. The intricacy of the brain's makeup and functioning, our inadequate comprehension of the pathophysiology behind the complicated nature of CNS disorders, and the existence of the blood–brain barrier (BBB) make developing novel medications for these conditions more difficult than for other illnesses. Techniques that utilize a subset of ML, an aspect of AI, tackle a variety of issues in CNS medication studies and discoveries (Figure 7.2). Advances in omics and HTS methods have made a vast amount of chemical activity information available over time. Several teams of researchers in CNS pharmaceutical discovery have focused on analyzing such huge amounts of information for speedier medication discovery (Ansari et al., 2022a).

Effectively speeding up the medication development procedure for brain diseases is possible by utilizing computer models, such as artificial intelligence and machine learning, to foresee and project BBB transit. AI/ML techniques have been applied in ways never seen before in the past several years to tackle a wide range of issues, from drug development to medical image analysis (Cherian et al., 2022). By reducing the number of laborious and time-consuming BBB permeability experiments, a variety of AI/ML-based algorithms have been proposed to help with the rapid

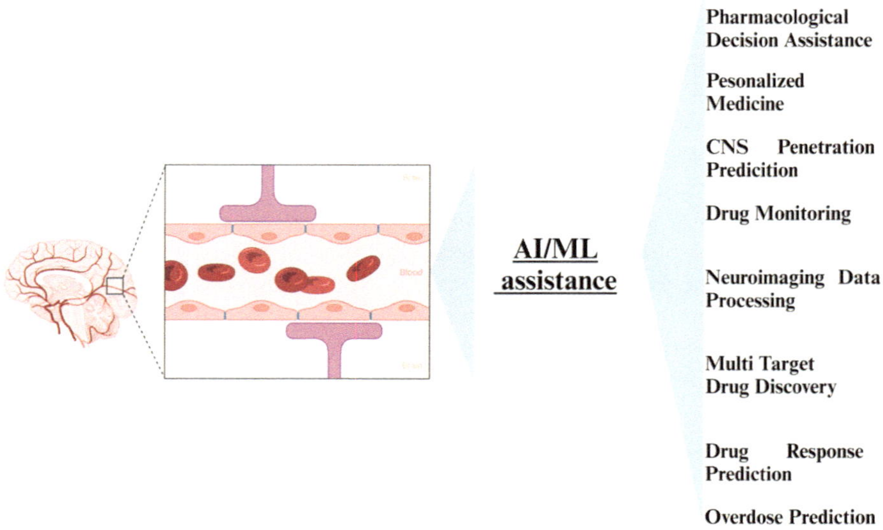

FIGURE 7.2 AI/ML-assisted challenges in CNS drug discovery.

identification and repurposing of CNS drugs. Numerous methods have been documented for determining and producing the essential molecular characteristics that contribute to BBB permeability (Alsenan et al., 2020).

AI/ML Targeting BBB

Even though our knowledge of CNS disorders has advanced significantly, there are still many obstacles to overcome before new treatments for CNS diseases may be developed. Not only is it challenging to identify CNS targets, but creating novel compounds that can cross the BBB is another significant challenge. The BBB shields the brain against circulating infections and changes in blood composition. It is composed of microglial cells, pericytes, astrocytic end feet, structural proteins, and capillary endothelial cells, which are bordered by the basal lamina. For peripheral medications, little penetration is preferred to reduce brain adverse reactions; nevertheless, for CNS-active agents, substantial penetration is required. Early attention to the BBB permeability of medication candidates is necessary to increase the efficacy rates of CNS drug development (Zeng et al., 2019).

In recent times, there have been proposals to reduce the quantity of tedious, expensive, and difficult BBB permeability studies required for CNS drug development by utilizing AI-based predictive models. Although ML and deep learning models might not offer a precise explanation for whether some medications pass or fail to penetrate the BBB, their practical uses make them significant and can be used as a first screening step in high-throughput chemical screening (Zeng et al., 2019).

Predictions have been made using ML techniques such as support vector machine (SVM), random forest (RF), k-nearest neighbors (kNN), multidimensional linear regression, linear discriminant analysis (LDA), and the naïve Bayesian classifier (NBC) (Ansari et al., 2022b). These techniques have been created to deal with various physical and chemical properties, such as molecular mass, topological polar surface area, hydrogen bond donors and acceptors, acidic and basic atom counts, hydrophilic properties (ClogP), lipophilicity (ClogD), water attainable volume, adaptability (rotatable bonds), van der Waals value, and ionization capacity. Most of the techniques have predictive power only for passive diffusional absorption and primarily depend on a small number of chemical descriptors. But a lot of chemicals, including insulin and glucose, cross the BBB through intricate processes involving interactions between the drug-transporter and drug-receptor (Ansari et al., 2022b). Furthermore, membrane-mediated transporters, including the ATP-binding channel and efflux transporter P-glycoprotein (P-gp), which facilitate drug efflux across the BBB, could pose a constraint on reaching therapeutic medication quantities in the CNS (Hindle et al., 2017).

Those transporters of efflux restrict the passage of neurotoxins into the nervous system, yet additionally restrict the entrance of numerous medications and could be involved in CNS chemoresistance. As a result, prediction techniques must consider numerous processes linked to medications that cross the BBB and persist in the brain, as well as get beyond the constraints of physicochemical properties (Hindle et al., 2017).

Numerous research studies revealed that SVM-based approaches were more widely used and more successful in determining BBB permeability. SVM may associate with greater dimensions to operate well with both nonlinear and linear information. For binary classification issues like biological barrier penetration forecasting, SVM has therefore been widely applied. By building a binary network with a decision node for whether classification or regression, the decision tree (DT) method has the flaw of overfitting, which is avoided by the method of ensemble learning known as RF (Wang et al., 2018). But unlike DTs, RF creates many DTs by randomly picking attributes, then averages the outcomes. A further key difference between RF and DTs is that RF prevents overfitting of modeling by randomly picking a characteristic subset and creating smaller DTs utilizing those subsets. The main drawback of RF is its degree of complexity, which makes creating DTs laborious (Berhane et al., 2018).

Another well-liked technique for predicting BBB permeability is kNN. Unlabeled datasets are categorized by allocating them to the most comparable neighboring class. Using methods like the Manhattan distance and Euclidean distance, kNN determines the distances among the data sources. The effectiveness of the created prediction model is affected by the choice of k, which is the ideal number of neighbors to be considered. Since there are very few legitimate datasets accessible to the BBB, LDA is a good fit for predicting permeability in the BBB and performs excellently on small datasets (Zhang, 2014). Among the more reliable, effective, and straightforward Bayesian classifiers is NBC. Its foundation is the ML Bayes hypothesis. NBC assumes that there exists no connection or link between the characteristics and that each of the factors is almost or totally independent. Despite giving conditionally independent class labels to every characteristic, the tradeoff curve yields intriguing result points. Regulation over accessibility to and use of medical information are two major issues with the uses of AI models. Even if most current models were tested using data that is easily accessible to the public, modifications of the same that involve secured patient information must be handled extremely carefully to preserve privacy and confidentiality. To handle the various security needs, it would likely be possible to employ several usage control approaches for that reason (Zhang, 2016).

AI/ML COMPONENTS AS NEUROLOGICAL DISORDERS THERAPEUTIC INTERVENTIONS

Academic and industrial investigators have generally taken a keen interest in AI/ML approaches in drug development. By June 2020, there were 230 start-ups in this sector, reflecting significant growth in this field (Schuhmacher et al., 2021). In addition, several pharmaceutical giants have made investments in their own AI-based research programs and have partnered with educational institutions and AI start-ups (Chan et al., 2019).

The AI/ML-powered subgroups and their function in the management and treatment of NDD are the main topics of the following section.

Deep Learning

Deep learning (DL) is a subfield of ML. This data abstraction technique abstracts data by applying various nonlinear modifications or processing layers of complicated structures. Artificial neural networks, also referred to as deep learning neural networks, mimic the functioning of the human brain by combining weights, biases, and data inputs. Collectively, these components enable efficient item recognition, classification, and characterization in the data. Predicting biomolecular targets, off-targets, and potentially dangerous effects of naturally occurring chemicals in food, household goods, and medications has been achieved through the application of DL. DL approaches may be divided into two categories: supervised and unsupervised (Lipinski et al., 2019).

Supervised deep learning (SDL) differs from learning about concepts in humans as well as animals because the former uses a computer circuit as the student. Deep learning frameworks that require supervision are trained on well-labeled data. They instruct the algorithm on how to apply in new scenarios and generalize from the training set. Artificial neural networks (ANNs), convolutional neural networks (CNNs), and recurrent neural networks (RNNs) are a few types of SDL techniques (Vemula et al., 2023).

An ANN is an analogous computer network to the CNS of humans. After being trained on an array of data, the neural network gains knowledge by contrasting its results to a given quantity. An effective neural network aims to minimize error by adjusting the importance and bias of its links. When assessing visual vision in deep learning, convolutional neural networks (CNNs/ConvNets) are deep neural networks. Convolution is the method used by it. With a pair of functions, convolution is a mathematical procedure that yields a third value that describes the way one function's appearance alters another's. CNNs have many layers of artificial neurons. Artificial neurons are mathematical operations that, like biological equivalents, determine the stimulation value by summing the weights of several inputs (Atak et al., 2021).

A supervised DL technique called recurrent neural network (RNN) uses the output from previous stages as input for the next phase. It consists of a hidden state with the data needed to commit an operation to memory. By transferring all outputs into the subsequent hidden layer, RNNs may convert independent inputs to dependent inputs, making it easier to increase parameters and retain each previous output (Lan & Corces, 2022). In theoretical terms, RNNs possess a pair of benefits over CNNs: they possess both short- and long-term memory, which should allow them to retain pertinent characteristics and employ these to relevant series elements anywhere across a chromosome; additionally, since the same core model segment is employed on each input component, RNNs are also resistant to translating switches in the listing of a source sequence of DNA. In action, nevertheless, multifaceted CNN frameworks have proved capable of competing or even outperforming RNN models in general (Lan & Corces, 2022).

DL ROLE IN GENOMICS

To determine if a model's prediction is consistent with actual biology, DL uses for genomics studies are utilized to gain an understanding of the molecular processes underlying the model. In silico mutagenesis (ISM) is a highly popular and most suitable method of analysis for sequence-based DL models. When bases of a DNA sequence source are changed and alterations are seen in the model output, like the degree of the transcription factor interaction estimated from the accessibility of chromatin data, this process is known as "in silico mutagenesis." By applying ISM, scientists can methodically determine which parts of the input DNA sequence cause the most modifications to the model's prediction, thus rendering them of the highest "importance" to the genomic relationships the model has learned (Lan & Corces, 2022). A streamlined version of ISM is frequently used for basic variant prioritization. Variant SNPs are disrupted, and the alteration in prediction is evaluated and scored. Higher scores are considered to have a bigger probable functional impact. ISM is still used in modern models as an essential interpretation strategy since it is the simplest means of evaluating the model (Zhou & Troyanskaya, 2015). It works best when combined with other, more specialized interpretability approaches. Nevertheless, comprehensive ISM is an extremely computationally costly procedure; to evaluate each SNP in an N-bp genome, a model must generate 3 N predictions, a number that gets unreasonably high when evaluating lengthy sequence inputs or several genomic loci (Avsec et al., 2019).

Kernel analysis is another straightforward interpretation technique in which investigators analyze model weights to extract important information. Scientists can examine the trends recorded in the convolutional filters that traverse the preceding layer by using CNN kernel analysis. CNN kernel analysis is very useful in bioinformatics for finding sequence motifs that symbolize biological events like TF binding sequences. When predicting the present area in transformers, the model focuses much of its energies on other parts of the order of inputs, which may be identified by kernel analysis (Baker et al., 2023). Yet, as sequencing motifs or attention connections may be distributed over several convolutional or attention adjustments on numerous layers, as models get more complicated and incorporate more layers, kernel analysis becomes less and less relevant (Vemula et al., 2023).

Neurodegeneration research is gradually embracing sequence-based ML genetics methods useful for variation effect assessment. In order to differentiate between transposase-accessible and transposase-inaccessible chromatin, gkm-SVM ML classifiers were developed for each of the 24 single-nucleus ATAC-seq groupings in a new investigation that aimed at assessing the epigenomic landscape of neural cell types for Alzheimer's and Parkinson's studies. Using a tiered multiomic analytic procedure, the gkm-SVM classifiers predicted the functional impacts of potential AD and PD SNPs in genome-wide association study (GWAS) locations and identified novel functional noncoding variations and related genes for additional study (Corces et al., 2020). DL frameworks have also been incorporated into the grading paradigm of several fine-mapping studies to score potential GWAS variations according to their propensity to cause neurological illness. DeepSEA and SpliceAI prediction scores of variable TF binding and splicing were included as one of numerous criteria for variant prioritization in a meta-analysis of noncoding variants from recent AD GWASs (Schwessinger et al., 2020). These preliminary applications highlight the growing significance of DL techniques for variant prioritization in the study of NDD. With the flood of additional risk loci following current GWASs, DL could support the variant fine-mapping operation on a wide scale by operating orthogonally to statistical approaches and standard analytical techniques for noncoding variant prioritization (Novikova et al., 2021b).

Even while the majority of genomics data analysis in the field of neurodegenerative studies still uses non-ML approaches, the ultimate inclusion of DL techniques will be critical for researching the impact of noncoding variations without being constrained by the duration or resource constraints of biological trials (Novikova et al., 2021a). Giant consortia dedicated to deciphering the human genome's transcriptional as well as gene regulatory circuits have produced enormous volumes of sequencing data for neuronal types connected to NDD. Some examples of such consortium efforts are the Allen Brain Atlas, the Human Cell Atlas, the Roadmap Epigenomics Consortium, and the ENCODE Project (Moore et al., 2020; Sunkin et al., 2013). Even while the majority of genomics data analysis in the field of neurodegenerative studies still uses non-ML approaches, the ultimate inclusion of DL techniques will be critical for researching the impact of noncoding variations without being constrained by the duration or resource constraints of biological trials. Giant consortia dedicated to deciphering the human genome's transcriptional as well as gene regulatory circuits have produced enormous volumes of sequencing data for neuronal types connected to NDD (Bellenguez et al., 2022).

Simultaneously, additional genetic loci linked to AD are being found by current GWASs, demonstrating the ongoing need for precise techniques for variant impact prediction and potential variant ranking. With data and use cases unique to AD, all the insights gained by the disease-independent DL sequencing techniques created in the past few years may be utilized (Vemula et al., 2023). Scientists may determine candidate causal variants from a wider variety of GWAS variants and recommend prospective biological mechanisms for NDD by training these models on pertinent cell type-specific sequencing data, implementing suitable downstream analysis algorithms, and conducting functional variant impact prediction (Turner et al., 2022).

PREDICTIVE MODELING

The relatively young discipline of computational psychiatry combines biological information, clinical information, and computer models to predict brain illnesses. The diagnosis of mental diseases, variability from subjective variables in measurement, absence of big systematic databases, and learning intricate technological features underlying various measurements all contribute to the enterprise's complexity (Polosecki et al., 2017). With the goal of expanding the transformation of biological data and large-scale clinical research into new therapies, computational psychiatry applies ML and computational methods to the investigation of neurological conditions. This is achieved by offering trustworthy biomarkers, recording the diversity in the clinical population, and proposing mechanisms of illness. Due to the rarity of patient brain cells for study, the prospective uses of human induced pluripotent stem cells (hiPSCs) are particularly pertinent to neurology (Huys et al., 2016). The use of hiPSCs from patients suffering from neuropsychiatric disorders or NDD has made it easier to conduct biological and multiomics research along with extensive chemical library screening. Finding prognostic and predictive biomarkers at pertinent substitute outcomes requires modeling illnesses in vitro. The creation of preventative therapies requires this information. To do this, it is imperative to develop the most interpretable and predictive in vitro cellular designs, which we think the precision healthcare paradigm may help with. Experts are having difficulty enhancing the descriptive disease modeling's scalability, repeatability, and reliability. In vitro experimental information and longitudinal clinical information will be correlated and integrated to create predictive disease modeling, an emerging field in hiPSC development that will begin with resolving these constraints (Rivetti et al., 2021).

Rather than adopting the reactive strategy that is currently the norm, precision healthcare is a relatively fresh operational philosophy that aims to maximize disease avoidance and treatment through predictive and preventative methods. Predicting each patient's unique illness trajectory and accurately intervening when a patient's diseases are avoidable or reversible are the ultimate goals of precision medicine. Predictive disease modeling using hiPSCs is thought to be best understood in relation to precision medicine regulations (Rivetti et al., 2021). It should include longitudinal research in which individuals and the cells they contain are stratified, as well as the application of computational models to combine data from various time points and produce data regarding the course of the disease.

Finding objectively quantified markers of pathological processes—henceforth referred to as biomarkers—is a major goal of computational psychiatry and is essential to redefining mental diseases (Younesi & Hofmann-Apitius, 2013). If biomarkers are adequately predictive of clinical advancement, one of the main objectives is to replace the existing standard, subjective clinical assessments with them (Ielapi et al., 2020). Thus, it is crucial to assess their biological significance to avoid making unintentional discoveries like epiphenomena that have some predictive value but no fundamental causal relationship to the illness. The neurobiology of mental disorders has proven particularly tricky when it comes to biomarkers.

However, significant progress has been made in bridging the gap between traditional psychiatry and neurobiology via ML-based modeling of brain imaging information. A few of these achievements include the recognition of pathological processes in schizophrenia. It has additionally been shown that the predictive power of functional magnetic resonance imaging (fMRI) for identifying signs of schizophrenia lies not merely in the commonly employed scale of this complex-valued message, but also in its timing (Huys et al., 2016). By taking advantage of the complex-valued nature of the fMRI signal through an array kernel depiction of the data, better localization of regions of the brain implicated in this mental illness and a more accurate classification of schizophrenia could be achieved. Attempts have been undertaken to lessen some of the widespread intersite heterogeneity found in multi-institutional structural MRI. In research, by identifying characteristics that offer reliable diagnostic ability across population specimens, aberrant alterations in patients' trends in gray matter composition that were consistent across site collection were detected (Ross et al., 2014). One of the only consistent structural anomaly results across schizophrenia studies—significant cortical shrinkage of the superior temporal gyrus—was found when this limitation was applied to feature identification (Castro et al., 2014a). Notwithstanding these specific accomplishments, crowdsourcing contests are having a greater cultural influence in the ML space. Organizers function as curators and supply participants with domain expertise to help them come up with innovative ideas (Castro et al., 2014b). Although ML holds great promise for offering biological understanding of mental illness, this research is hampered by an issue that is outside the purview of this area: the collection of data prior to the onset of sickness. This is a particularly complicated topic since most psychiatric diseases are only detectable after symptoms appear, which makes it impossible to have a thorough grasp of how mental illness develops.

Studies on beneficial medicines can be significantly advanced by creating predictive modeling frameworks that are useful for research studies in a this and by creating efficient and predictive in vitro systems for the modeling of human illnesses (Castro et al., 2014b).

These novel treatments necessitate the development of biomarkers that can forecast the onset and course of cognitive and motor decline, determine the best time to intervene therapeutically, and forecast the effectiveness of therapy across various patient populations (Ross et al., 2006). Since predictive modeling is based on inclusive elements of fundamental studies that are relevant to all disease modeling techniques, we believe it is also a legitimate strategy for other diseases, monogenic and otherwise (Rivetti et al., 2021).

SENSORS

As of now, one can use a variety of current sensors for identifying the molecular, cell-based, electrical, and behavioral alterations linked to the advancement of NDD because of the quick advancement of sensor technology (Gautam, 2022). Surprisingly, the development of nanomaterials, electrically charged polymers, and the Internet of Things (IoT) has made it possible to create extremely small sensors for neurological illnesses that function as well as conventional biosensors (Pateraki et al., 2020). To put it simply, a sensor is any device that can recognize any input signal and convert it into an apparent, measurable form. Each physiological movement or amount that might be monitored can be used as the signal being input. Biosensors have become state-of-the-art detection techniques in a variety of medical diagnostics, including those requiring immediate identification of human biological fluids (Khaksarinejad et al., 2022).

Whenever an NDD biosensor comes into contact with a pathogen that causes neurological infections or target analyte, which may be a patient's blood or cerebrospinal fluid, it generates an electrical signal. A physicochemical transmitter and a biological detection component are combined to generate these. Because of their interaction with the substances, biological molecules such as receptors, enzymes, and antibodies, can function as the recognizing or detector elements. However, transducers are typically composed of functionalized surfaces for electrodes and are used to translate biomolecular interactions into digital signals (Haleem et al., 2021).

The biosensor might be thermal, optical, mechanical, electrochemical, or based on its transducer information. Each kind is briefly described next: The functioning electrode serves as the transducer in an electrochemical biosensor. Electrodes are made of nonmetals like carbon, which is both conductive and semiconductive, metal oxides like titanium dioxide (TiO2), and metals like gold (Au) (Cesewski & Johnson, 2020). Glass, Teflon, or polyether ketone are examples of ceramic substrates or insulating polymers in which these conductive elements are inserted. An optical biosensor consists of an immobilized biorecognition component, an illumination source, an optical transmitting medium, and a signal-detecting gadget. These work in concert to quantify the interaction between a ligand and a target biomarker. Thermal biosensors measure the amount of heat energy received or emitted throughout an endogenous process. Heat is produced by the oxidation process and the production of free radicals, as oxidative stress is linked to most nondestructive diseases. By combining a thermal biosensor with an appropriate amplifier, a pH meter, and two thermistors, along with an operating system or output gadget, it is possible to identify these minuscule variations in thermal energy (Song et al., 2021). Moreover, the integration and inexpensive batch manufacture of small devices with effective heat isolation, minimal heat loss, and sample volume are made possible by microelectromechanical systems (MEMS) technologies. It is also feasible to measure many biological specimens at once with these manufactured instruments. The mechanical connections produced by adhesion to cells, motility, chemical connection, and transit inside a nerve cell serve as the foundation for mechanical biosensors. Outstanding opportunities exist for measuring physical properties (mass, speed, density, etc.) via cellular to subcellular operations in the mechanical arena using biological sensors. Integrating nanotechnology is now required to create biological detectors that are compatible with biomolecule sizes and have single-molecule sensitivities (Das et al., 2022).

Nowadays, there are implanted and wearable biosensors available in addition to the aforementioned biosensors that may be utilized by experienced scientists or medical professionals to identify modifications to biochemical components (Kumar et al., 2022). The operating concept for such wearable and implanted biosensors is nearly identical to that of conventional biosensors. They may be linked to the Internet of Things to provide immediate tracking data to anyone, including medical professionals and the patient's family. The most popular type of biosensors for monitoring the movement and bodily activity of NDD individuals is "wearable GPS with accelerometer" technology (Yin et al., 2021). Polymer composites and conjugated polymers are used as electrode materials in NDD wearable and implantable biosensors. For implanted and wearable biosensors, electrode–tissue mechanical fitting is made possible by the mechanical qualities of polymers (Zolfaghari et al., 2023)

Numerous inexpensive sensors may be used to monitor user location and movements, identify cognitive issues, and guarantee user security. As a result, the initial stage in any locomotion assessment technique is to gather movement information using sensory devices. The most popular wearable and non-wearable sensors for medical applications and have their benefits and drawbacks in relation to NDDs and abnormalities of locomotion. Still, many of those gadgets may be intrusive and noticeable, particularly the types that need constant wear from the user (Oguntala et al., 2018). Above all, there are significant privacy hazards associated with indoor location systems, given that the position of the user wants is always known, which the user can find undesirable.

Further, there are techniques that are gadget-based, requiring users to carry or wear a device or tag, while techniques that are device-free do not need the user to interact and are typically embedded in environments or artifacts (Liu et al., 2020).

Device-Based Localization Sensors

Wearable/mobile gadgets that assist in finding the user, including watches, bracelets, pendants, earbuds, and collars, are referred to as device-based location identification (Table 7.1).

Device-Free Localization Sensors

Device-free technologies, which mainly depend on the detection of motion and offer information in real time, enable the collection of the user's position without requiring them to keep any devices (Li et al., 2019). A variety of sensor techniques capable of being employed for this objective are summarized in Table 7.2.

Wearable Devices

Devices that are incorporated into clothing or created as wearable items are commonly referred to as wearables. Sensors like gyroscopes, magnetometers, and accelerometers that are integrated into

TABLE 7.1
Summary of the Features of a Few Device-Based Localization Sensors

Type	Characteristics	Reference
Magnetic-based technology	Uses a portable magnetic sensor—typically a smartphone—to gauge the magnetic fields in various interior locations. Using these measures, a map of the area is created that serves as a guide for client localization.	(de Oliveira et al., 2020)
Mechanical-based devices	An inertial measurement unit (IMU) is used to assess the user's position. The target's present position is calculated by the IMU using a known beginning point and a mix of gyroscopes, accelerometers, and sometimes magnetometers, together with previously calculated velocity and heading.	(Sanpechuda & Kovavisaruch, 2008)
Location-based radio-frequency identification (RFID)	Makes use of electromagnetic waves to identify tags attached to objects. A radio transponder, receiver, and transmitter are components of RFID technology. Whenever the reading device sends out an electromagnetic interrogator pulse, it sends information to the gadget.	(Kumar et al., 2017)
Ultra-wideband (UWB) systems	To find the gap between two things, they send data across a wide bandwidth and utilize metrics like the moment of arrival and the time variance of arrival. Excellent precision is provided by UWB systems.	(Delaunay & Guérin, 2017)
Wristband, watch, or smartphone	They included IMU methods. A smartphone, for instance, is a good example of an IMU. It is frequently used for NDD screening since it allows for remote monitoring of patients and provides more rapid access to care and providers. Additionally, it results in fewer hospital visits, improved interaction, and lower patient expenses. In addition, owing to their straightforward designs, widespread on-body location, and cell phone access, these wearable smartwatches and wristbands might be utilized in medical treatments.	(Sakpere & Adeyeye, 2015; Mazilu et al., 2016; Ganesan et al., 2019)

wearables enable constant, long-term tracking of physiological factors or motion patterns (Zhao et al., 2023). Wearable sensors are often positioned on various body regions that have no impact on normal mobility. Wireless links are used to transfer the data they gather, and built-in storage devices are used to store it. Wearable technology has developed into accessories, integrating apparel, body attachments, and body inserts due to the quick advancement of electronics, biologically compatible substances, and nanotechnology. Regarding the identification and surveillance of NDDs, a variety of sensors may be used to fully and constantly capture human activity data, including plantar pressure, posture of the lower limbs, joint flexing, and so forth (Ramgopal et al., 2014). Wearables in neuroscience provide new opportunities for objective, ongoing symptom assessment in both clinical and non-hospital contexts. The two major neurological disorders for which accelerometry-based innovation has been used most frequently are Parkinson's disease and stroke (van Uem et al., 2018). Additionally, there is considerable interest in the use of wearable technology to identify epileptic episodes (van Ejik et al., 2019; Shen et al., 2022).

Wearable sensors give the active element mechanical strength by being flexible or having a flexible platform, because of the compressive or tensile force sensors are separated into strain and pressure sensors. There are several different transformation mechanisms for pressure and strain sensors, including capacitive, triboelectric, piezoelectric, and piezoresistive (Sundar et al., 2015).

Piezoresistive sensors: The physical deformations brought on by exerted pressure or strain are converted to electrical resistance changes by piezoresistivity. Force-sensitive resistors (FSRs) are

TABLE 7.2
Summary of Sensor Techniques

Type	Characteristics	Reference
Infrared-based technologies	Used in both active and passive infrared detectors. People release thermal energy in the form of radiation, which is measured by the former device. For the second type, a beam of infrared light must be continually sent by a transmitter, and a receiver must sense whenever a moving item crosses the scan region to break the beam flow.	(Sharma et al., 2019)
Microelectromechanical systems (MEMS)	Pressure sensors assess foot striking and identify wandering tendencies and falls depending on the quantity and arrangement of activations on the ground or gait mat.	(Tsukagoshi et al., 2020)
Ultrasonic transmitters	Generate an ultrasonic pulse that is reflected by items in the scan region. Like microwave equipment, they can sense Doppler changes at low audio frequencies.	(Brena et al., 2017)
Computer vision-based sensors	The use of visual sensor systems or cameras in environmental sensing approaches to ubiquitous medical care shows promise. These methods allow for the comparison of successive images from recorded video streams, the tracking of 3D areas, as well as the detection of motion when sufficient changes occur among frames through the analysis of scenes and proximity approaches, as well as small on-node processing of images along with computer vision algorithms.	(Kearns et al., 2010)
Radio signals	One may determine a person's orientation and motion over time by looking at their own reflections. Additionally, when employing microwave messages, the motion of an object induces a period change in the radiation it emits, producing a low-frequency signal. This technique enables ultra-low-power radio signals in wall-mounted sensors and tomographic motion sensors to recognize radio waves at frequencies that pass through most walls and obstructions, allowing the user's location to be detected across a wide region.	(Johansson et al., 2018)

extremely tiny, inexpensive piezoresistive pressure sensors that are commonly employed. When collecting biosignals for the purpose of diagnosing and tracking illnesses, an FSR can be employed (Selzler et al., 2018). It may be inserted into gloves to track how frequently patients click as well as how hard they grasp. As the circular construction sensor works well for measuring foot pressure, FSRs are often radial in design. Gait evaluation can be utilized to diagnose non-disordered diabetes because the variance of vertical ground reaction force (vGRF) readings resulting from gait irregularities might reveal distinct force sequence alterations in contrast to healthy controls (Wang et al., 2022).

Piezoelectric sensors: Both negative and positive charges build up on the electrodes and create an electrostatic force whenever an element is crushed or bent, separating anionic and cationic groups from the center to the edge. This process is known as piezoelectricity. Owing to the influence of piezoelectricity on biological structures and its numerous biomedical uses, polyvinylidene fluoride (PVDF) may display a reasonably strong piezoelectric capability following polarization at elevated temperatures. PVDF is a material that has been extensively explored in this regard (Liu et al., 2022). It may be utilized for tracking a range of vital indicators, including blood pressure, respiration, and heart rate, in addition to providing a steady and long-term energy source. In order to identify motor abnormalities, a system of PVDF sensors was installed in the insoles to continuously measure plantar pressure (Homayounfar & Andrew, 2020).

Capacitive sensors: A regulated airspace between two electrodes or dielectric material positioned within each other makes up a capacitive sensor. By converting electrical power through an electrostatic field, it may be stored. It is frequently used in knee jerks, joint motions, wrist and finger flexing, neck motion, and other situations (Ullah et al., 2022) Furthermore, wearable and flexible capacitive electrophysiological sensors can be used for long-term wellness tracking. Capacitive sensors also offer promise for use in NDD surveillance and diagnostic applications. For tracking eye motion, a capacitive ocular tracker was created. Because PD patients have aberrant eye movements, this gadget can be used to diagnose and track PD patients (Mathew et al., 2021).

Triboelectric sensors: Triboelectric sensors use the so-called triboelectric effect, which involves electrification by contact accompanied by electrostatic induction, to convert mechanical distortion into electricity. TENG is used for healthcare surveillance through heart and breathing tracking, joint motion detecting, sweat and gait stage identification, hearing aid, and CO_2 detection. As recommended by the Unified PD Rating Scale (UPDRS), triboelectric transducers—alternative stretchables that are responsive to the movements of muscles and tendons—were put on the forearm to monitor PD by looking for stiffness and bradykinesia in addition to tremor (Fillekes et al., 2019a).

Since GPS and related technologies are widely integrated into consumer gadgets like cellphones, navigation apps, and wearables (like GPS watches), it has become easier to measure a person's location and movement objectively. These gadgets can pinpoint a person's position accurately and precisely within a few meters. Information from GPS gadgets can be employed as input data to produce mobility signals that define an individual's daily mobility habits. Although wearable GPS has become more popular in aging studies, most research has only provided a small number of mobility indices (Fillekes et al., 2019b). According to research, GPS-based technology might be able to distinguish between those with mild-to-moderate AD and those who do not. According to validation studies, the most reliable techniques of measuring living space mobility through conversation and diary are still quite accurate when using GPS-based devices that track position and movement. When worn, GPS-based devices have the benefit of passively monitoring a patient's movement with little effort, thus minimizing recall bias. Because cognitive function and memory may be compromised in the setting of NDD, wearable solutions might be quite helpful due to their simplicity (Breasail et al., 2021). The GPS and accelerometer technologies that can be placed at various anatomical locations and utilized by individuals with NDD are summarized in Table 7.3.

VIRTUAL REALITY AND AUGMENTED REALITY

Virtual reality (VR) is a technological advance that employs wearable screens to fully immerse an individual in a synthetic three-dimensional (3D) setting; augmented reality (AR), which is closely associated with VR, incorporates components of VR into the real-world setting by superimposing them onto live footage on an electronic device's display. Although AR is an emerging idea, VR has been developing for the last 50 years to provide users with an immersive experience that engages their visual and auditory senses (Cipresso et al., 2018). Both VR and AR technologies have recently improved in portability, realism, and real-time navigation because of technological advancements, introducing a sensory and occasionally olfactory aspect to the variety of sensations (Piromchai et al., 2015).

In addition, head-mounted gadgets are becoming easier to obtain. Since they first hit stores a few years ago, personalized VR and AR products have continued to advance. Integrated body tracking systems have been used recently, together with portable gadgets such as joysticks and keyboards, to allow the user to engage with the virtual world (Piromchai et al., 2015).

TABLE 7.3
GPS and Accelerometer Gadgets Utilized by NDD Patients

Neurological Disorder	Device	Device Type	Anatomical Site	Outcome Measures	Reference
PD	WIMU-GPS	Wearable GPS	Upper back	Frequency and length of trips outdoors	(van et al., 2018)
PD	DynaPort Minimod	Accelerometer based	Lower back	Duration of active and inactive periods	(Pradhan & Kelly, 2019)
PD	Fitbit Charge HR	Accelerometer based	Wrist	Step total, the amount of movement bouts, and the duration of activity bouts	(Mantri et al., 2019)
PD	Actigraph GT3X	Accelerometer based	Waist	Step count	(Sulzer et al., 2021)
PD	DynaPort Minimod	Accelerometer based	Lower back	Duration of motor behavior, quantity of behavioral episodes, and timing of movement intervals, step measure	(Moyle et al., 2017)
Dementia	SenseWear activity armbands	Accelerometer based	Arm	Steps taken, amount of time spent moving about, lying down, being awake, and sleeping	(Varma & Watts, 2017)
AD	Actigraph GT3X+	Accelerometer based	Hip	Type of exercise and intensity	(Kim et al., 2019)
PD	ActiGraph GT3X+	Accelerometer based	Wrist and waist	Step number, activity measure, and kind of activity time involved	(Klenk et al., 2016)
NDD	ActivPAL3	Accelerometer based	Thigh	Walk time, step measure, time spent walking, count of everyday sit-to-stand transitions	(van et al., 2019)
ALS	ActiGraph GT9X Link	Accelerometer based	Hip	Activity measure	(Azuma, 1997)

Source: Zhu et al. (2020).

There might be several advantages to adopting VR or AR over traditional treatment: It could make it possible to repeatedly perform simple activities in a clinical setting in an immersive environment without requiring continual supervision from medical professionals, which might significantly lower the expense of training facilities and qualified medical personnel. Moreover, head-mounted devices might be utilized securely in patients' homes, reducing the necessity for hospital stays, particularly for immobile individuals. By making VR and AR experiences visually appealing and easy to use, attrition rates of patients may be reduced and a more enjoyable atmosphere can be created. Patient safety might be significantly improved by using VR in surgical training since it can significantly lower the likelihood of surgical mistakes (Clare & Woods, 2004).

Specifically, computerized cognitive trainings (CCTs) include VR and AR. Researchers create computer-generated settings for VR participants to engage with, with the goal of controlling environmental factors and simulating multimodal sensations (Choi & Twamley, 2013). The kind of operation carried out and the technologies used to create an immersive atmosphere can distinguish VR standards. On 2D displays, non-immersive virtual reality (niVR) uses computer protocols to simulate a real-world experience. On the other hand, fully immersive virtual reality, or fiVR, is a computer simulation that substitutes a 3D virtual setting for the participants' external sensory world, allowing them to move around and engage with it much like they would in the real world. Head-mounted displays (HMDs) or specific spaces with the fake world displayed on the walls and recording motion gadgets, as in the cave automatic virtual environment (CAVE) technology (Paszkiel, 2020), can be used to create this immersion. VR communications tend to be facilitated through the use of gadgets that record motor actions (e.g., buttons or sensors that record body part movement (e.g., motion-capture cameras, accelerometers, and gaze tracking); however, the use of brain–computer interfaces as regulating gadgets in the near future may result in a greater degree of VR immersion.

In AR, extra digital items are superimposed on top of the actual world as seen through HMDs, specialty glasses, or other gadgets like smartphones. This encourages engagement with simulated things or integrates simulated elements on real items without having to create a whole virtual environment simulation (Teo et al., 2016).

Cognitive training (CT) therapies seek to rebuild individual global functioning by reducing the degeneration of cognitive processes, such as memory, attention, and problem-solving, which are primarily affected in AD (Paszkiel, 2020). Cognitive training interventions are beneficial for AD. Several investigations have demonstrated their effectiveness in postponing deficits and enhancing global cognitive performance. Other research did not find a benefit. However, given that subpar performance may be linked to decreased therapeutic involvement, which heightens dissatisfaction and sadness, cognitive training may be problematic (Gates et al., 2019).

Over the past ten years, increased accessibility to electronic gadgets and the demand for treatment from huge populations have led to the growing popularity of CCT as a scientific tool, particularly for in-home treatment (also known as telerehabilitation) for individuals who are bedridden or in an outbreak of a pandemic that necessitates social isolation. It is debatable whether to use CCT for mental illness therapy and cognitive rehabilitation (Cavallo and Angilletta, 2019). The wide range of methodologies and outcome metrics used in studies and the reality that it is still unknown whether training-related effects last a long time or only temporarily improve cognition (Barban et al., 2016) are further significant obstacles to proving the effectiveness of CCT. In telerehabilitation procedures, CT and CCT can be employed alone or in conjunction with other forms of therapy to maximize their efficacy. The advantages of this combination strategy, however, vary depending on the medicines used. In mild AD, mild cognitive impairment (MCI), and healthy participants, CCT for cognition and executive functioning in conjunction with recollection therapy (RT) had favorable results on verbal episodic recall. It also produced favorable results on MMSE assessments for individuals with mild AD. Nevertheless, three months after the program, there was no continued improvement in the daily abilities of individuals with mild AD, and mixtures with RT produced encouraging but weak data (Tárraga et al., 2006).

Integrating CCT and CT with neurostimulation, such as transcranial direct current stimulation (tDCS), pharmaceutical treatment, such as cholinesterase inhibitors (Peretz et al., 2011), and physical activity is a potential treatment method, according to other research. Surprisingly, when contrasted with groups completing nonspecific CCT standards, placebo-control categories, or classic computer games, CCT procedures given to patients and intended to improve targeted cognitive abilities customized to subjective demands have been associated with greater cognitive gains. As cognitive training may enhance the efficacy of neural pathways underpinning the learned cognitive function, CCT-related benefits may reflect brain neuroplasticity mechanisms (Meola et al., 2017).

AR has proven to be beneficial in physician education, particularly in the context of delicate and intricate surgical procedures such as trans-sphenoid or neurological surgery. To aid medical professionals in their practice of performing neurosurgery, an AR system has been developed. Surgeons are taught to operate on the enhanced robotics with the aid of this technology, and as a consequence, there has been a decrease in the misconduct ratio (Maggio et al., 2019). AR devices assist patients in carrying out their daily activities to enhance their quality of life. AR applications help patients with conditions like dementia, whose memory deteriorates daily, to carry out tasks such as preparing tea or identifying objects. For instance, AR systems are being used to alert people with intellectual disabilities to avoid hot objects (Guinet et al., 2021).

AR devices are mostly utilized in the treatment of motor disorders, which affect patients' ability to balance, move their limbs in unison, or coordinate their hand–eye motions. Another tendency that has been noticed is the employment of AR applications to preserve the brain's ability to process information and remember, particularly in certain areas of the cerebrum. When engaging in AR-enabled interactive games or tasks, a person with a disorder or degenerative condition loses functionality and becomes sluggish. AR apps that assist with mobility in conjunction with tangible equipment, such as a treadmill, have been utilized for the management of motor illnesses (Chan et al., 2019). AR-mounted headsets have been used in the rehabilitation of Level 3 GMFCS (Gross Motor Function Classification System) individuals for movement problems (Gorman & Gustafsson, 2022). Patients who have had strokes have benefited from AR treatment through the use of gamification to replace traditional medical procedures like physiotherapy with technologies that use color and form to enhance critical thinking (Matarazzo et al., 2019).

THERAPEUTIC APPLICATIONS OF AI/ML IN SPECIFIC NEUROLOGICAL DISORDERS

Parkinson's Disease

AI has already been utilized in PD research to uncover diagnostic biomarkers in blood and cerebrospinal fluid and to remotely track responses to therapy employing wearable technology. In order to identify potential inhibitors of the LRRK2 protein, a major genetic contributor to both familiar and sporadic PD, multiple machine learning approaches were employed to infer QSAR models. Additionally, the negative effects of drugs used to treat PD have been mitigated using AI-based technology, for example, inhibitors of the LRRK2 protein, which is a major hereditary risk factor for both PD and non-PD. Furthermore, AI-driven technologies have contributed to the reduction of medication adverse effects in PD therapy, such as levodopa-induced dyskinesia (LID) (Humpel, 2011).

In vitro and in vivo PD models have been developed for drug screening using AI/ML approaches, in addition to the recognition of new and repurposing possibilities. As an in vitro toxicity test, Monzel et al. (2020) developed a human midbrain organoid prototype for PD and constructed an RF classifier to forecast a compound's neurotoxic impact on organoids according to cellular characteristics. Hughes et al. (2020) created a zebrafish model of PD and an AI/ML-based technique to categorize movement problems in this model using high-resolution video recordings to construct an effective drug-testing approach. Positive findings from every study previously mentioned show how AI and ML technologies may be used to find effective, multitarget medications for PD.

Alzheimer's Disease

AI/ML-based approaches have gained popularity in AD studies during recent years, primarily utilized for prediction and diagnosis of AD when working with pictures and electronic health information. Conversely, there has not been much use of AI/ML methods in AD medication development. A few studies have, nonetheless, demonstrated the possible advantages of using AI and ML in the search for AD medications. The first step of drug development in AD, target identification and analysis, has benefited from the application of ML techniques. ML algorithms are trained and tested using neuroimaging data from the Alzheimer's Disease Neuroimaging Initiative (ADNI) database (Jack et al., 2008).

Two main parts make up the framework of DL methods: an output level for SoftMax regression that helps with AD diagnosis. In order to get deep illustrations of the initial input vectors, which can be expressed by neurons of every input layer, the first component, a sparse autoencoder, is composed of a neural network with several hidden layers. Numerous machine learning techniques have been proposed for the management of AD utilizing new biomarkers including total tau, phospho-tau-181, and β-amyloid (1–42) (Humpel, 2011). Since a biomarker's specificity and sensitivity determine how useful it is for diagnosing an illness, biomarkers serve as indicators. Accordingly, ML techniques use them in conjunction with the danger assessment of MCI to detect AD patients (Salvatore et al., 2015).

Epilepsy

Although seizure prediction is a highly difficult diagnostic assignment, machine learning approaches have made it possible to accomplish this quickly. Contemporary methods for predicting seizures allow for the acquisition of features from an individual's intracranial status electroencephalogram (EEG) and employ statistical analysis methods to draw conclusions. Using several electrodes, the intracranial EEG measures the local voltage current across brain cells. This potential is produced by the brain's many neuronal activities. Such voltage signals provide a wealth of information on the brain's neurological and behavioral disorders. The symptoms of epilepsy are evidently indicated by a significant electrical discharge (Geva & Kerem, 2017).

Additionally, as bivariate characteristics quantify the connections between nearby or distant EEG pathways, studies support their application in predicting seizures. An effective method for identifying epileptic episodes and categorizing EEG information uses a hybrid SVM that combines particle swarm optimization (PSO) and genetic algorithms (GAs) (Subasi et al., 2019). Appropriate kernel parameters in the SVM design are necessary for improved data categorization. A computer is trained using the categorized data to understand the nonlinear association between the characteristics and their corresponding labels. These skilled computers assist in automating the method and examining the EEG data more quickly and accurately (Subasi, 2007). As a result, PSO-SVM- and GA-SVM-based systems work well as real-time specialist diagnosis tools in clinical settings.

Amyotrophic Lateral Sclerosis

Brain–computer interface (BCI) software is a useful means of interaction available to individuals with ALS during the diagnosing process (Riccio et al., 2013; Ejik et al., 2019). BCI uses sophisticated machine learning algorithms to convert brain impulses into instructions with the objective of restoring mobility, just like the human brain does. A collection of electrodes is inserted into the brain's gray matter as part of the BCI technology, and these electrodes measure the voltage differential among neurons to interpret brain signals. By decoding signals—such as magnetoencephalography, functional magnetic resonance imaging, electroencephalogram, sensorimotor patterns, and slow cortical potentials—invasive and noninvasive BCI systems offer a potential way to interact for people with disabilities. For communication, an alternate BCI model that utilizes motor imagery that modifies oscillations generated in the motor cortex aids the individual with ALS. In addition to this, there is ALS.AI, a deep learning framework intended to create and utilize algorithms to identify potential illness causes and forecast patient reactions to potential medications. Since there are currently few diagnostic options, there is a pressing need to create novel ALS therapies utilizing DL algorithms (Schomer & Da Silva, 2012).

Depression

In neuropsychiatric research on drugs, AI/ML-based techniques have been applied, particularly for pharmacological decision assistance (Zhao and So, 2019). In an investigation on depression, scientists created a gradient boosting machine that uses the predictors found by the elastic net to determine if a patient taking citalopram would experience symptomatic relapse. An independent clinical trials escitalopram therapy group also benefited from the effective use of this model. Additionally, pharmacological responses to depression therapies have been predicted using data from fMRI and EEG (Chekroud et al., 2016).

CONCLUSION

CNS medication development is still a drawn-out, costly, ineffective, and difficult process with a low rate of new, effective therapy discoveries due to the intricacy of neurological illnesses. Scientists have employed AI/ML-based techniques to address the obstacles associated with CNS drug development, and these approaches have shown promise at every step of medication development for a range of illnesses. Academic and industrial researchers have generally taken a keen interest in AI/ML approaches in the development of pharmaceuticals. The use of AI technologies to improve CNS drug discovery is still at an early stage. Eventually, unconventional AI systems may be able to identify acute, complicated neurodegenerative crises. Chronic illnesses can also be identified early, allowing for more individualized and high-quality therapy at a notably lower expense. Drug development has been using ML for decades.

Modern ML techniques have expanded with the advent of new DL-based designs, which make it simple to include enormous volumes of biomedical data. Numerous virtual screening experiments have been conducted to find possible hits that might target and inhibit the specified protein. A traditional medical practitioner education curriculum, the scarcity of training in ML methods from an array of international data, and moral issues confronting specialists represent a few of the drawbacks that prevent the system from being universal. Other limitations include the absence of worldwide biomedical knowledge sharing network channels, a shortage of professionals who possess both clinical and programming skills, and a lack of trustworthy guidelines for interaction and information collaboration.

International organizations must also develop policies that raise awareness to encourage AI investment by compensating doctors and hospitals. The inability of AI-based models to be understood by the human mind is another drawback. With a growing number of investigations, mathematical model comprehension continues to be a difficult challenge for scholars. Deciphering such DL "black boxes" is therefore crucial to comprehending basic biological systems. Additionally, the insufficient data makes it difficult for AI/ML systems to effectively learn from large datasets. Laboratories and various scientific groups should cooperate to exchange data to prevent this. Addressing these obstacles may eventually allow for the reimbursement of the original financial outlay, opening the door for the use of AI in clinical settings.

ABBREVIATIONS

AD: Alzheimer's disease
ADME-T: absorption, distribution, metabolism, and excretion—toxicity
AL: artificial intelligence
ALS: amyotrophic lateral sclerosis
ANN: artificial neural network
AR: augmentedrReality
BBB: blood–brain barrier
CCT: computerized cognitive training
CNN: convolutional neural network
CNS: central nervous system
CT: cognitive training
DL: deep learning

DT: decision tree
FSR: force-sensitive resistor
GPU: graphics pprocessing unit
HD: Huntington's disease
HTA: high-throughput screening
ISM: in silico mutagenesis
kNN: K-nearest neighbors
LDA: linear discriminant analysis
LGVS: ligand-based virtual screening
ML: machine learning
MLR: multidimensional linear regression
NBC: naïve Bayesian classifiers
NDD: neurodegenerative disorder
PD: Parkinson's disease
QSAR: quantitative structure–activity relationship
RF: random forest
RNN: recurrent neural network
SBVS: structure-based virtual screening
SDL: supervised deep learning
SVM: support vector machine
TPU: tensor processing unit
VR: virtual reality
VS: virtual screening

REFERENCES

Alsenan S, Al-Turaiki I, Hafez A. A Recurrent Neural Network model to predict blood-brain barrier permeability. *Comput Biol Chem* (2020) 89:107377.

Ansari MY, Abdalla A, Ansari MY, Ansari MI, Malluhi B, Mohanty S, Mishra S, Singh SS, Abinahed J, Al-Ansari A, Balakrishnan S, Dakua SP. Practical utility of liver segmentation methods in clinical surgeries and interventions. *BMC Med Imaging* (2022a) 22(1):97.

Ansari MY, Chandrasekar V, Singh AV, Dakua SP. Re-routing drugs to blood brain barrier: A comprehensive analysis of Machine Learning approaches with fingerprint amalgamation and data balancing. *IEEE Access* (2022b) 11:9890–9906.

Atak ZK, Taskiran II, Demeulemeester J, Flerin C, Mauduit D, Minnoye L, Hulselmans G, Christiaens V, Ghanem GE, Wouters J, Aerts S. Interpretation of allele-specific chromatin accessibility using cell state-aware deep learning. *Genome Res* (2021) 31(6):1082–1096.

Avsec Ž, Kreuzhuber R, Israeli J, Xu N, Cheng J, Shrikumar A, Banerjee A, Kim DS, Beier T, Urban L, Kundaje A, Stegle O, Gagneur J. The Kipoi repository accelerates community exchange and reuse of predictive models for genomics. *Nat Biotechnol* (2019) 37(6):592–600.

Azuma RT. A survey of augmented reality. *Presence: Teleop Virt Envi* (1997) 6(4):355–385.

Baker N, Garrigan P, Phillips A, Kellman PJ. Configural relations in humans and deep convolutional neural networks. *Front Artif Intell* (2023) 5:961595.

Barban F, Annicchiarico R, Pantelopoulos S, Federici A, Perri R, Fadda L, et al. Protecting cognition from aging and Alzheimer's disease: a computerized cognitive training combined with reminiscence therapy. *Int J Geriatr Psychiatry* (2016) 31(4):340–348.

Bellenguez C, Küçükali F, Jansen IE, Kleineidam L, Moreno-Grau S, Amin N, et al. New insights into the genetic etiology of Alzheimer's disease and related dementias. *Nat Genet* (2022) 54(4):412–436.

Berhane TM, Lane CR, Wu Q, Autrey BC, Anenkhonov OA, Chepinoga VV, Liu H. Decision-Tree, Rule-Based, and Random Forest classification of high-resolution multispectral imagery for wetland mapping and inventory. *Remote Sens (Basel)* (2018) 10(4):580.

Breasail MÓ, Biswas B, Smith MD, Mazhar MKA, Tenison E, Cullen A, Lithander FE, Roudaut A, Henderson EJ. Wearable GPS and accelerometer technologies for monitoring mobility and physical activity in neurodegenerative disorders: a systematic review. *Sensors (Basel)* (2021) 21(24):8261.

Brena RF, García-Vázquez JP, Galván-Tejada CE, Muñoz-Rodriguez D, Vargas-Rosales C, Fangmeyer J. Evolution of indoor positioning technologies: a survey. *J Sens* (2017) 2017:1–21.

Castro E, Gómez-Verdejo V, Martínez-Ramón M, Kiehl KA, Calhoun VD. A multiple kernel learning approach to perform classification of groups from complex-valued fMRI data analysis: application to schizophrenia. *Neuroimage* (2014a) 87:1–17.

Castro E, Gupta CN, Martínez-Ramón M, Calhoun VD, Arbabshirani MR, Turner J. Identification of patterns of gray matter abnormalities in schizophrenia using source-based morphometry and bagging. *Annu Int Conf IEEE Eng Med Biol Soc* (2014b) 2014:1513--1516.

Cavallo M, Angilletta C. Long-lasting neuropsychological effects of a computerized cognitive training in patients affected by early stage Alzheimer's disease: are they stable over time? *J Appl Gerontol* (2019) 38(7):1035–1044.

Cesewski E, Johnson BN. Electrochemical biosensors for pathogen detection. *Biosens Bioelectron* (2020) 159:112214.

Chan HCS, Shan H, Dahoun T, Vogel H, Yuan S. Advancing drug discovery via artificial intelligence. *Trends Pharmacol Sci* (2019) 40(8):592–604.

Chan ZYS, MacPhail AJC, Au IPH, Zhang JH, Lam BMF, Ferber R, Cheung RTH. Walking with head-mounted virtual and augmented reality devices: Effects on position control and gait biomechanics. *PLoS One* (2019) 14(12):e0225972.

Chekroud AM, Zotti RJ, Shehzad Z, Gueorguieva R, Johnson MK, Trivedi MH, Cannon TD, Krystal JH, Corlett PR. Cross-trial prediction of treatment outcome in depression: a machine learning approach. *Lancet Psychiatry* (2016) 3(3):243–250.

Cherian Parakkal S, Datta R, Das D. DeepBBBP: high accuracy blood-brain-barrier permeability prediction with a mixed deep learning model. *Mol Inform* (2022) 41(10):e2100315.

Choi J, Twamley EW. Cognitive rehabilitation therapies for Alzheimer's disease: a review of methods to improve treatment engagement and self-efficacy. *Neuropsychol Rev* (2013) 23(1):48–62.

Cipresso P, Giglioli IAC, Raya MA, Riva G. The past, present, and future of virtual and augmented reality research: a network and cluster analysis of the literature. *Front Psychol* (2018) 9:2086.

Clare L, Woods RT. Cognitive training and cognitive rehabilitation for people with early-stage Alzheimer's disease: a review. *Neuropsychol rehabil* (2004) 14(4):385–401.

Corces MR, Shcherbina A, Kundu S, Gloudemans MJ, Frésard L, Granja JM, Louie BH, Eulalio T, Shams S, Bagdatli ST, Mumbach MR, Liu B, Montine KS, Greenleaf WJ, Kundaje A, Montgomery SB, Chang HY, Montine TJ. Single-cell epigenomic analyses implicate candidate causal variants at inherited risk loci for Alzheimer's and Parkinson's diseases. *Nat Genet* (2020) 52(11):1158–1168.

Das A, Dhillon P. Application of machine learning in measurement of ageing and geriatric diseases: a systematic review. *BMC Geriatr* (2023) 23(1):841.

Das SK, Nayak KK, Krishnaswamy PR, Kumar V, Bhat N. Electrochemistry and other emerging technologies for continuous glucose monitoring devices. *ECS Sen Plus* (2022) 1(3):031601.

Delaunay A, Guérin J. *Wandering detection within an embedded system for Alzheimer suffering patients.* In 2017 AAAI spring symposium series (2017).

de Oliveira Gondim ITG, de Souza CCB, Rodrigues MAB, Azevedo IM, de Sales Coriolano MDGW, Lins OG. Portable accelerometers for the evaluation of spatio-temporal gait parameters in people with Parkinson's disease: an integrative review. *Arch Gerontol Geriatr* (2020) 90:104097.

Dugger BN, Dickson DW. Pathology of neurodegenerative diseases. *Cold Spring Harb Perspect Biol* (2017) 9(7):a028035.

Fillekes MP, Giannouli E, Kim EK, Zijlstra W, Weibel R. Towards a comprehensive set of GPS-based indicators reflecting the multidimensional nature of daily mobility for applications in health and aging research. *Int J Health Geogr* (2019a) 18(1):17.

Fillekes MP, Röcke C, Katana M, Weibel R. Self-reported versus GPS-derived indicators of daily mobility in a sample of healthy older adults. *Soc Sci Med* (2019b) 220:193–202.

Ganesan B, Gowda T, Al-Jumaily A, Fong KNK, Meena SK, Tong RKY. Ambient assisted living technologies for older adults with cognitive and physical impairments: a review. *Eur Rev Med Pharmacol Sci* (2019) 23(23):10470–10481.

Gates NJ, Vernooij RW, Di Nisio M, Karim S, March E, Martínez G, Rutjes AW. Computerised cognitive training for preventing dementia in people with mild cognitive impairment. *Cochrane Database Syst Rev* (2019) 3(3):CD012279.

Gautam A. Towards modern-age advanced sensors for the management of neurodegenerative disorders: current status, challenges and prospects. *ECS Sens Plus* (2022) 1(4):042401.

Geva AB, Kerem DH. Brain state identification and forecasting of acute pathology using unsupervised fuzzy clustering of EEG temporal patterns. In H-NL Teodorescu, A Kandel, LC Jain (Eds.), *Fuzzy and neuro-fuzzy systems in medicine* 2017 Nov 22 (pp. 19–68). CRC Press.

Gorman C, Gustafsson L. The use of augmented reality for rehabilitation after stroke: a narrative review. *Disabil Rehabil Assist Technol* (2022) 17(4):409–417.

Guinet AL, Bouyer G, Otmane S, Desailly E. Validity of Hololens augmented reality head mounted display for measuring gait parameters in healthy adults and children with cerebral palsy. *Sensors (Basel)* (2021) 21(8):2697.

Haleem A, Javaid M, Singh RP, Suman R, Rab S. Biosensors applications in medical field: a brief review. *Sens Int* (2021) 2:100100.

Hindle SJ, Munji RN, Dolghih E, Gaskins G, Orng S, Ishimoto H, Soung A, DeSalvo M, Kitamoto T, Keiser MJ, Jacobson MP. Evolutionarily conserved roles for blood-brain barrier xenobiotic transporters in endogenous steroid partitioning and behavior. *Cell Rep* (2017) 21(5):1304–1316.

Homayounfar SZ, Andrew TL. Wearable sensors for monitoring human motion: a review on mechanisms, materials, and challenges. *SLAS Technol* (2020) 25(1):9–24.

Hughes GL, Lones MA, Bedder M, Currie PD, Smith SL, Pownall ME. Machine learning discriminates a movement disorder in a zebrafish model of Parkinson's disease. *Dis Model Mech* (2020) 13(10):dmm045815.

Humpel C. Identifying and validating biomarkers for Alzheimer's disease. *Trends Biotechnol* (2011) 29(1):26–32.

Huys QJ, Maia TV, Frank MJ. Computational psychiatry as a bridge from neuroscience to clinical applications. *Nat Neurosci* (2016) 19(3):404–413.

Ielapi N, Andreucci M, Licastro N, Faga T, Grande R, Buffone G, Mellace S, Sapienza P, Serra R. Precision medicine and precision nursing: the era of biomarkers and precision health. *Int J Gen Med* (2020) 13:1705–1711.

Jack CR Jr, Bernstein MA, Fox NC, Thompson P, Alexander G, Harvey D, et al. The Alzheimer's disease neuroimaging initiative (ADNI): MRI methods. *J Magn Reson Imaging* (2008) 27(4):685–691.

Johansson D, Malmgren K, Alt Murphy M. Wearable sensors for clinical applications in epilepsy, Parkinson's disease, and stroke: a mixed-methods systematic review. *J Neurol* (2018) 265(8):1740–1752.

Kearns WD, Nams VO, Fozard JL. Tortuosity in movement paths is related to cognitive impairment. Wireless fractal estimation in assisted living facility residents. *Methods Inf Med* (2010) 49(6):592–598.

Khaksarinejad R, Arabpour Z, RezaKhani L, Parvizpour F, Rasmi Y. Biomarker based biosensors: an opportunity for diagnosis of COVID-19. *Rev Med Virol* (2022) 32(5):e2356.

Kim DW, Hassett LM, Nguy V, Allen NE. A comparison of activity monitor data from devices worn on the wrist and the waist in people with Parkinson's disease. *Mov Disord Clin Pract* (2019) 6(8):693–699.

Klenk J, Srulijes K, Schatton C, Schwickert L, Maetzler W, Becker C, Synofzik M. Ambulatory activity components deteriorate differently across neurodegenerative diseases: a cross-sectional sensor-based study. *Neurodegener Dis* (2016)16(5–6):317–323.

Kumar A, Lau CT, Ma M, Chan S, Kearns W. *Trend analysis in the trajectory of the dementia patients.* In 2017 21st International Computer Science and Engineering Conference (ICSEC), IEEE (2017):1–5.

Kumar R, Aadil KR, Mondal K, Mishra YK, Oupicky D, Ramakrishna S, Kaushik A. Neurodegenerative disorders management: state-of-art and prospects of nano-biotechnology. *Crit Rev Biotechnol* (2022) 42(8):1180–1212.

Lan AY, Corces MR. Deep learning approaches for noncoding variant prioritization in neurodegenerative diseases. *Front Aging Neurosci* (2022) 14:1027224.

Li TY, Chien YW, Chou CC, Liao CF, Cheah WT, Fu LC, Chen CC, Chou CC, Chen IA. A fast and low-cost repetitive movement pattern indicator for massive dementia screening. *IEEE Trans Autom Sci Eng* (2019) 17(2):771–783.

Lionta E, Spyrou G, Vassilatis DK, Cournia Z. Structure-based virtual screening for drug discovery: principles, applications and recent advances. *Curr Top Med Chem* (2014) 14(16):1923–1938.

Lipinski CF, Maltarollo VG, Oliveira PR, da Silva ABF, Honorio KM. Advances and perspectives in applying deep learning for drug design and discovery. *Front Robot AI* (2019) 6:108.

Liu D, Zhang D, Sun Z, Zhou S, Li W, Li C, Li W, Tang W, Wang ZL. Active-matrix sensing array assisted with machine-learning approach for lumbar degenerative disease diagnosis and postoperative assessment. *Adv Funct Mater* (2022) 32(21):2113008.

Liu GX, Shi LF, Chen S, Wu ZG. Focusing matching localization method based on indoor magnetic map. *IEEE Sens J* (2020) 20(17):10012–10020.

Liu X, Chen K, Wu T, Weidman D, Lure F, Li J. Use of multimodality imaging and artificial intelligence for diagnosis and prognosis of early stages of Alzheimer's disease. *Transl Res* (2018) 194:56–67.

Maggio MG, Russo M, Cuzzola MF, Destro M, La Rosa G, Molonia F, Bramanti P, Lombardo G, De Luca R, Calabrò RS. Virtual reality in multiple sclerosis rehabilitation: a review on cognitive and motor outcomes. *J Clin Neurosci* (2019) 65:106–111.

Mantri S, Wood S, Duda JE, Morley JF. Comparing self-reported and objective monitoring of physical activity in Parkinson disease. *Parkinsonism Relat Disord* (2019) 67:56–59.

Matarazzo M, Arroyo-Gallego T, Montero P, Puertas-Martín V, Butterworth I, Mendoza CS, et al. Remote monitoring of treatment response in Parkinson's disease: the habit of typing on a computer. *Mov Disord* (2019) 34(10):1488–1495.

Mathew AA, Chandrasekhar A, Vivekanandan S. A review on real-time implantable and wearable health monitoring sensors based on triboelectric nanogenerator approach. *Nano Energy* (2021) 80:105566.

Mazilu S, Blanke U, Calatroni A, Gazit E, Hausdorff JM, Tröster G. The role of wrist-mounted inertial sensors in detecting gait freeze episodes in Parkinson's disease. *Pervasive Mob Comput* (2016) 33:1–6.

Meola A, Cutolo F, Carbone M, Cagnazzo F, Ferrari M, Ferrari V. Augmented reality in neurosurgery: a systematic review. *Neurosurg Rev* (2017) 40(4):537–548.

Moffat JG, Vincent F, Lee JA, Eder J, Prunotto M. Opportunities and challenges in phenotypic drug discovery: an industry perspective. *Nat Rev Drug Discovery* (2017) 16(8):531–543.

Monzel AS, Hemmer K, Kaoma T, Smits LM, Bolognin S, Lucarelli P, Rosety I, Zagare A, Antony P, Nickels SL, Krueger R, Azuaje F, Schwamborn JC. Machine learning-assisted neurotoxicity prediction in human midbrain organoids. *Parkinsonism Relat Disord* (2020) 75:105–109.

Moore JE, Purcaro MJ, Pratt HE, Epstein CB, Shoresh N, Adrian J, et al. Expanded encyclopaedias of DNA elements in the human and mouse genomes. *Nature* (2020) 583(7818):699–710.

Moyle W, Jones C, Murfield J, Draper B, Beattie E, Shum D, Thalib L, O'Dwyer S, Mervin CM. Levels of physical activity and sleep patterns among older people with dementia living in long-term care facilities: a 24-h snapshot. *Maturitas* (2017) 102:62–68.

Novikova G, Andrews SJ, Renton AE, Marcora E. Beyond association: successes and challenges in linking non-coding genetic variation to functional consequences that modulate Alzheimer's disease risk. *Mol Neurodegener* (2021a) 16(1):27.

Novikova G, Kapoor M, Tcw J, Abud EM, Efthymiou AG, Chen SX, Cheng H, Fullard JF, Bendl J, Liu Y, Roussos P, Björkegren JL, Liu Y, Poon WW, Hao K, Marcora E, Goate AM. Integration of Alzheimer's disease genetics and myeloid genomics identifies disease risk regulatory elements and genes. *Nat Commun* (2021b) 12(1):1610.

Oguntala G, Abd-Alhameed R, Jones S, Noras J, Patwary M, Rodriguez J. Indoor location identification technologies for real-time IoT-based applications: an inclusive survey. *Computer Sci Rev* (2018) 30:55–79.

Paszkiel S. *Analysis and classification of EEG signals for brain-computer interfaces*. Cham: Springer International Publishing (2020) pp. 87–91.

Pateraki M, Fysarakis K, Sakkalis V, Spanoudakis G, Varlamis I, Maniadakis M, Lourakis M, Ioannidis S, Cummins N, Schuller B, Loutsetis E. Biosensors and internet of things in smart healthcare applications: challenges and opportunities. In N DeyTechno, AS Ashour, SJ Fong, C Bhatt (Eds.), *Wearable and implantable medical devices*. Cambridge: Academic Press (2020) pp. 25–53.

Paulsen JS, Nance M, Kim JI, Carlozzi NE, Panegyres PK, Erwin C, Goh A, McCusker E, Williams JK. A review of quality of life after predictive testing for and earlier identification of neurodegenerative diseases. *Prog Neurobiol* (2013)110:2–28.

Peretz C, Korczyn AD, Shatil E, Aharonson V, Birnboim S, Giladi N. Computer-based, personalized cognitive training versus classical computer games: a randomized double-blind prospective trial of cognitive stimulation. *Neuroepidemiol* (2011) 36(2):91–99.

Piromchai P, Avery A, Laopaiboon M, Kennedy G, O'Leary S. Virtual reality training for improving the skills needed for performing surgery of the ear, nose or throat. *Cochrane Database Syst Rev* (2015) 2015(9):CD010198.

Polosecki P, Castro E, Wood A, Warner JH, Rish I, Cecchi GA. Computational psychiatry: Advancing predictive modeling of neurodegeneration with neuroimaging of Huntington's disease. *IBM J Res Dev* (2017) 61(2/3):4–1.

Pradhan S, Kelly VE. Quantifying physical activity in early Parkinson disease using a commercial activity monitor. *Parkinsonism Relat Disord* (2019) 66:171–175.

Ramgopal S, Thome-Souza S, Jackson M, Kadish NE, Sánchez Fernández I, Klehm J, Bosl W, Reinsberger C, Schachter S, Loddenkemper T. Seizure detection, seizure prediction, and closed-loop warning systems in epilepsy. *Epilepsy Behav* (2014) 37:291–307.

Riccio A, Simione L, Schettini F, Pizzimenti A, Inghilleri M, Belardinelli MO, Mattia D, Cincotti F. Attention and P300-based BCI performance in people with amyotrophic lateral sclerosis. *Front Hum Neurosci* (2013) 7:732.

Rivetti di Val Cervo P, Besusso D, Conforti P, Cattaneo E. hiPSCs for predictive modelling of neurodegenerative diseases: dreaming the possible. *Nat Rev Neurol* (2021) 17(6):381–392.

Ross CA, Aylward EH, Wild EJ, Langbehn DR, Long JD, Warner JH, Scahill RI, Leavitt BR, Stout JC, Paulsen JS, Reilmann R, Unschuld PG, Wexler A, Margolis RL, Tabrizi SJ. Huntington disease: natural history, biomarkers and prospects for therapeutics. *Nat Rev Neurol* (2014) 10(4):204–216.

Ross CA, Margolis RL, Reading SA, Pletnikov M, Coyle JT. Neurobiology of schizophrenia. *Neuron* (2006) 52(1):139–153.

Sakpere WE, Adeyeye MO. Can near field communication solve the limitations in mobile indoor navigation?. In In Lee (Ed.), *RFID technology integration for business performance improvement*. Pennsylvania: IGI Global (2015) pp. 52–79.

Salvatore C, Cerasa A, Battista P, Gilardi MC, Quattrone A, Castiglioni I; Alzheimer's Disease Neuroimaging Initiative. Magnetic resonance imaging biomarkers for the early diagnosis of Alzheimer's disease: a machine learning approach. *Front Neurosci* (2015) 1(9):307.

Sanpechuda T, Kovavisaruch LO. A review of RFID localization: Applications and techniques. In 2008 5th international conference on electrical engineering/electronics, computer, telecommunications and information technology. *IEEE* (2008) 2:769–772.

Schomer DL, Da Silva FL. *Niedermeyer's electroencephalography: basic principles, clinical applications, and related fields*. Philadelphia: Lippincott Williams & Wilkins; 2012 October 18.

Schuhmacher A, Gatto A, Kuss M, Gassmann O, Hinder M. Big techs and startups in pharmaceutical R&D - A 2020 perspective on artificial intelligence. *Drug Discov Today* (2021) 26(10):2226–2231.

Schwessinger R, Gosden M, Downes D, Brown RC, Oudelaar AM, Telenius J, Teh YW, Lunter G, Hughes JR. DeepC: predicting 3D genome folding using megabase-scale transfer learning. *Nat Methods* (2020) 17(11):1118–1124.

Selzler R, Green JR and Goubran R. Neurodegenerative disease prediction based on gait analysis signals acquired with force-sensitive resistors. In: *2018 IEEE life sciences conference (LSC)*, Montreal, QC, Canada, 28–30 October 2018, pp. 122–125. New York: IEEE.

Sharma R, Pillai L, Glover A, Virmani T. Objective impairment of tandem gait in Parkinson's disease patients increases with disease severity. *Parkinsonism Relat Disord* (2019) 68:33–39.

Sheikh S, Safia, Haque E, Mir SS. Neurodegenerative diseases: multifactorial conformational diseases and their therapeutic interventions. *J Neurodegener Dis* (2013) 2013:563481.

Shen Z, Liu F, Huang S, Wang H, Yang C, Hang T, Tao J, Xia W, Xie X. Progress of flexible strain sensors for physiological signal monitoring. *Biosens Bioelectron* (2022) 211:114298.

Song Z, Zhou Y, Han X, Qin J, Tang X. Recent advances in enzymeless-based electrochemical sensors to diagnose neurodegenerative diseases. *J Mater Chem B* (2021) 9(5):1175–1188.

Subasi A. EEG signal classification using wavelet feature extraction and a mixture of expert model. *Expert Syst Appl* (2007) 32(4):1084–1093.

Subasi A, Kevric J, Abdullah Canbaz M. Epileptic seizure detection using hybrid machine learning methods. *Neural Comput Appl* (2019) 31:317–325.

Sulzer P, Gräber S, Schaeffer E, van Lummel R, Berg D, Maetzler W, Liepelt-Scarfone I. Cognitive impairment and sedentary behavior predict health-related attrition in a prospective longitudinal Parkinson's disease study. *Parkinsonism Relat Disord* (2021) 82:37–43.

Summers MJ, Madl T, Vercelli AE, Aumayr G, Bleier DM, Ciferri L. Deep machine learning application to the detection of preclinical neurodegenerative diseases of aging. *DigitCult-Sci J Digi Cult* (2017) 2(2):9–24.

Sundar A, Das C, Deshmukh MK. *Novel applications of force sensing resistors in healthcare technologies*. In Proceedings of healthy world conference 2015 January.

Sunkin SM, Ng L, Lau C, Dolbeare T, Gilbert TL, Thompson CL, Hawrylycz M, Dang C. Allen Brain Atlas: an integrated spatio-temporal portal for exploring the central nervous system. *Nucleic Acids Res* (2013) 41(Database issue):D996–D1008.

Tárraga L, Boada M, Modinos G, Espinosa A, Diego S, Morera A, Guitart M, Balcells J, López OL, Becker JT. A randomised pilot study to assess the efficacy of an interactive, multimedia tool of cognitive stimulation in Alzheimer's disease. *J Neurol Neurosurg Psychiatry* (2006) 77(10):1116–1121.

Tăuţan AM, Ionescu B, Santarnecchi E. Artificial intelligence in neurodegenerative diseases: a review of available tools with a focus on machine learning techniques. *Artif Intell Med* (2021) 117:102081.

Teo WP, Muthalib M, Yamin S, Hendy AM, Bramstedt K, Kotsopoulos E, Perrey S, Ayaz H. Does a combination of virtual reality, neuromodulation and neuroimaging provide a comprehensive platform for neurorehabilitation? - a narrative review of the literature. *Front Hum Neurosci* (2016) 10:284.

Tsukagoshi S, Furuta M, Hirayanagi K, Furuta N, Nakazato S, Fujii M, Yuminaka Y, Ikeda Y. Noninvasive and quantitative evaluation of movement disorder disability using an infrared depth sensor. *J Clin Neurosci* (2020) 71:135–140.

Turner AW, Hu SS, Mosquera JV, Ma WF, Hodonsky CJ, Wong D, et al. Single-nucleus chromatin accessibility profiling highlights regulatory mechanisms of coronary artery disease risk. *Nat Genet* (2022) 54(6):804–816.

Ullah H, Wahab MA, Will G, Karim MR, Pan T, Gao M, Lai D, Lin Y, Miraz MH. Recent advances in stretchable and wearable capacitive electrophysiological sensors for long-term health monitoring. *Biosensors* (Basel) (2022) 12(8):630.

van Eijk RPA, Bakers JNE, Bunte TM, de Fockert AJ, Eijkemans MJC, van den Berg LH. Accelerometry for remote monitoring of physical activity in amyotrophic lateral sclerosis: a longitudinal cohort study. *J Neurol* (2019) 266(10):2387–2395.

van Uem JMT, Cerff B, Kampmeyer M, Prinzen J, Zuidema M, Hobert MA, Gräber S, Berg D, Maetzler W, Liepelt-Scarfone I. The association between objectively measured physical activity, depression, cognition, and health-related quality of life in Parkinson's disease. *Parkinsonism Relat Disord* (2018) 48:74–81.

Varma VR, Watts A. Daily physical activity patterns during the early stage of Alzheimer's disease. *J Alzheimers Dis* (2017) 55(2):659–667.

Vatansever S, Schlessinger A, Wacker D, Kaniskan HÜ, Jin J, Zhou MM, Zhang B. Artificial intelligence and machine learning-aided drug discovery in central nervous system diseases: state-of-the-arts and future directions. *Med Res Rev* (2021) 41(3):1427–1473.

Vemula D, Jayasurya P, Sushmitha V, Kumar YN, Bhandari V. CADD, AI and ML in drug discovery: a comprehensive review. *Eur J Pharm Sci* (2023) 181:106324.

Wal P, Dwivedi J, Wal A, Vig H, Singh Y. Detailed insight into the pathophysiology and the behavioral complications associated with the Parkinson's disease and its medications. *Futur J Pharm Sci* (2022) 8(1):1–4.

Wal P, Wal A, Vig H, Mahmood D, Kan MMU. Potential applications of mitochondrial therapy with a focus on Parkinson's disease and mitochondrial transplantation. *Adv Pharm Bull* (2023):1–25.

Wang Y, Yu Y, Wei X, Narita F. Self-powered wearable piezoelectric monitoring of human motion and physiological signals for the postpandemic era: a review. *Adv Mater Tech* (2022) 7(12):2200318.

Wang Z, Yang H, Wu Z, Wang T, Li W, Tang Y, Liu G. In silico Prediction of blood-brain barrier permeability of compounds by machine learning and resampling methods. *ChemMedChem* (2018) 13(20):2189–2201.

Yin R, Wang D, Zhao S, Lou Z, Shen G. Wearable sensors-enabled human–machine interaction systems: from design to application. *Adv Funct Mater* (2021) 31(11):2008936.

Younesi E, Hofmann-Apitius M. From integrative disease modeling to predictive, preventive, personalized and participatory (P4) medicine. *EPMA J* (2013) 4(1):23.

Zeng X, Zhu S, Liu X, Zhou Y, Nussinov R, Cheng F. deepDR: a network-based deep learning approach to in silico drug repositioning. *Bioinform* (2019) 35(24):5191–5198.

Zhang Z. Introduction to machine learning: K-nearest neighbors. *Ann Transl Med* (2016) 4(11):218.

Zhang Z. Too much covariates in a multivariable model may cause the problem of overfitting. *J Thorac Dis* (2014) 6(9):E196–E197.

Zhao H, Cao J, Xie J, Liao WH, Lei Y, Cao H, Qu Q, Bowen C. Wearable sensors and features for diagnosis of neurodegenerative diseases: a systematic review. *Digit Health* (2023) 9:20552076231173569.

Zhao K, So HC. Drug repositioning for schizophrenia and depression/anxiety disorders: a machine learning approach leveraging expression. *Data. IEEE J Biomed Health Inform* (2019) 23(3):1304–1315.

Zhou J, Troyanskaya OG. Predicting effects of noncoding variants with deep learning-based sequence model. *Nat Methods* (2015) 12(10):931–934.

Zhu L, Duval C, Boissy P, Montero-Odasso M, Zou G, Jog M, Speechley M. Comparing GPS-based community mobility measures with self-report assessments in older adults with Parkinson's disease. *J Gerontol A Biol Sci Med Sci* (2020) 75(12):2361–2370.

Zilian D, Sotriffer CA. SFCscore(RF): a random forest-based scoring function for improved affinity prediction of protein-ligand complexes. *J Chem Inf Model* (2013) 53(8):1923–1933.

Zolfaghari S, Suravee S, Riboni D, Yordanova K. Sensor-based locomotion data mining for supporting the diagnosis of neurodegenerative disorders: a survey. *ACM Comput Surv* (2023) 56(1):10:1–10:36.

8 Microglia, a New Avenue in the Cure and Treatment of Alzheimer's Disease

Dipanjan Karati[1] and Dileep Kumar[2,3,4]*

[1]Department of Pharmaceutical Technology, School of Pharmacy, Techno India University, Kolkata, West Bengal India

[2]Poona College of Pharmacy, Department of Pharmaceutical Chemistry, Bharati Vidyapeeth (Deemed to be University), Erandwane, Pune, Maharashtra, India

[3]Department of Pharm. Chemistry, Manipal College of Pharmaceutical Sciences, Manipal Academy of Higher Education, Manipal, Karnataka, India

[4]UC Davis Comprehensive Cancer Center, Department of Entomology and Nematology, University of California Davis, One Shields Avenue, Davis, CA, USA

*Corresponding author

INTRODUCTION

Alzheimer's disease (AD) is the most conventional type of dementia, characterized by gradual memory loss and cognitive impairment. Extracellular senile plaques made up of amyloid deposits, intracellular neurofibrillary tangles, and cerebral atrophy are all neuropathological signs of AD. AD is a neurodegenerative condition that affects the elderly and is characterized by chronic neuroinflammation and amyloid beta (Aβ) deposition in the brain parenchyma (Blennow et al., 2006; Perl, 2010; Benilova et al., 2012; Shimada et al., 2017). Microglia, the central nervous system's resident myeloid cells, perform an essential role in the pathogenesis of AD and have emerged as a possible therapeutic target for treating or preventing the disease. Microglial activation is essential in the fundamental reason for AD (Heneka et al., 2015). Microglia participate in creating usual neuronal connectivity and monitoring methods essential for growth of the central nervous system (CNS), such as synaptic pruning, which ensures the elimination of inappropriate synapses to reinforce the appropriate ones based on neuronal activity and experience, such as the brain's resident immune cells (Pierre et al., 2017). Microglia play a role in the development of cognitive activities, including learning and intellect in the mature brain (Parkhurst et al., 2013; Yao et al., 2019). Microglia are tasked with the maintenance of homeostasis by eradicating cellular waste, misfolded proteins, and dying cells (Hristovska et al., 2015; Paolicelli et al., 2011; Sierra et al., 2010; Stevens et al., 2007; Tremblay et al., 2010; Wake et al., 2009).

Similar to peripheral nerve macrophages, triggered microglia are divided into two polarized states: M1 and M2 phenotypes (Wes et al., 2016; Subramaniam et al., 2017; Xu et al., 2017). Even though a simple M1 or M2 phenotype does not encompass the entire standing of triggered microglia, this classification is an important exhibitor for microglia-directed care in Alzheimer's-affected persons (Ransohoff, 2016; Tang et al., 2016). The M1 phenotype is generally related to the development of pro-inflammatory cytokines such as tumor necrosis factor (TNF) and interleukin (IL)-1,

DOI: 10.1201/9781003506317-8

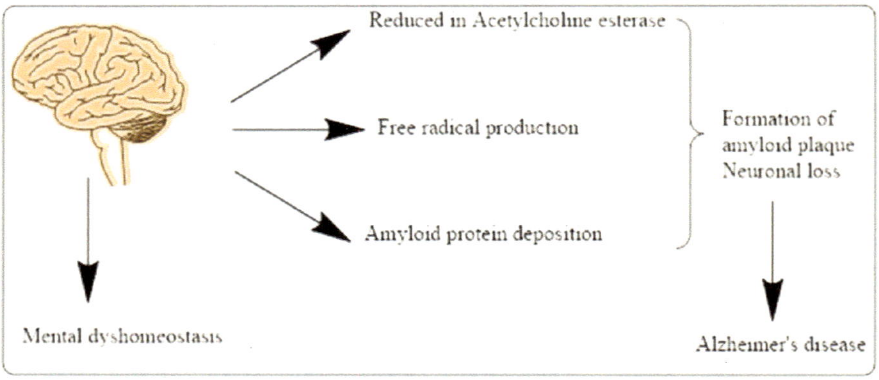

FIGURE 8.1 Pathophysiology of AD.

whereas the M2 phenotype is associated with the generation of anti-inflammatory components such as IL-10 and IL-4 (Jimenez et al., 2008; Malm et al., 2015). Microglia-targeted therapies should aim to change the microglia status, given the opposing states of microglial activation.

PATHOPHYSIOLOGY OF AD

The hippocampus, amygdala, entorhinal cortex, and cortical interaction regions of the frontal, temporal, and parietal cortices, as well as subcortical nuclei, including the dorsal serotonergic raphe, noradrenergic locus coeruleus, and basal cholinergic nucleus, can show neuronal failure. Tangles form in a specific fashion, beginning in the trans-entorhinal cortex and progressing through the entorhinal cortex, the CA1 region of the hippocampus, and, finally, the cortical interaction regions, which are significantly influenced in the frontal, parietal, and temporal lobes (Kumar et al., 2015). Tau protein accumulation is strongly linked to cognitive impairment and brain atrophy, including hippocampal atrophy. In the neuropathology of AD, there is a loss of neurons and atrophy in the temporal frontal cortex, which causes inflammation and the formation of amyloid plaques, an abnormal cluster of protein fragments, and tangled bundles of fibers, increasing the presence of monocytes and macrophages in the cerebral cortex, as well as activation of microglial cells in the parenchyma (Mishra et al., 2008). Figure 8.1 describes the pathophysiology of AD.

MOLECULAR INSIGHT INTO MICROGLIA

Microglia are the resident phagocytes and innate immune cells of the brain, protecting the substantia nigra pars compacta (SNC) from infections and regulating homeostasis. Since discovering microglia's role in synaptic plasticity dysfunctions a few decades ago, there has been a surge in interest in the cell. A change in the physiological functioning of these cells, in particular, can damage neuronal function, eventually leading to cognitive loss and behavioral disorders (Kang et al., 1987; Calabrò et al., 2020). Synaptic plasticity has also been confirmed as a crucial mechanism in establishing Aβ and the long-term potentiation (LTP) pathways in AD (Qi et al., 2014; Lo et al., 2013). A growing body of evidence suggests that reactive microglia in the brain can contribute to AD development. The brain's microglial cells are generally dormant. They can still execute some control functions in this state (Giulian et al., 1986; Kreutzberg, 1996; Mecha et al., 2016) but don't appear in AD processes. However, their activation can set off a chain of biological events linked to an elevated risk of AD.

When activated, microglia may trigger pro-inflammatory (M1 state) or anti-inflammatory (M2 state) processes (Henkel et al., 2009). The M2 state plays a role in neuroprotective-related processes. Microglial cells in the M1 state, on the other hand, are connected to the development of Alzheimer's disease because they increase neuroinflammation through the release of pro-inflammatory cytokines. As described before, neuroinflammation harms neurons and may lead to neurodegenerative

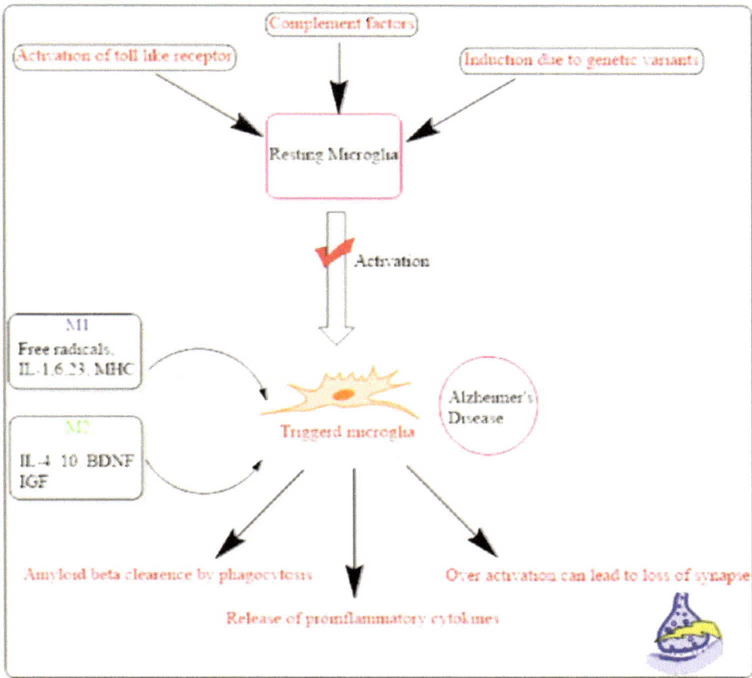

FIGURE 8.2 Activated microglia cause Alzheimer's disease.

processes. Also, cells in the M1 state may release reactive oxygen species (ROS), thus increasing the OS in the microenvironment (Stansley et al., 2012). M1 microglial cells can serve as a link between distinct AD processes. They have also been connected to plaque formation and hyperphosphorylation. However, microglial activities in early-stage AD, particularly their role in synaptic dysfunction, are poorly understood. The factors responsible for the activation of microglia and their involvement in AD are represented in Figure 8.2.

MICROGLIA-INDUCED NEUROINFLAMMATION IN AD

Microglial cells across the CNS make up around 15% of the brain's total cellular population (Cardona et al., 2006). Tau protein accumulation is strongly linked to cognitive decline and brain atrophy, especially hippocampal atrophy. There is a loss of neurons and atrophy in the temporal-frontal cortex in Alzheimer's disease neuropathology, which produces inflammation and deposits amyloid plaques, as well as an aberrant cluster of protein fragments and tangled bundles of fibers (Thakur et al., 2018). If stimulated, microglia can move to the injury site and generate an innate immune response. Microglia participate in the protection and remodeling of synaptic plasticity, which is necessary for neuronal circuits to function normally (Ji et al., 2013). Microglia play a crucial part in learning and memory functions by protecting and remodeling learning-related synapses. Aging, particularly microglia aging, plays a significant role during AD, as it is the essential factor linked to it. Aged microglia express large quantities of pro-inflammatory cytokines such as interleukin-1 beta (IL-1β), tumor necrosis factor-alpha (TNF-α), and IL-6 (Sierra et al., 2007) and suffer phenotypic and functional alterations (Wong, 2013). Dystrophic microglia are also present in the brains of the elderly and those with Alzheimer's disease. Recent research has discovered a direct link between inflammasome activation and age-related functional decline (Youm et al., 2013). In contrast, neuroinflammation induced by microglia has been proven to be a critical factor in the progression of Alzheimer's disease, implying a close relationship between immune aging, neuroinflammation, and AD. Through a series of intercellular exchanges, microglia contribute to neuroinflammation. Reactive microglia emit pro-inflammatory cytokines such as IL-1β, IL-6, IL-18, and TNF-α, as well as upregulate the production of chemokines like CCL2, CCR3, and CCR5, triggering local inflammatory reactions and neural cell death (Xia et al., 1998; Ishizuka et al., 1997). Microglia can contribute to neuroinflammation in Alzheimer's disease by activating cell surface receptors like toll-like receptors (TLRs), lipopolysaccharide (LPS) receptor cluster of differentiation 14 (CD14), and scavenger receptors such as SCARA1, CD36, and CD47 in response to Aβ oligomers and plaques (Stewart et al., 2010; Bamberger et al., 2003; Fassbender et al., 2004; Yamanaka et al., 2012; Mandrekarcolucci et al., 2012; Xu et al., 2015). Reduced Aβ clearance has been associated with poor microglial phagocytic ability in sporadic AD, which is characterized by Aβ phagocytosis receptor downregulation and elevated cytokine concentrations (Mawuenyega et al., 2010).

MICROGLIAL DRUG TARGETS IN THE TREATMENT OF AD

It is critical to show that possible targets are relevant to neuronal disorders and that modifying them will result in a medicinal advantage with a reasonable safety profile. Indications for a putative target's participation in the disease could incorporate:

a) The different appearances of illness
b) An integral relationship with illness pathogenesis
c) Hereditary interaction with the disorder
d) Having a mechanistic contribution to infection etiology
e) Accomplishment of a promising result when modified in a disease-relevant *in vitro* or *in vivo* model

Microglia have harmful and protective functions in the progression of Alzheimer's disease, depending on their polarized condition. Excessive neuroinflammation generates toxins like nitric oxide (NO) and pro-inflammatory cytokines/chemokines, which increase neuronal injury and, as a result, speed up the onset of Alzheimer's disease (Michaud et al., 2013). A close genetic link between a microglial substance and AD, such as CD33, complement receptor 1 (CR1), phospholipase C gamma 2 (PLCγ2), and triggering receptor expressed on myeloid cells 2 (TREM2) (Tato et al., 2016; Henstridge et al., 2019), indicates that it may be a drug target. If a mechanistic connection to a disease model can be established by *in vitro* or *in vivo* genetic or pharmacologic intervention studies (e.g., knocking out CD33) (Bradshaw et al., 2013) this would help to boost interest in the goal. TREM2 is needed for maintaining microglial metabolic fitness during stressful events, allowing for the development of fully mature disease-associated microglia (DAM) profiles and, ultimately, sustaining the microglial response to Aβ-plaque-induced pathology. Different factors make an excellent microglial target. Druggability, specificity, and safety are those significant factors that should be considered.

There are other, more realistic reasons for choosing a target besides relating it to the disease. To be sure, the term "druggable" is not well defined; it typically refers to the biochemical controllability of a ligand, but it may also denote the properties of those ligands in general, such as pharmacokinetics, delivery, metabolism (including pharmacological features of important metabolites), and therapeutic characteristics such as biochemical stability, and manufacturability. It is not enough to create an effective pharmacological compound against a drug target; it is also critical that the molecule modulates only this one target to prevent off-target effects that can make it difficult to interpret the experimental results. However, the most critical criterion for goal selection is safety. Many drugs fail during clinical development, not because of a lack of effectiveness, but because of adverse safety effects that were not adequately anticipated by preclinical studies. Toxicity can occur not only from modifying the target itself, implying a narrow therapeutic window, but also from off-target effects of a drug or from a toxic metabolite of the drug. As a result, it is essential to choose targets with the best chance of being safe as soon as possible (Biber et al.,

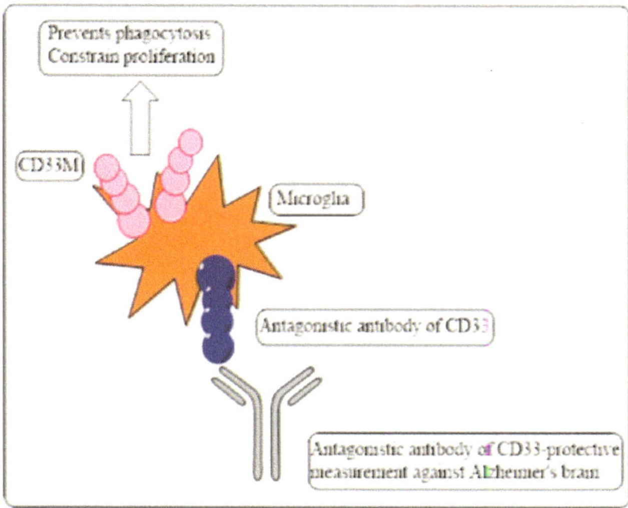

FIGURE 8.3 Antagonistic approach toward CD33.

2019). In Figure 8.3, it has been noted that the inhibitory action of CD33 is an essential approach to AD treatment.

PROMISING ROLE OF IMMUNE RECEPTORS OF MICROGLIA IMMUNE IN AD

Microglia, like other tissues' macrophages (Mφs), express a number of activating and inhibitory immune receptors, each with its own set of properties. Microglia's functional program is controlled by the balance of opposing signals transmitted by these receptors (Ulland et al., 2018). Recent large-scale genome-wide association studies (GWAS) have shown that people with unusual variants of certain immune receptors expressed in microglia, such as paired immunoglobulin-like type 2 receptor (PILR), CD3358, and TREM2, have a significantly higher risk of developing late-onset AD (Guerreiro et al., 2013b; Jonsson et al., 2013). The finding that microglial immune receptor genetic polymorphisms influence the likelihood of developing Alzheimer's disease confirms the critical role of microglia in regulating the development of AD pathology and identifies particular activating and inhibitory receptors as potential therapeutic targets. PILRα and PILRβ are two members of the PILR family that are expressed mainly on various myeloid cell types, including microglia, and have numerous roles in innate immune control in different organisms (Lu et al., 2014; Sun et al., 2012; Tato et al., 2012). Both PILRα and PILRβ are type I transmembrane proteins with identical extracellular domains but distinct intracellular signaling domains. By interacting with SHP-1 and SHP-2, PILRα produces two immunoreceptor tyrosine-based inhibitory motifs that block signal transduction (Mousseau et al., 2000). PILRβ, on the other hand, is associated with DAP12, a signaling adaptor with a tyrosine-based activation motif that recruits the protein tyrosine kinase Syk to deliver active signals (Shiratori et al., 2004). PILRα G78R (rs1859788) is a probable protective variant at the 7q21 locus (rs1476679) that reduces AD risk, according to a recent whole genome sequencing study of 1357 European ancestry samples. PILRα binds to HSV-1 glycoprotein B, which has been proposed as a key cause of AD (Satoh et al., 2008; Harris et al., 2015). HSV-1 binding to PILRα is impaired by G78R, and human monocyte-derived Mφs expressing R78/R78 are less susceptible to HSV-1 infection *in vitro* than G78/G78 Mφs, meaning that G78R can protect against AD by reducing HSV-1 infection. However, the ligands of PILRα and PILRβ during AD pathogenesis, as well as the mechanism by which PILR–ligand interaction reduces AD risk *in vivo*, are still unknown, and further research is required. CD33 is an immunoglobulin superfamily cell surface molecule that microglia generate in large quantities. By attracting the protein phosphatases SHP1 and SHP2, it binds sialic acid and delivers intracellular inhibitory signals. Individuals with a minor CD33 allele variant (rs3865444) that is associated with relatively high levels of CD33 expression have a higher risk of Alzheimer's disease, possibly due to CD33's inhibitory effects on microglial activation, especially phagocytosis of Aβ. Another CD33 variant associated with low levels of CD33 expression, on the other hand, may be protective against Alzheimer's disease. In APP/PS1 mice lacking CD33, insoluble Aβ1-42 and amyloid plaque burden in the brain are significantly reduced (Griciuc et al., 2013). Overall, human and mouse evidence supports the idea that CD33-mediated inhibitory signaling helps AD progression. TREM2 is an immunoglobulin superfamily cell surface receptor that is only expressed by microglia in the CNS. TREM2 binds phospholipids that are anionic or zwitterionic, lipoproteins such as HDL and LDL, DNA, and other polyanions (Daws et al., 2003; N'Diaye et al., 2009; Atagi et al., 2015; Butler et al., 2021). It primarily transmits intracellular signals through the DAP12 signaling adaptor (encoded by TYROBP) (Lanier, 2009). Src kinases phosphorylate DAP12 on cytoplasmic tyrosine residues as it binds ligands to TREM2. Phosphorylated DAP12 recruits the protein tyrosine kinase Syk, which activates phospholipase C gamma 2, PI3K, and MAPKs, among other downstream signaling mediators. The synthesis of inositol triphosphate and the release of inositol triphosphate-gated Ca^{2+} stores by phospholipase C gamma 2 promote intracellular Ca^{2+} elevation (Colonna, 2003; Ford et al., 2009). TREM2 or TYROBP gene loss-of-function mutations were first identified as the cause of Nasu–Hakola disease, a very rare type of autosomal recessive early-onset dementia, providing clear evidence that a lack of TREM2 expression or TREM2-activated signaling facilitates neurodegenerative diseases (Zhou et al., 2018; Song et al., 2018). T66M and Y38C homozygous mutations in the TREM2 gene, which cause protein misfolding and degradation, were later related to frontotemporal dementia (Guerreiro et al., 2013a; Thelen et al., 2014). GWAS have recently identified rare TREM2 variants as genetic risk factors for non-familial AD. The rs75932628 polymorphism, which causes an arginine-to-histidine transition at TREM2's amino acid 47 (R47H), increases the risk of AD2 by a factor of three (Abduljaleel et al., 2014; Pottier et al., 2013; Jin et al., 2014). TREM2 expression was found at both the transcriptional and protein levels in the brains of AD patients and mice with AD-like pathology, implying that TREM2 is directly involved in the microglial response to AD lesions. In the APPPS1-21 and 5XFAD mouse models of AD, TREM2 haploinsufficiency and complete deficiency prevented microglia proliferation and clustering around amyloid plaques (Wang et al., 2015; Ulrich et al., 2014). In TREM2-deficient mice, there was an abnormal accumulation of amyloid beta plaques in the late stages of AD progression, as well as the formation of dystrophic neurites around the plaques (Wang et al., 2016). Overexpression of TREM2 in cultured primary microglia, on the other hand, aided Aβ1-42 phagocytosis and suppressed Aβ1-42-induced pro-inflammatory responses (Jiang et al., 2014). Furthermore, in APP/PS1 mice, overexpression of TREM2 by intracerebral injection of lentiviral particles encoding mouse TREM2 reduced Aβ deposition, neuroinflammation, and neuronal and synaptic loss, resulting in better spatial cognitive function. Similarly, crossing 5XFAD mice with a BAC transgenic mouse expressing human TREM2 in microglia enhanced neuropathology and memory deficits. These findings indicate that TREM2 activation can help delay the progression of Alzheimer's disease.

DRUGS TARGETING MICROGLIA FOR EXHIBITING NEUROPROTECTIVE ACTIVITY

There are several drugs available that can be used in Alzheimer's treatment by directing the polarization of microglia (Wang et al., 2017; Ślusarczyk et al., 2018; Li et al., 2018; Pan et al., 2019; Park et al., 2019). These drugs are listed in Table 8.1.

TABLE 8.1

List of drugs used in Alzheimer's disease targeting microglia.

Drugs (Ref)	Chemical Structure	Pharmacological activities
Betulinic acid (Li et al., 2018)	{{{inline8.1}}} 9-hydroxy-5b,8,8,11a-tetramethyl-1-(prop-1-en-2-yl)icosahydro-3aH-cyclopenta(a)chrysene-3a-carboxylic acid	Endorsed M2 polarizing capability and reserved M1 polarization in LPS-triggered BV-2 microglial cells
Telmisartan (Xu et al., 2015)	{{{inline8.2}}} 4'-((1,7'-dimethyl-2'-propyl-1H,3'H-(2,5'-bibenzo(d)imidazol)-3'-yl)methyl)-(1,1'-biphenyl)-2-carboxylic acid	It inhibits glutamate-induced inflammation in cultured primary neurons and aided the transition from M1 to M2 polarization in LPS-stimulated BV2 and primary microglia cells.
5-Amino-4-imidazole carboxamide riboside (Giri et al., 2004)	{{{inline8.3}}} 5-amino-1H-imidazole-4-carboxamide compound with (*3R,4S,5R*)-5-(hydroxymethyl) tetrahydrofuran-2,3,4-triol (1:1)	It represses LPS-induced appearance of pro-inflammatory cytokines and i-NOS in primary rat astrocytes, and microglia.
Pioglitazone (Mandrekarcolucci et al., 2012)	{{{inline8.4}}} 5-(4-(2-(5-(1-hydroxyethyl)pyridin-2-yl)ethoxy) benzyl) thiazolidine-2,4-dione	In LPS-stimulated N9 microglial cells, reversed the imbalance of M1 and M2 inflammatory cytokines
DSP-8658 (Yamanaka et al., 2012)	{{{inline8.5}}} (*E*)-2,2-dimethyl-3-(4-(3-(2-(4-methylbenzoyl)-1H-pyrrol-1-yl)prop-1-en-1-yl)phenoxy)propanoic acid	In (APP/PS1) transgenic mice, increased Ap absorption by microglia, decreased Ap levels in the cortex and hippocampus, and improved spatial memory function.
Rutin (Pan et al., 2019)	{{{inline8.6}}} 2-(3,4-dihydroxyphenyl)-5,7-dihydroxy-3-((3,4,5-trihydroxy-6-(((3,4,5-trihydroxy-6-methyltetrahydro-2H-pyran-2-yl)oxy)methyl)tetrahydro-2H-pyran-2-yl)oxy)-4H-chromen-4-one	It increases synaptic plasticity, restored spatial learning and memory impairment, and promoted Mitochondrial Oxidative Phosphorylation System (OXPHOS) and microglia-mediated Aβ clearance.
Salidroside (Wang et al., 2017)	{{{inline8.7}}} 2-(hydroxymethyl)-6-(4-hydroxyphenethoxy)tetrahydro-2H-pyran-3,4,5-triol	It prohibits mTOR signals which leads to inhibition of the polarization of M1 phenotype.
Fasudil (Chen et al., 2014)	{{{inline8.8}}} 5-(piperazin-1-ylsulfonyl)isoquinoline	Antigen-reactive T cell proliferation is inhibited by inhibiting the Rho kinase enzyme.
Tianeptine (Ślusarczyk et al., 2018)	{{{inline8.9}}} 7-((3-chloro-6-methyl-5,5-dioxido-6,11-dihydrodibenzo(c,f)(1,2)thiazepin-11-yl)amino)heptanoic acid	Toll-like receptor 4 (TLR4)/NF-KB pathway inhibition.
N,N'-Diacetyl-p-Phenylenediamine (DAPPD) (Park et al., 2019)	{{{inline8.10}}} N,N'-(1,4-phenylene)diacetamide	It helps to reduce pro-inflammatory cytokines and promotes anti-inflammatory ones, which enhances spatial learning and memory, reduces Aβ aggregate deposition, and restores microglial activity.

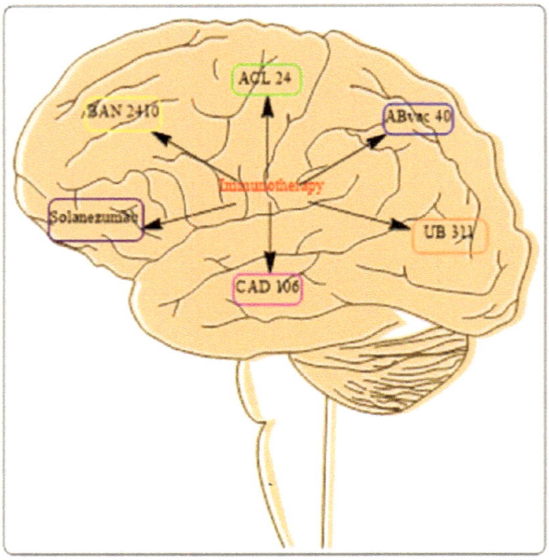

FIGURE 8.4 Presentation of Aβ therapeutics with rationale for AD prevention.

APPROACHES TOWARD AMYLOID ANTICIPATION FOR AD TREATMENT

According to the amyloid hypothesis, amyloid plaques, produced by aggregates of the Aβ peptide created by proteolytic cleavage of amyloid precursor protein (APP), are important to AD pathogenesis (Korczyn, 2008). An increase in the levels of Aβ42 or other factors that can affect the clearance of Aβ leads to the development of amyloidosis (Potter et al., 2013; Patterson et al., 2015). The accumulated Aβ amyloid fibrils form senile plaques, resulting in localized plaque-associated neurological disorders and then facilitate tau pathology (Guo et al., 2020). Multiple treatment techniques targeting Aβ in the brain have been investigated using the amyloid hypothesis. The methods comprise Aβ active or passive immunization (Gilman et al., 2005; Farlow et al., 2015; Pasquier et al., 2016; Salloway et al., 2014; Farlow et al., 2012; Doody et al., 2013), and β- and γ-secretase antagonists or modulators (Barão et al., 2016; Doody et al., 2015). Figure 8.4 represents the possible immunotherapy that can be used for AD treatment.

CONCLUSION

Despite numerous studies demonstrating efficacy in animal models, no disease-modifying medication for Alzheimer's disease has been approved by the Food and Drug Administration (FDA). Both Aβ and tau are essential components of Alzheimer's disease, and either might theoretically be a "disease-modifying" target. Multiple failures to stop AD by facilitating Aβ clearance or inhibiting production have piqued interest in other treatment options, such as neuroinflammation. Microglia were identified as essential actors in neurodegenerative disorders more than two decades ago. The dual role of microglial cells in Alzheimer's disease is a double-edged sword.

On the one hand, they play a vital role in reducing Aβ plaque in the CNS, but their excessive activation can also cause neurotoxicity. Microglia do not need robust provocative action to increase their Aβ clearance efficacy. It's tough to master inflammatory or anti-inflammatory phenotypes based on the AD stage and maintain efficient responses to remove Aβ. As a result, encouraging microglia phagocytosis while avoiding a long-term excessive inflammatory response could be a promising therapeutic method. CD33 and TREM2, microglial receptors that govern microglial pathogenesis and neuroinflammation, have surfaced as therapeutic development targets in AD. CD33 works against TREM2 signaling and is an appealing target since it may be suppressed, for example, by gene therapy, small drugs, or immunotherapy. In addition to the obstruction of Aβ clearance, additional biological processes by which CD33 contributes to the development of AD, such as its function in neuroinflammation and neuronal death, should be studied. TREM2 seems to be a viable therapeutic target, with various agonist antibodies causing receptor signaling to be activated. However, many scientists have verified and are seeking to expand on this theory, as several facets of this theory must be investigated. As a result, by creating strategies and medications that target these receptors (PILR, CD3358, and TREM2), it will be feasible to increase the positive actions of microglial cells, thus slowing the development of AD.

ACKNOWLEDGMENTS

Dipanjan Karati is thankful to Techno India University for its continuous support.

REFERENCES

Abduljaleel Z, Al-Allaf FA, Khan W, et al. Evidence of Trem2 variant associated with triple risk of Alzheimer's disease. *PLoS One* (2014) 9:1–11.

Atagi Y, Liu C-C, Painter MM, et al. Apolipoprotein E is a ligand for triggering receptor expressed on myeloid cells 2 (TREM2). *J Biol Chem* (2015) 290:26043–26050.

Bamberger ME, Harris ME, McDonald DR, Husemann J, Landreth GE. A cell surface receptor complex for fibrillar beta-amyloid mediates microglial activation. *J Neurosci* (2003) 23:2665–2674.

Barão S, Moechars D, Lichtenthaler SF, De Strooper B. BACE1 physiological functions may limit its use as therapeutic target for Alzheimer's disease. *Trends Neurosci* (2016) 39:158–169.

Benilova I, Karran E, De Strooper B. The toxic Aβ oligomer and Alzheimer's disease: an emperor in need of clothes. *Nature Neurosci* (2012) 15(3):349–357.

Biber K, Bhattacharya A, Campbell BM, et al. Microglial drug targets in AD: opportunities and challenges in drug discovery and development. *Front Pharmacol* (2019) 10:1–19.

Blennow K, Leon MJD, Zetterberg H. Alzheimer's disease. *Lancet* (2006) 368:387.

Bradshaw EM, Chibnik LB, Keenan BT, et al. CD33 Alzheimer's disease locus: altered monocyte function and amyloid biology. *Nat Neurosci* (2013) 16(7):848–850.

Butler CA, Thornton P, Brown GC. CD33M inhibits microglial phagocytosis, migration and proliferation, but the Alzheimer's disease protective variant CD33m stimulates phagocytosis and proliferation and inhibits adhesion. *J Neurochem* (2021) 00:1–14.

Calabrò M, Rinaldi C, Santoro G, Crisafulli C. The biological pathways of Alzheimer disease: a review. *AIMS Neurosci* (2020) 8(1):86–132.

Cardona AE, Huang D, Sasse ME, Ransohoff RM. Isolation of murine microglial cells for RNA analysis or flow cytometry. *Nat Protoc* (2006) 1:1947–1951.

Colonna M. TREMs in the immune system and beyond. *Nat Rev Immunol* (2003) 3:445–453.

Daws MR, Sullam PM, Niemi EC, Chen TT, Tchao NK, Seaman WE. Pattern recognition by TREM-2: binding of anionic ligands, *J Immunol* (2003) 171:594–599.

Doody RS, Raman R, Farlow M, et al. A phase 3 trial of semagacestat for treatment of Alzheimer's disease. *N Engl J Med* (2013) 369:341–350.

Doody RS, Raman R, Sperling RA, et al. Peripheral and central effects of γ-secretase inhibition by semagacestat in Alzheimer's disease. *Alzheimers Res Ther* (2015) 7:36.

Farlow M, Arnold SE, Van Dyck CH, et al. Safety and biomarker effects of solanezumab in patients with Alzheimer's disease. *Alzheimers Dement* (2012) 8:261–271.

Farlow MR, Andreasen N, Riviere ME, et al. Long-term treatment with active Aβ immunotherapy with CAD106 in mild Alzheimer's disease. *Alzheimers Res Ther* (2015) 7:1–13.

Fassbender K, Walter S, Kuhl S, et al. The LPS receptor (CD14) links innate immunity with Alzheimer's disease. *FASEB J* (2004) 18:203–205.

Ford JW, McVicar DW. TREM and TREM-like receptors in inflammation and disease. *Curr Opin Immunol* (2009) 21:38–46.

Gilman S, Koller M, Black RS, et al. Clinical effects of Aβ immunization (AN1792) in patients with AD in an interrupted trial. *Neurology* (2005) 64:1553–1562.

Giulian D, Baker TJ. Characterization of ameboid microglia isolated from developing mammalian brain. *J Neurosci* (1986) 6:2163–2178.

Griciuc A, Serrano-Pozo A, Parrado AR, et al. Alzheimer's disease risk gene cd33 inhibits microglial uptake of amyloid beta. *Neuron* (2013) 78:631–643.

Guerreiro R, Wojtas A, Bras J, et al. TREM2 variants in Alzheimer's disease. *N Engl J Med* (2013a) 368:117–127.

Guerreiro RJ, Lohmann E, Brás JM, et al. Using exome sequencing to reveal mutations in TREM2 presenting as a frontotemporal dementia-like syndrome without bone involvement. *JAMA Neurol* (2013b) 70:78–84.

Guo T, Zhang D, Zeng Y, Huang TY, Xu H, Zhao Y. Molecular and cellular mechanisms underlying the pathogenesis of Alzheimer's disease. Mol Neurodegener (2020) 15:40.

Harris SA, Harris EA. Herpes simplex virus type 1 and other pathogens are key causative factors in sporadic Alzheimer's disease. *J Alzheimer's Dis* (2015) 48:319–353.

Heneka MT, Carson MJ, El Khoury J, et al. Neuroinflammation in Alzheimer's disease. *The Lancet Neurol* (2015) 14(4):388–405.

Henkel JS, Beers DR, Zhao W, et al. Microglia in ALS: the good, the bad, and the resting. *J Neuroimmune Pharmacol* (2009) 4:389–398.

Henstridge CM, Hyman BT, Spires-Jones TL. Beyond the neuron-cellular interactions early in Alzheimer disease pathogenesis. *Nat Rev Neurosci* (2019) 20(2):94–108.

Hristovska I, Pascual O. Deciphering resting microglial morphology and process motility from a synaptic prospect. *Front Integr Neurosci* (2015) 9:73.

Ishizuka K, Kimura T, Igata-yi R, Katsuragi S, Takamatsu J, Miyakawa T. Identification of monocyte chemoattractant protein-1 in senile plaques and reactive microglia of Alzheimer's disease. *Psychiatry Clin Neurosci* (1997) 51:135–138.

Jiang T, Tan L, Zhu X-C, et al. Upregulation of TREM2 ameliorates neuropathology and rescues spatial cognitive impairment in a transgenic mouse model of Alzheimer's disease. *Neuropsycho Pharmacol* (2014) 39:2949–2962.

Ji K, Akgul G, Wollmuth LP, Tsirka SE. Microglia actively regulate the number of functional synapses. *PLoS One* (2013) 8:e56293.

Jimenez S, Baglietto-Vargas D, Caballero C, et al. Inflammatory response in the hippocampus of PS1M146L/APP751SL mouse model of Alzheimer's disease: age-dependent switch in the microglial phenotype from alternative to classic. *J Neurosci* (2008) 28:11650–11661.

Jin SC, Benitez BA, Karch CM, et al. Coding variants in TREM2 increase risk for Alzheimer's disease. *Hum Mol Genet* (2014) 23:5838–5846.

Jonsson T, Stefansson H, Steinberg S, et al. Variant of TREM2 associated with the risk of Alzheimer's disease. *N Engl J Med* (2013) 368:107–116.

Kang J, Lemaire HG, Unterbeck A, et al., The precursor of Alzheimer's disease amyloid A4 protein resembles a cell-surface receptor. *Nature* (1987) 325(6106):733–736.

Korczyn AD. The amyloid cascade hypothesis. *Alzheimers Dement* (2008) 4:176–178.

Kreutzberg GW. Microglia: a sensor for pathological events in the CNS. *Trends Neurosci* (1996) 19:312–318.

Kumar A, Singh A, Ekavali. A review on Alzheimer's disease pathophysiology and its management: an update. *Pharmacol Rep* (2015) 67(2):195–203.

Lanier LL. DAP10- and DAP12-associatedreceptors in innate immunity. *Immunol Rev* (2009) 227:150–160.

Li C, Zhang C, Zhou H, Feng Y, Tang F, Hoi MPM, He C, Ma D, Zhao C, Lee SMY. Inhibitory effects of betulinic acid on LPS-induced neuroinflammation involve M2 microglial polarization via CaMKK β-dependent AMPK activation. *Front Mol Neurosci* (2018) 11:98.

Lo AC, Iscru E, Blum D, et al. Amyloid and tau neuropathology differentially affect prefrontal synaptic plasticity and cognitive performance in mouse models of Alzheimer's disease. *J Alzheimers Dis* (2013) 37:109–125.

Lu Q, Lu G, Qi J, et al. PILRα and PILRβ have a siglec fold and provide the basis of binding to sialic acid. *Proc Natl Acad Sci USA* (2014) 111:8221–8226.

Malm TM, Jay TR, Landreth GE. The evolving biology of microglia in Alzheimer's disease. *Neurother* (2015) 12:81–93.

Mandrekarcolucci S, Karlo JC, Landreth GE. Mechanisms underlying the rapid peroxisome proliferator-activated receptor γ-mediated amyloid clearance and reversal of cognitive deficits in a murine model of Alzheimer's disease. *J Neurosci J Soc Neurosci* (2012) 32:10117–10128.

Mawuenyega KG, Sigurdson W, Ovod V, Munsell L, Kasten T, Morris JC, Yarasheski KE, Bateman RJ. Decreased clearance of CNS beta-amyloid in Alzheimer's disease. *Science* (2010) 330:1774.

Mecha M, Carrillo-Salinas FJ, Feliu A, et al. Microglia activation states and cannabinoid system: therapeutic implications. *Pharmacol Ther* (2016) 166:40–55.
Michaud JP, Hallé M, Lampron A, Thériault P, Préfontaine P, Filali M, Triboutjover P, Lanteigne AM, Jodoin R, Cluf C. Toll like receptor 4 stimulation with the detoxified ligand mono-phosphoryl lipid A improves Alzheimer's disease-related pathology. *Proc Natl Acad Sci USA* (2013) 110(5):1941–1946.
Mishra S, Palanivelu K. The effect of curcumin (turmeric) on Alzheimer's disease: an overview. *Ann Indian Acad Neurol* (2008) 11(1):13–19.
Mousseau DD, Banville D, L'Abbé D, Bouchard P, Shen S-H. PILRα, a novel immune receptor tyrosine-based inhibitory motif-bearing protein, recruitsSHP-1 upon tyrosine phosphorylation and is paired with the truncated counterpart PILRβ. *J Biol Chem* (2000) 275:4467–4474.
N'Diaye EN, Branda CS, Branda SS, et al. TREM-2 (triggering receptor expressed on myeloid cells 2) is a phagocytic receptor for bacteria. *J Cell Biol* (2009) 184:215–223.
Pan RY, Ma J, Kong X-X, et al. Sodium rutin ameliorates Alzheimer's disease–like pathology by enhancing microglial amyloid-β clearance. *Sci Adv* (2019) 5(2):6328.
Park MH, Lee M, Nam G, et al. N, N'-Diacetyl-p-phenylenediamine restores microglial phagocytosis and improves cognitive defects in Alzheimer's disease transgenic mice. *Proc Nat Acad Sci USA* (2019) 116(47):23426–23436.
Parkhurst CN, Yang G, Ninan I, et al. Microglia promote learning-dependent synapse formation through brain-derived neurotrophic factor. *Cell* (2013) 155(7):1596–1609.
Paolicelli RC, Bolasco G, Pagani F, et al. Synaptic pruning by microglia is necessary for normal brain development. *Science* (2011) 333(6048):1456–1458.
Pasquier F, Sadowsky C, Holstein A, et al. Two phase 2 multiple ascending-dose studies of vanutide cridificar (ACC-001) and QS-21 adjuvant in mild-to-moderate Alzheimer's disease. *J Alzheimers Dis* (2016) 51:1131–1143.
Patterson BW, Elbert DL, Mawuenyega KG, et al. Age and amyloid effects on human central nervous system amyloid-beta kinetics. *Ann Neurol* (2015) 78:439–453.
Perl DP. Neuropathology of Alzheimer's disease. *Mt Sinai J Med, NY* (2010) 77(1):32–42.
Pierre WC, Smith PL, Londono I, Chemtob S, Mallard C, Lodygensky GA. Neonatal microglia: The cornerstone of brain fate. *Brain Behav Immun* (2017) 59:333–345.
Potter R, Patterson BW, Elbert DL, et al. Increased in vivo amyloid-b42 production, exchange, and loss in presenilin mutation carriers. *Sci Transl Med* (2013) 5:189ra77.
Pottier C, Wallon D, Rousseau S, et al. TREM2 R47H variant as a risk factor for early-onset Alzheimer's disease. *J Alzheimers Dis* (2013) 35:45–49.
Qi Y, Klyubin I, Harney SC, et al. Longitudinal testing of hippocampal plasticity reveals the onset and maintenance of endogenous human Ass-induced synaptic dysfunction in individual freely behaving pre-plaque transgenic rats: rapid reversal by anti-Ass agents. *Acta Neuropathol Commun* (2014) 2:175.
Ransohoff RM. A polarizing question: do M1 and M2 microglia exist? *Nat Neurosci* (2016) 19:987–991.
Salloway S, Sperling R, Fox NC, et al. Two phase 3 trials of Bapineuzumab in mild-to-moderate Alzheimer's disease. *N Engl J Med* (2014) 370:322–333.
Satoh T, Arii J, Suenaga T, et al. PILRα is a herpes simplex virus-1 entry coreceptor that associates with glycoprotein B. *Cell* (2008) 132:935–944.
Shimada H, Kitamura S, Shinotoh H, et al. Association between Aβ and tau accumulations and their influence on clinical features in aging and Alzheimer's disease spectrum brains: A [(11)C] PBB3-PET study. *Alzheimer's Dement* (Amsterdam, Netherlands) (2017) 6:11–20.
Shiratori I, Ogasawara K, Saito T, Lanier LL, Arase H. Activation of natural killer cells and dendritic cells upon recognition of a novel CD99 like ligand by paired immunoglobulin-like type 2 receptors. *J Exp Med* (2004) 199:525–533.
Sierra A, Encinas JM, Deudero JJP, et al. Microglia shape adult hippocampal neurogenesis through apoptosis-coupled phagocytosis. *Cell Stem Cell* (2010) 7(4):483–495.
Sierra A, Gottfried-Blackmore AC, McEwen BS, Bulloch K. Microglia derived from aging mice exhibit an altered inflammatory profile. *Glia* (2007) 55:412–424.
Ślusarczyk J, Trojan E, Głombik K, Piotrowska A, Budziszewska B, Kubera M, Popiołekbarczyk K, Lasoń W, Mika J, Bastakaim A Targeting the NLRP3 inflammasome-related pathways via tianeptine treatment-suppressed microglia polarization to the M1 phenotype in lipopolysaccharide-stimulated cultures. *Int J Mol Sci* (2018) 19.
Song WM, Colonna M. The identity and function of microglia in neurodegeneration. *Nat Immunol* (2018) 19:1048–1058.

Stansley B, Post J, Hensley K. A comparative review of cell culture systems for the study of microglial biology in Alzheimer's disease. *J Neuroinflamm* (2012) 9:115.
Stevens B, Allen NJ, Vazquez LE, et al. The classical complement cascade mediates CNS synapse elimination. *Cell* (2007) 31(6):1164–1178.
Stewart CR, Stuart LM, Wilkinson K, et al. CD36 ligands promote sterile inflammation through assembly of a Toll-like receptor 4 and 6 heterodimer. *Nature Immunol* (2010) 11:155–161.
Subramaniam SR, Federoff HJ. Targeting microglial activation states as a therapeutic avenue in Parkinson's disease. *Front Aging Neurosci* (2017) 9:176.
Sun Y, Senger K, Baginski TK, et al. Evolutionarily conserved paired immunoglobulin-like receptor α (PILRα) domain mediates its interaction with diverse sialylated ligands. *J Biol Chem* (2012) 287:15837–15850.
Tang Y, L, W. Differential roles of M1 and M2 microglia in neurodegenerative diseases. *Mol Neurobiol* (2016) 53:1181–1194.
Tato CM, Joyce-Shaikh B, Banerjee A, et al. The myeloid receptor PILRβ mediates the balance of inflammatory responses through regulation of IL 27 production. *PLoS One* (2012) 7:e31680.
Tato PD, Holtman IR, Boddeke EW, Möller T, Eggen BJ. Next generation transcriptomics and genomics elucidate biological complexity of microglia in health and disease. *Glia* (2016) 64(2):197–213.
Thakur AK, Kamboj P, Goswami K, Ahuja K. Pathophysiology and management of Alzheimer's disease: an overview. *J Analy Pharm Res* (2018) 7(2):226–235.
Thelen M, Razquin C, Hernández I, et al. Investigation of the role of rare TREM2 variants in frontotemporal dementia subtypes. *Neurobiol Aging* (2014) 35:2657.
Tremblay MÈ, Lowery RL, Majewska AK. Microglial interactions with synapses are modulated by visual experience. *PLoS Biol* (2010) 8(11):e1000527.
Ulland TK, Colonna M. TREM2—a key player in microglial biology and Alzheimer disease. *Nat Rev Neurol* (2018) 14:667–675.
Ulrich JD, Finn M, Wang Y, et al. Altered microglial response to Aβ plaques in APPPS1-21 mice heterozygous for TREM2. *Mol Neurodegener* (2014) 9:1–9.
Wake H, Moorhouse AJ, Jinno S, Kohsaka S, Nabekura J. Resting microglia directly monitor the functional state of synapses in vivo and determine the fate of ischemic terminals. *J Neurosci* (2009) 29(13):3974–3980.
Wang C, Wang Q, Lou Y, Xu J, Feng Z, Chen Y, Tang Q, Zheng G, Zhang Z, Wu Y. Salidroside attenuates neuroinflammation and improves functional recovery after spinal cord injury through microglia polarization regulation. *J Cell Mol Med* (2017) 22:1148–1166.
Wang Y, Cella M, Mallinson K, et al. TREM2 lipid sensing sustains the microglial response in an Alzheimer's disease model. *Cell* (2015) 160:1061–1071.
Wang Y, Ulland TK, Ulrich JD, et al. TREM2-mediatedearlymicroglial response limits diffusion and toxicity of amyloid plaques. *J Exp Med* (2016) 213:667–675.
Wes PD, Sayed FA, Bard F, Gan L. Targeting microglia for the treatment of Alzheimer's disease. *Glia* (2016) 64:1710–1732.
Wong WT. Microglial aging in the healthy CNS: Phenotypes, drivers, and rejuvenation. *Front Cell Neurosci* (2013) 7:22.
Xia MQ, Qin SX, Wu LJ, Mackay CR, Hyman BT. Immunohistochemical study of the beta-chemokine receptors CCR3 and CCR5 and their ligands in normal and Alzheimer's disease brains. *Am J Pathol* (1998) 153:31–37.
Xu H, Wang Z, Li J, et al. The polarization states of microglia in TBI: a new paradigm for pharmacological intervention. *Neural Plast* (2017) 5405104. https://doi.org/10.1155/2017/5405104
Xu Y, Xu Y, Wang Y, Wang Y, He L, Jiang Z, Huang Z, Liao H, Li J, Saavedra JM. Telmisartan prevention of LPS-induced microglia activation involves M2 microglia polarization via CaMKK β-dependent AMPK activation. *Brain Behav Immun* (2015) 50:298–313.
Yamanaka M, Ishikawa T, Griep A, Axt D, Kummer MP, Heneka MT. PPARγ/RXRα-induced and CD36-mediated microglial amyloid-β phagocytosis results in cognitive improvement in amyloid precursor protein/presenilin 1 mouse. *J Neurosci* (2012) 32:17321–17331.
Yao K, Zu HB, Microglial polarization: novel therapeutic mechanism against Alzheimer's disease. *Infllammopharmacol* (2019) 28:95–110.
Youm YH, Grant RW, McCabe LR, et al. Canonical Nlrp3 inflammasome links systemic low-grade inflammation to functional decline in aging. *Cell Metabol* (2013) 18:519–532.
Zhou Y, Ulland TK, Colonna M. TREM2-dependent effects on microglia in Alzheimer's disease. *Front Aging Neurosci* (2018) 10:1–7.

9 Neuropharmacological Mechanisms of SARS-CoV-2

Swati Kaushik[1], Manisha Singh[1], Sonika Srivastava[2], Anupam K. Sachan[3], and Saroj Verma[1]*

[1]Department of Pharmacy, School of Medical and Allied Sciences, K.R. Mangalam University, Gurugram, India
[2]Department of Pharmacology, SGT College of Pharmacy, SGT University, Gurugram, Haryana, India
[3]Dayanand Dinanath College, Institute of Pharmacy, Kanpur, Uttar Pradesh, India
*Corresponding author

INTRODUCTION

Coronavirus is mainly responsible for respiratory illness in humans. The word "corona" originated from crown-like spikes present on the surface of the virus. This virus was initially detected in bats, cats and camels. From there, it further mutated and infected other species, one of which was humans. These viruses are RNA viruses (enveloped, positive-sense, single-stranded RNA viruses that were first isolated from humans) (Tyrrell and Bynoe, 1966). They have been studied extensively for decades and found to spread diseases like severe acute respiratory syndrome (SARS), Middle East respiratory syndrome (MERS) and the common cold (Chowdhury and Oommen, 2020).

The virus mutated over a period of time and recently emerged as the causative agent of a novel highly contagious viral disease, which was identified in humans in Wuhan, Hubei, a city in China. The disease was declared a pandemic in March 2020, and the virus was named the "2019 novel coronavirus" (2019-nCoV) or human coronavirus 2019 (HCoV-19/hCoV-19) (Zhu *et al.*, 2020). In February 2020, it was renamed "severe acute respiratory syndrome coronavirus 2" (SARS-CoV-2). This name was adopted as the official name by the International Committee on Taxonomy of Viruses (ICTV). SARS-CoV-2-induced infection was recognized by the World Health Organization (WHO), and it was termed "coronavirus disease 2019" or "COVID-19," and was subsequently declared a pandemic in March 2020 (WHO, 2020).

After originating in China, the first case reported was from Thailand, which was preceded by the spread of the infection globally within a few months, except Antarctica (WHO statement on novel coronavirus in Thailand, 2020). The virus spread at an alarming rate, with high morbidity and mortality among infected patients by the end of 2020 (WHO: COVID-19 Dashboard; Ministry of Health and Family Welfare). The SARS-CoV-2 virus has a complex physiological nature, and the behaviour of this viral organism should be understood in order to comprehend its pathophysiology. The next paragraph explains the disease caused by SARS-CoV-2 (COVID-19), along with its main signs and symptoms.

SARS CoV-2 is a novel β-coronavirus capable of causing severe respiratory illness in humans (Cascella *et al.*, 2022) and it is the main causative agent of COVID-19 disease. The symptomatic manifestation of COVID-19 includes fever, dry cough, breathlessness and severe infections like pneumonia (severe infection) (Wang *et al.*, 2020). Initial stages of infection may be identified by the loss of sensory responses, such as loss of taste and smell in the infected patient. Other symptoms

DOI: 10.1201/9781003506317-9

include diarrhoea (gastrointestinal distress), cardiovascular abnormalities, and minor symptoms such as headache, nausea, vomiting, impaired consciousness, encephalitis and acute cerebrovascular diseases (including stroke, venous sinus thrombosis and intracerebral haemorrhage) (Spudich and Nath, 2022). As the disease progresses and the patient gets proper rest and medication, the symptoms dampen, but the story doesn't end here. It comes with a number of post-disease complications. The most common and prevailing post-COVID symptoms involve respiratory diseases and neurological issues (ranging from mild headache and delirium to severe neurological disorders) in recently recovered individuals. Since the lungs were the most affected part of the body during the initial stages of the pandemic, doctors only focussed on monitoring the lung condition of the patient. The major goal of the physicians was to keep the patient alive by maintaining normal lung health. Although respiratory symptoms were maximally reported, brain-related symptoms were also observed and reported (Yu *et al.*, 2020). Over time, people became more aware of the changes occurring in the brain after the COVID pandemic, and awareness about this aspect of post-COVID complications increased (Wu et al., 2020). This chapter highlights the pathophysiological changes occurring in COVID patients post-recovery.

The reported symptoms of neurological changes post-COVID (post-SARS-CoV-2 infection) range from simple delirium to brain stroke, haemorrhage, memory loss and lifelong brain damage. The main problem currently faced by physicians is the lack of information regarding the basic underlying mechanisms leading to these neurological changes post-COVID. As there are a plethora of aspects that may be the cause of these neurological changes, there is a need to study the pathology of the virus along with the different pathophysiological mechanisms related to the neurological conditions faced by the patients. This chapter gives insight into the pathology of the virus as well as the different mechanisms related to neurological changes occurring after the pandemic hit the world.

SARS-COV-2

Classification

Family: Coronaviridae. Subfamily: Coronavirinae. Genera: *Alphacoronavirus*, *Betacoronavirus*, *Gammacoronavirus* and *Deltacoronavirus*. Genome size of CoV: 27–32 kb [a single-stranded positive-sense RNA (+ssRNA)] (Brian and Baric, 2005).

Structure

SARS-CoV-2 has a capsid made up of nucleocapsid (N) protein and contains an envelope that is made up of three structural proteins, namely, spike (S) protein, membrane (M) protein and envelope (E) protein (Brian and Baric, 2005).

Transmission of SARS-CoV-2

S protein is the entry target of the virus responsible for binding to the host angiotensin converting enzyme 2 (ACE-2). It is the antigenic component of the virus (Figure 9.1). S protein has two subunits. One is the S_1-membrane binding subunit which binds to the host cell receptor, and the other is the S_2-membrane fusion subunit which is responsible for the fusion of viral and host-cell membranes (Huang *et al.*, 2020; Letko *et al.*, 2020). After binding to the ACE-2 receptor, the S protein undergoes proteolytic cleavage catalyzed by a host membrane-anchored protein. It is a type 2 transmembrane (TM) serine protease enzyme (present on the surface of the host cell), i.e., TM protease serine 2 (TMPRSS2), which is responsible for the activation of the S protein. Activation of this protein facilitates the entry of the viral contents inside the host cell as the TMPRSS2 protein

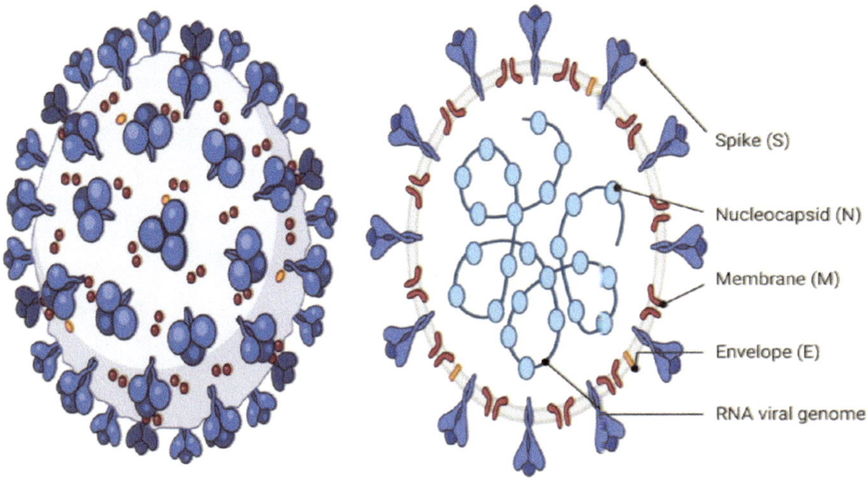

FIGURE 9.1 Structure of SARS-CoV-2: Double-layer enveloped viral genome along with nucleocapsid (N) and three membrane-bound structural proteins: spike (S) protein, membrane (M) protein and envelope (E) protein.

causes conformational changes in the structure of the S protein necessary for viral and host cell–membrane fusion (Jackson *et al.*, 2022).

After fusion of the membranes, viral RNA is released inside the host cell. The RNA strands replicate and form new proteins. These new proteins are then carried to the endoplasmic reticulum (ER) by the N protein. The nucleocapsids are formed and are further transported to the lumen, and then ultimately to the Golgi vesicles. In the end, the viral content moves to the adjacent cells through exocytosis and also to the upper respiratory tract for community spread of the COVID infection (Figure 9.2).

The early phases of viral infection in the human body remain confined to the lungs and upper respiratory tract. The interaction of immune cells with the infected epithelial cells decides the extent of spread of the viral infection as the immune cells activate cytokine signalling pathways. The transcription factors activated by the immune cells in the lungs after interacting with the infected epithelial cells include interferon regulatory factors (IRFs) and the nuclear factor-κB (NFκB). IRF activation further activates interferons (IFNs)—IFN I and II and IFN genes. IFN genes are responsible for host-cell defence system activation (activation of antiviral proteins of the defence system of host cell).

NFκB activates various pro-inflammatory cytokines and chemokines, which further activate specific types of leucocytes. The virus also activates pro-inflammatory cytokines by triggering inflammasomes. These inflammasomes are multiprotein complexes containing caspase-1 protein. Caspase-1 is responsible for the cleavage of precursor molecules and activation of ILs like IL-1β and IL-18. The viral gene has certain proteins, including non-structural protein-1 (NSP-1), accessory proteins (ORF 6 and ORF 3B) and the nucleocapsid. All these proteins of the viral genome inactivate the signal transducer and activator of transcription-1 protein (STAT-1), which decreases IFN synthesis in the host cell and thus the infection prevails and spreads.

As all these actions have been initiated after the binding of S protein to the host cell receptor protein (ACE-2) present on the cell surface. This special type of protein is responsible for the entry of the viral content into the human body and the different sites where this protein is present, making it feasible for the virus to invade different human body systems easily.

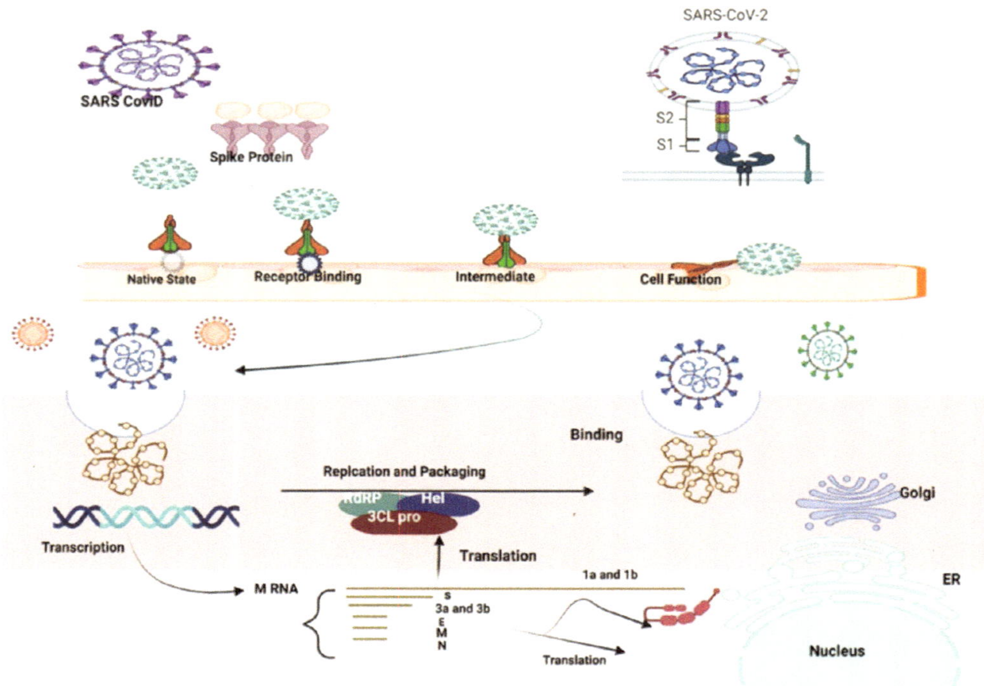

FIGURE 9.2 Transmission of SARS-CoV-2 inside the human body.

ACE-2

ACE-2 is a cell surface protein primarily expressed in the ciliated epithelium of the nasopharynx, respiratory tract (upper respiratory tract), epithelial layer of the bronchi and bronchioles, mast cells, macrophages, monocytes, and vascular endothelial cells. ACE-2 receptors are present in abundance in the lungs, followed by the brain, liver, kidney and heart (Cheng *et al.*, 2007).

These receptors are primarily present in the respiratory system, which is why lungs are the major site for infection due to COVID-19. The brain is the second most affected human body organ. This chapter includes a detailed insight into the neurological pathophysiology of SARS-CoV-2 (Zhang *et al.*, 2021).

PATHOLOGICAL ASPECTS UNDERLYING SARS-COV-2-INDUCED NEUROLOGICAL CHANGES

SARS-CoV-2 is selective in affecting the human body. It may enter the brain tissues using different mechanisms, including directly invading the CNS, by binding to the ACE-2 receptor proteins, through an outburst of ILs (hyperactivity of the immune system) by inducing hypoxia (leading to CNS invasion of the virus) and certain miscellaneous factors (Guadarrama-Ortiz *et al.*, 2020).

Direct CNS Invasion

SARS-CoV-2 virus has been reported to affect the brain directly as well as indirectly (Koyuncu *et al.*, 2013; Leber *et al.*, 2016). The direct invasion pathway includes two further types of subpathways: the haematogenous pathway (blood circulation pathway) and the neuronal pathway (olfactory pathway).

Haematogenous Route (Blood Circulation Pathway)

Primary infection is followed by invasion of the virus into the bloodstream. It is the phase when the virus is present in the bloodstream, and it finally moves to the brain (Keyhanian *et al.*, 2021).

This virus can potentially disrupt the epithelial barrier of the respiratory system and attack the bloodstream. As ACE-2 receptors are primarily housed on the surface of type II alveolar epithelial cells, these cells are the primary victims of SARS-CoV. The virus breaks through the alveolar cells and enters the blood circulation.

Gastrointestinal (GI) epithelial cells also contain ACE-2 receptors; thus, the GI tract may also play a crucial role in infecting the bloodstream with SARS-CoV-2 (Sia *et al.*, 2020).

Neuronal Pathway (Olfactory Nerve Pathway)

Invasion of the nerves of different areas of the human body is the easiest way to invade the brain and thus the CNS. Olfactory neurons are directly connected to the nasal cavity (van Riel *et al.*, 2015; Butowt and Bilinska, 2020) and many viruses target the brain using this very pathway, avoiding the blood–brain barrier (BBB) (Bryche *et al.*, 2020).

The olfactory neuron dendrites are projected in the nasal cavity, and the axons extend up to the olfactory bulb of the brain through the lamina cribrosa (Figure 9.3) (van Riel *et al.*, 2015). ACE-2 and TMPRSS2 proteins are present in the olfactory epithelial cells, thus they can also be a possible route of invasion of the virus into the CNS (Bilinska *et al.*, 2020; Kanageswaran *et al.*, 2015). Anosmia may be a symptom that surfaces due to the invasion of the SARS-CoV-2 virus through the olfactory pathway (Figure 9.3). The viral content has also been reported to be present in the trigeminal nerve branches (Figure 9.3).

The nerves of the nasal mucosa also connect to the trigeminal ganglion, and thus they connect to the trigeminal nerve. Viral invasion and damage to these nerves may be responsible for anosmia and headache in COVID-19 patients, and it may also lead to neuronal damage and necrosis, ultimately causing cognitive impairment (Figure 9.3).

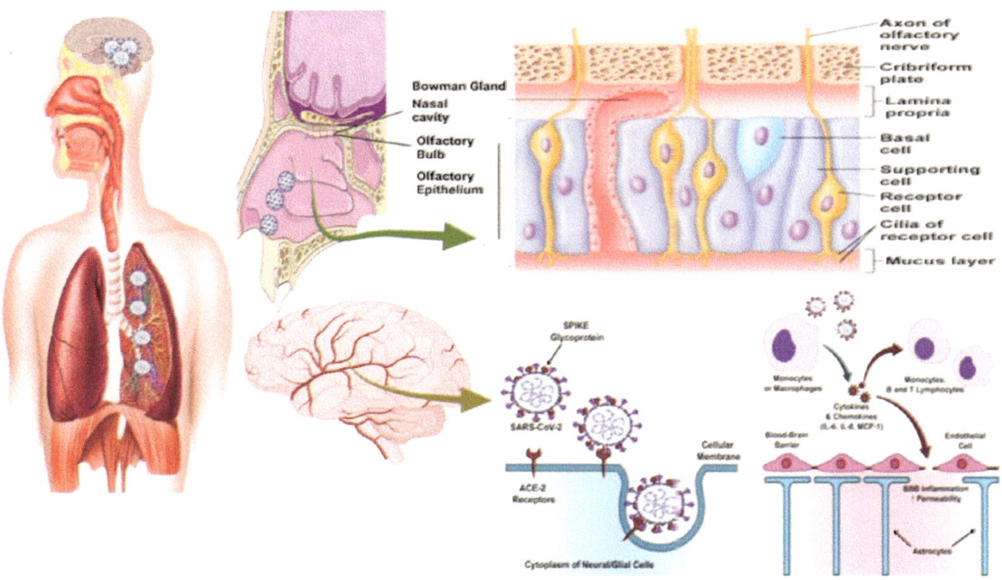

FIGURE 9.3 Different areas of the CNS affected by viral infection after invasion in the human body and the mechanism of neurological spread of the SARS-CoV-2 virus: (a) via ACE-2 receptor binding; and (b) via cytokine flood due to a hyperactive host immune system.

ACE-2-Mediated CNS Infection

It is well known that the initial infection occurs in the lung epithelial tissues and spreads further to adjacent tissues and other organs depending on the availability of receptors (ACE-2 in the case of SARS-CoV-2) required for the binding and fusion of the virus to the host cell. Recent studies have reported that brain tissues express ACE-2 receptors on the glial cells and neurons. These receptors are most prominent in the brainstem, paraventricular nucleus (PVN) and nucleus tractus solitarius (NTS) (Xia and Lazartigues, 2010). Thus, these receptors play a crucial role in providing entry to the viral genome inside the cell, and because they are reported to be found in brain areas too, it may be a plausible reason underlying the mild to severe neurological symptoms reported by patients.

Immune Cell Hyperactivity-Induced CNS Infection

As of now, it is clear that the presence of ACE-2 is very important for the invasion of SARS-CoV-2 inside human body organs, but it is not the sole player in spreading infection in the brain tissues. Cytokine overload also plays crucial role in increasing the concentration of SARS-CoV-2 viral components inside brain tissues. This action of cytokines is due to the hyperactivity of the immune system (in response to the initial infection of the virus in the respiratory system and further in the blood circulation). This cytokine storm may lead to systemic inflammatory syndrome (SIRS) or SIRS-like phenomena (Qin *et al.*, 2020).

The ILs, IL-6 and IL-8 are mainly found to be responsible for the widespread infection (Bell *et al.*, 1996; Kaplanski *et al.*, 1995). The monocyte chemoattractant protein-1 (MCP-1) also releases IL-6 and IL-18, and it is also involved in the transmigration of monocytes across the BBB. This may be the possible reason for inflammation of the CNS after hyperactivity of the immune system during or after COVID-19 infection (Sankowski *et al.*, 2015).

Hypoxia-Induced CNS Invasion

One more factor of concern is that recent studies have found that the SARS-CoV-2 virus can also cause damage to the CNS without invading the organ, i.e., by causing hypoxemia, ultimately predisposing various neurological symptoms like brain hypoxia (with vasodilation), hyperemia and brain edema, which may further lead to serious problems like unconsciousness and coma (Abdennour *et al.*, 2012).

After invasion of the CNS, the virus causes neurological changes in the brain which are diagnosed as mild to severe symptoms in patients post-COVID. This chapter further includes discussion about different neurological complications related to SARS-CoV-2.

NEUROLOGICAL COMPLICATIONS RELATED TO SARS-COV-2

The neurological manifestations due to SARS-CoV-2 infection range from minor infections to severe disease states in humans. Initial symptoms may start with mild dizziness, headache, dysgeusia or anosmia to severe disorders such as stroke, Guillain-Barré syndrome (GBS), acute haemorrhagic necrotizing encephalopathy, meningoencephalitis and cerebral venous thrombosis (Mao *et al.*, 2020; Lechien *et al.*, 2020; Yan *et al.*, 2020; Helms *et al.*, 2020). Other symptoms may include feelings of agitation and confusion, corticospinal tract signs, dysexecutive syndrome (Yan *et al.*, 2020), encephalopathy, autoimmune encephalitis, acute disseminated encephalomyelitis (ADEM) and GBS. The neurological manifestations due to SARS-CoV-2 can be categorized into two major categories, i.e., acute disorders and chronic disorders.

ACUTE DISORDERS DUE TO SARS-COV-2 INFECTION

Anosmia/Ageusia

Anosmia may be defined as a decreased ability to detect any kind of smell (loss of sense of smell). It may be due to a head injury, nasal blockage and infections like COVID-19. It is one of the cardinal symptoms that occurred in COVID-positive patients just after encountering the SARS-CoV-2 virus (Agyeman *et al.*, 2020). Recent reports have studied the underlying mechanism of anosmia and ageusia and concluded that these disorders occur in COVID-19 patients through the pathways of direct entry of the virus, infection and/or death of sustentacular cells (the most preferable pathway for viral entry and infection) (Butowt and Bartheld, 2021). Post-COVID, this disorder is also associated with other complications like cough, chest pain, congested nose, sore throat, GI disturbances, dysphagia, headache, insomnia, arthralgia and myalgia (Algahtani *et al.*, 2022).

Ageusia is defined as the loss of sense of taste. ACE-2 receptors are found to be distributed throughout the nasal and oral cavities. The tongue is mainly involved in the sensation of taste, which is lost because of the inhibition of the ACE-2 receptors by the SARS-CoV-2 virus in the oral cavity. This may be the principal mechanism of the induction of ageusia during COVID-19 infection in humans (Zhou *et al.*, 2020). Post-COVID, this disorder is also associated with other complications, including persistent anosmia, dyspnoea, chest pain, headache and hearing difficulties (Algahtani *et al.*, 2022).

Stroke

Increased risk of acute ischemic stroke was also being reported in COVID-19 patients, which also has a long-term effect on the human body. This risk is multifactorial as it activates a number of inflammatory as well as coagulation cascades, resulting in raised levels of different underlying factors such as erythrocyte sedimentation rate (ESR) levels, lymphopenia and increased fibrin D-dimer levels in the human body. The mechanism behind ischemic stroke may be endothelial cell damage and/or hypercoagulability of blood (Qureshi *et al.*, 2021).

Three main mechanisms appear to be responsible for the occurrence of ischemic strokes in COVID-19 (Oxley *et al.*, 2020; Beyrouti *et al.*, 2020): a hypercoagulable state, vasculitis and cardiomyopathy. As ACE-2 receptors are present in the endothelial layer (Hamming *et al.*, 2004) of the heart, binding of these receptors to the SARS-CoV-2 viral proteins may lead to severe inflammatory responses predisposing thrombotic complications like vasculitis (Varga *et al.*, 2020; Spence *et al.*, 2020).

Encephalitis

Encephalitis was reported to be a major complication in severely ill COVID-19 patients (Paterson *et al.*, 2020). It was reported during the infection period as well as a post-infection complication (Umapathi *et al.*, 2020; Pilotto *et al.*, 2020). The mechanism behind this disorder may be viral neuroinvasion, a disrupted BBB or activation of autoimmunity. Older patient groups are at higher risk of encephalitis as they also suffer from other diseases along with COVID-19 (Dogan *et al.*, 2020; Cao *et al.*, 2020). This disorder has also been reported to be present in patients with asymptomatic conditions of COVID-19 (Radmard *et al.*, 2020). Young children are found to be at higher risk of developing encephalitis after COVID-19 infection (Langley *et al.*, 2020; Zhang *et al.*, 2020). The cause of neuronal infection in children is multifactorial, as all the mechanisms have been reported in different cases of neurological changes in children's brains. The main causes include direct viral invasion (haematogenous pathway) (Pezzini and Padovani, 2020), autoimmune response activation post-COVID-19 infection and indirect viral attack, i.e., cytokine storm activation (Aghagoli *et al.*, 2021).

Encephalopathy

Encephalopathy is defined as any disease or disorder of the brain which alters the function or structure of the brain. It may be manifested as confusion, delirium and deep coma (Elkind, 2022). It may occur due to any infection (viral, bacterial, etc.), cellular malfunction (mitochondrial dysfunction), metabolic disorder, any kind of tumorous growth in the brain, increased pressure in the brain, elemental intoxication (exposure to toxic chemicals, solvents, drugs, radiation, paints, etc.), traumatic conditions, brain hypoxia, ischemia, etc. This condition may lead to altered brain activity and the mental state of the patient. It causes loss of memory and cognition, which may further lead to lethargic behaviour, concentration problems and other related issues (Akimoto *et al.*, 2022). Older COVID-19 patients are more prone to encephalopathy as there are other comorbidities associated with their health like Parkinson's disease, stroke and prior use of antipsychotic drugs. This may cause decreased immunity and allow invasion of the virus easier and infection more drastic (Elkind, 2022). Encephalopathy in children has been reported as a case of altered consciousness (the hallmark of encephalopathy in children) along with other alterations in brain scans (Elkind, 2022).

Peripheral Neuropathy

Peripheral neuropathy (PN) is a result of damage to the peripheral nerves, which can manifest as weakness, and pain in head and feet regions of the body. It also extends to other systems of the body, such as the gastrointestinal tract, urinary system and circulation. Post-COVID, PN has been reported as a common neurological complication. It is observed mostly in patients with comorbidities like diabetes since it affects the immune system. This disease may be caused by the side effects of COVID medications given to patients. The mechanism of PN is found to involve retrograde neuronal transmission through damaged or infected neurons. These neurons may be part of the olfactory system or the vascular endothelial system (Zubair *et al.*, 2020; Algahtani *et al.*, 2016).

Guillain-Barré syndrome is a type of peripheral neuropathy, as it is a disorder of inflamed neurons in the peripheral nervous system due to any kind of viral or bacterial infection of neurons (Levin, 2004). Manifestations related to GBS range from muscular paralysis (ascending) to sensory as well as autonomic neuronal disturbances/dysfunction and respiratory failure (Willison *et al.*, 2016). This disease may be caused by a number of microbes, with SARS-CoV-2 as the newest addition to the list. This disease is more prevalent post-COVID as the nervous system is damaged due to the COVID-19 infection in patients (Jacobs *et al.*, 1998). In children, post-COVID GBS has been reported as a severe complication with a common mechanism as in adults (Dugue *et al.*, 2020; El Mezzeoui *et al.*, 2021).

Myositis

Myositis is defined as inflammation of muscles, more precisely the musculoskeletal system. It is manifested as pain, swelling, tenderness and weakness of the muscles. This disease is more commonly reported in children with multisystem inflammatory syndrome (MIS) (Lindan *et al.*, 2021; Sánchez-Morales *et al.*, 2021). Myositis of the neck and face has been reported in children in the later phases of COVID infection as a complication. The underlying mechanism of the disease post-COVID involves either bioenergetic dysfunction or autoimmune inflammatory damage of the muscles (Mohkhedkar *et al.*, 2021).

Different muscles have been found to be affected post-COVID. Types of myositis include acute viral myositis, rhabdomyolysis, dermatomyositis and paraspinal myositis (Saud *et al.*, 2021).

CHRONIC DISORDERS DUE TO SARS-COV-2 INFECTION

Fatigue

Fatigue is the most prevalent symptom reported in patients recovering from COVID infection. It is a common and enervating sign in patients with post-COVID neurological complications. As fatigue doesn't have any specific cause, its mechanism is not clearly distinguished. Underlying causes of fatigue may be attributed to changes in concentration of neurotransmitters in the nervous system, inflammatory changes in the body, neurological disorders, psychological issues, cognitive impairment and issues with substrate metabolism. Respiratory and cardiac damage may also be counted as predisposing factors precipitating fatigue as a post-COVID complication (Rudroff *et al.*, 2020; de Medeiros Carvalho et al., 2020; Rogers et al., 2020).

Cognitive Impairment

Cognitive impairment is defined as improper functioning of the brain, thus affecting the activities related to the cognitive power of the brain. It is a complication that persists even after six months in patients recovered from COVID-19. The cases are found to be more severe in patients with psychological and cognition-related comorbidities. The pathophysiology underlying cognitive impairment includes microglial inflammation (leading to Alzheimer's disease) (Mandrekar-Colucci and Landreth, 2010) and cardiovascular changes (cognitive impairment and dementia) (Pendlebury and Rothwell, 2019; Solomon *et al.*, 2020).

Depression/Altered Mood

Depression, anxiety and insomnia are some of the most common post-COVID complications reported in the long COVID phase. These complications may occur as a result of psychological changes due to coronavirus invasion in the brain directly, or indirectly by activating immunomodulatory pathways in the body (Khubchandani *et al.*, 2022). Certain stressors like fear of illness, loneliness, future uncertainty, stigma and trauma due to long and drastic diseased conditions play a significant role in defining the underlying mechanism of direct activation of psychological illness, i.e., depression and mood swings. Old age and related comorbidities lead to more severe post-COVID stress-induced depression as well as anxiety in patients (Cucinotta and Vanelli, 2020).

CONCLUSION

SARS-CoV-2-induced COVID-19 infection is a life-threatening infection that caused severe infection in humans and was declared a pandemic by the WHO in 2020. This disease has been found to be associated with severe respiratory infection along with devastating effects on different organ systems of the human body. As the symptoms of COVID-19 included anosmia and/or ageusia, the role of neurological infection came into view, which was not highlighted and was neglected in the initial phase of the pandemic. The current chapter has highlighted the neuropathological aspects of SARS-CoV-2 infection, keeping in mind the different age groups and the severity of infection, either acute or chronic. From this data, it can be concluded that the neurological effects of SARS-CoV-2 are not only restricted to the acute phase of the disease but are also precipitating even after the recovery of the patient, which has been termed as post-COVID consequences or more precisely, long-COVID complications.

ACKNOWLEDGEMENTS

The authors thankfully acknowledge Mr. Debashish Paramanick for graphical assistance.

REFERENCES

Abdennour L, Zeghal C, Deme M, Puybasset L. Interaction brain-lungs. *Ann Fr Anesth Reanim* (2012) 31(6):e101–107.

Aghagoli G, Gallo Marin B, Katchur NJ, Chaves-Sell F, Asaad WF, Murphy SA. Neurological involvement in COVID-19 and potential mechanisms: a review. *Neurocritic Care* (2021) 34:1062–1071.

Agyeman AA, Chin KL, Landersdorfer CB, Liew D, Ofori-Asenso R. Smell and taste dysfunction in patients with COVID-19: a systematic review and meta-analysis. *Mayo Clin Proc* (2020) 95(8):1621–1631.

Akimoto T, Hara M, Tasaki K, Kurosawa Y, Nakamoto T, Hirose S, Mizoguchi T, Yokota Y, Ninomiya S, Nakajima H. Delayed encephalopathy after COVID-19: a case series of six patients. *Med* (2022) 101(42):e31029.

Algahtani H, Subahi A, Shirah B. Neurological complications of Middle East respiratory syndrome coronavirus: a report of two cases and review of the literature. *Case Rep Neurol Med* (2016) 2016:3502683.

Algahtani SN, Alzarroug AF, Alghamdi HK, Algahtani HK, Alsywina NB, Bin Abdulrahman KA. Investigation on the factors associated with the persistence of Anosmia and Ageusia in Saudi COVID-19 patients. *Int J Environ Res Public Health* (2022) 19(3):1047.

Bell MD, Taub DD, Perry VH. Overriding the brain's intrinsic resistance to leukocyte recruitment with intraparenchymal injections of recombinant chemokines. *Neurosci* (1996) 74(1):283–292.

Beyrouti R, Adams ME, Benjamin L, Cohen H, Farmer SF, Goh YY, Humphries F, Jäger HR, Losseff NA, Perry RJ, Shah S. Characteristics of ischaemic stroke associated with COVID-19. *J Neurol Neurosurg Psychiatry* (2020) 91(8):889–891.

Bilinska K, Jakubowska P, Von Bartheld CS, Butowt R. Expression of the SARS-CoV-2 entry proteins, ACE2 and TMPRSS2, in cells of the olfactory epithelium: identification of cell types and trends with age. *ACS Chem Neurosci* (2020) 11(11):1555–1562.

Brian DA, Baric RS. Coronavirus genome structure and replication. In Enjuanes L (ed.) *Coronavirus replication and reverse genetics.* Berlin: Springer (2005) 1–30.

Bryche B, Fretaud M, Saint-Albin Deliot A, Galloux M, Sedano L, Langevin C, Descamps D, Rameix-Welti MA, Eléouët JF, Le Goffic R, Meunier N. Respiratory syncytial virus tropism for olfactory sensory neurons in mice. *J Neurochem* (2020) 155(2):137–153.

Butowt R, Bilinska K. SARS-CoV-2: olfaction, brain infection, and the urgent need for clinical samples allowing earlier virus detection. *ACS Chem Neurosci* (2020) 11(9):1200–1203.

Butowt R, von Bartheld CS. Anosmia in COVID-19: underlying mechanisms and assessment of an olfactory route to brain infection. *NRO* (2021) 27(6):582–603.

Cao A, Rohaut B, Le Guennec L, Saheb S, Marois C, Altmayer V, Carpentier VT, Nemlaghi S, Soulie M, Morlon Q, Berthet-Delteil B. Severe COVID-19-related encephalitis can respond to immunotherapy. *Brain* (2020) 143(12):e102.

Cascella M, Rajnik M, Aleem A, Dulebohn SC, Di Napoli R. *Features, evaluation, and treatment of coronavirus (COVID-19).* Statpearls [internet]. 2022 February 5.

Cheng VC, Lau SK, Woo PC, Yuen KY. Severe acute respiratory syndrome coronavirus as an agent of emerging and reemerging infection. *Clin Microbio Rev* (2007) 20(4):660–694.

Chowdhury SD, Oommen AM. Epidemiology of COVID-19. *J Digestive Endoscopy* (2020) 11(01):3–7.

Cucinotta D, Vanelli M. WHO declares COVID-19 a pandemic: A review. *Acta Biomed* (2020) 91:157–160.

de Medeiros Carvalho PM, Moreira MM, de Oliveira MN, Landim JM, Neto ML. The psychiatric impact of the novel coronavirus outbreak. *Psychiatry Res* (2020) 286:112902.

Dogan L, Kaya D, Sarikaya T, Zengin R, Dincer A, Akinci IO, Afsar N. Plasmapheresis treatment in COVID-19–related autoimmune meningoencephalitis: case series. *Brain Behav Immun* (2020) 87:155–158.

Dugue R, Cay-Martínez KC, Thakur KT, Garcia JA, Chauhan LV, Williams SH, Briese T, Jain K, Foca M, McBrian DK, Bain JM. Neurologic manifestations in an infant with COVID-19. *Neurol* (2020) 94(24):1100–1102.

El Mezzeoui S, zahra Aftiss F, Aabdi M, Bkiyar H, Housni B. Guillan barre syndrome in post Covid-19 infection in children. *Ann Med Surg* (2021) 67(July):102524.

Elkind MSV. From the heart to the brain: Building bridges to a better future. *Stroke* (2022) 53:1037–1042.

Guadarrama-Ortiz P, Choreno-Parra JA, Sanchez-Martinez CM, Pacheco-Sanchez FJ, Rodriguez-Nava AI, Garcia-Quintero G. Neurological aspects of SARS-CoV-2 infection: mechanisms and manifestations. *Front Neurol* (2020) 11:1039.

Hamming I, Timens W, Bulthuis ML, Lely AT, Navis GV, van Goor H. Tissue distribution of ACE2 protein, the functional receptor for SARS coronavirus. A first step in understanding SARS pathogenesis. *J Pathol Soc GB Ire* (2004) 203(2):631–637.

Helms J, Kremer S, Merdji H, Clere-Jehl R, Schenck M, Kummerlen C Collange O, Boulay C, Fafi-Kremer S, Ohana M, Anheim M. Neurologic features in severe SARS-CoV-2 infection. *New Engl J Medi* (2020) 382(23):2268–2270.

Huang Y, Yang C, Xu XF, Xu W, Liu SW. Structural and functional properties of SARS-CoV-2 spike protein: potential antivirus drug development for COVID-19. *Acta Pharmacologica Sinica* (2020) 41(9):1141–1149.

Jackson CB, Farzan M, Chen B, Choe H. Mechanisms of SARS-CoV-2 entry into cells. *Nat Rev Mol Cell Biol* (2022) 23(1):3–20.

Jacobs BC, Rothbarth PH, Van der Meché FG, Herbrink P, Schmitz PI, De Klerk MA, Van Doorn PA. The spectrum of antecedent infections in Guillain-Barré syndrome: a case-control study. *Neurol* (1998) 51(4):1110–1115.

Kanageswaran N, Demond M, Nagel M, Schreiner BS, Baumgart S, Scholz P, Altmüller J, Becker C, Doerner JF, Conrad H, Oberland S. Deep sequencing of the murine olfactory receptor neuron transcriptome. *PLoS One* (2015) 10(1):e0113170.

Kaplanski G, Teysseire N, Farnarier C, Kaplanski S, Lissitzky JC, Durand JM, Soubeyrand J, Dinarello CA, Bongrand P. IL-6 and IL-8 production from cultured human endothelial cells stimulated by infection with Rickettsia conorii via a cell-associated IL-1 alpha-dependent pathway. *J Clin Invest* (1995) 96(6):2839–2844.

Keyhanian K, Umeton RP, Mohit B, Davoudi V, Hajighasemi F, Ghasemi M. SARS-CoV-2 and nervous system: from pathogenesis to clinical manifestation. *J Neuroimmunol* (2021) 350:577436.

Khubchandani J, Price JH, Sharma S, Wiblishauser MJ, Webb FJ. COVID-19 infection survivors and the risk of depression and anxiety symptoms: A nationwide study of adults in the United States. *Eur J Intern Med* (2022) 97:119–121.

Koyuncu OO, Hogue IB, Enquist LW. Virus infections in the nervous system. *Cell Host Microbe* (2013) 13(4):379–393.

Langley L, Zeicu C, Whitton L, Pauls M. Acute disseminated encephalomyelitis (ADEM) associated with COVID-19. *BMJ Case Rep CP* (2020) 13(12):e239597.

Leber AL, Everhart K, Balada-Llasat JM, Cullison J, Daly J, Holt S, Lephart P, Salimnia H, Schreckenberger PC, DesJarlais S, Reed SL. Multicenter evaluation of BioFire FilmArray meningitis/encephalitis panel for detection of bacteria, viruses, and yeast in cerebrospinal fluid specimens. *J Clin Microbiol* (2016) 54(9):2251–2261.

Lechien JR, Chiesa-Estomba CM, De Siati DR, Horoi M, Le Bon SD, Rodriguez A, Dequanter D, Blecic S, El Afia F, Distinguin L, Chekkoury-Idrissi Y. Olfactory and gustatory dysfunctions as a clinical presentation of mild-to-moderate forms of the coronavirus disease (COVID-19): a multicenter European study. *Eur Arch Oto-rhino-laryngol* (2020) 277(8):2251–2261.

Letko M, Marzi A, Munster V. Functional assessment of cell entry and receptor usage for SARS-CoV-2 and other lineage B betacoronaviruses. *Nat Microbiol* (2020) 5(4):562–569.

Levin KH. Variants and mimics of Guillain Barré syndrome. *Neuro* (2004) 10(2):61–74.

Lindan CE, Mankad K, Ram D, Kociolek LK, Silvera VM, Boddaert N, Stivaros SM, Palasis S, Akhtar S, Alden D, Amonkar S. Neuroimaging manifestations in children with SARS-CoV-2 infection: a multinational, multicentre collaborative study. *Lancet Child Adolesc Health* (2021) 5(3):167–177.

Mandrekar-Colucci S, Landreth GE. Microglia and inflammation in Alzheimer's disease. CNS Neurol Disord-Drug Targets. *Formerly Curr Drug Targets-CNS & Neurol Disord* (2010) 9(2):156–167.

Mao L, Jin H, Wang M, Hu Y, Chen S, He Q, Chang J, Hong C, Zhou Y, Wang D, Miao X. Neurologic manifestations of hospitalized patients with coronavirus disease 2019 in Wuhan, China. *JAMA Neurol* (2020) 77(6):683–690.

Ministry of Health and Family Welfare. https://Www.Mohfw.Gov.In/.

Mohkhedkar M, Venigalla SS, Janakiraman V. Untangling COVID-19 and autoimmunity: identification of plausible targets suggests multi organ involvement. *Mol Immunol* (2021) 137:105–113.

Oxley TJ, Mocco J, Majidi S, Kellner CP, Shoirah H, Singh IP, De Leacy RA, Shigematsu T, Ladner TR, Yaeger KA, Skliut M. Large-vessel stroke as a presenting feature of Covid-19 in the young. *New Engl J Medi* (2020) 382(20):e60.

Paterson RW, Brown RL, Benjamin L, Nortley R, Wiethoff S, Bharucha T, Jayaseelan DL, Kumar G, Raftopoulos RE, Zambreanu L, Vivekanandam V. The emerging spectrum of COVID-19 neurology: clinical, radiological and laboratory findings. *Brain* (2020) 143(10):3104–3120.

Pendlebury ST, Rothwell PM. Incidence and prevalence of dementia associated with transient ischaemic attack and stroke: analysis of the population-based Oxford Vascular Study. *Lancet Neurol* (2019) 18(3):248–258.

Pezzini A, Padovani A. Lifting the mask on neurological manifestations of COVID-19. *Nat Rev Neurol* (2020) 16(11):636–644.

Pilotto A, Masciocchi S, Volonghi I, del Zotto E, Magni E, De Giuli V, Caprioli F, Rifino N, Sessa M, Gennuso M, Cotelli MS. The clinical spectrum of encephalitis in COVID-19 disease: the ENCOVID multicentre study. *Medrxiv* (2020) 88(2):423–427.

Qin C, Zhou L, Hu Z, Zhang S, Yang S, Tao Y, Xie C, Ma K, Shang K, Wang W, Tian DS. Dysregulation of immune response in patients with coronavirus 2019 (COVID-19) in Wuhan, China. *Clin Infect Dis* (2020) 71(15):762–768.

Qureshi AI, Baskett WI, Huang W, Shyu D, Myers D, Raju M, Lobanova I, Suri MF, Naqvi SH, French BR, Siddiq F. Acute ischemic stroke and COVID-19: an analysis of 27 676 patients. *Stroke* (2021) 52(3):905–912.

Radmard S, Epstein SE, Roeder HJ, Michalak AJ, Shapiro SD, Boehme A, Wilson TJ, Duran JC, Bain JM, Willey JZ, Thakur KT. Inpatient neurology consultations during the onset of the SARS-CoV-2 New York City pandemic: a single center case series. *Front Neurol* (2020) 11:805.

Rogers JP, Chesney E, Oliver D, Pollak TA, McGuire P, Fusar-Poli P, Zandi MS, Lewis G, David AS. Psychiatric and neuropsychiatric presentations associated with severe coronavirus infections: a systematic review and meta-analysis with comparison to the COVID-19 pandemic. *Lancet Psychiatry* (2020) 7(7):611–627.

Rudroff T, Fietsam AC, Deters JR, Bryant AD, Kamholz J. Post-COVID-19 fatigue: potential contributing factors. *Brain Sci* (2020) 10(12):1012.

Sánchez-Morales AE, Urrutia-Osorio M, Camacho-Mendoza E, Rosales-Pedraza G, Dávila-Maldonado L, González-Duarte A, Herrera-Mora P, Ruiz-García M. Neurological manifestations temporally associated with SARS-CoV-2 infection in pediatric patients in Mexico. *Childs Nerv Syst* (2021) 37:2305–2312.

Sankowski R, Mader S, Valdés-Ferrer SI. Systemic inflammation and the brain: novel roles of genetic, molecular, and environmental cues as drivers of neurodegeneration. *Front Cellular Neurosci* (2015) 9:28.

Saud A, Naveen R, Aggarwal R, Gupta L. COVID-19 and myositis: what we know so far. *Curr Rheumatol Rep* (2021) 23:1–6.

Sia SF, Yan LM, Chin AW, Fung K, Choy KT, Wong AY, Kaewpreedee P, Perera RA, Poon LL, Nicholls JM, Peiris M. Pathogenesis and transmission of SARS-CoV-2 in golden hamsters. *Nature* (2020) 583(7818):834–848.

Solomon IH, Normandin E, Bhattacharyya S, Mukerji SS, Keller K, Ali AS, Adams G, Hornick JL, Padera Jr RF, Sabeti P. Neuropathological features of Covid-19. *New Engl J Medi* (2020) 383(10):989–992.

Spence JD, De Freitas GR, Pettigrew LC, Ay H, Liebeskind DS, Kase CS, Del Brutto OH, Hankey GJ, Venketasubramanian N. Mechanisms of stroke in COVID-19. *Cerebrovascul Dis* (2020) 49(4):451–458.

Spudich S, Nath A. Nervous system consequences of COVID-19. *Science* (2022) 375(6578):267–269.

Tyrrell DA, Bynoe ML. Cultivation of viruses from a high proportion of patients with colds. *Lancet* (1966) 76–77.

Umapathi T, Quek WM, Yen JM, Khin HS, Mah YY, Chan CY, Ling LM, Yu WY. Encephalopathy in COVID-19 patients; viral, parainfectious, or both? *Eneurologicalsci* (2020) 21:100275.

Van Riel D, Verdijk R, Kuiken T. The olfactory nerve: a shortcut for influenza and other viral diseases into the central nervous system. *J Pathol* (2015) 235(2):277–287.

Varga Z, Flammer AJ, Steiger P, Haberecker M, Andermatt R, Zinkernagel AS, Mehra MR, Schuepbach RA, Ruschitzka F, Moch H. Endothelial cell infection and endotheliitis in COVID-19. *Lancet* (2020) 395(10234):1417–1418.

Wang D, Hu B, Hu C, Zhu F, Liu X, Zhang J, Wang B, Xiang H, Cheng Z, Xiong Y, Zhao Y. Clinical characteristics of 138 hospitalized patients with 2019 novel coronavirus–infected pneumonia in Wuhan, China. *JAMA* (2020) 323(11):1061–1069.

WHO. COVID-19 Dashboard. https://covid19.who.int/.

WHO W. *Naming the coronavirus disease (COVID-19) and the virus that causes it.* WHO. 2020 February 11.

Willison HJ, Jacobs BC, van Doorn PA. Guillain-barre syndrome. *Lancet* (2016) 388(10045):717–727.

World Health Organization. *WHO statement on novel coronavirus in Thailand*. 2020.

Wu Y, Xu X, Chen Z, Duan J, Hashimoto K, Yang L, Liu C, Yang C. Nervous system involvement after infection with COVID-19 and other coronaviruses. *Brain Behav Immun* (2020) 87:18–22.

Xia H, Lazartigues E. Angiotensin-converting enzyme 2: central regulator for cardiovascular function. *Curr Hypertens Rep* (2010) 12:170–175.

Yan CH, Faraji F, Prajapati DP, Boone CE, DeConde AS. Association of chemosensory dysfunction and Covid-19 in patients presenting with influenza-like symptoms. *Int Forum Allergy Rhinol* (2020) 10 (7): 806–813.

Yu H, Sun T, Feng J. Complications and pathophysiology of COVID-19 in the nervous system. *Front Neurol* (2020) 11:573421.

Zhang Q, Xiang R, Huo S, Zhou Y, Jiang S, Wang Q, Yu F. Molecular mechanism of interaction between SARS-CoV-2 and host cells and interventional therapy. *Signal Transduct Target Ther* (2021) 6(1):233.

Zhang T, Rodricks MB, Hirsh E. COVID-19-associated acute disseminated encephalomyelitis–a case report. *MedRxiv* (2020) 2020(4).

Zhou P, Yang XL, Wang XG, Hu B, Zhang L, Zhang W, Si HR, Zhu Y, Li B, Huang CL, Chen HD. A pneumonia outbreak associated with a new coronavirus of probable bat origin. *Nature* (2020) 579(7798):270–273.

Zhu N, Zhang D, Wang W, Li X, Yang B, Song J, Zhao X, Huang B, Shi W, Lu R, Niu P. A novel coronavirus from patients with pneumonia in China, 2019. *New Eng J Med* (2020) Jan 24 82(8):727–733.

Zubair AS, McAlpine LS, Gardin T, Farhadian S, Kuruvilla DE, Spudich S. Neuropathogenesis and neurologic manifestations of the coronaviruses in the age of coronavirus disease 2019: a review. *JAMA Neurol* (2020) 77(8):1018–1027.

10 Current Perspectives on Pyrazole Scaffolds and Their Potential Impact on Alzheimer's Disease

Akshay Kumar[1], *Pragya Gawande*[1], *Souranava Jana*[1,2], *Anjali Murmu*[1], *Balaji Wamanrao Matore*[1], *Jagadish Singh*[1], *and Partha Pratim Roy*[1*]

[1]Department of Pharmacy, Guru Ghasidas Vishwavidyalaya (A Central University), Chhattisgarh, India
[2]Haldia Institute of Pharmacy, Haldia, West Bengal, India
*Corresponding author

INTRODUCTION

Alzheimer's disease (AD), named in honor of the German psychiatrist Alois Alzheimer, who initially identified this condition over a century ago, is a significant health concern. Statistical data published by AD International in the World Alzheimer Report 2023 revealed that the number of people living with AD across the world is expected to rise from 55 million in 2019 to 139 million in 2050. The Alzheimer's Association in Chicago stated that more than 6 million Americans are currently living with Alzheimerand's is projected to rise to 14 million by 2050 (Hebert *et al.*, 2013). AD is an irreversible, progressive neurodegenerative disorder that occurs gradually and results in memory loss, unusual behavior, personality changes, and a decline in thinking abilities (Ogbodo *et al.*, 2022). AD leads to the degeneration of brain cells and serves as the primary contributor to dementia. Dementia manifests as a decline in cognitive abilities and a loss of autonomy in daily personal tasks. It represents a serious problem for the elderly population. It involves the degeneration of brain cells, primarily contributing to dementia, which manifests as cognitive decline and diminished independence in daily tasks. This condition poses a significant challenge for the elderly population, characterized by progressive deprivation in daily activities, cognitive decline, various neuropsychiatric symptoms, and changes in thinking patterns and behavior (Breijyeh and Karaman, 2020). The Alzheimer's Association in Chicago (2023) proposed ten early signs and symptoms of AD, *viz.*, memory loss that disrupts daily life; challenges in planning or solving problems; difficulty completing familiar tasks; confusion with time or place; trouble understanding visual images and spatial relationships; new problems with words in speaking or writing; misplacing things and losing the ability to retrace steps; decreased or poor judgment; withdrawal from work or social activities; and changes in mood and personality.

Heterocyclic compounds are widely employed in drug discovery as well as drug treatment. The heterocycles are conjugated with different scaffolds to better treat complex disorders like cancer, Alzheimer's disease, and Parkinson's disease (Lakra *et al.*, 2022; Matore, Banjare, Sarthi, *et al.*, 2023; Matore *et al.*, 2022; Matore, Roy, *et al.*, 2023; Murmu, Banjare, Wamanrao Matore, *et al.*,

DOI: 10.1201/9781003506317-10

2023). One of the most studied scaffolds is pyrazole. Pyrazole, an organic compound belonging to the azole group with the chemical formula $C_3H_3N_2H_1$, features a unique heterocyclic structure composed of a five-member ring containing three carbon atoms and two neighboring nitrogen atoms in ortho substitution. One of these nitrogen atoms displays basic properties, while the other remains neutral. Functioning as a weak base, it exhibits a pKb of 11.5 (with a corresponding pKa of its conjugate acid at 2.49 at 25°C) (Faisal *et al.*, 2019).

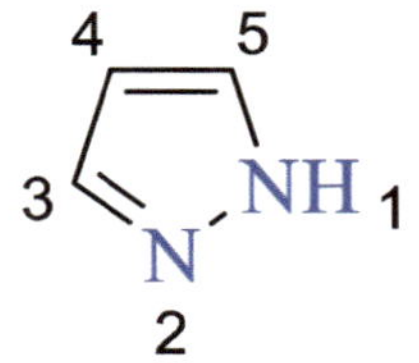

Pyrazole serves as a versatile foundation in organic chemistry and has been recognized as a crucial element in various pharmacological agents. Its significance in the realm of AD treatment stems from its prevalence in bioactive compounds and its adaptability in medicinal chemistry. With Alzheimer's disease associated with over 30 targets, numerous studies have highlighted pyrazole derivatives as potential anti-Alzheimer's agents due to their ability to interact with a maximum number of targets. These studies underscore that through appropriate modification and optimization, the pyrazole scaffold holds promise for the development of potent anti-Alzheimer's agents.

This chapter aims to offer a concise overview of AD, covering its etiology, pathology, and existing treatment approaches while also highlighting recent advancements in pyrazole derivatives as anti-AD agents. These advancements demonstrate the potential of pyrazole derivatives to address various pathological mechanisms associated with AD, thereby presenting opportunities for prevention and treatment.

Disease Progression: An Overview

AD generally advances gradually through three stages known as early stage, middle stage, and late stage, sometimes termed mild, moderate, and severe within medical circles. On average, an individual diagnosed with AD typically survives for 4 to 8 years post-diagnosis, though some live from the beginning as long as 20 years. Alterations within the brain associated with Alzheimer's commence years prior to the manifestation of any discernible symptoms of the illness, denoted as preclinical AD (Figure 10.1) ("2023 Alzheimer's Disease Facts and Figures," 2023).

Historical Development

The term "Alzheimer's disease" has been in use for more than a century, coined by Emil Kraepelin, a colleague of Alois Alzheimer, in 1910 (Kraepelin, 1910). As science and medical technologies have advanced significantly, so too have the methods for diagnosing and treating dementia (Möller and Graeber, 1998; Yang *et al.*, 2016). Alois Alzheimer first described the clinical and pathological verdicts of "a peculiar disease" (Dahm, 2006). He published his findings under the title "A characteristic serious disease of the cerebral cortex" and elaborated the background of the peculiar disease (Alzheimer, 1907). The first drug trial for the treatment of AD, tacrine, was studied in 1987 (Davis *et al.*, 1992). A historical development in early 1993 focused on the development of the naming and primary identification of risk factors of AD (Chyan *et al.*, 1999; Tolnay and Probst, 1999) (Figure 10.2).

The purpose of this review is to offer a brief summary of the causes, progression, and current therapies for AD. It also highlights recent progress in utilizing pyrazole compounds to potentially combat Alzheimer's by addressing key disease mechanisms (Livingston *et al.*, 2020; Yiannopoulou and Papageorgiou, 2020).

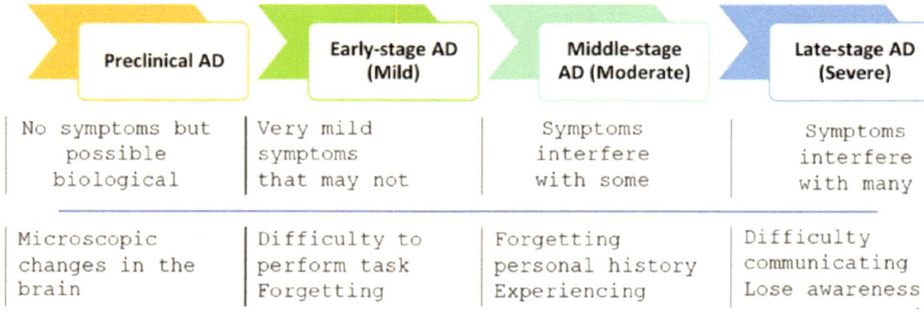

FIGURE 10.1 Stages of Alzheimer's disease and common difficulties.

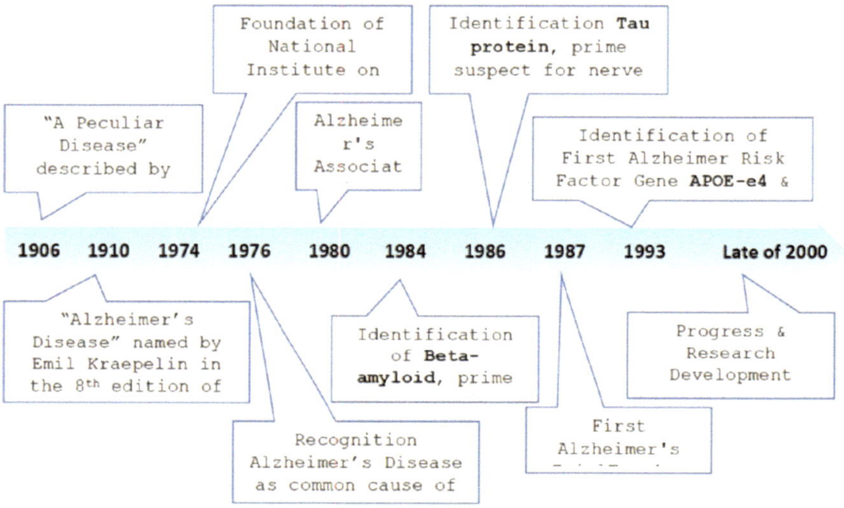

FIGURE 10.2 Historical development of Alzheimer's.

ETIOLOGY AND PATHOPHYSIOLOGY

Microscopic alterations in the brain are responsible for indicating the onset of AD. The brain consists of 100 billion nerve cells that are interconnected to form a communicative network. Scientists believe that AD interferes with the normal function of the brain (Figure 10.3). The primary agents responsible for damaging and ultimately causing the demise of nerve cells are plaques and tangles. Plaques form from accumulations of a protein fragment known as beta-amyloid (pronounced BAY-tuh AM-uh-loyd) in the gaps between nerve cells. Tangles consist of entangled fibers composed of another protein called tau (pronounced like "wow"), which accumulate within cells. Scientists posit that the genesis and aggregation of plaques and tangles are pivotal in impeding communication among nerve cells and disrupting essential cellular survival processes (Serrano-Pozo *et al.*, 2011; Spires-Jones and Hyman, 2014).

The formation of Alzheimer's is caused by neuronal cell death in the brain. Scientists believe that AD interferes with the normal function of the brain. Alterations in the structure of tau proteins lead to their aggregation into neurofibrillary tangles. In AD, tau proteins undergo alterations in structure and assemble into formations known as neurofibrillary tangles. These tangles impede the transportation system, leading to cellular damage. The fundamental reason behind the pathological transformations in AD remains unknown. The etiology of AD is supported by several hypotheses.

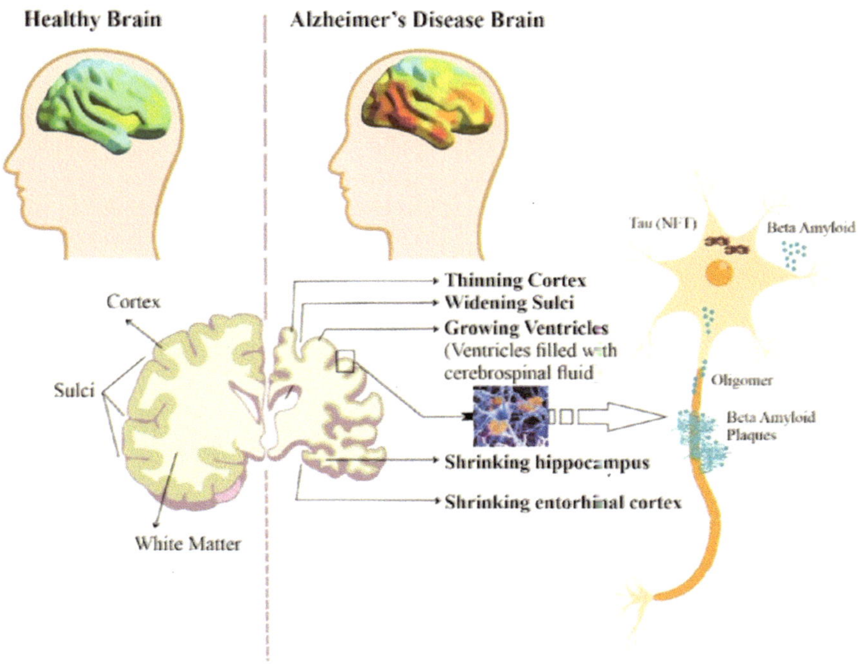

FIGURE 10.3 Healthy and Alzheimer's diseased brain.

One hypothesis describes a deficiency in cholinergic function as a key risk factor for AD, while another contends that changes in the production and processing of amyloid β (Aβ) protein serve as the principal instigating factor. However, there is no universally accepted hypothesis accepted for the etiology of pathogenesis (Armstrong, 2019; Anand and Singh, 2013). The premature onset of AD is linked to various genetic abnormalities, such as trisomy 21 (located on chromosome 21), a mutation in the presenilin-1 gene on chromosome 14, and an anomalous allele known as ε4 for the lipid-associated protein ApoE on chromosome 19. Unlike the standard ApoE ε2 variant, the ε4 variant promotes the accumulation of amyloid β deposits (Katzung *et al.*, 2012).

Risk Factors

The overwhelming majority of individuals who experience Alzheimer's dementia are aged 65 and older, categorized as late-onset alzheimer's. Experts suggest that Alzheimer's, similar to other prevalent chronic conditions, arises from a combination of various factors rather than a singular cause. Instances of Alzheimer's linked to specific genetic mutations are uncommon exceptions. AD has been linked to various risk factors, including aging, familial background, genetic predisposition, environmental influences, infections, head trauma, vascular disorders, and neurochemical imbalances (Breijyeh and Karaman, 2020).

Age: The majority of individuals with the disease are aged 65 and older. Beyond the age of 65 (Guerreiro and Bras, 2015), according to information available from the Alzheimer's Association, the risk of AD doubles approximately every five years. By the age of 85, the risk escalates to nearly one-third (Bekris *et al.*, 2010).

Family history: The history related to AD and dementia increases the likelihood of developing the disease. The risk is enhanced when a greater number of family members suffer from AD.

Genetic history: Scientists estimated that genes are involved in the development of AD. Studies revealed that 70% of AD is associated with genetic factors (Wang and Reddy, 2017), and most cases are due to mutations of dominant genes (Table 10.1). The existence of the APOE e4 gene represents a significant risk factor for AD.

Environmental factors: Environmental factors encompass a wide range of risk factors for the formation of AD. Air pollution, diet, heavy metals, trace metals, infections, and certain reported medical factors (e.g., cardiovascular diseases, obesity, diabetes) may contribute to the increased development of AD (Santos *et al.*, 2017; Stampfer, 2006).

Neurochemical factors: Neurochemicals such as acetylcholine, somatostatin, substance P, and norepinephrine play a role in the progression of AD.

The key areas associated with AD are categorized as physical risk reduction (diet), brain health (education), environmental and population level risk (air pollution), non-modifiable risk (genetic, age), preventive risk reduction (childhood), and post diagnostic risk reduction (modifiable risks to slow progression).

Pathogenesis and Pathophysiology

The AD has three distinguished features that are based on the accumulation of neuritic senile plaques (β-amyloid), the formation of numerous neurofibrillary tangles (NFTs), and loss of cortical neurons, especially the cholinergic neurons (Whalen, 2018). The pathogenesis of AD can be characterized by the following features.

TABLE 10.1
Major Dominant Genes Associated with AD

Gene	Chromosomal Location	Reported Year	Normal Function	Pathology in AD
Amyloid precursor protein (APP)	Chromosome 21	1987	Regulating synaptic function	Causes early-onset AD (Chartier-Harlin *et al.*, 1991; Goate *et al.*, 1991)
Presenilin-1 (PS-1)	Chromosome 14	1992	Associated with amyloid precursor protein and forming Aβ	Causes early-onset AD (Schellenberg *et al.*, 1992)
Presenilin-2 (PS-2)	Chromosome 1	1993	Associated with amyloid precursor protein and forming Aβ	Causes early-onset AD (Levy-Lahad *et al.*, 1995)
Apolipoprotein E4 (APOE4)	Chromosome 19	1993	Transporting cholesterol	Risk factor of AD (Corder *et al.*, 1993)
Clusterin gene (CLU)	Chromosome 8	2009	Preventive action by interacting with Aβ	Neurotoxic activity by decreasing Aβ clearance (Foster *et al.*, 2019)
Bridging integrator 1 (BIN1) or amphiphysin 2	Chromosome 2q14.3	2010	Netastasis suppressor and chemosensitizer in neuroblastoma	Aβ generation and NFT modulator (Tan *et al.*, 2013)
Evolutionarily conserved signaling intermediate in Toll pathway (ECSIT)	Chromosome 19	2013	Cell homeostasis by activating protective molecular mechanisms in response to signals of Aβ or oxidative damage	Interactions with other AD genes support oxidative stress, inflammation, and mitochondrial dysfunction in AD (Soler-López *et al.*, 2012; Tanzi, 2012)

Pathogenesis of AD

Formation of Neuritic Plaques

The peptide β-amyloid is produced from amyloid precursor protein (APP) with the assistance of either alpha- or beta-secretase, leading to formation of tiny fragments that are nontoxic to neurons. However, the consecutive breakdown of β-amyloid by beta-secretase followed by gamma-secretase is responsible for the formation of β-amyloid 42, which is responsible for inducing toxicity after accumulation. Plaques are small spherical lesions at the microscopic level characterized by a central core surrounded by swollen axonal endings. β-amyloid is a component of a larger protein. When these components aggregate, they seem to exert a harmful impact on neurons, which may hamper neuronal communications in brain cells. These aggregates coalesce to form larger accumulations known as amyloid plaques, which also include other cellular debris (Fiala, 2007; Mackenzie *et al.*, 1995).

Formation of Numerous Neurofibrillary Tangles (NFTs)

NFTs are filamentous structures found inside neurons, composed of a protein known as tau. Tau protein's fundamental role is to preserve the stability of axonal microtubules, crucial for intracellular transportation along neuronal axons. It binds and supports the assembly of microtubules. Tau proteins play a role in supporting the internal structure and transport system within brain cells, aiding in the movement of essential nutrients and materials. In AD, the accumulation of extracellular β-amyloid in the brain leads to enhanced excessive phosphorylation of tau, resulting in aggregation. These aggregates of tau protein appear as twisted paired spiral filaments, referred to as neurofibrillary tangles. At the primary stage, NFTs occur in the hippocampus, then subsequently spread throughout the cerebral cortex. These tau aggregates are deposited within neurons. These tangles obstruct the transport system, resulting in harm to the cells (Brion, 1998).

Pathophysiology of AD

Pathological changes of AD may arise due to several reasons that have been recognized by hypotheses. Among the various hypotheses, the cholinergic, oxidative, and amyloid hypotheses are considered to be the main pathophysiological explanations (Figures 10.4 and 10.5).

Cholinergic Hypothesis

In AD, there is a gradual decline in neurons, particularly cholinergic neurons, and cortical thinning is a prominent characteristic. The depletion of cholinergic neurons leads to a significant reduction in choline acetyltransferase (ChAT) and other indicators of cholinergic activity. In the brain, acetylcholine (ACh) participates in numerous essential physiological functions, including memory, attention, sensory perception, learning, and other vital processes. The decline of cholinergic neurons has been observed in AD, leading to disruptions in cognitive function and memory decline. It is theorized that β-amyloid impacts cholinergic neurotransmission, resulting in decreased choline uptake and the release of ACh (Babic *et al.*, 1999). It is assumed that inhibition of acetylcholinesterase (AChE) and butyrylcholinesterase (BuChE) within the CNS will improve cholinergic transmission (Darvesh *et al.*, 1998).

Amyloid Cascade Hypothesis

The amyloid hypothesis postulates that the breakdown of Aβ, produced from amyloid precursor protein through the action of β- and γ-secretase, is diminished with age or under pathological conditions. This reduction results in the accumulation of Aβ peptides (Bales *et al.*, 1999). In Alzheimer's disease, amyloid deposition takes place around the meningeal and cerebral vessels as well as within the gray matter, resulting in the introduction of neuronal toxicity in the central nervous system by disrupting cell function (Paroni *et al.*, 2019; Ricciarelli and Fedele, 2017).

Neurofibrillary Tangles Hypothesis

In AD, the buildup of extracellular β-amyloid leads to an overabundance of tau phosphorylation, resulting in the creation of tau aggregates (Goedert and Spillantini, 2001). These aggregates

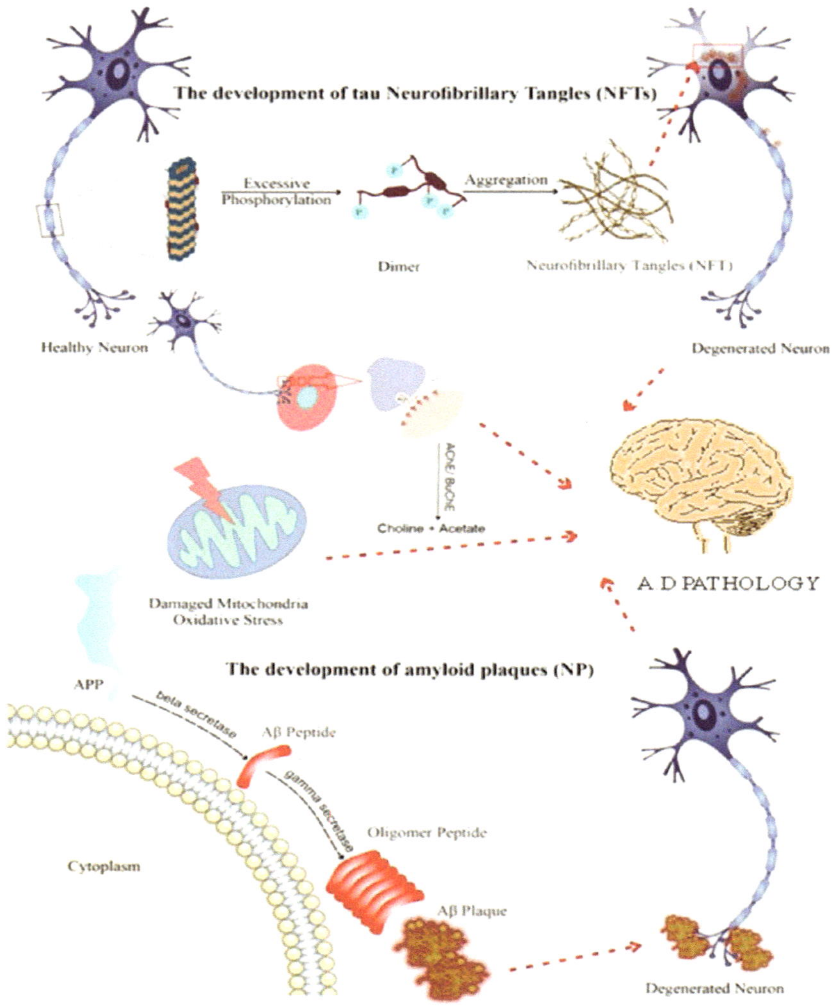

FIGURE 10.4 Schematic representation of the pathology of AD.

assemble into twisted paired helical filaments known as neurofibrillary tangles, leading to neuropathological lesions of neurons (Braak and Del Tredici, 2010; Götz *et al.*, 2001).

The formation of intracellular protein aggregates known as neurofibrillary tangles, comprising hyperphosphorylated tau protein, and the presence of extracellular protein deposits termed β-amyloid or senile plaques or neuritic amyloid plaques contribute to neurotoxicity, ultimately resulting in neuronal damage (Paroni *et al.*, 2019; Ricciarelli and Fedele, 2017).

Glutamatergic Hypothesis

Certain indications suggest that an excess of excitation triggered by glutamate may play a role in neuronal demise. Binding of glutamate with N-methyl-D-aspartate receptors (NMDARs) increases the availability of Ca^{2+} inside the neuron increases the intracellular Ca^{2+} resulting in irregularities in mitochondrial function that could also play a part in neuronal death. Excessive NMDAR activity can induce excitotoxicity, potentially fostering cell demise and serving as a mechanism of neurodegeneration in AD. It is established that NMDAR antagonists are recommended for moderate to severe cases of Alzheimer's disease.

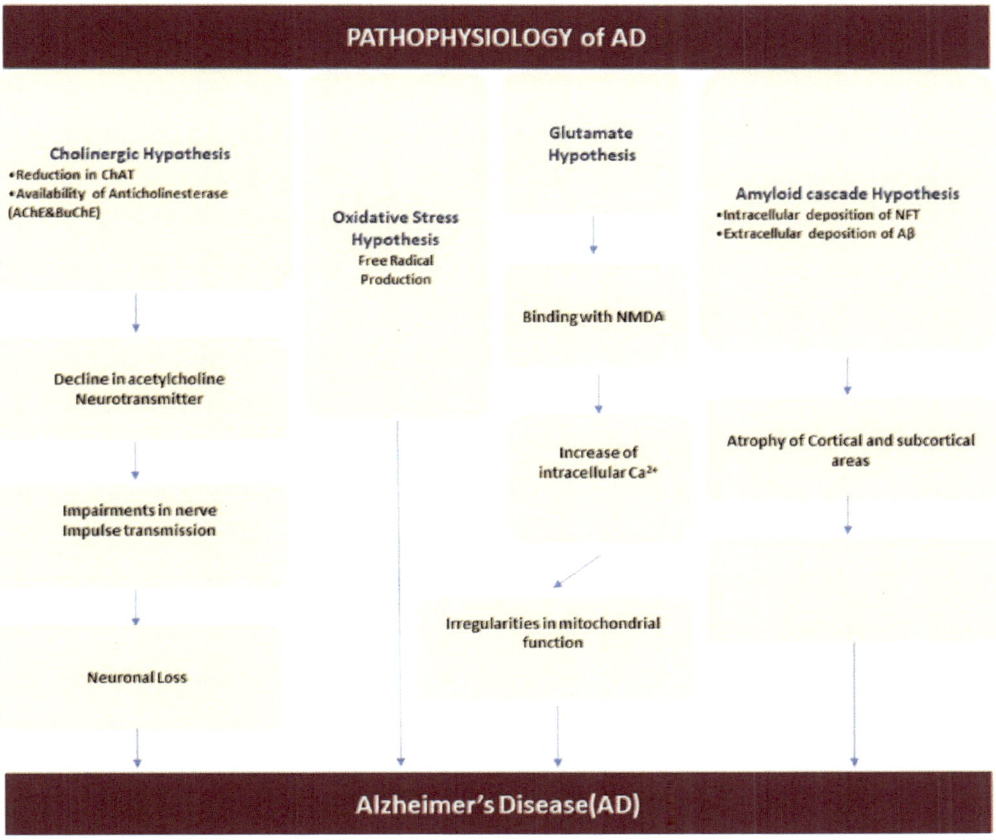

FIGURE 10.5 Pathophysiology of AD at a glance.

Oxidative Stress Hypothesis

Another indication assumed the oxidative stress hypothesis, which causes neuronal damage occurs due to the formation of free radicals (ROS) (Markesbery, 1997). It has been proposed that free radicals bind to and modify the structure of neuronal lipid molecules, leading to alterations in membrane fluidity and permeability. This disruption can interfere with various membrane functions, including transportation and barrier-like properties, leading to neuronal death. Pinpointing the specific origins of free radicals, including mitochondrial dysfunction and mechanisms underlying amyloid-β-related accumulation, transition metals, and genetic factors such as apolipoprotein E and presenilins, is imperative for comprehending the underlying redox imbalance in the pathogenesis of Alzheimer's disease (Chauhan and Chauhan, 2006; Smith *et al.*, 2000). The oxidative damage observed in Alzheimer's disease incorporates glycation end-products found in the process of lipid peroxidation, nitration, free carbonyls, and carbonyl-modified neurofilament protein (Smith *et al.*, 2000; Zhu *et al.*, 2004).

HETEROCYCLIC SCAFFOLDS REPORTED AS ANTI-ALZHEIMER'S AGENTS

Heterocycles have been significant in drug discovery, with various heterocyclic structures being utilized in crafting effective anti-AD compounds. The FDA approved drugs are summarized in Table 10.2. Systemic search revealed the prioritization of heterocyclic scaffolds for synthetic approaches to find new drugs with potential activity compared with existing available drugs. PubMed was used

TABLE 10.2

FDA-Approved Drugs for Management of Alzheimer's Disease

Drug	Application	Targets	Mechanism of Action
Memantine (Namenda) 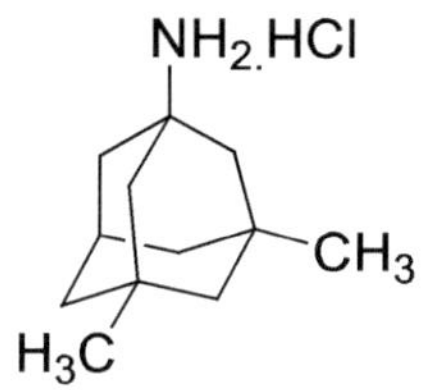	Alzheimer's disease (Molinaro *et al.*, 2009)	NMDA receptor (glutamate-gated ion channels) (Cacabelos *et al.*, 1999), 5-hydroxytryptamine receptor 3A (Nisijima *et al.*, 2004), neuronal acetylcholine receptor subunit alpha 7 (Pohanka, 2012), dopamine D2 receptor (Mancini *et al.*, 2016), glutamate receptor ionotropic NMDA (Kotermanski *et al.*, 2009), GABA receptor, glycine receptor (Molinaro *et al.*, 2009)	Received FDA approval in 2003 for the treatment of moderate-to-severe AD. Continuous NMDA receptor activation by glutamate contributes to Alzheimer's symptoms and neurotoxicity. Memantine, an uncompetitive NMDA receptor antagonist, lacks conclusive evidence in preventing neurodegeneration in Alzheimer's patients (Karimi Tari *et al.*, 2024).
Galantamine (Razadyne) 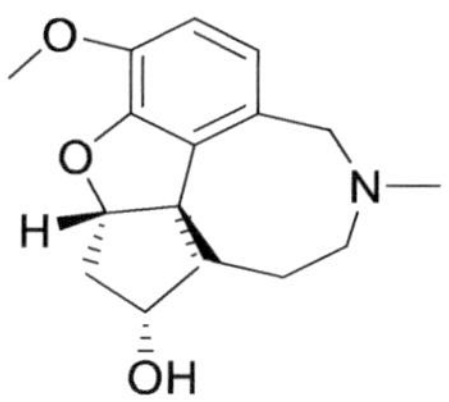	Mild to moderate dementia due to Alzheimer's disease Mild dementia of the Alzheimer's type	Nicotinic acetylcholine receptors (nAChRs) (Ludwig *et al.*, 2010), ACh, muscle nicotinic acetylcholine receptor (Akk and Steinbach, 2005), cholinesterase (Wilkinson, 2001)	Alzheimer's disease involves irreversible degeneration of acetylcholine-producing neurons, leading to cognitive decline and the accumulation of neurofibrillary tangles and amyloid plaques. The cholinergic system's role in memory underscores the potential benefits of drugs like galantamine, which inhibits acetylcholinesterase, preserving acetylcholine levels and memory function (Ferreira-Vieira *et al.*, 2016). Nicotinic acetylcholine receptors (nAChRs) in the CNS regulate neurotransmitter release. Alzheimer's patients exhibit decreased nAChR activity. Galantamine, binding to nAChRs, enhances acetylcholine release, promoting neurotransmission and potentially alleviating AD behavioral symptoms (Farlow, 2003).

Drug	Application	Targets	Mechanism of Action
Rivastigmine (Exelon) 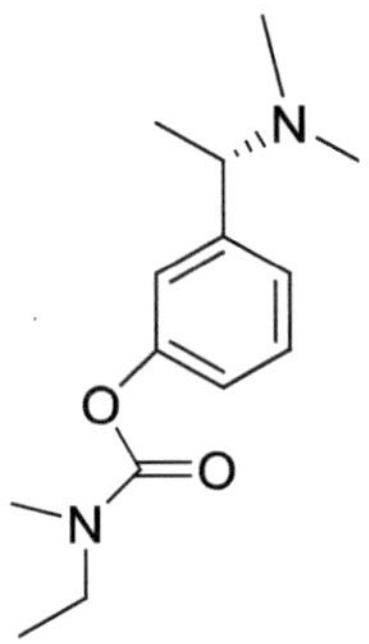	Mild to moderate dementia in Alzheimer's and Parkinson's (Smith, 2009)	Acetylcholinesterase (Birks *et al.*, 2009), cholinesterase (Birks *et al.*, 2009; Smith, 2009)	Pseudo-selective irreversible inhibitor of AChE and BuChE (Kennedy *et al.*, 1999). Rivastigmine, a carbamate derivative, likely inactivates cholinesterase, specifically in the brain, increasing acetylcholine concentration at cholinergic synapses, potentially improving cognitive functions (Birks *et al.*, 2009).
Donepezil (Aricept) 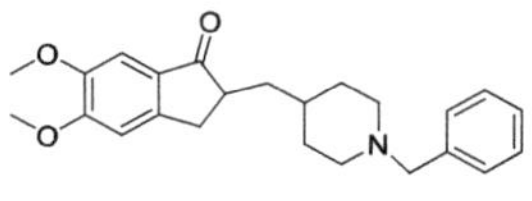	Used to treat the behavioral and cognitive effects of Alzheimer's disease and other types of dementia (Murphy and LeVine, 2010)	AChE, HTR2A (Hayslett and Tizabi, 2005; Repantis *et al.*, 2010), BuChE (Seltzer, 2005), NOS1 (Nakahata *et al.*, 2008), TNFAIP6 (Hwang *et al.*, 2010), interleukin-1 beta (Hwang *et al.*, 2010), NMDA (Shen *et al.*, 2010)	Donepezil selectively and reversibly inhibits the acetylcholinesterase enzyme, which normally breaks down acetylcholine. The main pharmacological actions of this drug are believed to occur as the result of this enzyme inhibition, enhancing cholinergic transmission, which relieves the symptoms of Alzheimer's dementia (Murphy and LeVine, 2010).
Brexpiprazole (Rexulti) 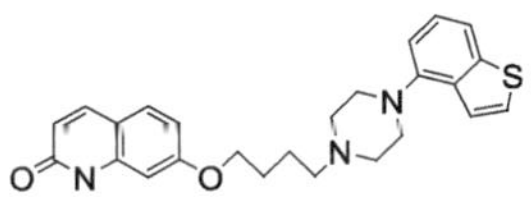	Schizophrenia and agitation associated with dementia due to Alzheimer's disease Works by balancing the levels of dopamine and serotonin in the brain (Stahl, 2016)	5-hydroxytryptamine receptor 1A (Maeda *et al.*, 2014), D (2) dopamine receptor, D (3) dopamine receptor (Greig, 2015), 5-hydroxytryptamine receptor 2A and 2B (Greig, 2015), alpha-1A adrenergic receptor, alpha-2C adrenergic receptor (Greig, 2015), alpha-1B adrenergic receptor, alpha-1D adrenergic receptor, alpha-2C adrenergic receptor (Oosterhof *et al.*, 2014), histamine H_1 receptor (Diefenderfer and Iuppa, 2017), muscarinic acetylcholine receptor M_1 (Das *et al.*, 2016)	Brexpiprazole's mechanism involves partial agonism at 5-HT1A and D2 receptors, antagonist activity at 5-HT2A receptors, and α-adrenergic receptor antagonism, aiding in psychiatric disorders like schizophrenia and depression. Influences dopamine and memory function (Brasso *et al.*, 2023; Das *et al.*, 2016).

Drug	Application	Targets	Mechanism of Action
Aducanumab (Aduhelm) $C_{6472}H_{10028}N_{1740}O_{2014}S_{46}$ (a human monoclonal antibody)	Mild dementia of the Alzheimer's type (Sevigny *et al.*, 2016)	Amyloid beta A4 protein (Arndt *et al.*, 2018; Uhlmann *et al.*, 2020)	Aducanumab act against aggregated Aβ and lowers the Aβ plaques in the brain. Aducanumab, an IgG1 antibody, binds Aβ in Alzheimer's, potentially slowing disease progression, supported by cognitive assessments and biomarkers (Ferrero *et al.*, 2016). It has reported positive results in phase III drug trials (Ali *et al.*, 2022).
Lecanemab (Lequembi) $C_{6544}H_{10088}N_{1744}O_{2032}S_{46}$	Early-stage of Alzheimer's disease (Swanson *et al.*, 2021)	Amyloid beta A4 protein (Logovinsky *et al.*, 2016; Swanson *et al.*, 2021)	An IV anti-AD drug was granted accelerated approval by the FDA in January 2023. Alzheimer's disease is characterized by extracellular Aβ plaques, contributing to neurodegeneration. Aβ, in various forms, impairs synaptic function and leads to dementia. Lecanemab, an antibody, targets soluble Aβ aggregates, showing promise in reducing Aβ plaques (Song *et al.*, 2022).

TABLE 10.3
Number of Scholarly Works during Last Decade on Heterocyclic Scaffolds

Scaffold	Papers
Xanthone	66
Donepezil hybrid	113
Tacrine hybrid	135
Benzofuran	238
Pyrazole	244
β-carboline	300
Coumarin	329
Imidazole	459
Benzothiazole	544

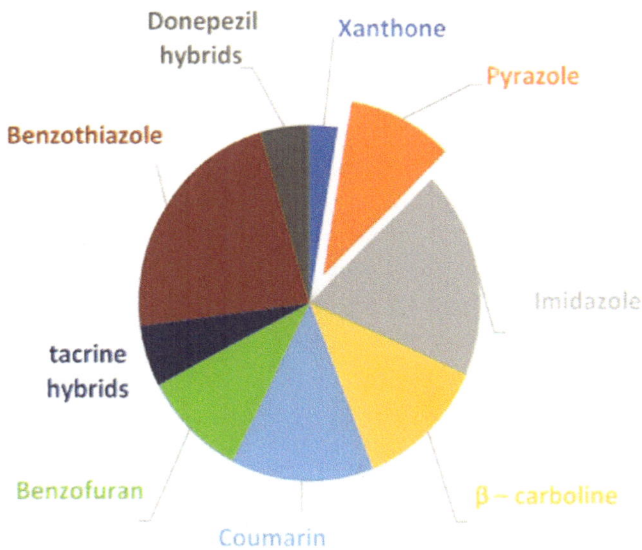

FIGURE 10.6 Graphical representation of heterocyclic scaffolds.

as a search engine for the meta-analysis of mentioned scaffolds and hybrids. The most research work was conducted on benzothiazole scaffolds during the last decade, and study of xanthone scaffolds saw the fewest number of articles. Consequently, recent reports on heterocyclic molecules have been collected and thoroughly assessed, with a comprehensive summary presented in Table 10.3 and Figure 10.6. The different scaffolds and their targets are reported in Table 10.4.

PYRAZOLE AS PROMISING SCAFFOLD FOR MULTI-TARGETED ANTI-ALZHIMER'S AGENTS

Pyrazole scaffold is one of the demanding scaffolds for synthetic approaches and *in silico* studies due to its multi-targeted nature in anti-Alzheimer's activity. The literature showed that the binding affinity of pyrazole derivatives toward the different targets of AD was higher than other heterocyclic scaffolds. To date, it has shown potential against 37 targets of AD. Some studies reported the

TABLE 10.4
Promising Heterocyclic Scaffolds for AD with Targets

Heterocyclic Scaffold	Targets	References
Xanthone scaffold 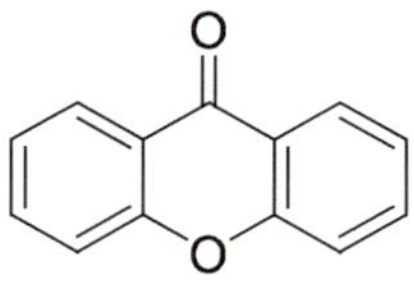	AChE, BuChE, MAO, amyloid β, H3 receptor, α-mangostin, GSK-3β, hSIRT2	(Akawa *et al.*, 2021; Dinakar *et al.*, 2022; Guglielmi *et al.*, 2022; Łażewska *et al.*, 2020; Loh *et al.*, 2021; Malafaia *et al.*, 2021; Yao *et al.*, 2016)
Pyrazole scaffold 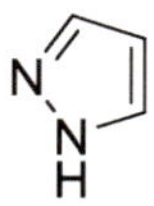	AChE, BuChE, MAO, BACE1, COX-2, THP-1, PC12, GSK-3β, TAU2N4R, α-synuclein, SHIP1, BMPR2, protein kinase C, PDE2, TTBK1, SIRT3, PDE9A, GLP1R, SHIP2, CB2, AMPK, αβ aggregation, CDK9, phospodiestrase 9A, TAU aggregation, 5-HT6, PDE1, NMDA, liver X receptor, KCa3.1, HDAC, PDE5, 15 LOX, NOS, carbonic anhydrase, DMT1, HSPA5	(Amin *et al.*, 2018; Amrhein *et al.*, 2023; Booth *et al.*, 2015; Cuadrado-Tejedor *et al.*, 2017; De, 2024; del Cerro *et al.*, 2018; Elbatrawy *et al.*, 2024; Gameiro *et al.*, 2017; Govindarajulu *et al.*, 2021; Halkina *et al.*, 2021; Hussain *et al.*, 2017; Ivachtchenko *et al.*, 2017; Jesudason *et al.*, 2023; Lan *et al.*, 2018; Lim *et al.*, 2018; Liu *et al.*, 2020; Maccallini *et al.*, 2016; Majd *et al.*, 2018; Montalbetti *et al.*, 2015; Okuda *et al.*, 2017; Pourabdi *et al.*, 2016; Tamborini *et al.*, 2016; Yi *et al.*, 2016; Yu *et al.*, 2017; Zhang *et al.*, 2020; Zhou *et al.*, 2021; De, 2023)
β-carboline scaffold 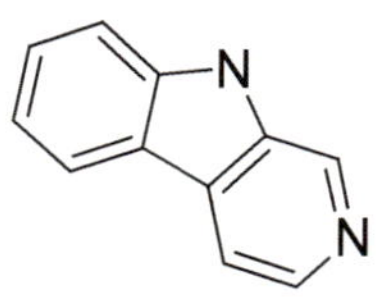	AChE, BuChE, GSK-3β, amyloid-β, HDAC6, NMDA, BACE1, MAO, PDE5, DYRK1 kinase	(Drake *et al.*, 2019; Du *et al.*, 2023; Narang *et al.*, 2019; Ni *et al.*, 2018; Qiu *et al.*, 2024; Rüben *et al.*, 2015; Schwarthoff *et al.*, 2021; Wang *et al.*, 2021)
Benzothiazole scaffold 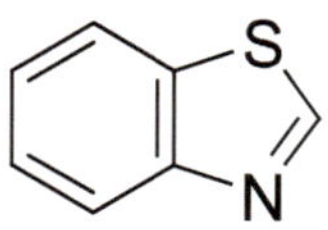	AChE, BACE1, TAU2N4R and α-dynuclein, MAO-A, 17β-hydroxysteroid dehydrogenase type 10 (HSD10), MAO-B, BuChE, NMDAR, β-amyloid, DYRK1A kinase, α-synuclein, MEK1/2, GSK-3	(Elbatrawy *et al.*, 2024; Hafez *et al.*, 2023; He *et al.*, 2023; Schmidt *et al.*, 2023)
Coumarin scaffold 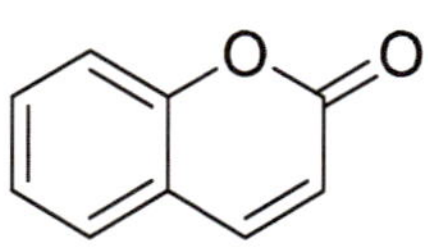	AChE, BuChE, MAO, BACE1	(Annunziata *et al.*, 2020; Liu *et al.*, 2022; Mellado *et al.*, 2022; Piazzi *et al.*, 2008; Stefanachi *et al.*, 2018)
Benzofuran scaffold 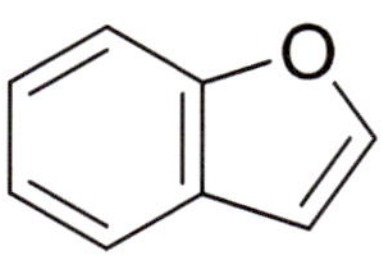	AChE, BuChE, BACE1, TAU, amyloid, H3, H4, DYRK1A, NMDA, α-glucosidase, β-amyloid, 17β-HSD10, CK1, ABAD, MAO, 5-HT6, σ1 protein	(Agis-Torres *et al.*, 2014; Aitken *et al.*, 2019; Cabrera-Pardo *et al.*, 2020; Fancellu *et al.*, 2020; Goyal *et al.*, 2018; Hasanvand *et al.*, 2022; Kaur *et al.*, 2023; Montanari *et al.*, 2019; Narlawar *et al.*, 2006; Ono *et al.*, 2006; Watanabe *et al.*, 2014)

Heterocyclic Scaffold	Targets	References
Tachrine hybrids 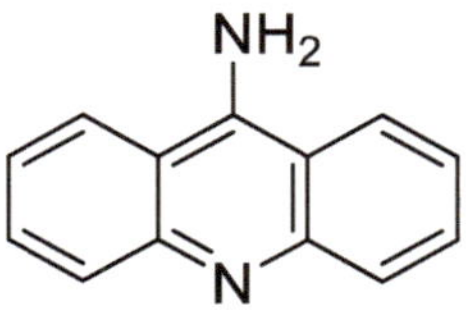	AChE, BuChE, BACE1, MAO-B, ROCK2, NMDA, TAU	(Bubley *et al.*, 2023; Moreira *et al.*, 2022; Oset-Gasque and Marco-Contelles, 2020; Remya *et al.*, 2021; Remya *et al.*, 2022; Sadafi Kohnehshahri *et al.*, 2022; Xie *et al.*, 2015)
Benzothiazole scaffold 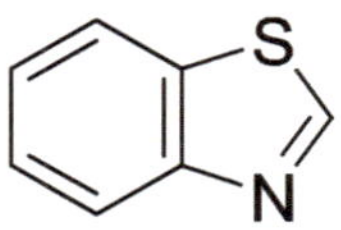	AChE, BuChE, TAU-2N4R, α-synuclein, GSK-3, MAO, MEK1/2, NMDAR, β-amyloid, DYRK1A kinase, HSD10	(Aitken *et al.*, 2019; Hafez *et al.*, 2023; Kamal *et al.*, 2012; Kramer *et al.*, 2012; Ono *et al.*, 2009; Rajeshwari *et al.*, 2019; Ramirez *et al.*, 2023; Rosales Hernández *et al.*, 2022; Rothweiler *et al.*, 2016; Shahroz *et al.*, 2022)
Donepezil hybrids 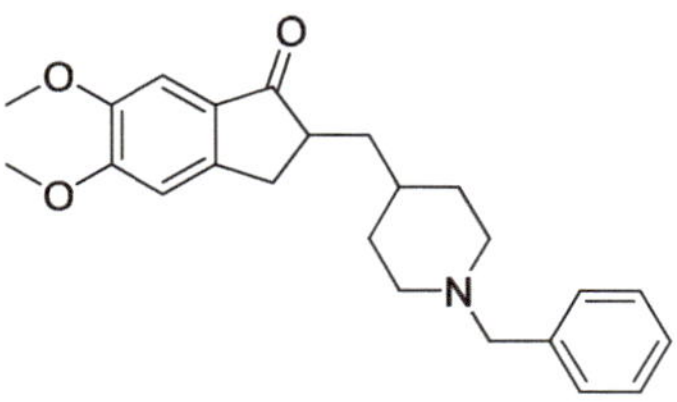	AChE, BACE1, ROCK2, COX, NMDAR, 5HT4, MAO-A/B	(Fang *et al.*, 2022; Maiuolo *et al.*, 2023; Moreira *et al.*, 2022; Piemontese *et al.*, 2018; Wang *et al.*, 2020; Zagórska and Jaromin, 2020; Zang *et al.*, 2021)

TABLE 10.5

Number of Scholarly Works on Pyrazole Scaffold in the Last Decade

Year	2014	2015	2016	2017	2018	2019	2020	2021	2022	2023	2024
Number of studies	26	29	29	41	28	28	28	26	16	24	5

potential and selective pyrazole derivatives for their inhibitory activity, while the major studies demonstrated dual or multi-targeted action of pyrazole derivatives. The compounds acting on dual or multiple targets had a great impact on anti-Alzheimer's activity. A systematic search on PubMed for the last ten years (2014–2024) with the keywords "pyrazole" and "AD" revealed the gaining attention to pyrazole in AD. Table 10.5 represents the yearwise count of research work on pyrazole scaffold in the last decade.

The table clearly indicates the potential of this heterocyclic scaffold in AD therapy. The number of researchers' works reported in the last decade on the pyrazole scaffold is 244, with an average of 20 articles per year. Further studies revealed the association of pyrazole with various targets related to AD (Figure 10.7). Pyrazole has a wide spectrum of affinity for more than 35 targets. A bar diagram (Figure 10.7) also highlights the number of studies done so far with respective targets. These targets involve HDAC, NOS, α-sinuclein, 5-HT6, CB2, NMDA, CA, PDE9A, GSK-3β, BACE1, COX-A/B, MAO-A/B, BuChE, TAU, AChE, and β-amyloid.

In continuation of literature studies, bibliometric network analysis was performed to find the correlation between pyrazole and different AD targets using the VOSviewer tool (Figure 10.8).

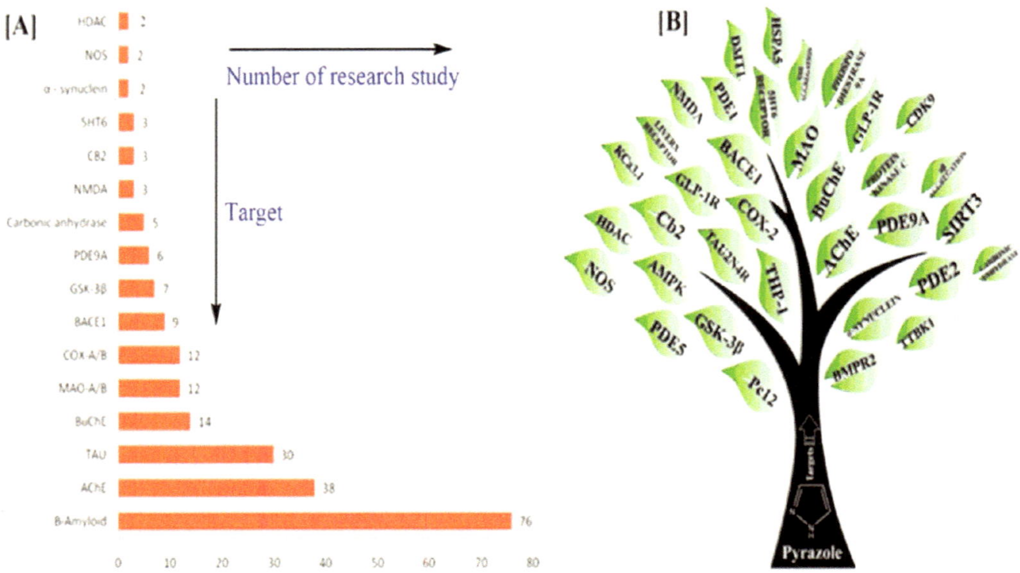

FIGURE 10.7 (A) The number of studies on important targets of pyrazole in the last decade. (B) Different AD targets associated with the pyrazole scaffold in the last decade.

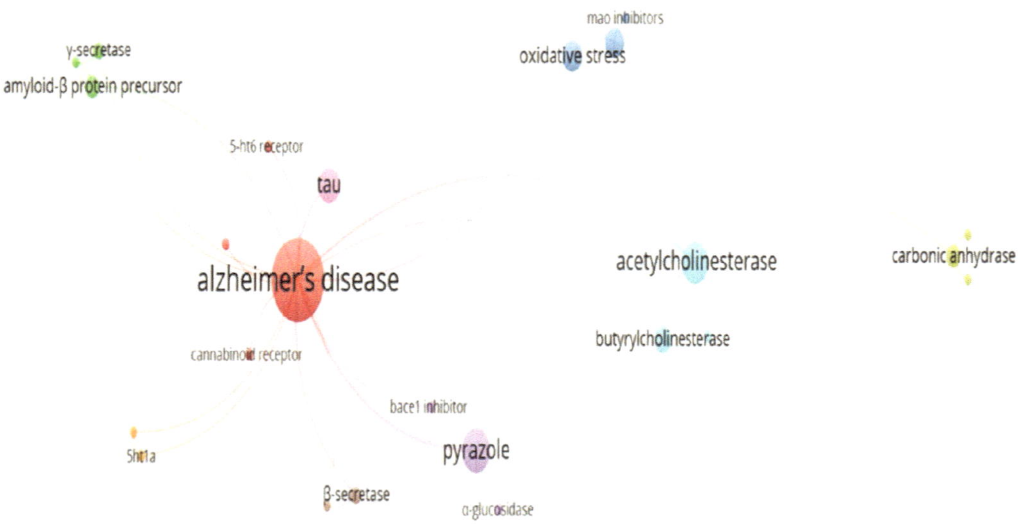

FIGURE 10.8 Bibliometric network connection between pyrazole, AD, and its targets.

The bibliometric network represented in Figure 10.8 shows the direct and indirect involvement of different targets with AD, while the pyrazole scaffold is directly linked or connected with AD. Along with AD, there are some important targets that also show a direct connection with BACE1, β-secretase, AChE, BuChE, and α-glucosidase receptor. Overall, this analysis shows the strong connection between pyrazole, AD, and its crucial targets.

Upon conducting the literature search and bibliometric network analysis, some important pyrazole-linked targets were identified and are highlighted next.

AChE

AChE is an enzyme that breaks down acetylcholine, a key neurotransmitter in the brain that is involved in memory and learning (Tabet, 2006). Acetylcholinesterase inhibitors (AChEIs) have been observed to protect cells from free radical toxicity and β-amyloid-induced injury, and increase the production of antioxidants (Li, 2022). In one study, a series of pyrazolo-benzothiazole analogs were systematically planned, synthesized, and biologically tested for their AChE activities. In another study, 18 pyrazolines were synthesized and evaluated for their inhibitory effects against AChE *in vitro*.

BuChE

The research topic focuses on the identification and synthesis of pyrazole scaffold derivatives as potential therapeutic agents for AD, with a particular emphasis on butyrylcholinesterase (BuChE) inhibition. BuChE is a target for symptomatic treatment in AD due to its role in cholinergic neurotransmission. A study revealed pyrazolopyrimidinone–rivastigmine hybrids that showed remarkable inhibitory activities against both PDE9A and BuChE (Li *et al.*, 2021). Another study revealed pyrazole derivatives with various substituents have been identified as selective and potent BuChE inhibitors, with some compounds showing submicromolar inhibitory activity and reversible, mixed-type, non-competitive inhibition (Li *et al.*, 2022; Pourabdi *et al.*, 2016).

MAOA/B

Monoamine oxidase (MAO) has surfaced as a promising focus for treating AD because of its link to the irregular production of γ-aminobutyric acid (GABA) in reactive astrocytes (Ajala *et al.*, 2023). In the context of the pyrazole scaffold, research has been conducted on the synthesis and evaluation of pyrazolo-benzothiazole derivatives as possible therapeutic agents for the treatment of Alzheimer's disease (Kumar *et al.*, 2022). Novel pyrazole derivatives have been synthesized with dual MAO-B inhibitory activity (Al-Hazmy *et al.*, 2023). A study revealed that pyrazolines have been found to inhibit AChE, Aβ plaques, and MAO enzymes, which are all targets in managing these diseases (Ucar *et al.*, 2005). This suggests that the pyrazole scaffold, when appropriately modified and optimized, could potentially be used in the development of effective anti-AD agents targeting MAO.

TAU

Tau protein, particularly its hyperphosphorylated form, is a significant target in AD treatment. It forms the neurofibrillary tangles that are a hallmark of AD. Tau pathology is associated with cognitive dysfunction and is accompanied by neuroinflammation (Chen and Yu, 2023). Studies exposing pyrazolotriazines targeting GSK-3 show potential as brain-permeable leads for AD treatment with potential neuroprotective properties against tau hyperphosphorylation (Sciú *et al.*, 2019).

β-Amyloid

Aβ, the primary constituent of neuritic plaques in AD, is viewed as the fundamental molecular force behind the pathogenesis and advancement of Alzheimer's disease due to its buildup (Zhang *et al.*, 2023). The advancement of therapeutic treatments has predominantly centered on inhibiting the enzymatic cleavage of APP by β- and γ-secretases, stimulating an immune response through active immunotherapy, and employing passive immunotherapeutic methods that aim monoclonal antibodies at Aβ aggregates (Behl *et al.*, 2020).

Moreover, the pyrazole scaffold, when appropriately modified and optimized, could potentially be explored in the direction of establishing potent anti-Alzheimer's agents.

HDAC

Histone deacetylases (HDACs) are enzymes that regulate the expression of genes involved in learning and memory, making them important targets for AD research. In preclinical studies, HDAC inhibitors have been found to have neuroprotective properties and inhibit the hyperphosphorylation of tau protein induced by Aβ (Xu *et al.*, 2011). However, identifying the specific role of individual HDACs in AD is critical to developing targeted HDAC inhibitors for treating AD (Yang *et al.*, 2017). HDAC6 is a cytosolic enzyme that has been found to be significantly increased in the cortex and hippocampus of AD brains compared to normal brains, indicating its involvement in AD pathogenesis (Xu *et al.*, 2011).

NOS

Nitric oxide synthesizing enzyme inhibitors show potential as a therapeutic strategy to delay AD progression by reducing amyloid beta peptide production and cholinergic deficits (Fernandez *et al.*, 2010). Nitric oxide synthase (NOS), an enzyme that catalyzes the creation of nitric oxide (NO), is a signaling molecule involved in various physiological and pathological processes. NO plays a crucial role in the pathogenesis of AD. NO can modulate the activity of APP, the precursor of amyloid beta peptide, which forms the senile plaques in the brains of AD patients. NO can also affect the phosphorylation of tau protein, which forms the senile plaques in the brains of AD patients. NO can also affect the phosphorylation of tau protein, which forms the neurofibrillary tangles in AD. Therefore, modulating the levels of NO by targeting NOS enzymes could be a potential strategy for treating AD (Li *et al.*, 2021).

A-Synuclein

Recent evidence suggests that α-synuclein (αSyn), a presynaptic protein primarily associated with synucleinopathies, may also play a role in AD pathophysiology (Twohig and Nielsen, 2019). Lewy-related pathology, mainly comprised of αSyn, has been identified in autopsied AD brains. Additionally, higher levels of αSyn in the cerebrospinal fluid of patients with mild reasoning damage and AD correlate with cognitive decline. Interestingly, the accumulation of Aβ plaques is associated with elevated CSF αSyn levels in individuals at risk of sporadic AD and those carrying autosomal dominant AD mutations (Kargbo, 2020; Paik, 2022).

5-HT6

5-HT6 is a type of G protein-coupled receptor (GPCR) that is responsible for modulating adenylate cyclase activity that enhances cyclic adenosine monophosphate (cAMP) formation upon agonist activation. Reported preclinical studies have consistently supported the idea that 5-HT6 receptor antagonists could improve memory and alleviate behavioral symptoms associated with AD (Ferrero *et al.*, 2017; Ramírez, 2013). It is established that AD is produced by a decrease in striatal GABA activity (Ferrero *et al.*, 2017). Two antagonists, idalopirdine and RVT-101, are currently being evaluated in ongoing phase III clinical trials (Ferrero *et al.*, 2017).

CB2

CB2 receptors belong to the family of GPCRs and are primarily expressed in peripheral tissues, including immune organs, nerve tissues, and various organs. Accumulated evidence suggests that CB2 receptors play a crucial role in modulating neuroinflammation, oxidative stress, and neuronal survival (Wu *et al.*, 2022). CB2 receptors are implicated in Aβ clearance mechanisms. Enhancing

Aβ phagocytosis by microglia via CB2 activation may help reduce Aβ burden in the brain, a hallmark of AD pathology (Aso and Ferrer, 2016).

N-Methyl-D-aspartate (NMDA)

NMDARs, a subtype of ionotropic glutamate receptors, have recently been implicated in AD-related synaptic deficits. Early synaptic dysfunction in AD is linked to increased oligomeric Aβ peptide, which causes NMDAR-dependent synaptic depression and spine elimination (Camps *et al.*, 2009).

Carbonic anhydrase (CA)

Carbonic anhydrase activators have been identified as potential therapeutic agents for enhancing synaptic efficacy, spatial learning, and memory, which could be beneficial in the treatment of AD and other memory-related disorders (Solesio *et al.*, 2018; Supuran and Scozzafava, 2002). Carbonic anhydrase inhibitors (CAIs) have been shown to prevent amyloid β-induced mitochondrial toxicity, which is associated with AD, suggesting a protective role against neurovascular mitochondrial damage (Provensi *et al.*, 2019). A study evaluated thiazolyl-pyrazoline derivatives for their potential as both AChE and human carbonic anhydrase (hCA) inhibitors, offering another avenue for AD treatment.

Phosphodiesterase 9A (PDE9A)

Phosphodiesterase-9A (PDE9A), a cGMP-specific phosphodiesterase found in neurons throughout the brain, has been considered a potential therapeutic target. Inhibition of PDE9A can enhance cGMP signaling, which plays a role in synaptic plasticity and memory formation (Shao *et al.*, 2014). Pyrazolopyrimidinone derivatives, in particular, have shown promise due to their excellent inhibition of PDE9A and additional beneficial properties. For instance, one study reported the synthesis of 14 pyrazolopyrimidinone derivatives that demonstrated potent PDE9A inhibition at nanomolar concentrations, along with favorable metabolic stability and solubility (Yu *et al.*, 2017). Another study revealed pyrazolopyrimidinone–rivastigmine hybrids that showed remarkable inhibitory activities against both PDE9A and BuChE (Li *et al.*, 2021).

Glycogen synthase kinase-3β (GSK-3β)

Glycogen synthase kinase 3 beta (GSK-3β) is a key enzyme implicated in the pathogenesis of AD, particularly in the hyperphosphorylation of tau protein, which is a hallmark of the disease. Inhibiting GSK-3β has therefore become a promising strategy for the development of therapeutic agents aimed at treating AD. Pyrazolotriazines and thieno[3,2-c]pyrazol-3-amine derivatives have been designed and synthesized as potent GSK-3β inhibitors, showing the ability to cross the blood–brain barrier and exhibit neuroprotective properties against tau hyperphosphorylation in cell-based models (Noh *et al.*, 2013; Sciú *et al.*, 2019). Another study revealed that pyrazolotriazines targeting GSK-3 show potential as brain-permeable leads for AD treatment with potential neuroprotective properties against tau hyperphosphorylation (Sciú *et al.*, 2019).

Beta-site amyloid precursor protein cleaving enzyme 1 (BACE1)

Targeting β-secretase (BACE1) has become a significant strategy in developing treatments for AD, with various inhibitors being explored to reduce Aβ production and deposition (Haghighijoo *et al.*, 2020). A study designed pyrano[3,2-c]quinoline–6-chlorotacrine hybrids, which not only inhibited AChE but also showed significant activity against beta-amyloid aggregation and BACE1, suggesting an approach to AD treatment (Yu *et al.*, 2017).

Cyclooxygenase (COX)

The enzyme COX is responsible for neuroinflammation and the development of AD. Derivative pyrrolidine cores have been developed, targeting COX-2, with compounds exhibiting nanomolar range COX-2 inhibition and balanced inhibitory activity against all tested targets (Machado *et al.*, 2022).

ARTIFICIAL INTELLIGENCE (AI)-BASED TARGET PRIORITIZATION

The selection of an appropriate target is an essential step in any drug discovery, and it has more importance in complex disorders like AD due to the availability of more than half a century targets (Banjare *et al.*, 2023; Matore, Banjare, Singh, *et al.*, 2022; Murmu, Matore, *et al.*, 2023). Reports showed pyrazole derivatives having an affinity toward more than 30 AD receptors. Before designing newer pyrazole derivatives, it is necessary to prioritize a set of promising targets to expedite the drug discovery process. Based on literature reports, we have collected different targets of AD that show a strong affinity for pyrazole derivatives. Further, we have generated an AI-based protein–protein interaction and gene neighborhood network using the STRING server to analyze the gene correlation and identify promising targets (https://string-db.org/). The protein–protein interaction network is presented in Figure 10.9. The targets shown in the figure are well explored, and some of them are very promising due to their crucial role in AD pathophysiology. ACHE, BCHE, BACE1, and APP cover the center of the network and are the first-choice targets. GSK3B, SNCA, MAO-A, MAPT, and PTGS2 are the second-line targets. There are some good targets that still need to be explored and are placed in third-line targets: PTGS1, HDAC4, CA1, NOS2, KMO, CNR2, and KMT2E. During the selection of a single or selective target, it is good to choose from first-choice targets. All the first-choice targets have connections with second-line targets, suggesting that we can pick dual or multiple targets from two different categories of targets to get better results. Based on reports

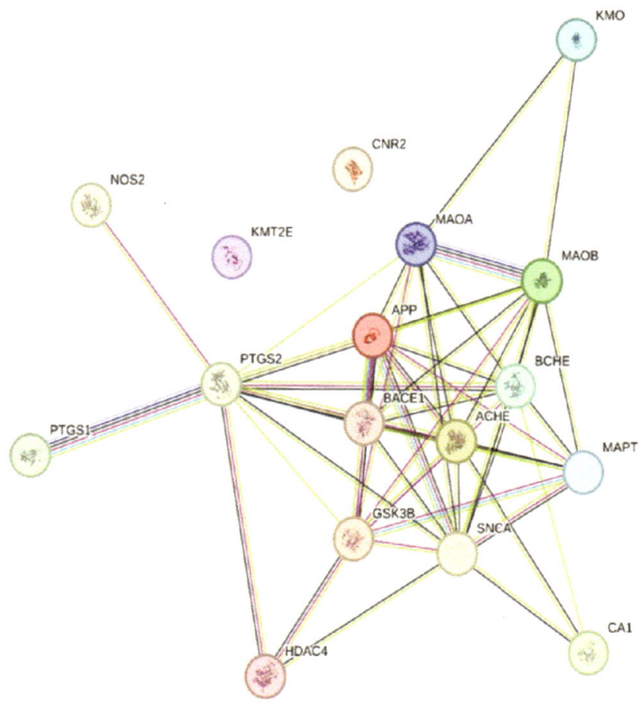

FIGURE 10.9 AI based protein-protein interaction network for different targets.

and our analysis, we have proposed some dual and multiple target combinations. When picking a dual target always choose the neighborhood proteins. For example, AChE–BuChE, AChE–BACE1, AChE–SNCA, AChE–GSK3B, AChE–MAPT, BuChE–MAOA, BuChE–MAOB, BuChE–MAPT, BACE1–APP, BACE1–GSK3B, APP–MAOA, and SNCA–GSK3B. The triple-target combination can be easily picked by identifying a three-point triangle of neighborhood proteins such as AChE–BuChE–APP, APP–BACE1–AChE, and MAOB–BuChE–APP. Similarly, four or multiple targets can be prioritized from the network shown in Figure 10.9. These are the most promising targets available for pyrazole-containing molecules to tackle AD.

PYRAZOLE DERIVATIVES AS ANTI-ALZHEIMER'S AGENTS

Researchers are continuously working on finding of curable drug molecules to tackle AD. One of them is the pyrazole scaffold. It is a five-membered azole compound containing nitrogen at the 1 and 2 position and three carbon atoms. It is an interesting scaffold because of its availability of four different substitution positions to modify the drug molecules. Pyrazole has wide pharmacological properties and has been widely studied due to its potency, safety profile, and synthetic accessibility. To date, 16 commercial drugs are available with pyrazole as the base framework and used to treat cancer, heart diseases, arthritis, sleep disorders, and leukemia. This defines the potential and priority of pyrazole scaffolds. In the last decade, more than 200 articles were published on AD treatment or management studies, showing the potency of pyrazole derivatives over the 40 targets of AD.

Umar *et al.* (2019) designed and synthesized a new series of 2-(piperazin-1-yl)-N-(1H-pyrazolo[3,4-b]pyridin-3-yl)acetamides as selective AChE and β-amyloid aggregation inhibitors. Primarily, all the synthesized compounds were evaluated for their potential AChE and BuChE inhibitory activity. Further, these compounds were tested for β-amyloid aggregation and Cu(II) mediated α β-aggregation. Compounds 1 and 2 showed similar types of AChE inhibitory activity with IC_{50} of 4.8 and 4.9 nM, respectively. Both compounds showed more than tenfold activity compared to the standard donepezil (IC_{50} = 49 nM). The substitution of piperidine and piperazine rings in compound 1 and 2 are responsible for the potent AChE inhibitory activity. The substitution of a methyl group in compound 2 showed slightly lower activity than compound 1. It has been reported that all synthesized compounds had lower BuChE inhibitory activity, showing IC_{50} > 100 μM. This data shows the AChE selectivity of these compounds. Compound 3 showed maximum α β-aggregation inhibitory activity with 81.64% due to the bulky substitution on the piperazine ring. It is disclosed that all designed compounds showed inhibition against β-amyloid. Synthesized compounds were also estimated for metal chelating activity and antioxidant activity (Umar *et al.*, 2019).

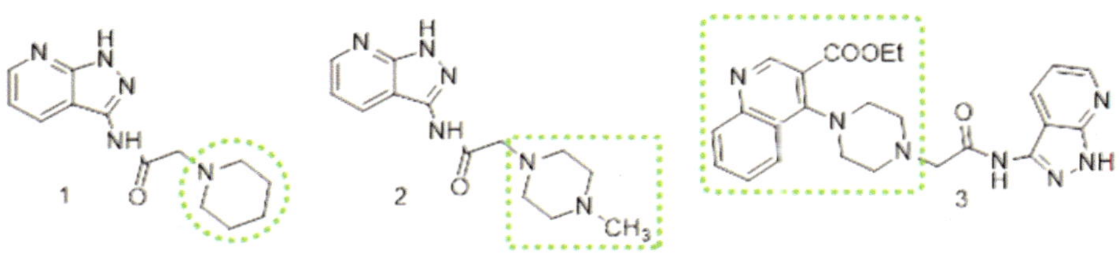

Chigurupati and coworkers (2016) synthesized some novel indolopyrazoline derivatives and estimated their AChE inhibitory and antioxidant activity. It was revealed that compound 4 exhibits maximum AChE inhibition (IC_{50} = 0.68 μM) and has strong DPPH and ABTS radical scavenging activity (IC_{50} = 13.77 μM and IC_{50} = 12.59 μM, respectively). Compound 5 also showed similar types of AChE inhibitory activity and is the second most potent compound with IC_{50} = 0.74 μM. It showed IC_{50} of 13.77 μM and 12.59 μM, respectively, by DPPH and ABTS radical scavenging assay. The N-substituted carbonyl phenyl ring at position 2 and phenyl ring at position 3 are essential for all type of activity. Substitution at the carbonyl phenyl ring leads to a decrease in AChE inhibitory and antioxidant activity. The docking studies also revealed the importance of 2, 3 substitutions for key interactions. Moreover, *in silico* prediction of ADME revealed that the designed derivatives exhibit properties of CNS-acting drugs (Chigurupati *et al.*, 2016).

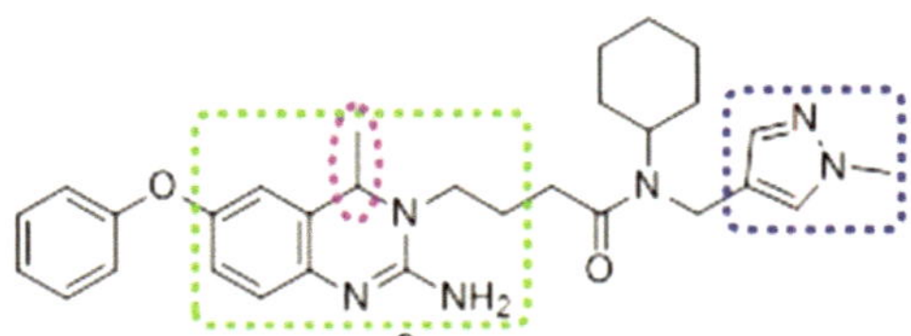

9

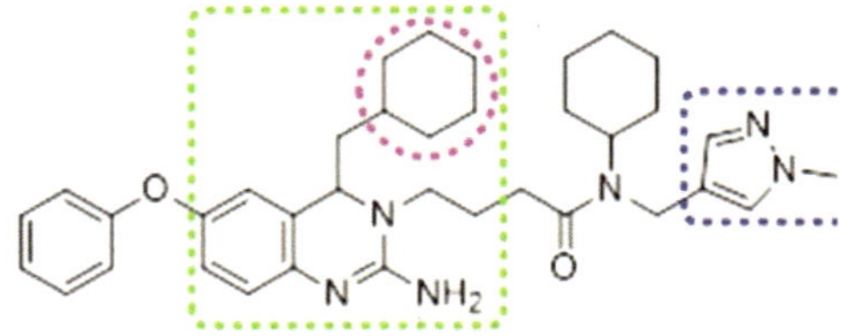

10

Mohd Faudzi and group (2021) designed and synthesized a series of pyrazole analogs. All the synthesized compounds were investigated for AChE and BuChE inhibitory activity. Compound 6 showed the most potent AChE and BuChE inhibitory activity with IC_{50} values of 0.3 and 0.11 μM, respectively. The structure–analysis relationship (SAR) analysis showed that the substation of 2-hydroxy phenyl at the third and unsubstituted phenyl ring at the fourth position of pyrazole enhanced both the AChE and BuChE inhibitory activity. The favorable docking interactions shown by both the phenyl rings also confirmed their potential in compound 6 (Mohd Faudzi *et al.*, 2021).

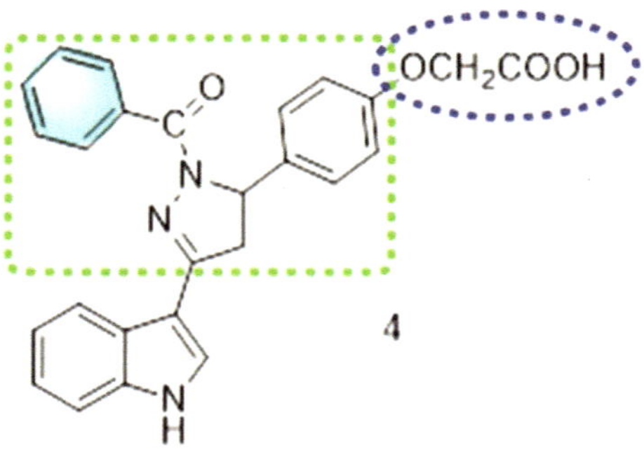

4

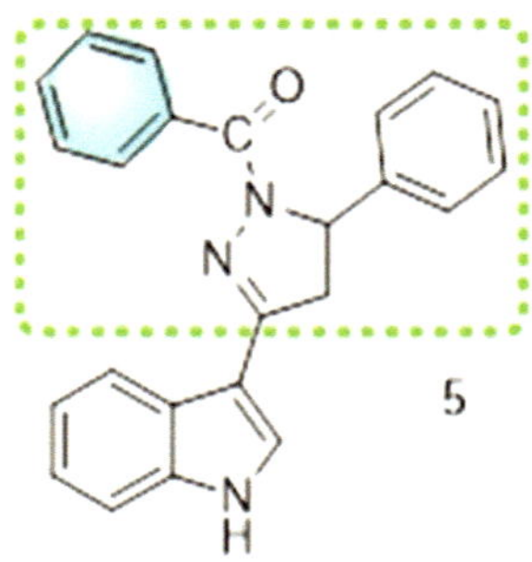

5

Shaikh *et al.* (2020) designed and synthesized N-substituted pyrazole α-aminophosphate derivatives and evaluated them for AChE and BuChE inhibitory activity. Compounds 7 and 8 were identified as the most potent compounds with IC_{50} of 0.017 and 0.055 μM, respectively, demonstrating AChE inhibitory activity greater than the available drugs tacrine (IC_{50} = 0.21 μM), rivastigmine (IC_{50}= 3.15 μM), and galantamine (IC_{50}= 2.63 μM). The synthesized compounds also showed good to moderate BuChE inhibitory activity. Similar to the AChE, compound 7 also exhibited potent activity against BuChE with IC_{50} of 6.331 μM, surpassing the standard rivastigmine (IC_{50} = 6.901 μM). The substitution of a phenyl ring at the second and fifth positions of pyrazole are essential for anti-AD activity. Chloro substitution at the phenyl ring at the fifth position enhances AChE inhibitory activity. The substitution of a bulky group at the fourth position of pyrazole is essential for its potency. Additionally, the synthesized compounds also exhibited strong antioxidant activity by the DPPH and H_2O_2 scavenging methods. Molecular docking revealed that compounds 7 and 8 expressed dual types of inhibition (Shaikh *et al.*, 2020).

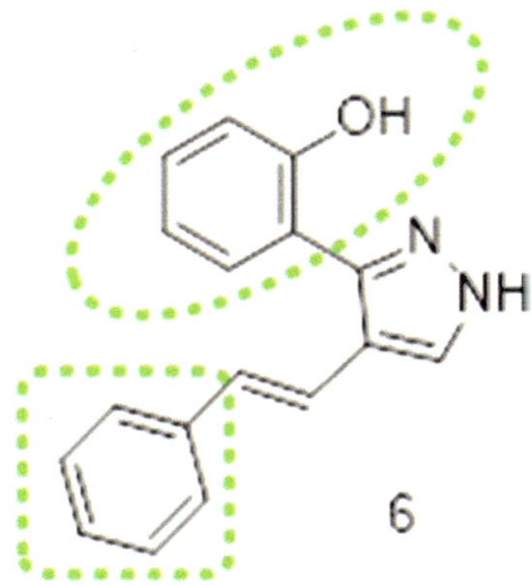

6

Jagtap *et al.* (2020) synthesized a series of conjugated pyrazole and 2-amino-3,4-dihydroquinazoline derivatives as novel inhibitors targeting BACE1. The bulky substitution of 2-amino-3,4-dihydroquinazoline was carried out at the fourth position of pyrazole with an intermediate linker amide moiety. The study revealed that the pyrazole conjugation exerted potent activity against the BACE1 receptor. The report suggested that compound 9, a 4-methyl-substituted derivative, exhibited maximum inhibitory activity with an IC_{50} value of 0.38 μM (cell-free assay) and 0.14 μM (whole cell assay), while compound 10, a 4-cyclohexylmethyl-substituted derivative, exhibited an IC_{50} value of 0.49 μM (cell-free assay) and 0.14 μM (whole cell assay).

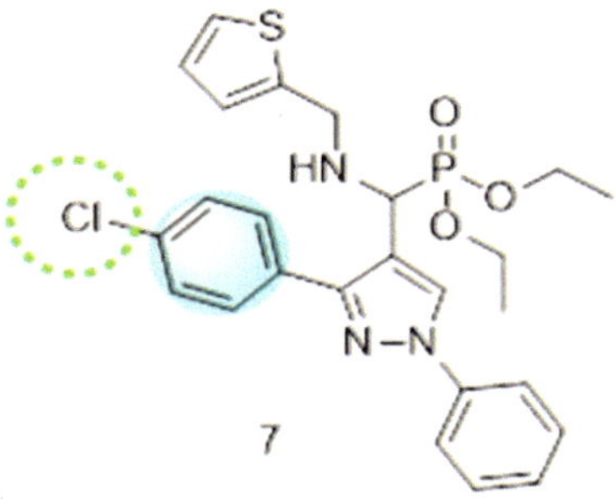

7

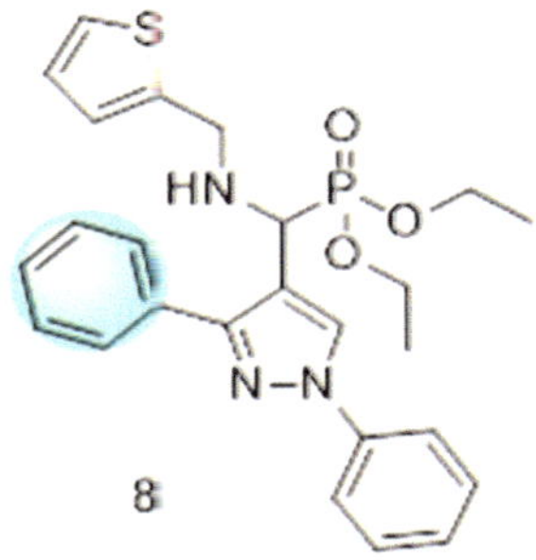

8

The study revealed the relationship between structure and BACE1 inhibitory activity, indicating that the N-cyclohexyl substituent at C3 and C4 binds with the S1 and S1′ subsites of BACE1 to enhance inhibitory potency. Also, the addition of the pyrazole moiety facilitates further binding with the S2 subsite of BACE1, with interaction with Thr 232 in the S2 subsite being responsible for the improvement of inhibitory activity. The pharmacophore 2-amino-3,4-dihydroquinazoline facilitates the formation of hydrogen bonds with Asp 32 and Asp 228. Moreover, the presence of α-hydrophobic properties in the substituent and the pure (S)-2 enantiomer contribute to the observed activity (Figure 10.10) (Jagtap *et al.*, 2020).

Huan-Huan Li *et al.* (2022) synthesized fluorosulfate-appended pyrazole derivatives targeting selective BuChE inhibitors for the effective treatment of AD. They devised a series of pyrazole-5-fluorosulfate derivatives and tested them against cholinesterase enzyme. They reported that compound 11 exhibited potential inhibitory activity against BuChE with an IC_{50} of 0.79 μM and hBuChE with 6.59 μM, while it displayed inhibitory activity against hAChE at 31.65 μM (Figure 10.11).

The potent activity can be elucidated with the assistance of the SAR of pyrazole-5-fluorosulfates. It is presumed that a phenyl ring at the N_1 position with a p-chloro substituent is essential for inhibitory activity. Replacement of chlorine with another electronegative atom, such as fluorine or bromine, decreases inhibitory activity. Phenyl substitution with 4-H or 4-methyl does not

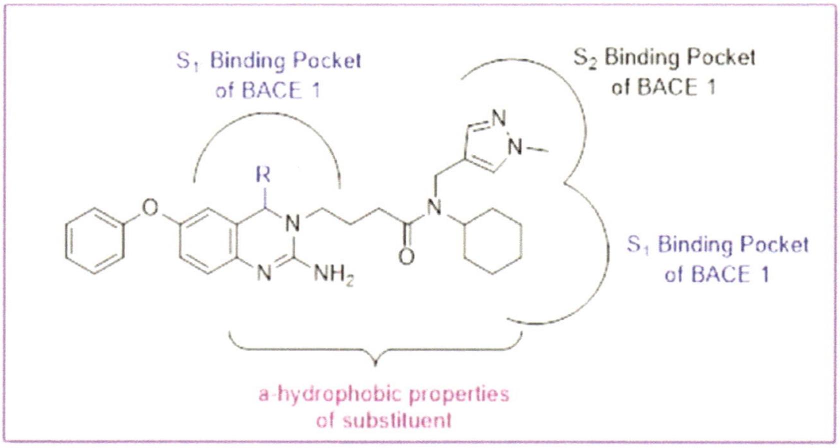

FIGURE 10.10 Binding interactions with BACE1.

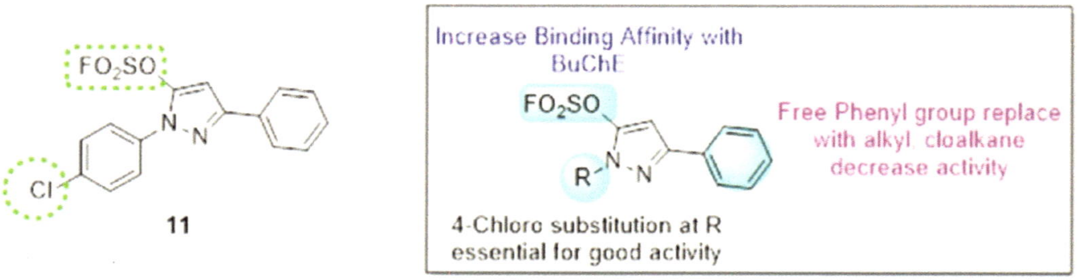

FIGURE 10.11 Chemical structure of compound 11 and SAR characteristics.

enhance the activity. Disubstitution of the phenyl ring at N_1, such as 3,4-dimethyl, also reduces the potency of the compound as a BuChE inhibitor. Removing the phenyl ring and replacing it with a smaller non-aromatic group decreases activity. The study established that *p*-chloro phenyl substitution at N1 of the pyrazole is essential for optimum activity. Another substitution at C-3 of the pyrazole, indicating the availability of a free phenyl ring, exhibits the best inhibition among the series of compounds. However, a substituent at C-3 with an isopropyl group also possesses potent activity in this series, whereas substitution with cyclopropyl at the same position decreases activity on BuChE. The docking study confirmed that the 5-fluorosulfate of pyrazole increases binding affinity with BuChE through pi-sulfur interaction and van der Waals forces (Li *et al.*, 2022).

Mroueh *et al.* (2020) established derivatives of conjugated pyrazole and quinoline derivatives as *h*AChE and COX-2 inhibitors. In this study, a series of 1H-pyrazolo[3,4-b]quinoline compounds were developed and screened for anti-cholinesterase and anti-inflammatory evaluation. The study established compounds for *in vitro* AChE inhibitory activity, whereas an *in vivo* study revealed almost ten compounds with greater inhibitory activities when compared to tacrine. The common compound that shows good inhibition for both AChE and COX-2 is based on a thiazolidinone derivative, compound 12, with an *in vitro* increase in contraction of 65.5% and *in vivo* AChE activity showing 18.43% inhibition.

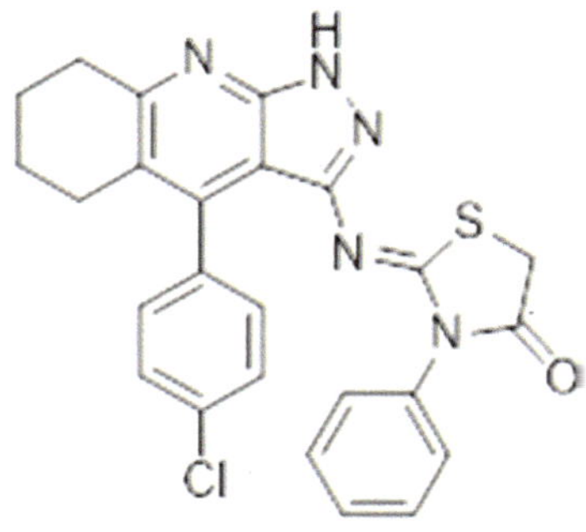

12

The designed structure has been developed rationally based on the condensation of a free amino group and aromatic aldehydes, developing a Schiff base that has good AChE inhibitory activity. Further developments of the amino group into thiourea possess good inhibitory activity for both AChE and COX-2. The study deals with cyclization of thiourea leading to formation of thiazolidine, which developed a compact pyrazole-conjugated quinoline ring with thiazolidine and thiazolidinone moieties, showing good *in vitro* and *in vivo* inhibitory activity. The structure–activity relationship demonstrated that the 1H-pyrazolo[3,4-b]quinoline ring is essential to exert activity and the substitution of the cyclized thiazolidinone amino moiety at the C-3 position of pyrazole increases the activity over the open-chain thiourea moiety (Figure 10.12) (Mroueh *et al.*, 2020).

McKenzie *et al.* (2019) carried out the synthesis of pyrazole derivatives as potential inhibitors of human THP-1 receptors, aiming for the medication of various neurodegenerative diseases by harnessing their neuroprotective and anti-inflammatory properties. The research involved the synthesis and evaluation of several trisubstituted pyrazole derivatives. The top three compounds were identified as effective inhibitors of human THP-1 monocyte cells across concentrations ranging from 1 to 100 μM. It was observed that exposure of THP-1 cells to stimulating agents generally reduced their viability to around 40–60% compared to unstimulated cells. Notably, compounds 13, 14, and 15 exhibited significant inhibition at both 50 and 100 μM concentrations. The N-substituted phenyl, CN at the fourth position, and amide chain at the third position of pyrazole are essential for potency. At the amide chain end, primary or tertiary amine substitution increases the activity.

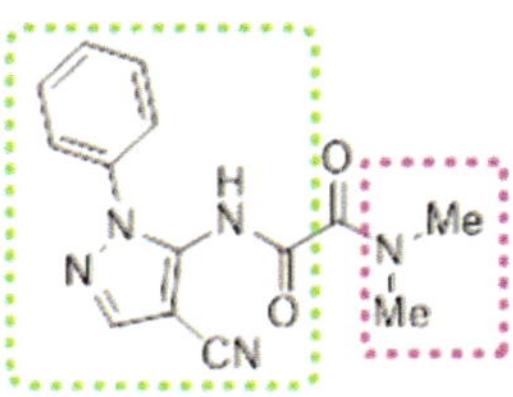

13

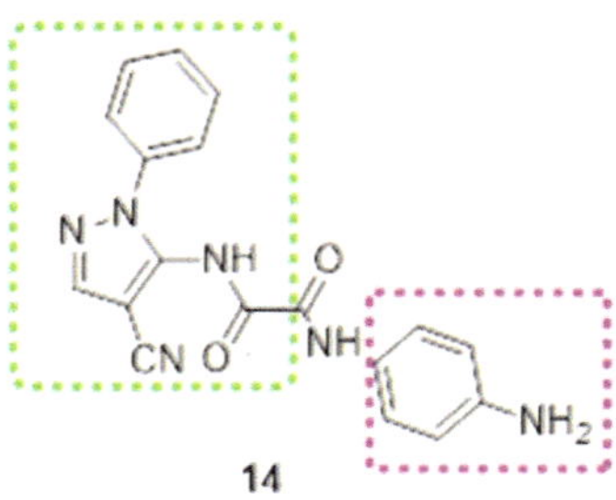

14

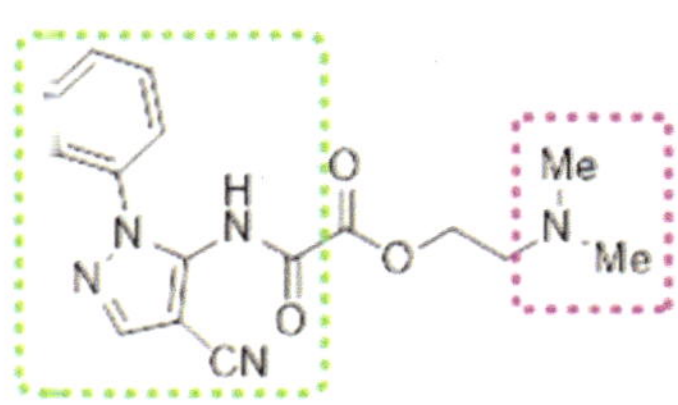

15

The new derivatives outlined in this research demonstrated moderate efficacy in inhibiting the neurotoxic secretions of microglia-like cells. Although additional refinement is needed to decrease their IC_{50} values as inhibitors of microglial neurotoxicity, compound 15 lays the groundwork for identifying more potent neuroprotective medications that can inhibit the harmful activation of microglia. The study emphasizes future investigations with pyrazole analogs with smaller substituents that possess electrophilic sites to be available as potent THP-1 inhibitors (McKenzie *et al.*, 2019).

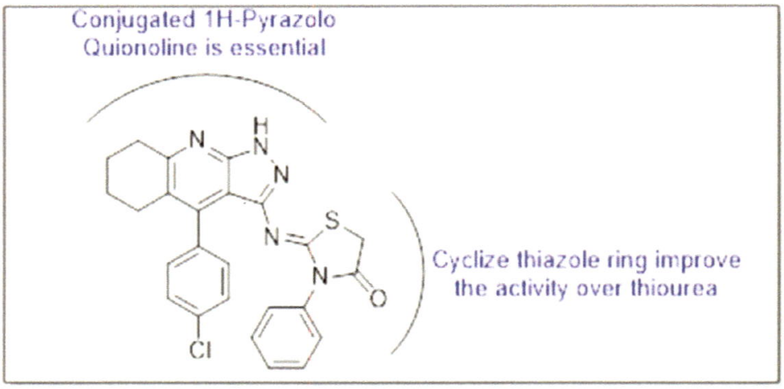

FIGURE 10.12 SAR characteristics of pyrazole-conjugated quinoline nuclei.

Baleh Leili *et al.* (2017) employed a rational design approach to transform tacrine into fused pyrazolo[1,2-b]phthalazine derivatives, aiming to create potent AChE inhibitors with high selectivity and minimal hepatocyte and neuroprotective toxicity. Among these derivatives, compound 17 emerged as particularly promising, exhibiting an IC_{50} value of 49 μM, making it seven times more potent than tacrine. Notably, unlike tacrine, compound 18 displayed significant selectivity against AChE over BuChE. Furthermore, assessments using cell-based assays against hepatocytes (HepG2) and a neuronal cell line (PC12) revealed the lower toxicity of this compound.

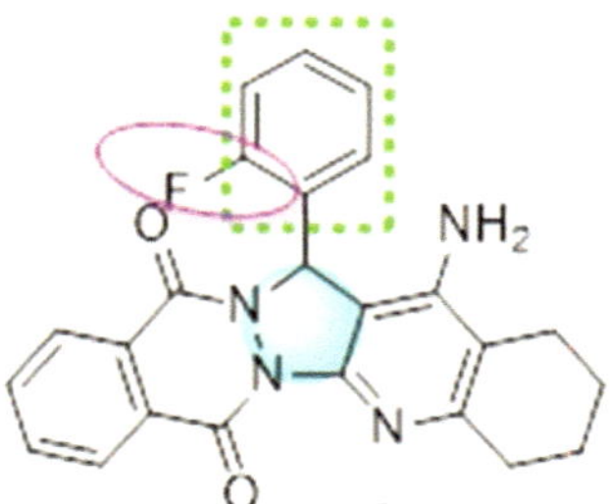

17

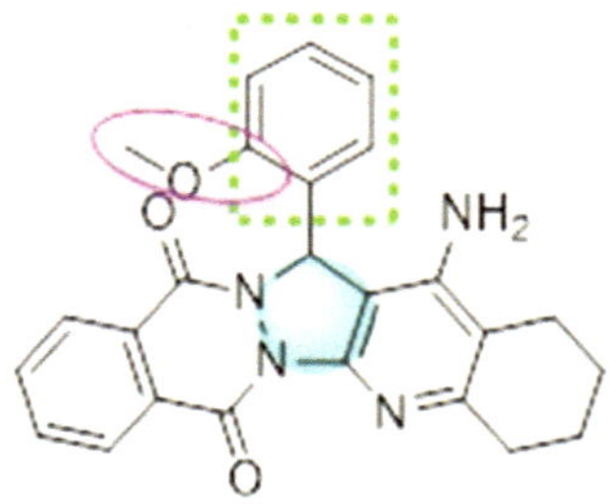

18

In the developed series, substitutions with R demonstrated that halogen and electron-withdrawing groups, such as the NO_2 group, when placed at the para position of the phenyl ring, decrease inhibitory activity, whereas halogen substitution at the ortho position enhances inhibitory activity (Figure 10.13). Additionally, replacing aromatic substitutions with smaller aliphatic alkyl groups, such as methyl or propyl, decreases activity, whereas substitution with a pentyl group enhances activity (Jalili-Baleh *et al.*, 2017).

Chigurrupati *et al.* (2016) synthesized indolopyrazoline derivatives and examined and assessed for their potential as anti-Alzheimer's agents by means of *in vitro* AChE inhibition. Compound 19 demonstrates superior AChE inhibition with an IC_{50} value of 0.68 ± 0.13 μM.

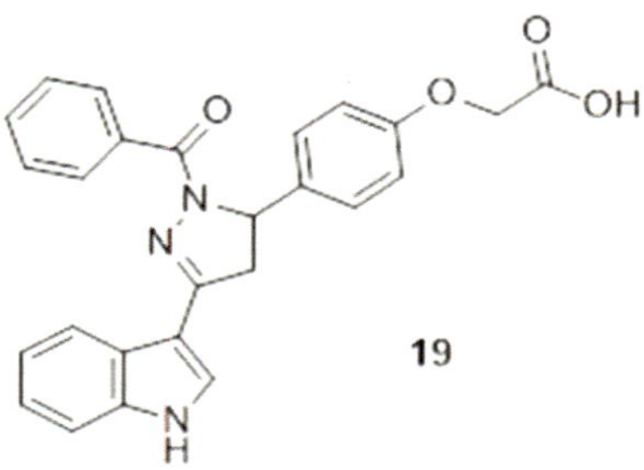

19

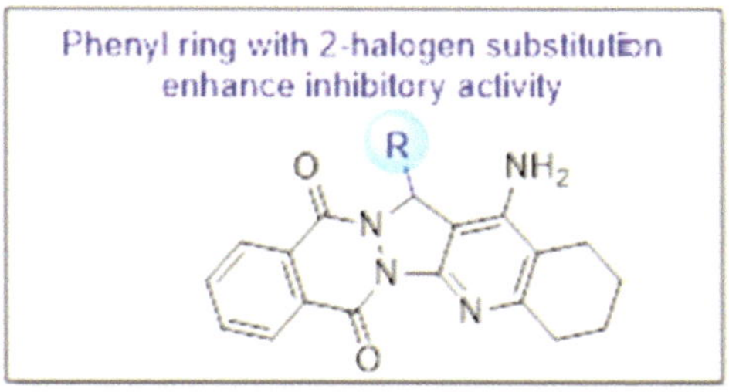

FIGURE 10.13 SAR characteristics of fused pyrazolo[1,2-b]phthalazine derivatives.

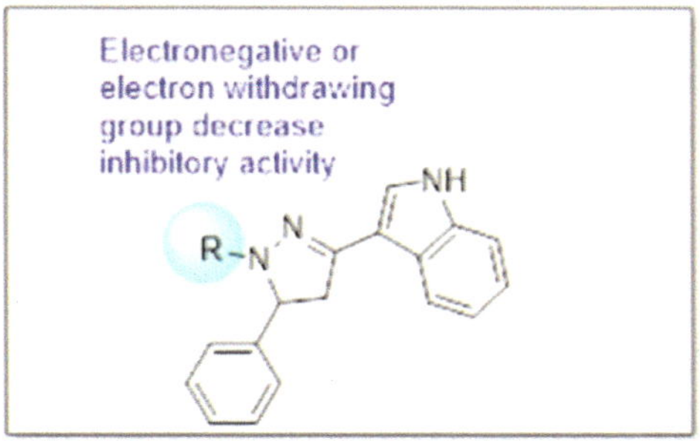

FIGURE 10.14 SAR features of indolopyrazoline derivatives.

In this series, the substitution at the imido nitrogen of imidazole ranges from a heterocyclic pyridine moiety to an aromatic moiety with either electron-withdrawing groups or no substitution. It has been resolved that aromatic substitution with an electronegative atom or an electron-withdrawing group diminishes inhibitory activity, whereas a phenoxy moiety with no substitution enhances inhibitory activity (Figure 10.14). The indole ring gives stable pi–pi (π–π) stacking interaction with Trp 84 and His 440 to exhibit the inhibitory activity (Chigurupati *et al.*, 2016).

Gutti *et al.* (2019) aimed at developing multifunctional agents for the treatment of Alzheimer's disease focused on the synthesis of 3,5-diarylpyrazoles and spiropyrazoline analogs as acetylcholinesterase inhibitors. The study identified a potent compound 20 with an AChE IC_{50} value of 1.937 μM, which was further modified to produce compound 21, exhibiting an AChE IC_{50} value of 0.464 μM. Both compounds demonstrated promising activity in an *in vivo* study using a scopolamine-induced amnesia model. Additionally, these two compounds were found to decrease metal-induced Aβ aggregation.

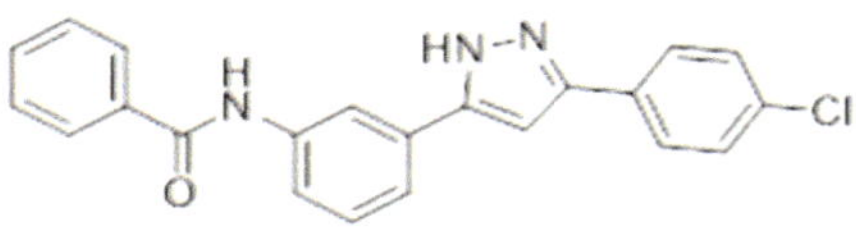

20

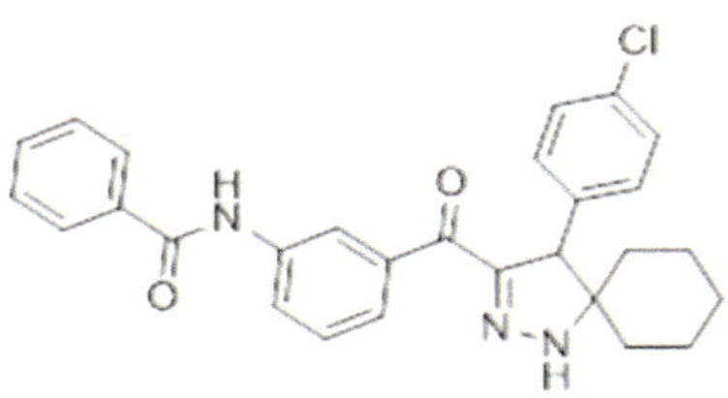

21

The structure–activity relationship analysis revealed that the presence of the N-phenylbenzamide moiety is crucial for inhibitory activity, as the benzamide segment establishes polar interactions with HIS447 and SER203 residues. In the case of 3,5-diarylpyrazole derivatives, the N-phenylbenzamide substitution at the 3-position of the pyrazole ring enhances binding affinity. The attachment of a p-chlorophenyl ring interacts hydrophobically with TYR124 and forms π–π stacking interactions with TRP286 and TYR341 residues of PAS-AChE. Additionally, an extra hydrogen on the heterocycle facilitates hydrogen bond formation in another compound (Gutti *et al.*, 2019).

Turkan *et al.* (2019) designed substituted pyrazole compounds that have emerged as effective inhibitors of cytosolic carbonic anhydrase I and II isoforms (hCA I and II), as well as AChE enzymes, with compound 22 demonstrating potency, displaying IC_{50} values of 0.83 μM for hCA I and 1.36 μM for hCA II, respectively. The Ki value for evaluated AChE compound 23 exhibited 48.94 μM. Docking interactions revealed that compound 23 binds in the active site through π–π interactions with Trp 286 and Phe 297 residues and forms hydrogen bonds with Tyr 72 and Phe 295 residues. The study confirmed the significant roles of Tyr 72 and Trp 286 residues in the inhibition of the AChE enzyme (Turkan *et al.*, 2019).

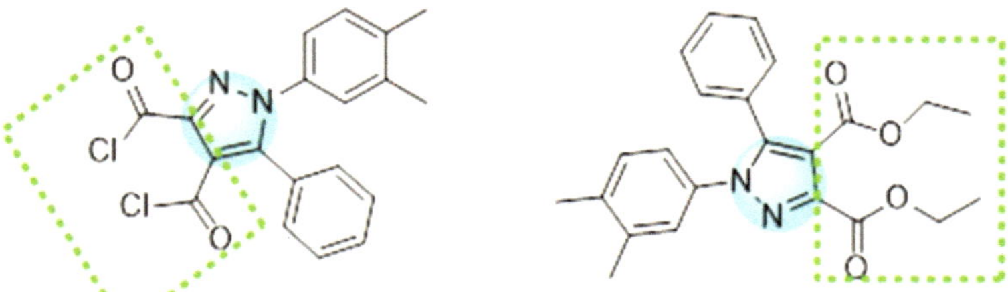

22 23

Lan-Jin *et al.* (2018) designed new hybrid compounds incorporating both MAO and Aβ self-aggregation inhibitory properties synthesized using an indazole moiety and resveratrol. Biological screening revealed that the majority of these compounds exhibited strong inhibitory effects on Aβ self-aggregation, along with notable selective inhibition of MAO-B. Compound 24 emerged as the most potent inhibitor, displaying significant activity against both hMAO-B with an IC_{50} of 1.14 μM and Aβ self-aggregation with 58.9% inhibition at 20 μM.

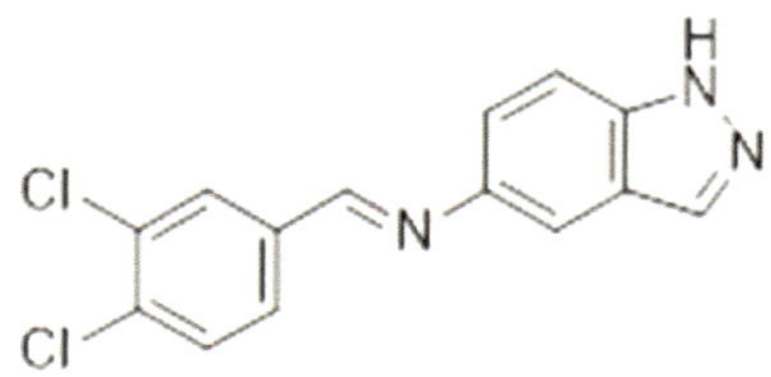

24

The structure–activity relationship revealed that the inclusion of an imidazole moiety enhances activity against MAO-B, likely attributed to the electronic characteristics introduced into the phenyl ring. Substituting the phenyl group with another group such as –OH or $–OCH_3$ diminishes activity against both hMAO-B and Aβ self-aggregation (Figure 10.15) (Lan *et al.*, 2018).

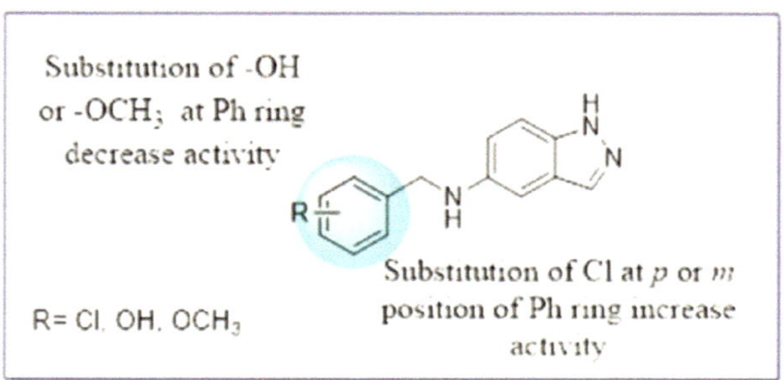

FIGURE 10.15 SAR characteristics of imidazole–resveratrol hybrid structure.

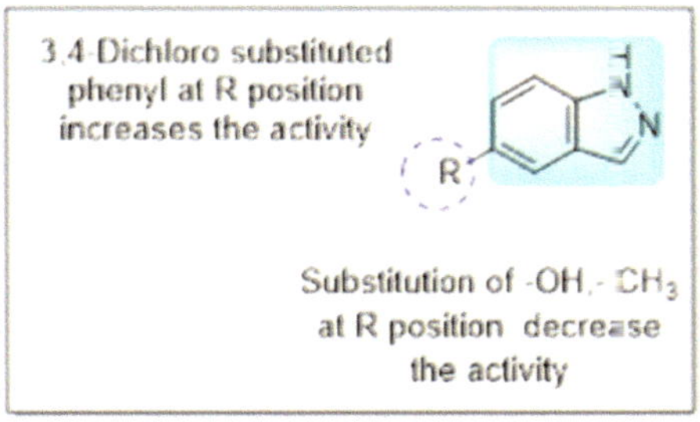

FIGURE 10.16 SAR features of bisindole-substituted aminopyrazoles

Liu-Jian *et al.* (2020) has identified bisindole-substituted aminopyrazoles as novel inhibitors of GSK-3β, showing promise for AD treatment. Within a series of newly designed disubstituted aminopyrazole derivatives aimed at targeting GSK-3β, compound 25 exhibited the most notable GSK-3β inhibition, displaying moderate activity with an IC_{50} value of 1.76 μM.

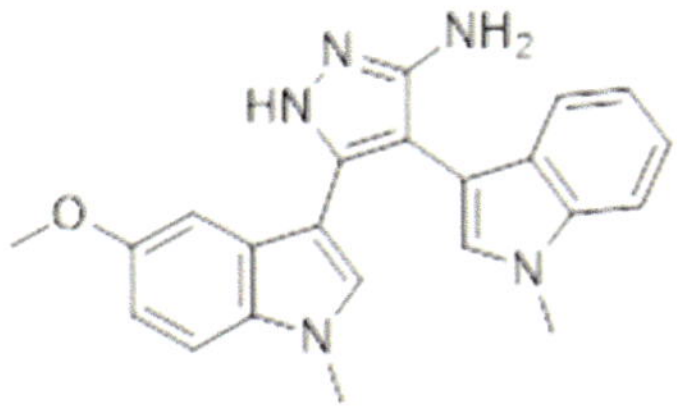

25

Upon examining the structure of compound 25, it was discovered that the inclusion of a methoxy group at the fifth position of the indole moiety, which is substituted at the third position of the pyrazole core ring, enhances activity compared to halogen substitution at the fifth or sixth position (Figure 10.16). This uniquely structured bisindole-substituted aminopyrazole could serve as a valuable prototype in the search for new therapeutic agents to address Alzheimer's disease (Liu *et al.*, 2020).

General Structure–Activity Relationship (SAR)

Throughout the chapter we have discussed the efficiency of the pyrazole scaffold. It has great affinity toward different essential AD targets. Pyrazole yielded potential drugs for various diseases and continuously produces drug-like molecules due to the wide scope of substitutions. It provides freedom for modification and optimization of drug-like molecules at four different positions. The general SAR of pyrazole scaffold with respect to anti-Alzheimer's action is represented in Figure 10.10. Most researchers substituted the second, third, and fifth positions of pyrazole with different functional groups, rings, and bulky groups. Substitution of a phenyl ring at the third position and heterocycic ring at the fifth position showed enhanced action. Substitution of a bulky group or long chain, amide chain, or carbonyl chain at the fourth position showed improved action. The detailed SAR is shown in Figure 10.17.

CONCLUSION

Pyrazole scaffold has emerged as one of the promising scaffolds for multi-target drug design in the recent era for AD. Pyrazole scaffold derivatives designed for anti-AD drugs are an important choice for researchers since various scaffolds and derivatives of pyrazole have better multi-targeted activities. Researchers are showing interest in pyrazole scaffold for its multi-target activity, less

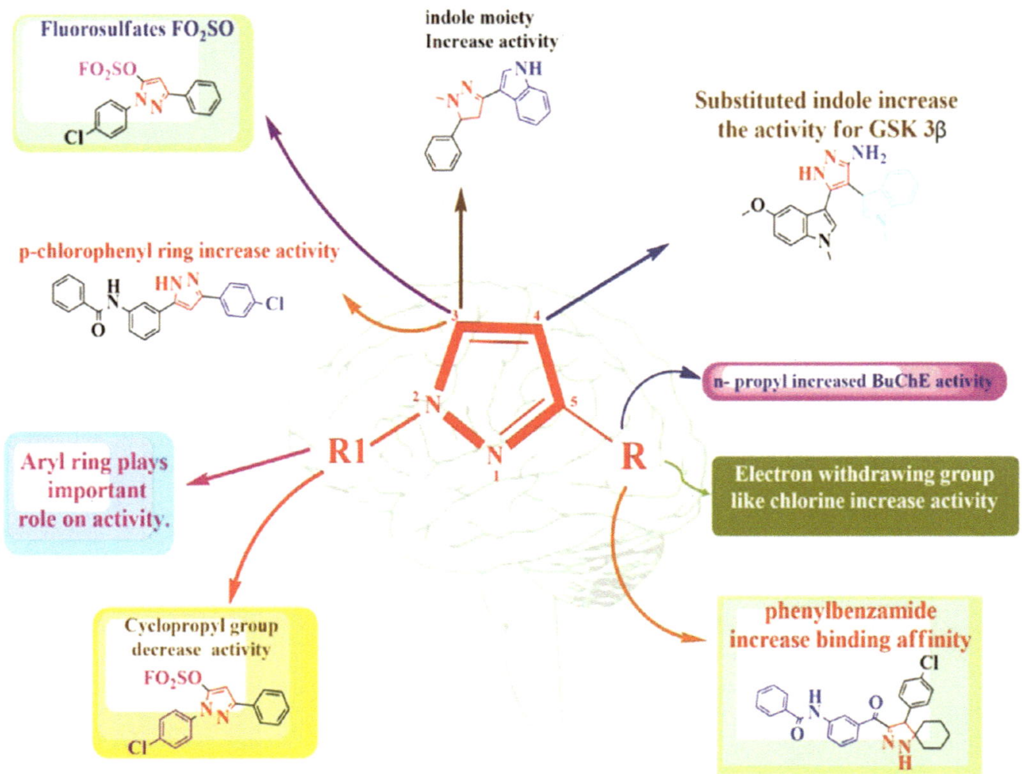

FIGURE 10.17 General SAR of pyrazole derivatives.

cytotoxicity, and neurotoxicity. The multiple target affinity of pyrazole scaffold should be handled carefully. It is anticipated to have the desired selectivity for optimum activity. For design purposes, it is noteworthy to mention the selection of targets by which compartments of an entire pathway can be achieved. The STRING strategy and predefined knowledge with the aid of *in silico* tools can solve the problem. Although the majority of pyrazole derivatives target AChE and BuChE, showing their inhibitory activity against Alzheimer's, other possibilities (targets) should be explored in a rational, stepwise manner. Looking at the current status of pyrazole scaffolds, we can conclude that showing interest in synthetic pyrazole scaffolds will give the desired output for anti-Alzheimer's activity.

REFERENCES

Agis-Torres A, Sollhuber M, Fernandez M, Sanchez-Montero JM. Multi-target-directed ligands and other therapeutic strategies in the search of a real solution for Alzheimer's isease. *Curr Neuropharmacol* (2014) 12(1):2–36.

Aitken L, Benek O, McKelvie BE, Hughes RE, Hroch L, et al. Novel Benzothiazole-based Ureas as 17β-HSD10 inhibitors, a potential Alzheimer's disease treatment. *Molecules* (2019) 24(15):2757.

Ajala A, Uzairu A, Shallangwa GA, Abechi SE. QSAR, simulation techniques, and ADMET/pharmacokinetics assessment of a set of compounds that target MAO-B as anti-Alzheimer agent. *Fut J Pharm Sci* (2023) 9(1):4.

Akawa OB, Subair TI, Soremekun OS, Olotu FA, Soliman MES. Structural alterations in the catalytic core of hSIRT2 enzyme predict therapeutic benefits of Garcinia mangostana derivatives in Alzheimer's disease: molecular dynamics simulation study. *RSC Adv* (2021) 11(14):8003–8018.

Akk G, Steinbach JH. Galantamine activates muscle-type Nicotinic Acetylcholine receptors without binding to the Acetylcholine-binding site. *J Neurosci* (2005) 25(8):1992–2001.

Al-Hazmy S, Zouaghi M, Amri N, Arfaoui Y, Alhagri I, Hamdi N. (2023). DFT study of regio- and stereoselective 13DC reaction between Diazopropane and substituted Chalcone derivatives: molecular docking of novel Pyrazole derivatives as anti-Alzheimer's agents. *Molecules* (2023) 28(4):1899.
Ali R, Gupta GD, Chawla PA. Aducanumab: a new hope in Alzheimer's disease. *Health Sci Rev* (2022) 4:100039.
ALZHEIMER, A. Uber eigenartige Erkrankung der Hirnrinde. *All Z Psychiatr* (1907) 64:146–148.
Alzheimer's Association. Alzheimer's disease facts and figures. *Alzheimer Dement* (2023) 19(4):1598–1695.
Amin SA, Bhargava S, Adhikari N, Gayen S, Jha T. Exploring pyrazolo[3,4- d]pyrimidine phosphodiesterase 1 (PDE1) inhibitors: a predictive approach combining comparative validated multiple molecular modelling techniques. *J Biomol Stru Dyn* (2018) 36(3):590–608.
Amrhein JA, Wang G, Berger B-T, Berger LM, Kalampaliki AD, Krämer A, Knapp S, Hanke T. Design and synthesis of Pyrazole-based macrocyclic Kinase inhibitors targeting BMPR2. *ACS Med Chem Lett* (2023) 14(6):833–840.
Anand P, Singh B. A review on cholinesterase inhibitors for Alzheimer's disease. *Arch Pharm Res* (2013) 36(4):375–399.
Annunziata F, Pinna C, Dallavalle S, Tamborini L, Pinto A. An overview of Coumarin as a versatile and readily accessible scaffold with broad-ranging biological activities. *Int J Mol Sci* (2020) 21(13):4618. https://doi.org/10.3390/ijms21134618
Armstrong RA. Risk factors for Alzheimer's disease. *Folia Neuropathol* (2019) 57(2):87–105. https://doi.org/10.5114/fn.2019.85929
Arndt JW, Qian F, Smith BA, Quan C, Kilambi KP, et al. Structural and kinetic basis for the selectivity of aducanumab for aggregated forms of amyloid-β. *Sci Rep* (2018) 8(1):6412. https://doi.org/10.1038/s41598-018-24501-0
Aso E, Ferrer I. CB2 Cannabinoid receptor as potential target against Alzheimer's disease. *Front Neurosci* (2016) 10:243. https://doi.org/10.3389/fnins.2016.00243
Babic T, Francis PT, Palmer AM, Snape M, Wilcock GK. The cholinergic hypothesis of Alzheimer's disease: a review of progress. *J Neurol Neurosurg Psychiatry* (1999) 67(4):558–558. https://doi.org/10.1136/jnnp.67.4.558
Bales KR, Verina T, Cummins DJ, Du Y, Dodel RC, et al. (1999). Apolipoprotein E is essential for amyloid deposition in the APP^{V717F} transgenic mouse model of Alzheimer's disease. *Proc Nat Acad Sci* (1999) 96(26):15233–15238. https://doi.org/10.1073/pnas.96.26.15233
Banjare P, Matore WB, Murmu A, Kumar V, Singh J, Roy PP. In silico strategy: a promising implement in the development of multitarget drugs against neurodegenerative diseases. *Curr Topics Med Chem* (2023) 23(29):2765–2791. https://doi.org/10.2174/1568026623666230811113231
Behl T, Kaur I, Fratila O, Brata R, Bungau S. Exploring the potential of therapeutic agents targeted towards mitigating the events associated with Amyloid-β cascade in Alzheimer's disease. *Int J Mol Sci* (2020) 21(20):7443. https://doi.org/10.3390/ijms21207443
Bekris LM, Yu C-E, Bird TD, Tsuang DW. Review article: genetics of Alzheimer disease. *J Geriatr Psychiatry* Neurol (2010) 23(4):213–227.
Birks J, Grimley Evans J, Iakovidou V, Tsolaki M. Rivastigmine for Alzheimer's disease. In J. Birks (Ed.), *Cochrane Database of Systematic Reviews* (2009) (p. CD001191). John Wiley & Sons, Ltd.
Booth L, Roberts, JL, Dent P. HSPA5/Dna K may be a useful target for human disease therapies. *DNA Cell Biol* (2015) 34(3):153–158.
Braak H, Del Tredici K. Neurofibrillary tangles. In K. Kompoliti and L. Verhagen Metman (Eds.), *Encyclopedia of Movement Disorders* (2010) (pp. 265–269). Elsevier.
Brasso C, Colli G, Sgro R, Bellino S, Bozzatello P, et al. Efficacy of Serotonin and Dopamine activity modulators in the treatment of negative symptoms in Schizophrenia: a rapid review. *Biomedicines* (2023) 11(3):921.
Breijyeh Z, Karaman R. Comprehensive review on Alzheimer's disease: causes and treatment. *Molecules* (2020) 25(24):5789.
Brion J-P. Neurofibrillary tangles and Alzheimer's disease. *Eur Neurol* (1998) 40(3):130–140.
Bubley A, Erofeev A, Gorelkin P, Beloglazkina E, Majouga A, Krasnovskaya O. Tacrine-based hybrids: past, present, and future. *Int J Mol Sci* (2023) 24(2):1717.
Cabrera-Pardo JR, Fuentealba J, Gavilán J, Cajas D, Becerra J, Napiórkowska M. Exploring the multi–target neuroprotective chemical space of Benzofuran scaffolds: a new strategy in drug development for Alzheimer's disease. *Front Pharmacol* (2020) 10:1679.
Cacabelos R, Takeda M, Winblad B. The glutamatergic system and neurodegeneration in dementia: preventive strategies in Alzheimer's disease. *Int J Geriatr Psychiatry* (1999) 14(1):3–47.

Camps P, Formosa X, Galdeano C, Muñoz-Torrero D, Ramírez L, et al. Pyrano[3,2- c]quinoline–6-Chlorotacrine hybrids as a novel family of Acetylcholinesterase- and β-Amyloid-directed anti-Alzheimer compounds. *J Med Chem* (2009) 52(17):5365–5379.

Chartier-Harlin M-C, Crawford F, Houlden H, Warren A, Hughes D, et al. Early-onset alzheimer's disease caused by mutations at codon 717 of the β-amyloid precursor protein gene. *Nature* (1991) 353(6347):844–846.

Chauhan V, Chauhan A. Oxidative stress in Alzheimer's disease. *Pathophysiol* (2006) 13(3): 195–208.

Chen Y, Yu Y. Tau and neuroinflammation in Alzheimer's disease: interplay mechanisms and clinical translation. *J Neuroinflammation* (2023) 20(1):165.

Chigurupati S, Selvaraj M, Mani V, Selvarajan KK, Mohammad JI, et al. Identification of novel acetylcholinesterase inhibitors: Indolopyrazoline derivatives and molecular docking studies. *Bioorg Chem* (2016) 67:9–17.

Chyan Y-J, Poeggeler B, Omar RA, Chain DG, Frangione B, Ghiso J, Pappolla MA. Potent neuroprotective properties against the Alzheimer β-Amyloid by an endogenous Melatonin-related Indole structure, Indole-3-propionic acid. *J Biological Chem* (1999) 274(31):21937–21942.

Corder EH, Saunders AM, Strittmatter WJ, Schmechel DE, Gaskell PC, et al. (1993). Gene dose of Apolipoprotein E type 4 Allele and the risk of Alzheimer's disease in late onset families. *Science* (1993) 261(5123):921–923.

Cuadrado-Tejedor M, Garcia-Barroso C, Sánchez-Arias JA, Rabal O, Pérez-González M, et al. A first-in-class small-molecule that acts as a dual inhibitor of HDAC and PDE5 and that rescues Hippocampal synaptic impairment in Alzheimer's disease mice. *Neuropsychopharmacol* (2017) 42(2):524–539.

Dahm R. Alzheimer's discovery. *Curr Biology* (2006) 16(21):R906–R910.

Darvesh S, Grantham DL, Hopkins DA. Distribution of butyrylcholinesterase in the human amygdala and hippocampal formation. *J Comparative Neurol* (1998) 393(3):374–390.

Das S, Barnwal P, Winston AB, Mondal S, Saha I. Brexpiprazole: so far so good. *Therapeutic Adv Psychopharmacol* (2016) 6(1):39–54.

Davis KL, Thal LJ, Gamzu ER, Davis CS, Woolson RF, et al. A double-blind, placebo-controlled multicenter study of Tacrine for Alzheimer's disease. *New Eng J Med* (1992) 327(18):1253–1259.

De SK. Novel Pyrimidin-4-yl-3-amino-pyrrolo[3,4-c]pyrazoles as protein kinase C inhibitors for treating diseases. *Curr Med Chem* (2023) 31(8):1036–1039.

Del Cerro P, Alquézar C, Bartolomé F, González-Naranjo P, Pérez C, et al. Activation of the Cannabinoid type 2 receptor by a novel Indazole derivative normalizes the survival pattern of Lymphoblasts from patients with late-onset alzheimer's Disease. *CNS Drugs* (2018) 32(6):579–591.

Diefenderfer LA, Iuppa C. Brexpiprazole: A review of a new treatment option for schizophrenia and major depressive disorder. *Ment Health Clin* (2017) 7(5):207–212.

Dinakar S, Gurubarath M, Dhananjayan K. Prediction of binding affinity of 1,2-diphenyline ketone analogues at adenosine triphosphate binding site of glycogen synthase kinase-3β: a molecular docking and dynamic simulation study. *J Biomol Stru Dyn* (2022) 1–16.

Drake LR, Pham JM, Desmond TJ, Mossine AV, Lee SJ, Kilbourn MR, Koeppe RA, Brooks AF, Scott PJH. Identification of AV-1451 as a weak, nonselective inhibitor of Monoamine Oxidase. *ACS Chem Neurosci* (2019) 10(8):3839–3846.

Du H, Song J, Ma F, Gao H, Zhao X, Mao R, He X, Yan Y. Novel harmine derivatives as potent acetylcholinesterase and amyloid beta aggregation dual inhibitors for management of Alzheimer's disease. *J Enzyme Inhib Med Chem* (2023) 38(1):2281893.

Elbatrawy AA, Ademoye TA, Alnakhala H, Tripathi A, Zami A, Ostafe R, Dettmer U, Fortin JS. Discovery of small molecule benzothiazole and indole derivatives tackling tau 2N4R and α-synuclein fibrils. *Bioorg Med Chem* (2024) 100:117613.

Faisal M, Saeed A, Hussain S, Dar P, Larik FA. Recent developments in synthetic chemistry and biological activities of pyrazole derivatives. *J Chem Sci* (2019) 131(8):70.

Fancellu G, Chand K, Tomás D, Orlandini E, Piemontese L, et al. Novel tacrine–benzofuran hybrids as potential multi-target drug candidates for the treatment of Alzheimer's Disease. *J Enzyme Inhib Med Chem* (2020) 35(1): 211–226.

Fang L, Shen S, Liu Q, Liu Z, Zhao J. Combination of NSAIDs with donepezil as multi-target directed ligands for the treatment of Alzheimer's disease. *Bioorg Med Chem Lett* (2022) 75:128976.

Farlow MR. Clinical Pharmacokinetics of Galantamine. *Clin Pharmacokinet* (2003) 42(15):1383–1392.

Fernandez AP, Pozo-Rodrigalvarez A, Serrano J, Martinez-Murillo R. Nitric Oxide: target for therapeutic strategies in Alzheimers disease. *Curr Pharm Design* (2010) 16(25):2837–2850.

Ferreira-Vieira TH, Guimaraes IM, Silva FR, Ribeiro FM. Alzheimer's disease: targeting the Cholinergic system. *Curr Neuropharmacol* (2016) 14(1):101–115.

Ferrero H, Solas M, Francis PT, Ramirez MJ. Serotonin 5-HT6 receptor antagonists in Alzheimer's disease: therapeutic rationale and current development status. *CNS Drugs* (2017) 31(1):19–32.

Ferrero J, Williams L, Stella H, Leitermann K, Mikulskis A, O'Gorman J, Sevigny J. First-in-human, double-blind, placebo-controlled, single-dose escalation study of aducanumab (BIIB037) in mild-to-moderate Alzheimer's disease. *Alzheimer Dement* (N Y): *Trans Res Clin Interventions* (2016) 2(3):169–176.

Fiala JC. Mechanisms of amyloid plaque pathogenesis. *Acta Neuropathol* (2007) 114(6):551–571.

Foster EM, Dangla-Valls A, Lovestone S, Ribe EM, Buckley NJ. Clusterin in Alzheimer's disease: mechanisms, genetics, and lessons from other pathologies. *Front Neurosci* (2019) 13:164.

Gameiro I, Michalska P, Tenti G, Cores Á, Buendia I, et al. Discovery of the first dual GSK3β inhibitor/Nrf2 inducer. A new multitarget therapeutic strategy for Alzheimer's disease. *Sci Rep* (2017) 7(1):45701.

Goate A, Chartier-Harlin M-C, Mullan M, Brown J, Crawford F, et al. Segregation of a missense mutation in the amyloid precursor protein gene with familial Alzheimer's disease. *Nature* (1991) 349(6311):704–706.

Goedert M, Spillantini MG. Tau gene mutations and neurodegeneration. *Biochem Soc Symp* (2001) 67:59–71.

Götz J, Chen F, van Dorpe J, Nitsch RM. Formation of neurofibrillary tangles in P301L Tau transgenic mice induced by Aβ42 fibrils. *Science* (2001) 293(5534):1491–1495.

Govindarajulu M, Ramesh S, Neel L, Fabbrini M, Buabeid M, et al. Nutraceutical based SIRT3 activators as therapeutic targets in Alzheimer's disease. *Neurochem Int* (2021) 144:104958.

Goyal D, Kaur A, Goyal, B. Benzofuran and Indole: promising scaffolds for drug development in Alzheimer's disease. *ChemMedChem* (2018) 13(13):1275–1299.

Greig SL. Brexpiprazole: first global approval. *Drugs* (2015) 75(14):1687–1697.

Guerreiro R, Bras J. The age factor in Alzheimer's disease. *Genome Med* (2015) 7(1):106.

Guglielmi P, Carradori S, D'Agostino I, Campestre C, Petzer JP. An updated patent review on monoamine oxidase (MAO) inhibitors. *Expert Opin Ther Pat* (2022) 32(8):849–883.

Gutti G, Kumar D, Paliwal P, Ganeshpurkar A, Lahre K, Kumar A, Krishnamurthy S, Singh SK. Development of pyrazole and spiropyrazoline analogs as multifunctional agents for treatment of Alzheimer's disease. *Bioorg Chem* (2019) 90:103080.

Hafez DE, Dubiel M, La Spada G, Catto M, Reiner-Link D, et al. Novel benzothiazole derivatives as multitargeted-directed ligands for the treatment of Alzheimer's disease. *J Enzyme Inhib Med Chem* (2023) 38(1):2175821.

Haghighijoo Z, Akrami S, Saeedi M, Zonouzi A, Iraji A, et al. N-Cyclohexylimidazo[1,2-a]pyridine derivatives as multi-target-directed ligands for treatment of Alzheimer's disease. *Bioorg Chem* (2020) 103:104146.

Halkina T, Henderson JL, Lin EY, Himmelbauer MK, Jones JH, et al. Discovery of potent and brain-penetrant Tau Tubulin Kinase 1 (TTBK1) inhibitors that lower Tau Phosphorylation in vivo. *J Med Chem* (2021) 64(9):6358–6380.

Hasanvand Z, Motahari R, Nadri H, Moghimi S, Foroumadi R, et al. Novel 3-aminobenzofuran derivatives as multifunctional agents for the treatment of Alzheimer's disease. *Front Chem* (2022) 10:882191.

Hayslett RL, Tizabi Y. Effects of donepezil, nicotine and haloperidol on the central serotonergic system in mice: Implications for Tourette's syndrome. *Pharmacol Biochem Behavior* (2005) 81(4):879–886.

He Z, Jiang BSK, Sun X, Tang M, Liu Y, Guan L, Wu D. Benzothiazole-Propanamide linker Pyrrolidine (Morpholine) as Monoamine Oxidase-B and Butyrylcholinesterase inhibitors. *Chem Biodiver* (2023) 20(12):e202301271.

Hebert LE, Weuve J, Scherr PA, Evans DA. Alzheimer disease in the United States (2010–2050) estimated using the 2010 census. *Neurol* (2013) 80(19):1778–1783.

Hussain A, Verma CK, Chouhan U. Identification of novel inhibitors against Cyclin Dependent Kinase 9/ Cyclin T1 complex as: anti cancer agent. *Saudi J Biol Sci* (2017) 24(6):1229–1242.

Hwang J, Hwang H, Lee H-W, Suk K. Microglia signaling as a target of donepezil. *Neuropharmacol* (2010) 58(7):1122–1129.

Ivachtchenko AV, Okun I, Aladinskiy V, Ivanenkov Y, Koryakova A, et al. AVN-492, a novel highly selective 5-HT6R antagonist: preclinical evaluation. *J Alzheimers Dis* (2017) 58(4):1043–1063.

Jagtap AD, Kondekar NB, Hung P-Y, Hsieh, C-E, Yang C-R, Chen GS, Chern J-W. 4-Substituted 2-amino-3,4-dihydroquinazolines with a 3-hairpin turn side chain as novel inhibitors of BACE-1. *Bioorg Chem* (2020) 95:103135.

Jalili-Baleh L, Nadri H, Moradi A, Bukhari SNA, Shakibaie M, et al. New racemic annulated pyrazolo[1,2-b]phthalazines as tacrine-like AChE inhibitors with potential use in Alzheimer's disease. *Eur J Med Chem* (2017) 139:280–289.

Jesudason CD, Mason ER, Chu S, Oblak AL, Javens-Wolfe J, et al. SHIP1 therapeutic target enablement: Identification and evaluation of inhibitors for the treatment of late-onset alzheimer's disease. *Alzheimer Dement (N Y): Trans Res Clin Interventions* (2023) 9(4):e12429.

Kamal A, Mallareddy A, Janaki Ramaiah M, Pushpavalli SNCVL, Suresh P, et al. Synthesis and biological evaluation of combretastatin-amidobenzothiazole conjugates as potential anticancer agents. *Eur J Med Chem* (2012) 56:166–178.

Kargbo RB. PROTAC compounds targeting α-Synuclein protein for treating neurogenerative disorders: Alzheimer's and Parkinson's diseases. *ACS Med Chem Lett* (2020) 11(6):1086–1087.

Karimi Tari P, Parsons CG, Collingridge GL, Rammes G. Memantine: updating a rare success story in pro-cognitive therapeutics. *Neuropharmacol* (2024) 244:109737.

Katzung BG, Masters SB, Trevor AJ, Eds. *Basic and clinical pharmacology.* 12th Edition (2012). McGraw-Hill Medical.

Kaur R, Narang SS, Singh P, Goyal B. Structural and molecular insights into tacrine-benzofuran hybrid induced inhibition of amyloid-β peptide aggregation and BACE1 activity. *J Biomol Stru Dyn* (2023) 41(22):13211–13227.

Kennedy JS, Polinsky RJ, Johnson B, Loosen P, Enz A, et al. Preferential Cerebrospinal fluid Acetylcholinesterase inhibition by Rivastigmine in humans. *J Clin Psychopharmacol* (1999) 19(6):513–521.

Kotermanski SE, Wood JT, Johnson JW. Memantine binding to a superficial site on NMDA receptors contributes to partial trapping. *J Physiol* (2009) 587(19):4589–4604.

Kraepelin E. Das senile und prasenilarlrrsein. In E. Kraepelin (Ed.), *Psychiatrie : Ein Lehrbuch Fur Studierende Und Alzte* (1910) (pp. 533;593–632). Barth.

Kramer T, Schmidt B, Lo Monte F. Small-molecule inhibitors of GSK-3: structural insights and their application to Alzheimer's disease models. *Int J Alzheimers Dis* (2012) 2012:1–32.

Kumar H, Goyal A, Kumar N, Garg P. Design, synthesis, and biological evaluation of pyrazolo-benzothiazole derivatives as a potential therapeutic agent for the treatment of Alzheimer's disease. *Med Chem Res* (2022) 31(11):1931–1947.

Lakra N, Matore BW, Banjare P, Singh R, Singh J, Roy PP. Pharmacophore based virtual screening of cholinesterase inhibitors: search of new potential drug candidates as antialzheimer agents. *Silico Pharmacol* (2022) 10(1):18.

Lan J-S, Liu Y, Hou J, Yang J, Zhang X-Y, Zhao Y, Xie S-S, Ding Y, Zhang T. Design, synthesis and evaluation of resveratrol-indazole hybrids as novel monoamine oxidases inhibitors with amyloid- β aggregation inhibition. *Bioorg Chem* (2018) 76, 130–139.

Łażewska D, Bajda M, Kaleta M, Zaręba P, Doroz-Płonka A, et al. Rational design of new multitarget histamine H3 receptor ligands as potential candidates for treatment of Alzheimer's disease. *Eur J Med Chem* (2020) 207:112743.

Levy-Lahad E, Wasco W, Poorkaj P, Romano DM, Oshima J, et al. Candidate gene for the Chromosome 1 familial Alzheimer's disease locus. *Science* (1995) 269(5226):973–977.

Li H-H, Wu C, Zhang S-L, Yang J-G, Qin H-L, Tang W. Fluorosulfate-containing pyrazole heterocycles as selective BuChE inhibitors: structure-activity relationship and biological evaluation for the treatment of Alzheimer's disease. *J Enzyme Inh Med Chem* (2022) 37(1):2099–2111.

Li X, Yu Y, Tu Z. Pyrazole Scaffold Synthesis, Functionalization, and Applications in Alzheimer's Disease and Parkinson's Disease Treatment (2011–2020). *Molecules* (2021) 26(5):1202.

Li X-T. Alzheimer's disease therapy based on acetylcholinesterase inhibitor/blocker effects on voltage-gated potassium channels. *Metabolic Brain Dis* (2022) 37(3):581–587.

Lim JW, Kim SK, Choi SY, Kim DH, Gadhe CG, et al. Identification of crizotinib derivatives as potent SHIP2 inhibitors for the treatment of Alzheimer's disease. *Eur J Med Chem* (2018) 157:405–422.

Liu J-G, Zhao D, Gong Q, Bao F, Chen W-W, Zhang H, Xu M-H. Development of Bisindole-substituted Aminopyrazoles as novel GSK-3β inhibitors with suppressive effects against microglial inflammation and oxidative neurotoxicity. *ACS Chem Neurosci* (2020) 11(20):3398–3408.

Liu W, Wu L, Liu W, Tian L, Chen H, et al. Design, synthesis and biological evaluation of novel coumarin derivatives as multifunctional ligands for the treatment of Alzheimer's disease. *Eur J Med Chem* (2022) 242:114689.

Liu Y, Xu Y, Li M, Pan D, Li Y, Wang Y, Wang L, Wu Q, Yang M. Multi-target PET evaluation in APP/PS1/tau mouse model of Alzheimer's disease. *Neurosci Lett* (2020) 728:134938.

Livingston G, Huntley J, Sommerlad A, Ames D, Ballard C, et al. Dementia prevention, intervention, and care: 2020 report of the Lancet Commission. *Lancet* (2020) 396(10248):413–446.

Logovinsky V, Satlin A, Lai R, Swanson C, Kaplow J, Osswald G, Basun H, Lannfelt L. Safety and tolerability of BAN2401 - a clinical study in Alzheimer's disease with a protofibril selective Aβ antibody. *Alzheimers Res Ther* (2016) 8(1):14.

Loh ZH, Kwong HC, Lam KW, Teh SS, Ee GCL, Quah CK, Ho ASH, Mah SH. New 3- O -substituted xanthone derivatives as promising acetylcholinesterase inhibitors. *J Enzyme Inhib Med Chem* (2021) 36(1):627–639.

Ludwig J, Höffle-Maas A, Samochocki M, Luttmann E, Albuquerque EX, Fels G, Maelicke A. Localization by site-directed mutagenesis of a galantamine binding site on α7 nicotinic acetylcholine receptor extracellular domain. *J Recept Signal Transduct Res* (2010) 30(6):469–483.

Maccallini C, Di Matteo M, Vullo D, Ammazzalorso A, Carradori S, et al. Indazole, Pyrazole, and Oxazole derivatives targeting Nitric Oxide synthases and Carbonic Anhydrases. *ChemMedChem* (2016) 11(16):1695–1699.

Machado V, Cenci AR, Teixeira KF, Sens L, Tizziani T, et al. Pyrazolines as potential anti-Alzheimer's agents: DFT, molecular docking, enzyme inhibition and pharmacokinetic studies. *RSC Med Chem* (2022) 13(12):1644–1656.

Mackenzie IRA, Hao C, Munoz DG. Role of microglia in senile plaque formation. *Neurobio Aging* (1995) 16(5):797–804.

Maeda K, Sugino H, Akazawa H, Amada N, Shimada J, et al. Brexpiprazole I: in vitro and in vivo characterization of a novel Serotonin-Dopamine activity modulator. *J Pharmacol Exp Ther* (2014) 350(3):589–604.

Maiuolo J, Costanzo P, Masullo M, D'Errico A, Nasso R, et al. Hydroxytyrosol–Donepezil hybrids play a protective role in an in vitro induced Alzheimer's disease model and in neuronal differentiated human SH-SY5Y Neuroblastoma cells. *Int J Mol Sci* (2023) 24(17):13461.

Majd S, Koblar S, Power J. Compound C enhances tau phosphorylation at Serine396 via PI3K activation in an AMPK and rapamycin independent way in differentiated SH-SY5Y cells. *Neurosci Lett* (2018) 670:53–61.

Malafaia D, Oliveira A, Fernandes PA, Ramos MJ, Albuquerque HMT, Silva AMS. Chromeno[3,4-b]xanthones as first-in-class AChE and Aβ aggregation dual-inhibitors. *Int J Mol Sci* (2021) 22(8):4145.

Mancini M, Ghiglieri V, Bagetta V, Pendolino V, Vannelli A, et al. Memantine alters striatal plasticity inducing a shift of synaptic responses toward long-term depression. *Neuropharmacol* (2016) 101:341–350.

Markesbery WR. Oxidative stress hypothesis in Alzheimer's disease. *Free Radic Biol Med* (1997) 23(1):134–147.

Matore BW, Banjare P, Guria T, Roy PP, Singh J. Oxadiazole derivatives: Histone deacetylase inhibitors in anticancer therapy and drug discovery. *Eur J Med Chem Rep* (2022) 5:100058.

Matore BW, Banjare P, Sarthi AS, Roy PP, Singh J. Phthalimides represent a promising scaffold for multi-targeted anticancer agents. *Chemistry Select* (2023) 8(9):e202204831.

Matore BW, Banjare P, Singh J, Roy PP. In silico selectivity modeling of pyridine and pyrimidine based CYP11B1 and CYP11B2 inhibitors: a case study. *J Mol Graph Model* (2022) 116:108238.

Matore BW, Roy PP, Singh J. Discovery of novel VEGFR2-TK inhibitors by phthalimide pharmacophore based virtual screening, molecular docking, MD simulation and DFT. *J Biomol Stru Dyn* (2023) 41(22):13056–13077.

McKenzie JA, Barghash RF, Alsaggaf AT, Kulkarni O, Boudreau K, Menard F, Neeland EG, Klegeris A. Synthesis and evaluation of novel Pyrazole Ethandiamide compounds as inhibitors of human THP-1 Monocytic cell neurotoxicity. *Cells* (2019) 8(7):655.

Mellado M, González C, Mella J, Aguilar LF, Celik I, et al. Coumarin-Resveratrol-Inspired hybrids as Monoamine Oxidase B inhibitors: 3-Phenylcoumarin versus trans-6-Styrylcoumarin. *Molecules* (2022) 27(3):928.

Mohd Faudzi SM, Leong SW, Auwal FA, Abas F, Wai LK, et al. In silico studies, nitric oxide, and cholinesterases inhibition activities of pyrazole and pyrazoline analogs of diarylpentanoids. *Archiv Der Pharmazie* (2021) 354(1):e2000161.

Molinaro G, Battaglia G, Riozzi B, Di Menna L, Rampello L, Bruno V, Nicoletti F. Memantine treatment reduces the expression of the K+/Cl– cotransporter KCC2 in the hippocampus and cerebral cortex, and attenuates behavioural responses mediated by GABAA receptor activation in mice. *Brain Res* (2009) 1265:75–79.

Möller H-J, Graeber MB. The case described by Alois Alzheimer in 1911. *Eur Arch Psychiatry Clin Neurosci* (1998) 248(3):111–122.

Montalbetti N, Simonin A, Simonin C, Awale M, Reymond J-L, Hediger MA. Discovery and characterization of a novel non-competitive inhibitor of the divalent metal transporter DMT1/SLC11A2. *Biochem Pharmacol* (2015) 96(3):216–224.

Montanari S, Mahmoud AM, Pruccoli L, Rabbito A, Naldi M, et al. Discovery of novel benzofuran-based compounds with neuroprotective and immunomodulatory properties for Alzheimer's disease treatment. *Eur J Med Chem* (2019) 178:243–258.

Moreira NC dos S, Tamarozzi ER, Lima JEB de F, Piassi L de O, Carvalho I, Passos GA, Sakamoto-Hojo ET. Novel dual AChE and ROCK2 inhibitor induces neurogenesis via PTEN/AKT pathway in Alzheimer's disease model. *Int J Mol Sci* (2022) 23(23):14788.

Mroueh M, Faour WH, Shebaby WN, Daher CF, Ibrahim TM, Ragab HM. Synthesis, biological evaluation and modeling of hybrids from tetrahydro-1H-pyrazolo[3,4-b]quinolines as dual cholinestrase and COX-2 inhibitors. *Bioorg Chem* (2020) 100:103895.

Murmu A, Banjare P, Wamanrao Matore B, Pratim Roy P, Singh J. 1,3,4-Oxadiazole: an emerging scaffold to inhibit the Thymidine Phosphorylase as an anticancer agent. *Curr Med Chem* (2023) 31(38):6227–6250.

Murmu A, Matore BW, Banjare P, Pandey NK, Chhabra N, Banjare L, Basak S, Singh J, Roy PP. Role of target fishing in discovery of novel anti-Alzheimer's agents: in silico applications. In D. Kumar, V. M. Patil, D. Wu, N. Thorat (Eds.), *Deciphering Drug Targets for Alzheimer's Disease* (2023) (pp. 245–284). Springer Nature Singapore.

Murphy MP, LeVine H. Alzheimer's disease and the Amyloid-β Peptide. *J Alzheimers Dis* (2010) 19(1):311–323.

Nakahata K, Kinoshita H, Hama-Tomioka K, Ishida Y, Matsuda N, Hatakeyama N, Haba M, Kondo T, Hatano Y. Cholinesterase inhibitor Donepezil dilates Cerebral Parenchymal Arterioles via the activation of Neuronal Nitric Oxide Synthase. *Anesthesiol* (2008) 109(1):124–129.

Narang SS, Goyal D, Goyal B. Molecular insights into the inhibitory mechanism of bi-functional bis-tryptoline triazole against β-secretase (BACE1) enzyme. *Amino Acids* (2019) 51(10–12):1593–1607.

Narlawar R, Pérez Revuelta BI, Haass C, Steiner H, Schmidt B, Baumann K. Scaffold of the Cyclooxygenase-2 (COX-2) inhibitor Carprofen provides Alzheimer γ-Secretase modulators. *J Med Chem* (2006) 49(26):7588–7591.

Ni W, Wang H, Li X, Zheng X, Wang M, Zhang J, Gong Q, Ling D, Mao F, Zhang H, Li J. Novel Tadalafil derivatives Ameliorates Scopolamine-induced cognitive impairment in mice via inhibition of Acetylcholinesterase (AChE) and Phosphodiesterase 5 (PDE5). *ACS Chem Neurosci* (2018) 9(7):1625–1636.

Nisijima K, Yoshino T, Takano K, Kato SKS. Memantine, an NMDA antagonist, prevents the development of hyperthermia in an animal model for Serotonin syndrome. *Pharmacopsychiatry* (2004) 37(02):57–62.

Noh M-Y, Chun K, Kang BY, Kim H, Park J-S, Lee H-C, Kim Y-H, Ku S, Kim SH. Newly developed glycogen synthase kinase-3 (GSK-3) inhibitors protect neuronal cells death in amyloid-beta induced cell model and in a transgenic mouse model of Alzheimer's disease. *Biochem Biophys Res Commun* (2013) 435(2):274–281.

Ogbodo JO, Agbo CP, Njoku UO, Ogugofor MO, Egba SI, et al. Alzheimer's disease: pathogenesis and therapeutic interventions. *Curr Aging Sci* (2022) 15(1):2–25.

Okuda M, Fujita Y, Hijikuro I, Wada M, Uemura T, et al. PE859, a novel Curcumin derivative, inhibits Amyloid-β and Tau aggregation, and Ameliorates cognitive dysfunction in Senescence-accelerated mouse Prone 8. *J Alzheimers Dis* (2017) 59(1):313–328.

Ono M, Hayashi S, Kimura H, Kawashima H, Nakayama M, Saji H. Push–pull benzothiazole derivatives as probes for detecting β-amyloid plaques in Alzheimer's brains. *Bioorg Med Chem* (2009) 17(19):7002–7007.

Ono M, Kawashima H, Nonaka A, Kawai T, Haratake M, et al. Novel Benzofuran derivatives for PET imaging of β-Amyloid plaques in Alzheimer's disease brains. *J Med Chem* (2006) 49(9):2725–2730.

Oosterhof CA, El Mansari M, Blier P. Acute effects of Brexpiprazole on Serotonin, Dopamine, and Norepinephrine systems: an in vivo electrophysiologic characterization. *J Pharmacol Exp Ther* (2014) 351(3):585–595.

Oset-Gasque MJ, Marco-Contelles JL. Tacrine-natural-product hybrids for Alzheimer's disease therapy. *Curr Med Chem* (2020) 27(26):4392–4400.

Paik J. Targeting toxic forms of α-synuclein with immunotherapy could alter the progression of Parkinson's disease. *Drugs Ther Perspect* (2022) 38(11):467–471.

Paroni G, Bisceglia P, Seripa D. *Understanding the Amyloid hypothesis in Alzheimer's disease. J Alzheimers Dis* (2019) 68(2):493–510.

Piazzi L, Cavalli A, Colizzi F, Belluti F, Bartolini M, Mancini F, Recanatini M, Andrisano V, Rampa A. Multi-target-directed coumarin derivatives: hAChE and BACE1 inhibitors as potential anti-Alzheimer compounds. *Bioorg Med Chem Lett* (2008) 18(1):423–426.

Piemontese L, Tomás D, Hiremathad A, Capriati V, Candeias E, et al. Donepezil structure-based hybrids as potential multifunctional anti-Alzheimer's drug candidates. *J Enzyme Inhib Med Chem* (2018) 33(1):1212–1224.

Pohanka M. Alpha7 Nicotinic Acetylcholine receptor is a target in pharmacology and toxicology. *Int J Mol Sci* (2012) 13(2):2219–2238.

Pourabdi L, Khoobi M, Nadri H, Moradi A, Moghadam FH, et al. (2016). Synthesis and structure-activity relationship study of tacrine-based pyrano[2,3-c]pyrazoles targeting AChE/BuChE and 15-LOX. *Eur J Med Chem* (2016) 123:298–308.

Provensi G, Carta F, Nocentini A, Supuran CT, Casamenti F, Passani MB, Fossati S. A new kid on the block? Carbonic Anhydrases as possible new targets in Alzheimer's disease. *Int J Mol Sci* (2019) 20(19):4724.

Qiu J, Feng X, Chen H, Liu W, Liu W, et al. Discovery of novel harmine derivatives as GSK-3β/DYRK1A dual inhibitors for Alzheimer's disease treatment. *Archiv Der Pharmazie* (2024) 357(2): e2300404.

Rajeshwari R, Chand K, Candeias E, Cardoso S, Chaves S, Santos M. New multitarget hybrids bearing Tacrine and Phenylbenzothiazole motifs as potential drug candidates for Alzheimer's disease. *Molecules* (2019) 24(3):587.

Ramirez E, Ganegamage SK, Elbatrawy AA, Alnakhala H, Shimanaka K, et al. 5-Nitro-1,2-benzothiazol-3-amine and N -Ethyl-1-[(ethylcarbamoyl)(5-nitro-1,2-benzothiazol-3-yl)amino]formamide modulate α-Synuclein and Tau aggregation. *ACS Omega* (2023) 8(22):20102–20115.

Ramírez MJ. 5-HT6 receptors and Alzheimer's disease. *Alzheimers Res Ther* (2013) 5(2):15.

Remya C, Dileep KV, Koti Reddy E, Mantosh K, Lakshmi K, et al. Neuroprotective derivatives of tacrine that target NMDA receptor and acetyl cholinesterase – Design, synthesis and biological evaluation. *Comput Struct Biotechnol J* (2021) 19:4517–4537.

Remya RS, Ramalakshmi N, Nalini CN, Niraimathi V, Amuthalakshmi S. Design synthesis and in vitro evaluation of Tacrine-flavone hybrids as multifunctional Cholinesterase inhibitors for Alzheimer's disease. *Curr Comp-Aided Drug Design* (2022) 18(4):271–292.

Repantis D, Laisney O, Heuser I. Acetylcholinesterase inhibitors and memantine for neuroenhancement in healthy individuals: a systematic review. *Pharmacol Res* (2010) 61(6):473–481.

Ricciarelli R, Fedele E. The Amyloid cascade hypothesis in Alzheimer's disease: it's time to change our mind. *Curr Neuropharmacol* (2017) 15(6):926–935.

Rosales Hernández MC, Fragoso Morales LG, Correa Basurto J, Olvera Valdez M, García Báez EV, et al. In silico and in vitro studies of Benzothiazole-Isothioureas derivatives as a multitarget compound for Alzheimer's disease. *Int J Mol Sci* (2022) 23(21):12945.

Rothweiler U, Stensen W, Brandsdal BO, Isaksson J, Leeson FA, Engh RA, Svendsen JSM. Probing the ATP-binding pocket of protein kinase DYRK1A with Benzothiazole fragment molecules. *J Med Chem* (2016) 59(21):9814–9824.

Rüben K, Wurzlbauer A, Walte A, Sippl W, Bracher F, Becker W. Selectivity profiling and biological activity of novel β-Carbolines as potent and selective DYRK1 kinase inhibitors. *PLoS One* (2015) 10(7):e0132453.

Sadafi Kohnehshahri M, Chehardoli G, Bahiraei M, Akbarzadeh T, Ranjbar A, Rastegari A, Najafi Z. Novel tacrine-based acetylcholinesterase inhibitors as potential agents for the treatment of Alzheimer's disease: Quinolotacrine hybrids. *Mol Divers* (2022) 26(1):489–503.

Santos CY, Snyder PJ, Wu W, Zhang M, Echeverria A, Alber J. Pathophysiologic relationship between Alzheimer's disease, cerebrovascular disease, and cardiovascular risk: a review and synthesis. *Alzheimers Dement (Amst): Diagn, Assess Dis Monit* (2017) 7(1):69–87.

Schellenberg GD, Bird TD, Wijsman EM, Orr HT, Anderson L, et al. Genetic linkage evidence for a familial Alzheimer's disease locus on chromosome 14. *Science* (1992) 258(5082):668–671.

Schmidt M, Vaskova M, Rotterova A, Fiandova P, Miskerikova M, Zemanova L, Benek O, Musilek K. Physiologically relevant fluorescent assay for identification of 17β -hydroxysteroid dehydrogenase type 10 inhibitors. *J Neurochem* (2023) 167(2):154–167.

Schwarthoff S, Tischer N, Sager H, Schätz B, Rohrbach MM, et al. Evaluation of γ-carboline-phenothiazine conjugates as simultaneous NMDA receptor blockers and cholinesterase inhibitors. *Bioorg Med Chem* (2021) 46:116355.

Sciú ML, Sebastián-Pérez V, Martinez-Gonzalez L, Benitez R, Perez DI, et al. Computer-aided molecular design of pyrazolotriazines targeting glycogen synthase kinase 3. *J Enzyme Inhib Med Chem* (2019) 34(1):87–96.

Seltzer B. Donepezil: a review. *Expert Opin Drug Metab Toxicol* (2005) 1(3):527–536.

Serrano-Pozo A, Frosch MP, Masliah E, Hyman BT. Neuropathological alterations in Alzheimer disease. *Cold Spring Harb Perspect Med* (2011) 1(1):a006189–a006189.

Sevigny J, Chiao P, Bussière T, Weinreb PH, Williams L, et al. The antibody aducanumab reduces Aβ plaques in Alzheimer's disease. *Nature* (2016) 537(7618):50–56.

Shahroz MM, Sharma HK, Altamimi ASA, Alamri MA, Ali A, et al. Novel and potential small molecule scaffolds as DYRK1A inhibitors by integrated molecular docking-based virtual screening and dynamics simulation study. *Molecules* (2022) 27(4):1159.

Shaikh S, Dhavan P, Pavale G, Ramana MMV, Jadhav BL. Design, synthesis and evaluation of pyrazole bearing α-aminophosphonate derivatives as potential acetylcholinesterase inhibitors against Alzheimer's disease. *Bioorg Chem* (2020) 96:103589.

Shao Y, Huang M, Cui W, Feng L-J, Wu Y, et al. Discovery of a Phosphodiesterase 9A inhibitor as a potential Hypoglycemic agent. *J Med Chem* (2014) 57(24):10304–10313.

Shen H, Kihara T, Hongo H, Wu X, Kem W, Shimohama S, Akaike A, Niidome T, Sugimoto H. Neuroprotection by donepezil against glutamate excitotoxicity involves stimulation of α7 nicotinic receptors and internalization of NMDA receptors. *British J Pharmacol* (2010) 161(1):127–139.

Smith DA. Treatment of Alzheimer's disease in the long-term-care setting. *Am J Health Syst Pharm* (2009) 66(10):899–907.

Smith MA, Rottkamp CA, Nunomura A, Raina AK, Perry G. Oxidative stress in Alzheimer's disease. *Biochim Biophys Acta* (2000) 1502(1):139–144.

Soler-López M, Badiola N, Zanzoni A, Aloy P. Towards Alzheimer's root cause: ECSIT as an integrating hub between oxidative stress, inflammation and mitochondrial dysfunction. *BioEssays* (2012) 34(7):532–541.

Solesio ME, Peixoto PM, Debure L, Madamba SM, de Leon MJ, Wisniewski T, Pavlov EV, Fossati S. Carbonic anhydrase inhibition selectively prevents amyloid β neurovascular mitochondrial toxicity. *Aging Cell* (2018) 17(4):e12787.

Song C, Shi J, Zhang P, Zhang Y, Xu J, Zhao L, Zhang R, Wang H, Chen H. Immunotherapy for Alzheimer's disease: targeting β-amyloid and beyond. *Transl Neurodegener* (2022) 11(1):18.

Spires-Jones TL, Hyman BT. The intersection of Amyloid beta and Tau at synapses in Alzheimer's disease. *Neuron* (2014) 82(4):756–771.

Stahl SM. Mechanism of action of brexpiprazole: comparison with aripiprazole. *CNS Spectrums* (2016) 21(1):1–6.

Stampfer MJ. Cardiovascular disease and Alzheimer's disease: common links. *J Intern Med* (2006) 260(3):211–223.

Stefanachi A, Leonetti F, Pisani L, Catto M, Carotti A. Coumarin: A natural, privileged and versatile scaffold for bioactive compounds. *Molecules* (2018) 23(2):250.

Supuran CT, Scozzafava A. Applications of carbonic anhydrase inhibitors and activators in therapy. *Expert Opin Ther Pat* (2002) 12(2):217–242.

Swanson CJ, Zhang Y, Dhadda S, Wang J, Kaplow J, et al. A randomized, double-blind, phase 2b proof-of-concept clinical trial in early Alzheimer's disease with lecanemab, an anti-Aβ protofibril antibody. *Alzheimers Res Ther* (2021) 13(1):80.

Tabet N. Acetylcholinesterase inhibitors for Alzheimer's disease: anti-inflammatories in acetylcholine clothing! *Age and Ageing* (2006) 35(4):336–338.

Tamborini L, Chen Y, Foss CA, Pinto A, Horti AG, et al. Development of radiolabeled ligands targeting the glutamate binding site of the N -Methyl- <scp>d</scp> -aspartate receptor as potential imaging agents for brain. *J Med Chem* (2016) 59(24):11110–11119.

Tan M-S Yu J-T, Tan L. Bridging integrator 1 (BIN1): form, function, and Alzheimer's disease. *Trends Mol Med* (2013) 19(10):594–603.

Tanzi RE. A brief history of Alzheimer's disease gene discovery. *J Alzheimers Dis* (2012) 33(s1):S5–S13.

Tolnay M, Probst A. Review: tau protein pathology in Alzheimer's disease and related disorders. *Neuropath Appl Neurobio* (1999) 25(3):171–187.

Turkan F, Cetin A, Taslimi P, Karaman M, Gulçin İ. Synthesis, biological evaluation and molecular docking of novel pyrazole derivatives as potent carbonic anhydrase and acetylcholinesterase inhibitors. *Bioorg Chem* (2019) 86:420–427.

Twohig D, Nielsen HM. α-synuclein in the pathophysiology of Alzheimer's disease. *Mol Neurodegenerat* (2019) 14(1):23.

Ucar G, Gokhan N, Yesilada A, Bilgin AA. 1-N-Substituted thiocarbamoyl-3-phenyl-5-thienyl-2-pyrazolines: a novel cholinesterase and selective monoamine oxidase B inhibitors for the treatment of Parkinson's and Alzheimer's diseases. *Neurosci Lett* (2005) 382(3):327–331.

Uhlmann RE, Rother C, Rasmussen J, Schelle J, Bergmann C, et al. Acute targeting of pre-amyloid seeds in transgenic mice reduces Alzheimer-like pathology later in life. *Nat Neurosci* (2020) 23(12):1580–1588.

Umar T, Shalini S, Raza MK, Gusain S, Kumar J, Seth P, Tiwari M, Hoda N. A multifunctional therapeutic approach: Synthesis, biological evaluation, crystal structure and molecular docking of diversified 1H-pyrazolo[3,4-b]pyridine derivatives against Alzheimer's disease. *Eur J Med Chem* (2019) 175:2–19.

Wang R, Reddy PH. Role of glutamate and NMDA receptors in Alzheimer's disease. *J Alzheimers Dis* (2017) 57(4):1041–1048.

Wang X-B, Yin F-C, Huang M, Jiang N, Lan J-S, Kong L-Y. Chromone and donepezil hybrids as new multipotent cholinesterase and monoamine oxidase inhibitors for the potential treatment of Alzheimer's disease. *RSC Med Chem* (2020) 11(2):225–233.

Wang X-X, Xie F, Jia C-C, Yan N, Zeng Y-L, Wu J-D, Liu Z-P. Synthesis and biological evaluation of selective histone deacetylase 6 inhibitors as multifunctional agents against Alzheimer's disease. *Eur J Med Chem* (2021) 225:113821.

Watanabe H, Ono M, Iikuni S, Yoshimura M, Matsumura K, Kimura H, Saji H. A 68Ga complex based on benzofuran scaffold for the detection of β-amyloid plaques. *Bioorg Med Chem Lett* (2014) 24(20):4834–4837.

Whalen K. *Lippincott® illustrated reviews: pharmacology* (2018). Wolters kluwerindia Pvt Ltd.

Wilkinson DG. Galantamine: a new treatment for Alzheimer's disease. *Expert Rev Neurother* (2001) 1(2):153–159.

Wu Y, Tang J, Zhang W, Zhuang C, Shi Y. Rational drug design of CB2 receptor ligands: from 2012 to 2021. *RSC Adv* (2022) 12(54):35242–35259.

Xie S-S, Wang X, Jiang N, Yu W, Wang KDG, Lan J-S, Li Z-R, Kong L-Y. Multi-target tacrine-coumarin hybrids: Cholinesterase and monoamine oxidase B inhibition properties against Alzheimer's disease. *Eur J Med Chem* (2015) 95:153–165.

Xu K, Dai X-L, Huang H-C, Jiang Z-F. Targeting HDACs: A promising therapy for Alzheimer's disease. *Oxid Med Cell Longev* (2011) 2011:1–5.

Yang HD, Kim DH, Lee SB, Young LD. History of Alzheimer's disease. *Dement Neurocogn Disord* (2016) 15(4):115.

Yang S, Zhang R, Wang G, Zhang Y. The development prospection of HDAC inhibitors as a potential therapeutic direction in Alzheimer's disease. *Transl Neurodegener* (2017) 6(1):19.

Yao L, Gu X, Song Q, Wang X, Huang M, Hu M, Hou L, Kang T, Chen J, Chen H, Gao X. Nanoformulated alpha-mangostin ameliorates Alzheimer's disease neuropathology by elevating LDLR expression and accelerating amyloid-beta clearance. *J Contr Release* (2016) 226:1–14.

Yi M, Yu P, Lu Q, Geller HM, Yu Z, Chen H. KCa3.1 constitutes a pharmacological target for astrogliosis associated with Alzheimer's disease. *Mol Cell Neurosci* (2016) 76:21–32.

Yiannopoulou KG, Papageorgiou SG. Current and future treatments in Alzheimer disease: an update. *J Cent Nerv Syst Dis* (2020) 12:117957352090739.

Yu Y-F Huang Y-D, Zhang C, Wu X-N, Zhou Q, Wu D, Wu Y, Luo H-B. Discovery of novel Pyrazolopyrimidinone derivatives as Phosphodiesterase 9A inhibitors capable of inhibiting Butyrylcholinesterase for treatment of Alzheimer's disease. *ACS Chem Neurosci* (2017) 8(11):2522–2534.

Zagórska A, Jaromin A. perspectives for new and more efficient multifunctional ligands for Alzheimer's disease therapy. *Molecules* (2020) 25(15):3337.

Zang Y, Liu K, Wang W, Li C, Ma J, Yang J, Chen X, Wang X, Zhang D. Claulansine F–Donepezil hybrids as anti-Alzheimer's disease agents with Cholinergic, free-radical scavenging, and neuroprotective activities. *Molecules* (2021) 26(5):1303.

Zhang P, Jiang M-Y, Le M-L, Zhang B, Zhou Q, Wu Y, Zhang C, Luo H-B. Design, synthesis and evaluation of pyrazolopyrimidinone derivatives as novel PDE9A inhibitors for treatment of Alzheimer's disease. *Bioorg Med Chem Lett* (2020) 30(14):127254.

Zhang Y, Chen H, Li R, Sterling K, Song W. Amyloid β-based therapy for Alzheimer's disease: challenges, successes and future. *Signal Transduct Target Ther* (2023) 8(1):248.

Zhou Y, Li J, Yuan H, Su R, Huang Y, Huang Y, Li Z, Wu Y, Luo H, Zhang C, Huang L. Design, synthesis, and evaluation of Dihydropyranopyrazole derivatives as novel PDE2 inhibitors for the treatment of Alzheimer's disease. *Molecules* (2021) 26(10):3034.

Zhu X, Raina AK, Lee H, Casadesus G, Smith MA, Perry G. Oxidative stress signalling in Alzheimer's disease. *Brain Res* (2004) 1000(1–2):32–39.

11 Role and Therapeutic Perspectives of Oxidative Stress in Neurodegenerative Diseases

Rakhi Mishra[1], Prem Shankar Mishra[2], Vaishali Manikrao Patil[3], and Preeti Kumari[1]*

[1]Noida Institute of Engineering and Technology (Pharmacy Institute), Greater Noida, India
[2]Department of Pharmacy, Galgotias University, Greater Noida, India
[3]Charak School of Pharmacy, Chaudhary Charan Singh University, Meerut, Uttar Pradesh, India
*Corresponding author

INTRODUCTION

Neurodegeneration is the progressive degeneration of neurons and their death, leading to various diseases that may be characterized as neurodegenerative diseases (Katsnelson et al., 2016). Various mechanisms involved in neurodegeneration and therapeutic strategies have been studied to search for potential treatments (Sukanya et al., 2023; Gupta and Patil, 2020). Oxidative stress plays an important role in neurodegeneration and the pathophysiology of related disorders. Despite various reported studies, the etiology of many neurodegenerative diseases is not well understood (Li et al., 2013). Some common findings highlight the role of mitochondrial function and oxidative stress as major contributors to neurodegeneration. Various mechanisms have shown the role of oxidative stress, which includes nucleic acid, oxidation of protein and lipid, production of advanced glycation end products, activation of glial cells, apoptosis, defective mitochondrial function and ubiquitin-proteasome system, protein oligomerization, production of cytokines and inflammation, dysfunction of the proteasome, and disrupted blood–brain barrier (Figure 11.1) (Yaribeygi et al., 2018; Leo and Campos, 2019; Pegoretti et al., 2020). Oxidative stress is related to higher levels of reactive oxygen species (ROS), which cause damage to lipids, proteins, and DNA (Singh et al., 2019). ROS accumulation causes defective energy production, metal homeostasis, and the buildup of toxic protein aggregates. This leads to triggering apoptosis, necrosis, autophagy, and thus neurodegenerative disorders. The ROS causing neurodegeneration include H_2O_2, the superoxide radical (O_2^-), and the hydroxyl radical (OH), and along with the nitric oxide radical (NO) and peroxynitrite ($ONOO^-$), have deleterious effects on neurons (Singh et al., 2019). The present chapter aims to describe various free radicals/ROS, their physiological functions, and their role in neurodegeneration, which can aid in developing understanding to develop therapeutic strategies.

DOI: 10.1201/9781003506317-11

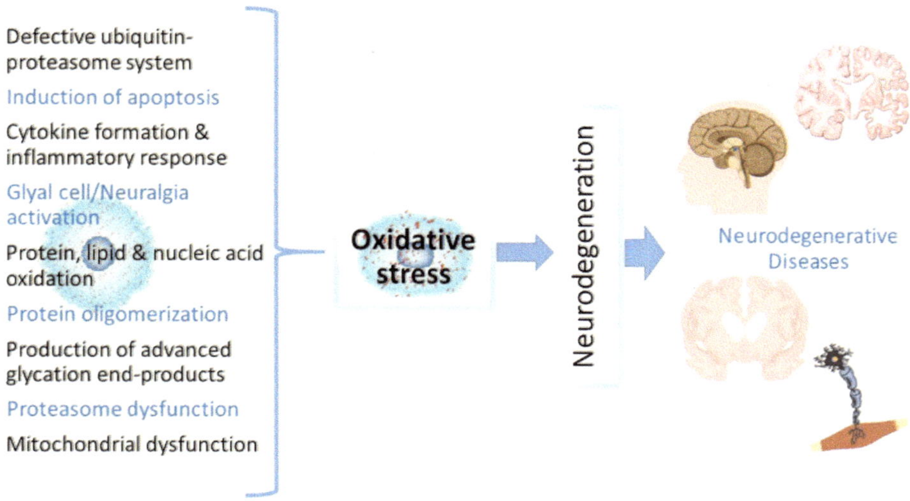

FIGURE 11.1 Various mechanisms of oxidative stress causing neurodegeneration and thus leading to neurodegenerative diseases.

REACTIVE OXYGEN SPECIES (ROS)

Free radicals are molecules or portions thereof that possess one or more unpaired electrons in their outer orbital, a state that greatly increases their reactivity (Bayr, 2005; Salvemini and Cuzzocrea, 2002). The most well-known reactive species that are produced when oxygen is present are peroxynitrite, hydroxyl radicals, and superoxide anion. Superoxide dismutase (SOD), an enzyme that is widely distributed, catalyzes the enzymatic reduction of superoxide to hydrogen peroxide (Lushchak, 2014). The term "reactive oxygen species (ROS)" refers to a group of highly reactive molecules that comprise oxygen radicals such as superoxide, hydroxyl, and hydroperoxyl (HO_2) radicals (Madkour, 2019; Krumova, 2016). They are formed as natural byproducts of normal cellular metabolism and have an important role in cell signaling and homeostasis (He et al., 2017). However, excessive ROS production can lead to oxidative stress, causing damage to cellular components such as proteins, lipids, and DNA, which is implicated in various diseases and aging processes (Cecarini et al., 2007). Following are some types of ROS and their characteristics (Villalpando-Rodriguez and Gibson, 2021):

Superoxide anion (O2•–): These are generated primarily in mitochondria during the electron transport chain and by enzymes such as NADPH oxidase. They are highly reactive and can readily initiate chain reactions (Pervaiz and Clement, 2007)

Hydrogen peroxide (H_2O_2): This is formed by the dismutation of the superoxide anion or by various enzymatic reactions. It is more stable than superoxide and can diffuse across membranes, allowing it to act as a signaling molecule (Habibi, 2014).

Hydroxyl radical (•OH): These are considered the most reactive ROS and are generated through the Fenton reaction, where hydrogen peroxide reacts with transition metals such as iron or copper. They can cause severe damage to biomolecules due to their high reactivity (Gutteridge, 1994).

Singlet oxygen ($^1O_2$): It is produced during photosynthesis and by certain enzyme-catalyzed reactions. It exhibits unique chemical properties, leading to oxidative damage to lipids, proteins, and nucleic acids (Vašková et al., 2012).

Alkoxyl radicals (RO•) and *peroxyl radicals* (ROO•): These are formed through the reaction of lipid peroxides with other radicals and particularly damage cell membranes, leading to lipid peroxidation (Yaman and Ayhanci, 2021).

Hypochlorous acid (HOCl): This is generated by the enzyme myeloperoxidase in neutrophils during the respiratory burst. It exhibits antimicrobial properties but can also damage host tissues if produced in excess (Winterbourn and Kettle, 2013).

Some of the characteristics of ROS include:

Reactivity: ROS possess high reactivity due to the presence of unpaired electrons, allowing them to readily react with other molecules (Juan et al., 2021).

Short lifespan: ROS have a shorter lifespan because they react rapidly with nearby biomolecules or undergo dismutation (Shields et al., 2021).

Membrane permeability: While some ROS, such as the superoxide anion, are charged and cannot cross membranes easily, others like hydrogen peroxide are neutral and can diffuse across membranes (Azadmanesh and Borgstahl, 2018).

Redox activity: ROS can participate in redox reactions, either as oxidants or reductants, influencing the oxidation state of cellular components (Okoh et al., 2011).

Role in signaling: Despite their damaging potential, ROS also serve as important signaling molecules for various physiological processes such as cell proliferation, apoptosis, and immune response (Waszczak et al., 2018).

PHYSIOLOGICAL FUNCTIONS AND ROLE OF ROS IN NEURODEGENERATION

ROS are highly reactive molecules containing oxygen that are generated as natural byproducts of cellular metabolism (Mansoor et al., 2022). While ROS serve important physiological functions in cellular signaling, host defense, and regulation of cellular processes, excessive ROS production or impaired antioxidant defense mechanisms can lead to oxidative stress, a condition implicated in the pathogenesis of neurodegenerative diseases (Chen et al., 2012; Simpson et al., 2020). A breakdown of ROS physiological functions and their role in neurodegeneration are elaborated next.

Physiological Functions

Cell signaling: ROS, such as hydrogen peroxide and superoxide, can act as signaling molecules in various cellular pathways. They modulate the activity of signaling proteins and transcription factors, influencing processes such as cell proliferation, differentiation, and apoptosis (Lennicke and Cochemé, 2021).

Host defense: ROS are essential components of the immune system's defense mechanisms against pathogens. Phagocytic cells, such as neutrophils and macrophages, produce ROS to kill engulfed bacteria and viruses through a process called the respiratory burst (Dickson and Zhou, 2020).

Redox regulation: ROS are involved in redox signaling, where they regulate the redox state of proteins by the reversible oxidation of cysteine residues. This redox modulation controls protein function and cellular processes, including enzyme activity, gene expression, and cell survival (Lennicke and Cochemé, 2021).

Role in Neurodegeneration

Oxidative damage: Excessive ROS production can lead to oxidative damage to cellular components, including DNA, proteins, and lipids. This oxidative damage contributes to the neuronal dysfunction and death observed in neurodegenerative diseases such as Alzheimer's disease (AD), Parkinson's disease (PD), and amyotrophic lateral sclerosis (ALS) (Buccellato et al., 2021).

Protein misfolding and aggregation: ROS have been implicated in promoting the misfolding and aggregation of proteins associated with neurodegenerative diseases. For example, ROS can induce the aggregation of beta-amyloid and tau proteins in AD and alpha-synuclein in PD, leading to the formation of toxic protein aggregates and neuronal damage (Soto and Estrada, 2008).

Mitochondrial dysfunction: Mitochondria are major sources and targets of ROS within cells. Excessive ROS production can damage mitochondrial DNA, proteins, and membranes, impairing mitochondrial function and bioenergetics. Mitochondrial dysfunction further exacerbates ROS generation, creating a vicious cycle of oxidative stress and cellular damage implicated in neurodegeneration (Mancuso et al., 2006).

Neuroinflammation: ROS can activate inflammatory pathways in the central nervous system, leading to neuroinflammation characterized by the release of pro-inflammatory cytokines, chemokines, and reactive gliosis. Chronic neuroinflammation contributes to the neuronal injury and degeneration in various neurodegenerative diseases (Oliver, 2020).

In summary, while ROS serve important physiological functions in cellular signaling and host defense, their dysregulation can contribute to neurodegeneration through oxidative damage, protein misfolding, mitochondrial dysfunction, and neuroinflammation (Aguilera et al., 2018). Understanding the complex roles of ROS in neurodegenerative processes may offer insights into the development of therapeutic strategies targeting oxidative stress for the treatment of these debilitating diseases (Rekatsina et al., 2020).

OXIDATIVE STRESS

ROS are molecules derived from oxygen that possess high reactivity due to the presence of unpaired electrons (Bergamini et al., 2004). Common ROS include superoxide anion, hydroxyl radical, and hydrogen peroxide (Madkour, 2019). While ROS play essential roles in various physiological processes such as cell signaling and host defense, excessive ROS production or impaired antioxidant defenses can lead to oxidative stress (Liu, 2017).

Sources of ROS

Mitochondria: Mitochondria are a major source of ROS during cellular respiration. Leakage of electrons from the electron transport chain can generate superoxide radicals (Tabassum et al., 2020).

NADPH oxidase: This enzyme complex generates superoxide radicals as part of the immune response to pathogens, known as the respiratory burst (Panday et al., 2015).

Inflammation: Immune cells produce ROS to destroy pathogens during inflammation (Agita and Alasagaff, 2017).

Environmental factors: Exposure to ionizing radiation, pollutants, toxins, and certain drugs can also induce ROS production (Mena et al., 2009).

Antioxidant Defense

Cells have defense mechanisms to neutralize ROS and prevent oxidative damage. These include enzymes like superoxide dismutase (SOD), catalase, and glutathione peroxidase, as well as non-enzymatic antioxidants such as vitamins C and E, glutathione and flavonoids (Ighodaro and Akinloye, 2018).

Oxidative Stress

Cellular damage: ROS can damage various cellular components, including lipids, proteins, and DNA. Lipid peroxidation can lead to membrane damage, while protein oxidation can impair enzyme function and structural integrity. DNA damage by ROS can lead to mutations and genomic instability (Pisoschi and Pop, 2015).

Mitochondrial dysfunction: ROS can damage mitochondrial DNA, proteins, and membranes, leading to impaired mitochondrial function and bioenergetics. This can further exacerbate ROS production, creating a vicious cycle of oxidative stress and mitochondrial dysfunction (Zhu et al., 2018).

Cellular signaling: Moderate levels of ROS can act as signaling molecules in cellular pathways, regulating processes such as cell proliferation, differentiation, and apoptosis. However, excessive ROS can disrupt signaling pathways, leading to aberrant cell behavior and dysfunction (Martindale and Holbrook, 2002).

Neurodegeneration: Oxidative stress has been implicated in the pathogenesis of neurodegenerative diseases such as Alzheimer's disease, Parkinson's disease, and amyotrophic lateral sclerosis. Accumulation of oxidative damage in neurons contributes to the neuronal dysfunction and death observed in these diseases (Reynolds et al., 2007).

CONSEQUENCES OF OXIDATIVE STRESS

Aging is considered one of the hallmarks of oxidative stress, with cumulative damage contributing to age-related decline in cellular function and tissue integrity (Luo et al., 2020). Oxidative stress is implicated in the pathogenesis of various diseases, including cardiovascular diseases, cancer, diabetes, and neurodegenerative diseases (Dhalla et al., 2000).

Overall, while ROS are essential for cellular function, maintaining redox balance is crucial to prevent oxidative stress and its associated detrimental effects on cellular health and function. Balancing ROS production with antioxidant defenses is essential for cellular homeostasis and overall health (Liu et al., 2017).

BIOMARKERS

ROS play significant roles in various physiological processes and pathological conditions, including aging, cancer, neurodegenerative diseases, and cardiovascular diseases (Jomova et al., 2023). Due to their involvement in these processes, ROS levels and associated biomarkers have been studied extensively. Here are some ROS biomarkers commonly used in research and clinical settings (Umeno et al., 2017): Malondialdehyde (MDA) is a byproduct of lipid peroxidation, which occurs when ROS attack lipids in cell membranes (Jadoon and Malik, 2017). Elevated levels of MDA in biological samples, such as plasma or urine, are indicative of oxidative stress and lipid peroxidation (Tsikas, 2017). 8-Hydroxy-2′-deoxyguanosine (8-OHdG) is a marker of oxidative DNA damage. It is formed when ROS attack DNA, leading to the oxidation of guanine residues (Urbaniak et al., 2020). Measurement of 8-OHdG levels in urine or tissue samples is commonly used to assess oxidative stress-induced DNA damage (Valavanidis et al., 2009). Protein carbonyls are produced by protein carbonylation when ROS oxidize amino acid side chains in proteins (Suzuki et al., 2010). Detection of protein carbonyls in plasma, serum, or tissue samples is indicative of oxidative damage to proteins (Dalle-Donne et al., 2003). Glutathione (GSH) is an important antioxidant that scavenges ROS and maintains cellular redox balance (Raj Rai et al., 2021). The ratio of reduced GSH to oxidized glutathione, i.e., glutathione disulfide (GSSG), in cells or tissues is a marker of oxidative stress. A decreased GSH/GSSG ratio indicates oxidative stress and impaired antioxidant capacity (Baeza et al., 2010).

Superoxide dismutase is an enzyme that catalyzes the dismutation of the superoxide radicals into hydrogen peroxide and oxygen, thereby reducing oxidative stress (Ighodaro and Akinloye, 2018). Measurement of SOD activity or expression levels in tissues or cells provides insights into antioxidant defense mechanisms (Ho et al., 1998). Catalase is an enzyme that catalyzes the decomposition of hydrogen peroxide into water and oxygen, protecting cells from oxidative damage (Nelson et al.,

2006). Assessment of catalase activity or expression levels serves as a biomarker of antioxidant capacity (Demirci-Cekic et al., 2022). ROS-sensitive fluorescent probes, such as dihydroethidium (DHE) and dichlorofluorescein (DCFH), are used to measure intracellular ROS levels in real-time (Dikalov and Harrison, 2014). These probes can be used in live-cell imaging and flow cytometry experiments to monitor ROS production and oxidative stress responses (Jávega et al., 2023). F2-isoprostanes are prostaglandin-like compounds formed by the peroxidation of arachidonic acid in cell membranes (Halliwell and Lee, 2010). Measurement of F2-isoprostane levels in urine or plasma serves as a biomarker of oxidative stress and lipid peroxidation (Basu, 2004). These biomarkers provide valuable insights into the oxidative stress status of cells, tissues, and organisms. Monitoring ROS levels and oxidative damage biomarkers can help in understanding the role of oxidative stress in disease pathogenesis and evaluating the efficacy of antioxidant therapies (Liguori et al., 2018).

ROLE OF PRO-OXIDANTS

Pro-oxidants, often referred to as ROS, play significant roles in biological systems, despite their reputation for causing cellular damage (Vertuani et al., 2004). Here are some key roles of pro-oxidants:

Cell signaling: ROS serve as important signaling molecules in various cellular pathways. They can modulate the activity of proteins involved in signaling cascades, such as protein kinases, phosphatases, and transcription factors. ROS-mediated signaling regulates processes such as cell proliferation, differentiation, apoptosis, and stress response (Averill-Bates et al., 2023).

Immune response: ROS are essential components of the immune system's defense mechanisms against pathogens. Immune cells, such as neutrophils, macrophages, and monocytes, produce ROS as part of the oxidative burst to kill engulfed microorganisms. ROS play a crucial role in phagocytosis and the destruction of pathogens within phagosomes (Yang et al., 2013).

Mitochondrial function: Mitochondria are both a source and a target of ROS. Controlled production of ROS within mitochondria serves as signaling molecules involved in processes such as mitochondrial biogenesis, apoptosis, and adaptation to stress. However, excessive ROS production can damage mitochondrial DNA, proteins, and lipids, leading to mitochondrial dysfunction and cell death (Bolisetty and Jaimes, 2013).

Cellular homeostasis: ROS are involved in maintaining cellular redox balance. They participate in the regulation of antioxidant defense systems, such as the activation of antioxidant enzymes like superoxide dismutase, catalase, and glutathione peroxidase. ROS also contribute to the turnover of damaged cellular components through processes like autophagy and proteasomal degradation (Espinosa-Diez et al., 2015).

Hormesis: ROS can induce hormetic responses in cells and organisms. Hormesis refers to a phenomenon where exposure to low levels of stressors, including ROS, induces adaptive responses that confer resistance to subsequent stress. The hormetic effects of ROS include enhanced stress tolerance, improved cellular function, and increased longevity (Schirrmacher, 2021).

Cellular signaling in aging: ROS are implicated in the aging process and age-related diseases. While excessive ROS production contributes to cellular damage and senescence, moderate levels of ROS are involved in signaling pathways that regulate lifespan and aging-related processes. ROS-mediated signaling influences cellular senescence, genomic stability, and age-related diseases (Papa Constantinou, 2019).

Redox regulation: ROS participate in redox signaling by modifying the redox state of proteins and lipids. Oxidative modifications, such as cysteine oxidation and lipid peroxidation, regulate protein function, gene expression, and cellular responses to stress. Redox-sensitive signaling pathways mediated by ROS include MAPK pathways, NF-κB signaling, and Nrf2-mediated antioxidant responses (Biswas et al., 2006).

Overall, while excessive ROS production can lead to oxidative stress and cellular damage, pro-oxidants also serve important physiological functions in cellular signaling, immune defense, mitochondrial function, and redox regulation (Rahal et al., 2014). Understanding the complex roles of ROS in biological systems is essential for developing strategies to modulate ROS levels and mitigate their detrimental effects on health and disease (Sies et al., 2022).

ROLE OF HEAVY METALS

Heavy metals are metallic elements with high atomic weights and densities (Duffus, 2002). While some heavy metals, such as iron and zinc, are essential micronutrients required for various biological functions, others can be toxic to living organisms at higher concentrations (Peraza et al., 1998). Here are some key roles of heavy metals in biological systems:

Enzyme cofactors: Some heavy metals serve as essential cofactors for enzymes involved in critical biochemical processes. For example, zinc is a cofactor for hundreds of enzymes and is essential for DNA synthesis, cell division, and immune function (Jomova et al., 2022). Iron is a cofactor for enzymes involved in oxygen transport (hemoglobin), energy metabolism (cytochromes), and DNA synthesis (ribonucleotide reductase) (Crichton, 2001).

Structural functions: Heavy metals can have structural roles in proteins and enzymes. For instance, iron is a component of heme groups in hemoglobin and myoglobin, which are involved in oxygen transport and storage in red blood cells and muscle cells, respectively (Permyakov, 2021).

Signal transduction: Some heavy metals, such as zinc and copper, are involved in cellular signal transduction pathways. They can modulate the activity of signaling proteins, transcription factors, and ion channels, influencing processes such as cell proliferation, differentiation, and apoptosis (Paithankar et al., 2021).

Oxidative stress: Certain heavy metals, such as lead, cadmium, and mercury, can induce oxidative stress by generating ROS and disrupting antioxidant defense mechanisms. Oxidative stress caused by heavy metals can lead to damage to cellular components, including DNA, proteins, and lipids, contributing to various diseases (Mansoor et al., 2023).

Neurotransmission: Heavy metals like zinc and copper play roles in neurotransmission and synaptic function in the central nervous system. They are involved in the regulation of neurotransmitter release, synaptic plasticity, and neuronal excitability (Marchetti, 2014).

Nutrient transport: Some heavy metals, such as iron and copper, are involved in the transport and metabolism of nutrients (Williams et al., 2000). Iron is essential for the transport of oxygen by hemoglobin and the storage of iron in ferritin, while copper is required for the transport of iron across cell membranes and the synthesis of hemoglobin (Arredondo and Núñez, 2005).

Immune function: Heavy metals can influence immune function by modulating the activity of immune cells and cytokine production. Zinc, for example, is essential for the development and function of immune cells, including T cells and macrophages, and plays a role in regulating inflammatory responses (Wessels et al., 2017).

While heavy metals have important biological functions at physiological concentrations, exposure to excessive levels of certain heavy metals can be toxic and pose significant health risks. Chronic exposure to toxic heavy metals, such as lead, mercury, cadmium, and arsenic, is associated with a wide range of adverse health effects, including neurological disorders, cardiovascular disease, kidney damage, and cancer. Therefore, it is essential to monitor and regulate exposure to heavy metals to minimize their harmful effects on human health and the environment (Rehman et al., 2018; Balali-Mood, 2021).

OXIDATIVE STRESS AND NEURODEGENERATIVE DISEASES

Alzheimer's Disease

Oxidative stress is widely implicated in the pathogenesis of various neurodegenerative diseases, including Alzheimer's disease (Chauhan and Chauhan, 2006). AD is characterized by the progressive loss of cognitive function and memory due to the accumulation of beta-amyloid plaques and tau protein tangles in the brain (Agostinho et al., 2010). Oxidative stress occurs when there is an imbalance between the production of ROS and the ability of the body's antioxidant defenses to neutralize them (Ozougwu, 2016). ROS, such as superoxide radicals, hydroxyl radicals, and hydrogen peroxide, are highly reactive molecules that can damage cellular components like proteins, lipids, and DNA (Juan et al., 2021). In the context of AD, oxidative stress contributes to several pathological processes:

Beta-amyloid accumulation: Oxidative stress can promote the aggregation of beta-amyloid peptides into toxic oligomers and plaques, which are hallmark features of AD pathology. These aggregates induce neuronal injury and synaptic dysfunction, leading to cognitive decline (Zhao and Zhao, 2013).

Tau protein hyperphosphorylation: Oxidative stress can also trigger abnormal phosphorylation of tau proteins, leading to the formation of neurofibrillary tangles. Tau pathology disrupts microtubule stability and impairs axonal transport, further contributing to neuronal dysfunction and degeneration (Alavi Naini and Soussi-Yanicostas, 2015).

Neuronal damage: ROS can directly damage neurons by inducing lipid peroxidation, protein oxidation, and DNA damage. Neuronal oxidative damage impairs cellular function and can ultimately lead to cell death (Poh Loh et al., 2006).

Inflammation: Oxidative stress can activate inflammatory pathways in the brain, leading to the production of pro-inflammatory cytokines and chemokines. Chronic neuroinflammation exacerbates neuronal damage and contributes to the progression of Alzheimer's pathology (Agostinho et al., 2010).

Mitochondrial dysfunction: Mitochondria are major sources of ROS production, and oxidative stress can impair mitochondrial function, further exacerbating ROS generation. Mitochondrial dysfunction compromises energy production and cellular resilience, contributing to neuronal vulnerability in AD (Moreira et al., 2010).

Given the detrimental effects of oxidative stress on AD pathology antioxidant therapies have been investigated as potential treatment strategies (Teixeira et al., 2019). These include dietary antioxidants, such as vitamins C and E, as well as synthetic antioxidants and compounds with antioxidant properties (Gulcin, 2020). While some studies have shown promising results in animal models, the efficacy of antioxidant therapy in human AD patients remains inconclusive and requires further investigation. Nonetheless, targeting oxidative stress remains a promising avenue for therapeutic intervention in AD (Pritam et al., 2022).

Parkinson's Disease

Oxidative stress is also strongly implicated in the pathogenesis of Parkinson's disease, another neurodegenerative disorder characterized by the progressive loss of dopaminergic neurons in the substantia nigra region of the brain (Trist et al., 2019). Similar to Alzheimer's disease, oxidative stress plays a multifaceted role in the development and progression of PD:

Mitochondrial dysfunction: Mitochondrial dysfunction is a prominent feature of PD pathology, leading to impaired energy production and increased generation of ROS (Bhatia and Sharma, 2021). Dysfunctional mitochondria are less efficient at ATP synthesis and are more prone to leakage of electrons, resulting in ROS production. Excessive ROS production can further damage mitochondrial DNA, proteins, and lipids, exacerbating mitochondrial dysfunction and oxidative stress (Madamanchi and Runge, 2007).

Protein aggregation: Misfolded proteins, such as α-synuclein, aggregate to form Lewy bodies, a pathological hallmark of PD. Oxidative stress can promote the misfolding and aggregation of α-synuclein through mechanisms such as protein oxidation, nitration, and glycation. These aggregates are toxic to neurons and disrupt cellular function, contributing to neurodegeneration in PD (Tan et al., 2009).

Neuronal damage: Dopaminergic neurons in the substantia nigra are particularly vulnerable to oxidative stress due to their high metabolic activity and low antioxidant defense (Trist et al., 2019). ROS can damage cellular components, including proteins, lipids, and DNA, leading to neuronal dysfunction and death. Oxidative damage to dopaminergic neurons contributes to the motor symptoms characteristic of PD, such as tremors, rigidity, and bradykinesia (Guo et al., 2018).

Inflammation: Oxidative stress can activate inflammatory pathways in the brain, leading to the production of pro-inflammatory cytokines and chemokines. Chronic neuroinflammation exacerbates neuronal damage and contributes to the progression of PD pathology (Teleanu et al., 2022). Inflammatory mediators can also induce oxidative stress through the activation of NADPH oxidase and other ROS-generating enzymes (Lee and Yang, 2012).

Impaired autophagy and proteostasis: Autophagy is a cellular process responsible for removing damaged organelles and protein aggregates (Joshi et al., 2020). Oxidative stress can impair autophagic flux, leading to the accumulation of dysfunctional mitochondria and protein aggregates in PD. Disruption of proteostasis further contributes to the neuronal dysfunction and degeneration (Lehtonen et al., 2019).

The central role of oxidative stress in Parkinson's disease and antioxidant therapies has been explored as potential neuroprotective strategies (Jiang et al., 2016). Preclinical studies have shown that antioxidants, such as coenzyme Q10, vitamin E, and N-acetylcysteine, can mitigate oxidative damage and protect against dopaminergic neurodegeneration in animal models of PD (Moosmann and Behl, 2002). However, clinical trials investigating the efficacy of antioxidant therapy in PD patients have yielded mixed results, highlighting the complexity of antioxidant interventions in a multifactorial disease like PD (Deus et al., 2022). Nonetheless, targeting oxidative stress remains a promising therapeutic approach for slowing the progression of Parkinson's disease and improving patients' quality of life (Melo et al., 2011).

Amyotrophic Lateral Sclerosis

Oxidative stress also plays a significant role in the pathogenesis of amyotrophic lateral sclerosis, a progressive neurodegenerative disease characterized by the degeneration of motor neurons in the brain and spinal cord (Barber and Shaw, 2010). Here's how oxidative stress is involved in ALS:

Mitochondrial dysfunction: Similar to other neurodegenerative diseases, mitochondrial dysfunction is a key feature of ALS pathology (Golpich et al., 2017). Mitochondria are both a source and a target of oxidative stress. Oxidative damage to mitochondrial DNA, proteins, and lipids impairs mitochondrial function, leading to decreased energy production and increased ROS generation, which further exacerbates oxidative stress (Bhat et al., 2015).

Glutamate excitotoxicity: Dysregulation of glutamate signaling and excitotoxicity are implicated in ALS (Verche et al., 2010). Excessive glutamate release leads to calcium influx and activation of downstream pathways, including ROS production. ROS, in turn, can potentiate glutamate excitotoxicity, creating a vicious cycle of neuronal damage (Balkhi et al., 2014).

Protein aggregation: Misfolded protein aggregates, including mutant forms of superoxide dismutase 1 (SOD1) and TAR DNA-binding protein 43 (TDP-43), are found in ALS motor neurons (Prasad et al., 2019). These aggregates can induce oxidative stress directly by generating ROS and indirectly by impairing protein quality control mechanisms such as the ubiquitin-proteasome system and autophagy (Van Dam and Dansen, 2020).

Inflammation: Neuroinflammation is a prominent feature of ALS pathology, characterized by the activation of microglia and astrocytes, which release pro-inflammatory cytokines and ROS. Chronic inflammation contributes to motor neuron degeneration and disease progression in ALS (Geloso et al., 2017).

Axonal transport defects: Impaired axonal transport, a process critical for the delivery of essential molecules and organelles to distal neuronal compartments, is observed in ALS (Guo et al., 2020). Oxidative stress disrupts axonal transport mechanisms, causing the accumulation of toxic aggregates and mitochondrial dysfunction in motor neurons (Reddy, 2011).

Antioxidant therapies have been explored as potential treatment strategies for the role of oxidative stress in ALS pathogenesis (Hemerková and Vališ, 2021). Preclinical studies have shown some promising results with antioxidants such as coenzyme Q10, vitamin E, and N-acetylcysteine in ALS animal models (Kamat et al., 2008). However, clinical trials targeting oxidative stress in ALS patients have yielded mixed results, suggesting the need for further research to optimize antioxidant interventions and better understand their therapeutic potential in ALS (Cunha-Oliveira et al., 2020).

Huntington's Disease

Huntington's disease (HD) is another neurodegenerative disorder in which oxidative stress plays a significant role in its pathogenesis (Manoharan et al., 2016). HD is characterized by progressive motor dysfunction, cognitive decline, and psychiatric disturbances caused by a mutation in the huntingtin (HTT) gene, leading to the production of mutant huntingtin protein (mHTT) (Tabrizi et al., 2020). Here's how oxidative stress contributes to HD:

Mitochondrial dysfunction: Similar to other neurodegenerative diseases, mitochondrial dysfunction is a hallmark feature of HD (Golpich et al., 2017). Mutant huntingtin disrupts mitochondrial function, leading to impaired electron transport chain activity, decreased ATP production, and increased ROS generation. ROS-mediated oxidative damage further exacerbates mitochondrial dysfunction, creating a vicious cycle of oxidative stress (Jodeiri Farshbaf and Ghaedi, 2017; Cui, 2012; Bhat et al., 2015).

Excitotoxicity: Dysregulation of glutamate signaling and excitotoxicity also play a role in HD pathogenesis (Dong, 2009). Mutant huntingtin disrupts the balance of calcium homeostasis, leading to excessive glutamate release and subsequent excitotoxic neuronal damage. Excitotoxicity contributes to ROS production and oxidative stress in HD (Sepers and Raymond, 2014).

Protein aggregation: Aggregation of mHTT protein into intracellular inclusion bodies is a pathological hallmark of HD (Cisbani and Cicchetti, 2012). These aggregates can induce oxidative stress directly by generating ROS and indirectly by impairing protein quality control mechanisms, leading to further accumulation of damaged proteins and oxidative damage (Cenini and Voos, 2016).

Dysregulation of cellular signaling pathways: Mutant huntingtin disrupts various cellular signaling pathways involved in oxidative stress response and antioxidant defense mechanisms

(Paul and Snyder, 2019). This dysregulation impairs the ability of cells to cope with oxidative stress and exacerbates neuronal vulnerability in HD (Gorman, 2008).

Neuroinflammation: Chronic neuroinflammation is observed in HD, characterized by microglial activation and the release of pro-inflammatory cytokines and ROS. Neuroinflammation exacerbates neuronal damage and contributes to disease progression in HD (Subhramanyam et al., 2019).

ANTIOXIDANT THERAPEUTICS

For the treatment and prevention of neurodegenerative disorders, some of the important therapeutic strategies include the regulation of the generation of free radicals and protection from their effects (Yaribeygi et al., 2018). Antioxidants play an important role in the treatment of neurodegenerative disorders by protecting from harmful effects and reducing damage caused by oxidative stress (Rekatsina et al., 2020). Drawbacks of antioxidant therapies include blood–brain barrier (BBB) permeability, which has limited their application for the treatment of neurodegenerative disorders (Leo and Campos, 2019; Singh et al., 2019; Jiang et al., 2016). Details of the antioxidants and their mechanisms of action from clinical studies are summarized in Table 11.1.

CONCLUSION

Neurodegenerative diseases are associated with multiple etiologies and pathophysiology, and oxidative stress is observed as a major part of the pathophysiology. The detrimental effects of oxidative stress and ROS have been demonstrated, and they contribute toward the initiation and progression of various pathologies, including neurodegeneration.

The role of oxidative stress in HD pathogenesis and antioxidant therapies have been investigated as potential treatment strategies. Preclinical studies have shown promising results with antioxidants such as coenzyme Q10, vitamin E, and resveratrol in HD animal models, demonstrating neuroprotective effects and improvement in motor function. However, clinical trials targeting oxidative stress in HD patients have shown mixed results, highlighting the complexity of antioxidant interventions in human disease and the need for further research in this area.

ABBREVIATIONS

3-NT: 3-nitrotyrosine
8-OHdG: 8-hydroxydeoxyguanosine
AD: Alzheimer's disease
ALS: amyotrophic lateral sclerosis
BBB: blood–brain barrier
CNS: central nervous system
DCFH: dichlorofluorescein
DHE: dihydroethidium
GPx: glutathione peroxidase
GSH: reduced glutathione
HD: Huntington's disease
MDA: malondialdehyde
mHTT: mutated huntingtin (HTT) protein
PD: Parkinson's disease
ROS: reactive oxygen species
SOD: superoxide dismutase

TABLE 11.1
Details of Antioxidant Agents Evaluated in Clinical Trials

Disease Condition	Antioxidant Agent	Mechanism of Action	Reference
AD	Acetyl-L-carnitine	Reduces cytotoxicity, protein oxidation, and lipid peroxidation. It improves the concentration of GSH.	Mota et al., 2021
	Lipoic acid	Increases antioxidant action by reducing MDA. It increases glutathione reductase concentration.	Kaur et al., 2021
	Omega-3 acid	Improves activity of SOD, catalase, and GPx activity. It reduces lipid peroxidation.	Canhada et al., 2018
	Rasagilina, Selegilina	Improves iron chelator and antioxidant properties.	Behl et al., 2021
	Resveratrol	Decreases production of MDA and nitrile, activity of SOD, and helps to restore concentration of GSH.	Islam et al., 2022
PD	N-acetylcysteine	Decreases lipid peroxidation and SOD activity. It improves GPx and GSH action.	Niedzielska et al., 2016
	Zonisamide	It obstructs release of glutamate and improves 8-OHdG levels in patient's urine.	
	Vitamin E	Inhibits lipid peroxidation and reduces ROS effects.	Shoulson, 1998
	Deferiprone	Reduces generation of ROS and thus oxidative stress due to its ability to rescue iron-overloaded cells.	Devos et al., 2014
HD	Safranal	Inhibits 3-NT and MDA concentration and reduces activity of SOD, catalase, and GSH.	Fotoohi et al., 2021
	Epigallocatechin gallate	Decreases ROS and chelation of bonding metal ions.	Sebastiani et al., 2021
ALS	Riluzol	Acts by decreasing cytosolic calcium and attenuates excess release of glutamate.	Orrell et al., 2007
	Edaravone	Removes lipid peroxidase and hydroxyl radicals and reduces 3-NT concentration.	Yoshino and Kimura, 2006
	Curcumin	Reduces oxidative damage and dysfunction of mitochondria.	Chico et al., 2018
AD, PD, HD	Coenzyme Q10 and derivatives	Decreases levels of MDA and inhibits lipid peroxidation. It improves activity of mitochondrial complex I and SOD.	McManus et al., 2011; Zhu et al., 2017; McGarry et al., 2017

Abbreviations: 3-NT, 3-nitrotyrosine; 8-OHdG, 8-hydroxydeoxyguanosine; AD, Alzheimer's disease; HD, Huntington's disease; ALS, amyotrophic lateral sclerosis; PD, Parkinson's disease; GPx, glutathione peroxidase; GSH, reduced glutathione; MDA, malondialdehyde; ROS, reactive oxygen species; SOD, superoxide dismutase.

REFERENCES

Agita A, Alsagaff MT. Inflammation, immunity, and hypertension. *Acta Med Indones* (2017) 49(2):158.

Agostinho P, A Cunha R, Oliveira C. Neuroinflammation, oxidative stress and the pathogenesis of Alzheimer's disease. *Curr Pharm Design* (2010) 16(25):2766–2778.

Aguilera G, Colín-González AL, Rangel-López E, Chavarria A, Santamaria A. Redox signaling, neuroinflammation, and neurodegeneration. *Antioxid Redox Signal* (2018) 28(18):1626–1651.

Alavi Naini SM, Soussi-Yanicostas N. Tau hyperphosphorylation and oxidative stress, a critical vicious circle in neurodegenerative tauopathies? *Oxidative Med Cell Longev* (2015) 2015: 151979.

Arredondo M, Núñez MT. Iron and copper metabolism. *Mol Aspects Med* (2005) 26(4-5):313–327.

Averill-Bates D. Reactive oxygen species and cell signaling. Review. *Biochim Biophysic Acta (BBA)-Mol Cell Res* (2023) 1871:119573.

Azadmanesh J, Borgstahl GEO. Deciphering the catalytic mechanism of Human Manganese Superoxide Dismutase. *Antioxidants* (2018) 7(2):25.

Baeza I, Fdez-Tresguerres J, Ariznavarreta C, De la Fuente M. Effects of growth hormone, melatonin, oestrogens and phytoestrogens on the oxidized glutathione (GSSG)/reduced glutathione (GSH) ratio and lipid peroxidation in aged ovariectomized rats. *Biogerontol* (2010) 11:687–701.

Balali-Mood M, Naseri K, Tahergorabi Z, Khazdair MR, Sadeghi M. Toxic mechanisms of five heavy metals: mercury, lead, chromium, cadmium, and arsenic. *Front Pharmacol* (2021) 2021:227.

Balkhi HM, Gul T, Banday MZ, Haq E. Glutamate excitotoxicity: an insight into the mechanism. *Int J Adv Res* (2014) 2(7):361–373.

Barber SC, Shaw PJ. Oxidative stress in ALS: key role in motor neuron injury and therapeutic target. *Free Radic Bio Med* (2010) 48(5):629–641.

Basu S. Review isoprostanes: novel bioactive products of lipid peroxidation. *Free Radic Res* (2004) 38(2):105–122.

Bayr H. Reactive oxygen species. *Critic Care Med* (2005) 33(12): S498–501.

Behl T, Kaur D, Sehgal A, Singh S, Sharma N, Zengin G, et al. Role of monoamine oxidase activity in Alzheimer's disease: an insight into the therapeutic potential of inhibitors. *Molecules* (2021) 26:3724.

Bergamini CM, Gambetti S, Dondi A, Cervellati C. Oxygen, reactive oxygen species and tissue damage. *Curr Pharm Design* (2004) 10(14):1611–1626.

Bhat AH, Dar KB, Anees S, Zargar MA, Masood A, Sofi MA, Ganie SA. Oxidative stress, mitochondrial dysfunction and neurodegenerative diseases; a mechanistic insight. *Biomed Pharmacother* (2015) 74:101–110.

Bhatia V, Sharma S. Role of mitochondrial dysfunction, oxidative stress and autophagy in progression of Alzheimer's disease. *J Neurol Sci* (2021) 421:117253.

Biswas S, Chida AS, Rahman I. Redox modifications of protein–thiols: emerging roles in cell signaling. *Biochem Pharmacol* (2006) 71(5):551–564.

Bolisetty S, Jaimes EA. Mitochondria and reactive oxygen species: physiology and pathophysiology. *Int J Mol Sci* (2013) 14(3):6306–6344.

Buccellato FR, D'Anca M, Fenoglio C, Scarpini E, Galimberti D. Role of oxidative damage in alzheimer's disease and neurodegeneration: from pathogenic mechanisms to biomarker discovery. *Antioxidants* (2021) 10(9):1353.

Canhada S, Castro K, Perry IS, Luft VC. Omega-3 fatty acids' supplementation in Alzheimer's disease: a systematic review. *Nutr Neurosci* (2018) 21:529–538

Cecarini V, Gee J, Fioretti E, Amici M, Angeletti M, Eleuteri AM, Keller JN. Protein oxidation and cellular homeostasis: emphasis on metabolism. *Biochim Biophysic Acta (BBA)-Molecular Cell Res* (2007) 773(2):93–104.

Cenini G, Voos W. Role of mitochondrial protein quality control in oxidative stress-induced neurodegenerative diseases. *Curr Alzheimer Res* (2016) 13(2):164–173.

Chauhan V, Chauhan A. Oxidative stress in Alzheimer's disease. *Pathophysiol* (2006) 13(3):195–208.

Chen X, Guo C, Kong J. Oxidative stress in neurodegenerative diseases. *Neural Regen Res* (2012) 7(5):376.

Chico L, Ienco EC, Bisordi C, Gerfo AL, Petrozzi L, Petrucci A, et al. Amyotrophic lateral sclerosis and oxidative stress: a double-blind therapeutic trial after curcumin supplementation. *CNS Neurol Disord Drug Targets* (2018) 17:767–779.

Cisbani G, Cicchetti F. An in vitro perspective on the molecular mechanisms underlying mutant huntingtin protein toxicity. *Cell Death Dis* (2012) 3(8):e382.

Crichton R. "The importance of iron for biological systems", *Inorg Chem Iron Metabolism: From Molecular Mechanisms to Clinical Consequences,* John Wiley & Sons, Ltd, 2001, pp. 17–48.

Cui H, Kong Y, Zhang H. Oxidative stress, mitochondrial dysfunction, and aging. *J Signal Transduct* (2012) 2012:646354.

Cunha-Oliveira T, Montezinho L, Mendes C, Firuzi O, Saso L, Oliveira PJ, Silva FSG. Oxidative stress in amyotrophic lateral sclerosis: pathophysiology and opportunities for pharmacological intervention. *Oxid Med Cell Longev* (2020) 2020:5021694.

Dalle-Donne I, Rossi R, Giustarini D, Milzani A, Colombo R. Protein carbonyl groups as biomarkers of oxidative stress. *Clinica Chimica Acta* (2003) 329(1–2):23–38.

Demirci-Cekic S, Özkan G, Avan AN, Uzunboy S, Çapanoğlu E, Apak R. Biomarkers of oxidative stress and antioxidant defense. *J Pharm Biomed Analysis* (2022) 209:114477.

Deus CM, Teixeira J, Raimundo N, Tucci P, Borges F, Saso L, Oliveira PJ. Modulation of cellular redox environment as a novel therapeutic strategy for Parkinson's disease. *Eur J Clin Investigation* (2022) 52(10):e13820.

Devos D, Moreau C, Devedjian JC, Kluza J, Petrault M, Laloux C, et al. Targeting chelatable iron as a therapeutic modality in Parkinson's disease. *Antioxid Redox Signal* (2014) 21:195–210.

Dhalla NS, Temsah RM, Netticadan T. Role of oxidative stress in cardiovascular diseases. *J Hypertension* (2000) 18(6):655–673.

Dickson KB, Zhou J. Role of reactive oxygen species and iron in host defense against infection. *Front Biosci-Landmark* (2020) 25(8):1600–1616.

Dikalov SI, Harrison DG. Methods for detection of mitochondrial and cellular reactive oxygen species. *Antioxid Redox Signal* (2014) 20(2):372–382.

Dong XX, Wang Y, Qin ZH. Molecular mechanisms of excitotoxicity and their relevance to pathogenesis of neurodegenerative diseases. *Acta Pharmacol Sin* (2009) 30(4):379–387.

Duffus JH. "Heavy metals" a meaningless term? (IUPAC Technical Report). *Pure Applied Chem* (2002) 74(5):793–807.

Espinosa-Diez C, Miguel V, Mennerich D, Kietzmann T, Sánchez-Pérez P, Cadenas S, Lamas S. Antioxidant responses and cellular adjustments to oxidative stress. *Redox Biol* (2015) 6:183–197.

Fotoohi A, Moloudi MR, Hosseini S, Hassanzadeh K, Feligioni M, Izadpanah E. A novel pharmacological protective role for safranal in an animal model of Huntington's disease. *Neurochem Res* (2021) 46:1372–1379.

Geloso MC, Corvino V, Marchese E, Serrano A, Michetti F, D'Ambrosi N. The dual role of microglia in ALS: mechanisms and therapeutic approaches. *Front Aging Neurosci* (2017) 9:242.

Golpich M, Amini E, Mohamed Z, Azman Ali R, Mohamed Ibrahim N, Ahmadiani A. Mitochondrial dysfunction and biogenesis in neurodegenerative diseases: pathogenesis and treatment. *CNS Neurosci Ther* (2017) 23(1):5–22.

Gorman AM. Neuronal cell death in neurodegenerative diseases: recurring themes around protein handling. *J Cell Mol Med* (2008) 12(6a):2263–2280.

Gulcin İ. Antioxidants and antioxidant methods: an updated overview. *Archiv Toxicol* (2020) 94(3):651–715.

Gupta SP, Patil VM. Recent studies on design and development of drugs against Alzheimer's disease (AD) based on inhibition of BACE-1 and other AD-causative agents. *Curr Topics Med Chem* (2020) 20(13):1195–1213.

Guo JD, Zhao X, Li Y, Li GR, Liu XL. Damage to dopaminergic neurons by oxidative stress in Parkinson's disease. *Int J Mol Med* (2018) 41(4):1817–1825.

Guo W, Dittlau KS, Van Den Bosch L. Axonal transport defects and neurodegeneration: molecular mechanisms and therapeutic implications. *Semin Cell Dev Biol* (2020) 99:133–150.

Gutteridge JM. Biological origin of free radicals, and mechanisms of antioxidant protection. *Chem-Biol Interact* (1994) 91(2–3):133–140.

Habibi G. Hydrogen peroxide (H2O2) generation, scavenging and signaling in plants, in *Oxidative Damage to Plants*, ed G Habibi, Academic Press, 2014, pp. 557–584.

Halliwell B, Lee CY. Using isoprostanes as biomarkers of oxidative stress: some rarely considered issues. *Antioxid Redox Signal* (2010) 13(2):145–156.

He L, He T, Farrar S, Ji L, Liu T, Ma X. Antioxidants maintain cellular redox homeostasis by elimination of reactive oxygen species. *Cell Physiol Biochem* (2017) 44(2):532–553.

Hemerková P, Vališ M. Role of oxidative stress in the pathogenesis of amyotrophic lateral sclerosis: antioxidant metalloenzymes and therapeutic strategies. *Biomolecules* (2021) 11(3):437.

Ho YS, Magnenat JL, Gargano M, Cao J. The nature of antioxidant defense mechanisms: a lesson from transgenic studies. *Environ Health Perspect* (1998) 106(suppl 5):1219–1228.

Ighodaro OM, Akinloye OA. First line defence antioxidants-superoxide dismutase (SOD), catalase (CAT) and glutathione peroxidase (GPX): their fundamental role in the entire antioxidant defence grid. *Alexandria J Med* (2018) 54(4):287–293.

Islam F, Nafady MH, Islam MR, Saha S, Rashid S, Akter A, et al. Resveratrol and neuroprotection: an insight into prospective therapeutic approaches against Alzheimer's disease from bench to bedside. *Mol Neurobiol* (2022) 59:4384–4404.

Jadoon S, Malik A. A review article on the formation, mechanism and biochemistry of MDA and MDA as a biomarker of oxidative stress. *Int J Adv Res* (2017) 5:811–818.

Jávega B, Herrera G, Martínez-Romero A, O'Connor JE. Flow cytometry of oxygen and oxygen-related cellular stress. *Oxygen* (2023) 3(2):222–255.

Jiang T, Sun Q, Chen S. Oxidative stress: a major pathogenesis and potential therapeutic target of antioxidative agents in Parkinson's disease and Alzheimer's disease. *Prog Neurobiol* (2016) 147:1–19.

Jodeiri Farshbaf M, Ghaedi K. Huntington's disease and mitochondria. *Neurotoxicity Res* (2017) 32:518–529.

Jomova K, Makova M, Alomar SY, Alwasel SH, Nepovimova E, Kuca K, Rhodes CJ, Valko M. Essential metals in health and disease. *Chem-Biol Interact* (2022) 367:110173.

Jomova K, Raptova R, Alomar SY, Alwasel SH, Nepovimova E, Kuca K, Valko M. Reactive oxygen species, toxicity, oxidative stress, and antioxidants: chronic diseases and aging. *Archiv Toxicol* (2023) 97(10):2499–2574.

Joshi V, Upadhyay A, Prajapati VK, Mishra A. How autophagy can restore proteostasis defects in multiple diseases? *Med Res Rev* (2020) 40(4):1385–1439.

Juan CA, Pérez de la Lastra JM, Plou FJ, Pérez-Lebeña E. The chemistry of reactive oxygen species (ROS) revisited: outlining their role in biological macromolecules (DNA, lipids and proteins) and induced pathologies. *Int J Mol Sci* (2021) 22(9):4642.

Kamat CD, Gadal S, Mhatre M, Williamson KS, Pye QN, Hensley K. Antioxidants in central nervous system diseases: preclinical promise and translational challenges. *J Alzheimer's Dis* (2008) 15(3):473–493.

Katsnelson, A, Strooper, BD, Zoghbi, HY. Neurodegeneration: from cellular concepts to clinical applications. *Sci Transl Med* (2016) 8:364ps18.

Kaur D, Behl T, Sehgal A, Singh S, Sharma N, Chigurupati S, et al. Decrypting the potential role of α-lipoic acid in Alzheimer's disease. *Life Sci* (2021) 284:119899.

Krumova K, Cosa G. Overview of reactive oxygen species, in *Singlet Oxygen: Applications in Biosciences and Nanoscience*, ed S. Nonell, C. Flors, S. Nonell, and C. Flors, The Royal Society of Chemistry, 2016, Ch. 1, pp. 1–21.

Le Verche V, Ikiz B, Jacquier A, Przedborski S, Re DB. Glutamate pathway implication in amyotrophic lateral sclerosis: what is the signal in the noise? J Receptor, Ligand and Channel Research (2010) 2011:1–22.

Lee IT, Yang CM. Role of NADPH oxidase/ROS in pro-inflammatory mediators-induced airway and pulmonary diseases. *Biochem Pharmacol* (2012) 84(5):581–590.

Lehtonen Š, Sonninen TM, Wojciechowski S, Goldsteins G, Koistinaho J. Dysfunction of cellular proteostasis in Parkinson's disease. *Front Neurosci* (2019) 13:457.

Lennicke C, Cochemé HM. Redox metabolism: ROS as specific molecular regulators of cell signaling and function. *Mol Cell* (2021) 81(18):3691–3707.

Leo, EEM, Campos, MRS. Systemic oxidative stress: a key point in neurodegeneration – a review. *J Nutr Health Aging* (2019) 23:694–699.

Li J, O W, Li W, Jiang ZG, Ghanbari HA. Oxidative stress and neurodegenerative disorders. *Int J Mol Sci* (2013) 14:24438–24475.

Liguori I, Russo G, Curcio F, Bulli G, Aran L, Della-Morte D, Gargiulo G, Testa G, Cacciatore F, Bonaduce D, Abete P. Oxidative stress, aging, and diseases. *Clin Interven Aging* (2018) 2018:757–772.

Luo J, Mills K, le Cessie S, Noordam R, van Heemst D. Ageing, age-related diseases and oxidative stress: what to do next? *Ageing Res Rev* (2020) 57:100982.

Lushchak VI. Free radicals, reactive oxygen species, oxidative stress and its classification. *Chem-Biol Interact* (2014) 224:164–175.

Madamanchi NR, Runge MS. Mitochondrial dysfunction in atherosclerosis. *Circulation Res* (2007)100(4):460–473.

Madkour LH. Function of reactive oxygen species (ROS) inside the living organisms and sources of oxidants. *Pharm Sci Anal Res* (2019) 2:180023.

Mancuso M, Coppede F, Migliore L, Siciliano G, Murri L. Mitochondrial dysfunction, oxidative stress and neurodegeneration. *J Alzheimer's Dis* (2006) 10(1):59–73.

Manoharan S, Guillemin GJ, Abiramasundari RS, Essa MM, Akbar M, Akbar MD. The role of reactive oxygen species in the pathogenesis of Alzheimer's disease, Parkinson's disease, and Huntington's disease: a mini review. *Oxid Med Cell Longev* (2016) 2016:8590578.

Mansoor S, Ali A, Kour N, Bornhorst J, AlHarbi K, Rinklebe J, Abd El Moneim D, Ahmad P, Chung YS. Heavy metal induced oxidative stress mitigation and ROS scavenging in plants. *Plants* (2023) 12(16):3003.

Mansoor S, Ali Wani O, Lone JK, Manhas S, Kour N, Alam P, Ahmad A, Ahmad P. Reactive oxygen species in plants: from source to sink. *Antioxid* (2022) 11(2):225.

Marchetti C. Interaction of metal ions with neurotransmitter receptors and potential role in neurodiseases. *Biometals* (2014) 27:1097–1113.

Martindale JL, Holbrook NJ. Cellular response to oxidative stress: signaling for suicide and survival. *J Cell Physiol* (2002) 192(1):1–5.

McGarry A, McDermott M, Kieburtz K, de Blieck EA, Beal F, Marder K, et al. Huntington study group 2CARE investigators and coordinators. A randomized, double-blind, placebo-controlled trial of coenzyme Q10 in Huntington disease. *Neurol* (2017) 88: 152–159.

McManus MJ, Murphy MP, Franklin JL. The mitochondria-targeted antioxidant MitoQ prevents loss of spatial memory retention and early neuropathology in a transgenic mouse model of Alzheimer's disease. *J Neurosci* (2011) 31:15703–15715.

Melo A, Monteiro L, Lima RM, de Oliveira DM, de Cerqueira MD, El-Bachá RS. Oxidative stress in neurodegenerative diseases: mechanisms and therapeutic perspectives. *Oxida Med Cell Longev* (2011) 2011:467180.

Mena S, Ortega A, Estrela JM. Oxidative stress in environmental-induced carcinogenesis. *Mutat Res/Genet Toxicol Environ Mutagen* (2009) 674(1–2):36–44.

Moosmann B, Behl C. Antioxidants as treatment for neurodegenerative disorders. *Exp Opin Invest Drugs* (2002) 11(10):1407–1435.

Moreira PI, Carvalho C, Zhu X, Smith MA, Perry G. Mitochondrial dysfunction is a trigger of Alzheimer's disease pathophysiology. *Biochim Biophysic Acta (BBA)-Mol Basis Dis* (2010) 1802(1):2–10.

Mota SI, Pita I, Águas R, Tagorti S, Virmani A, Pereira FC, et al. Mechanistic perspectives on differential mitochondrial-based neuroprotective effects of several carnitine forms in Alzheimer's disease in vitro model. *Arch Toxicol* (2021) 95:2769–2784.

Nelson SK, Bose SK, Grunwald GK, Myhill P, McCord JM. The induction of human superoxide dismutase and catalase in vivo: a fundamentally new approach to antioxidant therapy. *Free Radic Biol Med* (2006) 40(2):341–347.

Niedzielska E, Smaga I, Gawlik M, Moniczewski A, Stankowicz P, Pera J, et al. Oxidative stress in neurodegenerative diseases. *Mol Neurobiol* (2016) 53:4094–4125.

Okoh V, Deoraj A, Roy D. Estrogen-induced reactive oxygen species-mediated signalings contribute to breast cancer. *Biochim Biophysic Acta (BBA)-Rev Cancer* (2011) 1815(1):115–133.

Orrell RW, Lane RJM, Ross M. Antioxidant treatment for amyotrophic lateral sclerosis/motor neuron disease. *Cochrane Database Syst Rev* (2007) 35:CD002829.

Ozougwu JC. The role of reactive oxygen species and antioxidants in oxidative stress. *Int J Res* (2016) 1(8).

Paithankar JG, Saini S, Dwivedi S, Sharma A, Chowdhuri DK. Heavy metal associated health hazards: an interplay of oxidative stress and signal transduction. *Chemosphere* (2021) 262:128350.

Panday A, Sahoo MK, Osorio D, Batra S. NADPH oxidases: an overview from structure to innate immunity-associated pathologies. *Cell Mol Immunol* (2015) 12(1):5–23.

Papaconstantinou J. The role of signaling pathways of inflammation and oxidative stress in development of senescence and aging phenotypes in cardiovascular disease. *Cells* (2019) 8(11):1383.

Paul BD, Snyder SH. Impaired redox signaling in Huntington's disease: therapeutic implications. *Front Mol Neurosci* (2019) 12:68.

Pegoretti, V, Swanson, KA, Bethea, JR, Probert, L, Eisel, ULM, Fischer, R. Inflammation and oxidative stress in multiple sclerosis: consequences for therapy development. *Oxid Med Cell Longev* (2020) 2020:7191080.

Peraza MA, Ayala-Fierro F, Barber DS, Casarez E, Rael LT. Effects of micronutrients on metal toxicity. *Environ Health Perspect* (1998) 106(suppl 1):203–216.

Permyakov EA. Metal binding proteins. *Encyclopedia* (2021) 1(1):261–292.

Pervaiz S, Clement MV. Superoxide anion: oncogenic reactive oxygen species? *Int J Biochem Cell Biol* (2007) 39(7–8):1297–1304.

Pisoschi AM, Pop A. The role of antioxidants in the chemistry of oxidative stress: a review. *Eur J Med Chem* (2015) 97:55–74.

Poh Loh K, Hong Huang S, De Silva R, H Tan BK, Zhun Zhu Y. Oxidative stress: apoptosis in neuronal injury. *Curr Alzheimer Res* (2006) 3(4):327–337.

Prasad A, Bharathi V, Sivalingam V, Girdhar A, Patel BK. Molecular mechanisms of TDP-43 misfolding and pathology in amyotrophic lateral sclerosis. *Front Mol Neurosci* (2019) 12:25.

Pritam P, Deka R, Bhardwaj A, Srivastava R, Kumar D, Jha AK, Jha NK, Villa C, Jha SK. Antioxidants in Alzheimer's disease. *Curr Ther Significance Future Prospect Bio* (2022) 11(2):212.

Rahal A, Kumar A, Singh V, Yadav B, Tiwari R, Chakraborty S, Dhama K. Oxidative stress, prooxidants, and antioxidants: the interplay. *BioMed Res Int* (2014) 2014:761264.

Raj Rai S, Bhattacharyya C, Sarkar A, Chakraborty S, Sircar E, Dutta S, Sengupta R. Glutathione: role in oxidative/nitrosative stress, antioxidant defense, and treatments. *Chem Select* (2021) 6(18):4566–4590.

Reddy PH. Abnormal tau, mitochondrial dysfunction, impaired axonal transport of mitochondria, and synaptic deprivation in Alzheimer's disease. *Brain Res* (2011) 30(1415):136–148.

Rehman K, Fatima F, Waheed I, Akash MS. Prevalence of exposure of heavy metals and their impact on health consequences. *J Cell Biochem* (2018) 119(1):157–184.

Rekatsina M, Paladini A, Piroli A, Zis P, Pergolizzi JV, Varrassi G. Pathophysiology and therapeutic perspectives of oxidative stress and neurodegenerative diseases: a narrative review. *Adv Therapy* (2020) 37:113–139.

Reynolds A, Laurie C, Mosley RL, Gendelman HE. Oxidative stress and the pathogenesis of neurodegenerative disorders. *Int Rev Neurobio* (2007) 82:297–325.

Salvemini D, Cuzzocrea S. Oxidative stress in septic shock and disseminated intravascular coagulation. *Free Radical Bio Med* (2002) 33(9):1173–1185.

Schirrmacher V. Less can be more: the hormesis theory of stress adaptation in the global biosphere and its implications. *Biomedicines* (2021) 9(3):293.

Sebastiani G, Almeida-Toledano L, Serra-Delgado M, Navarro-Tapia E, Sailer S, Valverde O, et al. Therapeutic effects of catechins in less common neurological and neurodegenerative disorders. *Nutrients* (2021) 13:2232.

Sepers MD, Raymond LA. Mechanisms of synaptic dysfunction and excitotoxicity in Huntington's disease. *Drug Discov Today* (2014) 19(7):990–996.

Shields HJ, Traa A, Van Raamsdonk JM. Beneficial and detrimental effects of reactive oxygen species on lifespan: a comprehensive review of comparative and experimental studies. *Front Cell Develop Bio* (2021) 9:181.

Shoulson I. DATATOP: a decade of neuroprotective inquiry. Parkinson study group. Deprenyl and tocopherol antioxidative therapy of parkinsonism. *Ann Neurol* (1998) 44:S160–166.

Sies H, Belousov VV, Chandel NS, Davies MJ, Jones DP, Mann GE, Murphy MP, Yamamoto M, Winterbourn C. Defining roles of specific reactive oxygen species (ROS) in cell biology and physiology. *Nature Rev Mol Cell Biol* (2022) 23(7):499–515.

Simpson DS, Oliver PL. ROS generation in microglia: understanding oxidative stress and inflammation in neurodegenerative disease. *Antioxid* (2020) 9(8):743.

Singh, A, Kukreti, R, Saso, L, Kukreti, S. Oxidative stress: a key modulator in neurodegenerative diseases. *Molecules* (2019) 24:1583–1602.

Soto C, Estrada LD. Protein misfolding and neurodegeneration. *Archiv Neurol* (2008) 65(2):184–189.

Subhramanyam CS, Wang C, Hu Q, Dheen ST. Microglia-mediated neuroinflammation in neurodegenerative diseases. *Semin Cell Dev Biol* (2019) 94:112–120.

Sukanya, Prajapati BG, Patil VM, Choudhary BS. Dual specificity Tyrosine Phosphorylation-regulated Kinase 1A (DYRK1A) inhibitors: the quest for a disease-modifying treatment for Alzheimer's disease, in *Deciphering Drug Targets for Alzheimer's Disease*, ed Kumar D, Patil VM, Wu D and Thorat N, Springer, Singapore 2023, pp. 69–94.

Suzuki YJ, Carini M, Butterfield DA. Protein carbonylation. *Antioxid Redox Signal* (2010) 12(3):323–325.

Tabassum N, Kheya IS, Asaduzzaman S, Maniha S, Fayz AH, Zakaria A, Noor R. A review on the possible leakage of electrons through the electron transport chain within mitochondria. *Life Sci* (2020) 6:105–113.

Tabrizi SJ, Flower MD, Ross CA, Wild EJ. Huntington disease: new insights into molecular pathogenesis and therapeutic opportunities. *Nature Rev Neurol* (2020) 16(10):529–546.

Tan JM, Wong ES, Lim KL. Protein misfolding and aggregation in Parkinson's disease. *Antioxid Redox Signal* (2009) 11(9):2119–2134.

Teixeira JP, de Castro AA, Soares FV, da Cunha EF, Ramalho TC. Future therapeutic perspectives into the Alzheimer's disease targeting the oxidative stress hypothesis. *Molecules* (2019) 24(23):4410.

Teleanu DM, Niculescu AG, Lungu II, Radu CI, Vladâcenco O, Roza E, Costăchescu B, Grumezescu AM, Teleanu RI. An overview of oxidative stress, neuroinflammation, and neurodegenerative diseases. *Int J Mol Sci* (2022) 23(11):5938.

Trist BG, Hare DJ, Double KL. Oxidative stress in the aging substantia nigra and the etiology of Parkinson's disease. *Aging Cell* (2019) 18(6):e13031.

Tsikas D. Assessment of lipid peroxidation by measuring malondialdehyde (MDA) and relatives in biological samples: Analytical and biological challenges. *Anal Biochem* (2017) 524:13–30.

Umeno A, Biju V, Yoshida Y. In vivo ROS production and use of oxidative stress-derived biomarkers to detect the onset of diseases such as Alzheimer's disease, Parkinson's disease, and diabetes. *Free Radical Res* (2017) 51(4):413–427.

Urbaniak SK, Boguszewska K, Szewczuk M, Kaźmierczak-Barańska J, Karwowski BT. 8-Oxo-7, 8-dihydro-2′-deoxyguanosine (8-oxodG) and 8-hydroxy-2′-deoxyguanosine (8-OHdG) as a potential biomarker for gestational diabetes mellitus (GDM) development. *Molecules* (2020) 25(1):202.

Valavanidis A, Vlachogianni T, Fiotakis C. 8-hydroxy-2′-deoxyguanosine (8-OHdG): a critical biomarker of oxidative stress and carcinogenesis. *J Environ Sci Health Part C* (2009) 27(2):120–139.

van Dam L, Dansen TB. Cross-talk between redox signalling and protein aggregation. *Biochem Soc Trans* (2020) 48(2):379–397.

Vašková J, Vaško L, Kron I. Oxidative processes and antioxidative metaloenzymes. *Antioxidant Enzyme* (2012) 2:19–58.

Vertuani S, Angusti A, Manfredini S. The antioxidants and pro-antioxidants network: an overview. *Curr Pharm Des* (2004) 10(14):1677–1694.

Villalpando-Rodriguez GE, Gibson SB. Reactive oxygen species (ROS) regulates different types of cell death by acting as a rheostat. *Oxid Med Cell Longev* (2021) 2021:9912436.

Waszczak C, Carmody M, Kangasjärvi J. Reactive oxygen species in plant signaling. *Ann Rev Plant Bio* (2018) 69:209–236.

Wessels I, Maywald M, Rink L. Zinc as a gatekeeper of immune function. *Nutrients* (2017) 9(12):1286.

Williams LE, Pittman JK, Hall JL. Emerging mechanisms for heavy metal transport in plants. *Biochimic Biophysic Acta (BBA)-Biomembranes* (2000) 1465(1–2):104–126.

Winterbourn CC, Kettle AJ. Redox reactions and microbial killing in the neutrophil phagosome. *Antioxid Redox Signal* (2013) 18(6):642–660.

Yaman SO, Ayhanci A. Lipid peroxidation, In *Accenting Lipid Peroxidation*, ed Pınar Atukeren, BoD – Books on Demand (2021) (pp.1–12).

Yang Y, Bazhin AV, Werner J, Karakhanova S. Reactive oxygen species in the immune system. *Int Rev Immunol* (2013) 32(3):249–270.

Yaribeygi, H, Panahi, Y, Javadi, B, Sahebkar, A. The underlying role of oxidative stress in neurodegeneration: a mechanistic review. *CNS Neurol Disord Drug Targets* (2018) 17:207–215.

Yoshino H, Kimura A. Investigation of the therapeutic effects of edaravone, a free radical scavenger, on amyotrophic lateral sclerosis (phase II study). *Amyotroph Lateral Scler* (2006) 7: 241–245

Zhao Y, Zhao B. Oxidative stress and the pathogenesis of Alzheimer's disease. *Oxid Med Cell Longev* (2013) 2013:316523.

Zhu MJ, Wang X, Shi L, Liang LY, Wang Y. Senescence, oxidative stress and mitochondria dysfunction. *Med Res Innov* (2018) 2(4):1–5.

Zhu ZG, Sun MX, Zhang WL, Wang WW, Jin YM, Xie CL. The efficacy and safety of coenzyme Q10 in Parkinson's disease: a meta-analysis of randomized controlled trials. *Neurol Sci* (2017) 38:215–224.

12 Herbs and Natural Products for Neurodegenerative Disorders

Rashmi Saxena Pal[1], Sharma Vedika Vijaypal[2], Yogendra Pal[3], MVNL Chaitanya[1], Priyanka[2], and Deepti Katiyar[4]*

[1]School of Pharmaceutical Sciences, Lovely Professional University, Phagwara, Punjab, India
[2]CT College of Pharmacy Jalandhar, Punjab, India
[3]School of Pharmaceutical Sciences, RIMT University, Mandi Gobindgarh, Punjab, India
[4]KIET School of Pharmacy, KIET Group of Institutions, Ghaziabad, Uttar Pradesh, India
*Corresponding author

INTRODUCTION

A class of illnesses known as neurodegenerative disorders results in damage to neural cells, which impairs motility and/or cognitive function (Tomoshige and Ishikawa, 2021). Protein aggregation causes neurodegenerative disorders like Alzheimer's disease (AD), Parkinson's disease (PD), Huntington's disease (HD), amyotrophic lateral sclerosis (ALS), and multiple sclerosis (MS). These diseases have distinct pathogenic features and causative components (Kabir et al., 2022). Neurodegenerative disorders significantly reduce the quality of life for both patients and those who care for them. These disorders are frequently deadly and incurable. Epidemiologists predict that by 2050, the cumulative prevalence of these diseases will triple due to the aging of the Western population (Valori et al., 2021). The highly efficient blood–brain barrier (BBB) nevertheless poses a significant obstacle to the proper management of neurodegenerative disorders. Although numerous successful procedures and extremely invasive approaches have been established, their clinical adoption is limited because of differing concerns regarding their long-term value due to possible brain barrier injury (Lamptey et al., 2022). Due to their unique symptoms and targeting of different brain regions, neurodegenerative disorders are typically regarded as discrete clinical entities. Nonetheless, there are a number of genetic, molecular, or physiological characteristics that they have in common (such as early vascular dysfunction, misfolded protein aggregation and dissemination, selective susceptibility of specific neurons, and immune response activation) (Bordoni et al., 2020). Transporting macromolecules or highly polar solutes over the BBB and into the brain capillaries is severely restricted by the BBB (Birajdar et al., 2023).

A few of the traditional herbs' neuroprotective methods include reducing inflammatory responses, improving antioxidative qualities, and suppressing the functional aspects of pro-inflammatory cytokines like tumors. Plants' defense mechanisms against these diseases heavily rely on variations in transcription and transduction pathways (Luthra and Roy, 2022). The Indian subcontinent has a rich traditional knowledge of nootropic plants, which are renowned for their multifaceted utility in diverse circumstances, and is widely employed in the ancient Ayurvedic medical system (Sharma et al., 2021).

Macamides have both *in vitro* and *in vivo* neuroprotective properties. Macamides may prove beneficial in the management of neurodegenerative illnesses, including Alzheimer's disease, as they presumably reduce the mitochondrial toxicity caused by Mn in glioma U-87 MG (magnesium) cells through

DOI: 10.1201/9781003506317-12

their interaction with the CB1 (cannabinoid receptor 1) receptor (Leuci et al., 2020). Nutraceuticals are becoming more in demand as the urgent need to stick to healthy solutions to counter neurological interventions grows. There is evidence linking dietary habits and lifestyle choices to neurodegeneration, which suggests the use of nutraceuticals as a preventative measure for neurological therapies (Makkar et al., 2020; Singh et al., 2021). According to recent research, withaferin A may be able to stop viral endocytosis by containing the host transmembrane protease TMPRSS2 without changing the expression of ACE-2. It is thought that this multiring compound's potential for conducting minor structural alterations will increase its pharmacotherapeutic applications. Known for its natural therapeutic properties and low occurrence of adverse effects, herbal remedies have been utilized for treating neurodegenerative diseases since ancient times and are today considered traditional therapy options. Phytochemicals such as flavonoids, phenols, terpenes, saponins, and alkaloids are important players in the regulation of most neurodegenerative disorders (Kumar et al., 2024).

SIGNIFICANCE OF HERBS AND NATURAL PRODUCTS IN NEURODEGENERATIVE DISORDERS

Chinese medicine has long utilized traditional herbs, such as *Schisandra chinensis* and *Evodia rutaecarpa*, to treat neurodegenerative diseases. Different ratios of the herbs can have different therapeutic effects, allowing Chinese medicine to adapt to the needs of different patients (Cao et al., 2023). The secondary metabolites, flavonoids, are generated from plants and have different biological functions. It was discovered in earlier studies that a variety of naturally occurring flavonoids might effectively cure the neuroinflammation linked to Alzheimer's (Zhang and Yan, 2023). Additionally, it is well acknowledged that they play a crucial role in the treatment and prevention of various neurodegenerative illnesses, including epilepsy, Parkinson's disease and Alzheimer's disease (Rahman et al., 2021b). A considerable and rising share of morbidity and death in the industrialized world is caused by neurodegenerative diseases, including Alzheimer's and Parkinson's disease. For thousands of years, clinical applications have been performed using *Ganoderma lucidum* (lingzhi), one of the most nourishing and potent medicinal herbs (Quan et al., 2019).

CLASSIFICATION

Neurodegenerative illnesses are conditions marked by a gradual loss of neurons in the brain and other peripheral organs, along with the accumulation of proteins exhibiting changed physicochemical characteristics. Neurodegenerative diseases are classified molecularly using protein-based criteria (Kovacs, 2018). Diseases in which the central nervous system's cells stop functioning or die are known as neurodegenerative disorders. These include Parkinson's, Alzheimer's, epilepsy, stroke, and other conditions. With no known treatment, neurodegenerative diseases usually get worse over time (Chaki and Woźniak, 2023).

Neurodegenerative diseases are prevalent disorders of the neurological system that develop over time and exhibit complex clinical patterns. The oscillations in gait are a reflection of pathological and physiological changes in the locomotor control system. Clinical trials and healthcare monitoring both depend on the use of gait fluctuations for illness condition evaluation. In patients with neurodegenerative diseases, the categorization of gait irregularities improves clinical diagnosis capabilities and quality of life (Yan et al., 2020). TAR DNA-binding protein 43 (TDP-43) proteopathy, a-synucleinopathy, tauopathies, and amyloidosis are the most prevalent neurodegenerative diseases (Yan et al., 2020).

COMMON FEATURES AND SYMPTOMS

The common factor across neurodegenerative disorders is that they are caused by altered proteins that unfold, form β-structures, and have a pathological propensity to self-aggregate in neuronal cells.

Tau protein in Alzheimer's disease, tau unfolding-related tauopathies, α-synuclein in Parkinson's disease, and huntingtin in Huntington's disease all exhibit this trait (Guzman-Martinez et al., 2019). Anxiety, dysphoric mood, apathy, disinhibition, and euphoric mood, along with their corresponding neuronal circuitry, are typical symptoms that arise in both neurodegenerative illness and psychopathology (Levenson et al., 2014). Less than half of cases of psychosis are caused by primary psychiatric problems, which include delusions, hallucinations, and disordered thought or behavior. Neurodegenerative diseases account for up to 60% of late-life psychotic symptoms, as do other systemic medical or neurological disorders (Seritan, 2023).

PREVALENCE AND IMPACT ON GLOBAL HEALTH PATHOPHYSIOLOGY OF NEURODEGENERATIVE DISORDERS

After a thorough search of the literature, the majority of studies assessing the impact of nutritional support, vitamins, EPA and DHA, and the Mediterranean diet on memory and cognitive impairment were found (Bianchi et al., 2021). The majority of neurological disorders have a dramatic rise in prevalence as people age, and aging also affects how risk factors, clinical presentation, and the natural course of these diseases interact (Dumurgier and Tzourio, 2020). A number of pathophysiological components are shared by neurodegenerative disorders. Inflammatory reactions and oxidative stress (OS) are the shared denominators. Regretfully, there is no easy way to address these ailments (Rekatsina et al., 2020). There is growing evidence that the conditions "ferroptosis in AD," "pyroptosis in AD," and "necroptosis in AD" are closely associated with AD, and these studies are becoming increasingly common (Yeerlan et al., 2024).

CELLULAR AND MOLECULAR MECHANISMS

To discover new treatment targets for neurodegenerative illnesses, it is crucial to comprehend the fundamental molecular processes that underpin aging. NAD+ precursors, mitophagy inducers, and cellular senescence inhibitors are a few examples of potential treatment approaches (Azam et al., 2021). Researchers must examine the most current findings about the pathways linked to Aβ and tau, and discuss how glial dysfunction occurs in AD, focusing on the neuronal and glial sensors that regulate Aβ-induced toxicity (Guo et al., 2020a).

In order to serve as a guide for the prevention of aging and diseases associated with it, a number of key mechanisms of neurodegenerative disorders, such as DNA damage and repair, free radical oxidation, telomeres and telomerase, mitochondrial damage, inflammation, and their functions in neurodegenerative diseases, are discussed (Zhao et al., 2024). Depression in the entorhinal cortex can affect the hippocampus's function, which depends on entorhinal inputs; additional degeneration in the hippocampus can exacerbate this effect and cause severe cognitive disruption. Degeneration in the entorhinal cortex and hippocampus during AD has been linked to disabilities in memory and cognitive function (Olajide et al., 2021). The cellular and molecular mechanisms of neurodegenerative disorders are explained in Figure 12.1.

ROLE OF OXIDATIVE STRESS AND INFLAMMATION

Although neuroinflammation and oxidative stress are two entirely distinct pathogenic phenomena, they are interconnected and have an impact on each other. Antioxidants have the ability to modify oxidative stress and mitigate the symptoms associated with neurodegenerative disorders (Teleanu et al., 2022). Neuroinflammation can involve the activation of mast cells, contributing to oxidative stress, in addition to other sources of reactive oxygen species (ROS). Antioxidants can powerfully neutralize reactive oxygen species and free radicals, decreasing oxidative damage. Antioxidant genes, like the manganese superoxide dismutase enzyme, can undergo epigenetic changes that reduce their expression, thus increasing oxidative stress in tissue (Houldsworth, 2024). The buildup

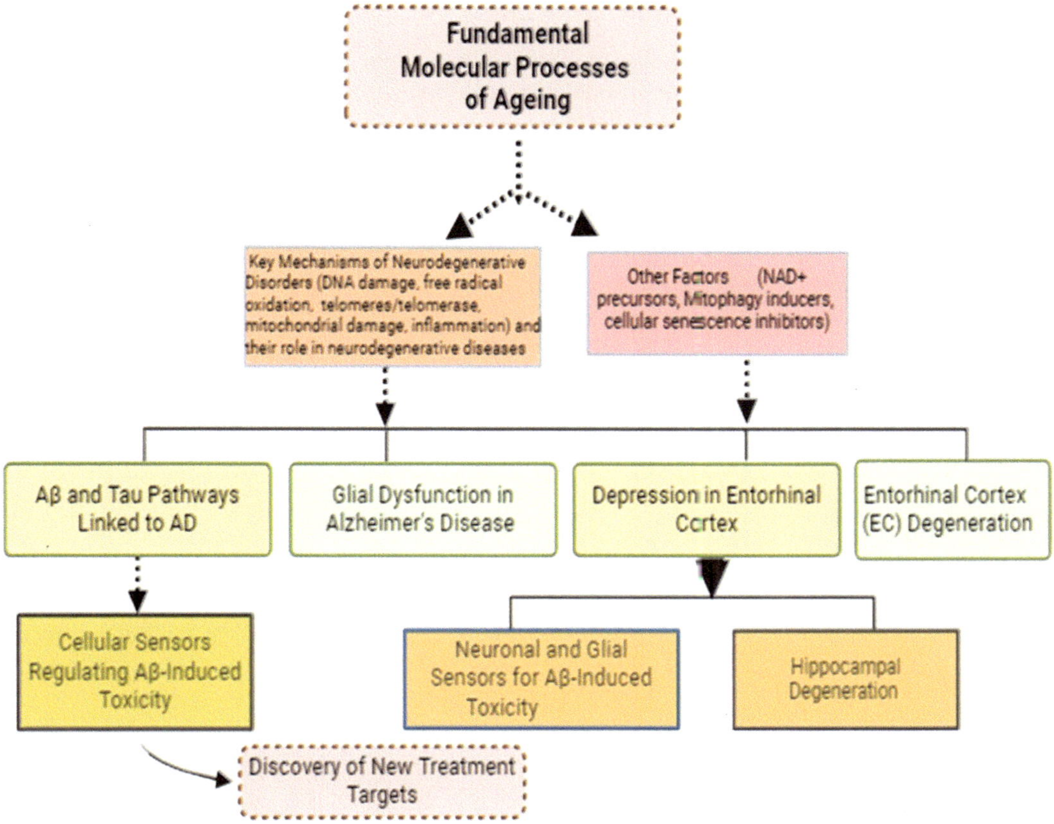

FIGURE 12.1 Cellular and molecular mechanisms of neurodegenerative disorders.

of its pro-oxidative and neuroactive metabolites results from the kynurenine pathway (KP) being activated, which is brought on by the pro-inflammatory variables that coexist with neurodegenerative processes. The cellular energy metabolism is disrupted and mitochondrial damage is increased as a result of this, raising the degree of oxidative stress (Mor et al., 2021).

NEUROTRANSMITTER IMBALANCE

Neurodegenerative Disorders: An In-Depth Molecular Perspective

Among the leading causes of lost years of healthy life, neurodegenerative diseases like Alzheimer's and Parkinson's place a heavy burden on open healthcare systems. At the molecular level, traditional neurodegeneration research is still tainted by narrow-sense tackles focused on amyloid β (Aβ), tau, and α-synuclein (α-syn) (Lam et al., 2020). When compared to the single-target action of the majority of the medications now used to treat Alzheimer's disease, natural compounds offer the advantage of a multitarget strategy, tagging different molecular locations in the human brain, given that Alzheimer's disease is a multifactorial disease (Bhat et al., 2022).

Parkinson's disease is a multifactorial illness that results in nigral dopaminergic depletion and motor impairment due to a variety of pathogenic mechanisms. Comprehensive knowledge of these intricate and potentially convergent pathways will facilitate the development of neuroprotective measures to avert disease presentations and treatment modalities to reduce disease symptoms (Pirooznia et al., 2021).

Genetic and Epigenetic Factors in Neurodegeneration

The study of heritable modifications to gene expression that take place without changing the DNA sequence and connect the genome to its environment is known as epigenetics. Finding epigenetic biomarkers to increase the effectiveness of diagnosis in patients with neurodegenerative illnesses was the goal of the current investigation (Martínez-Iglesias et al., 2021). It has been revealed that epigenetic mechanisms, which alter gene activity without changing the DNA bases, affect synaptic plasticity and cognition as well as mature neurons, having long-lasting impacts. In neural cells, the acetylation and methylation of histones play a role in coordinating protein activity and gene expression (Mohd Murshid et al., 2022). Neuronal loss is preceded by a long period of neuronal dysfunction, during which glial cells also undergo major changes, including a neuroinflammatory response. Those dramatic changes affecting both neuronal and glial cells are associated with epigenetic and transcriptional dysregulations characterized by defined cell-type-specific signatures (Alcalá-Vida et al., 2021). Thus, epigenetic imbalance has been shown to influence the progression of many neurological disorders independent of aberrations in the genetic code (Bertogliat et al., 2020).

Role of Genetic Mutations

A disease condition that primarily affects neurons is referred to as neurodegeneration (Zhang et al., 2023). Widespread and even fatal human multisystem illnesses are typically linked to genetic abnormalities that disrupt cellular pathways essential to basic biological activities (Deneubourg et al., 2022). The etiology and pathophysiology of numerous neurodegenerative diseases are influenced by genetic alterations as well as environmental variables like smoking, excessive alcohol intake, and aging (Luo et al., 2020). An expanding list of genetic variations linked to diseases is being produced by genome-wide association studies; the next step is to determine how these variants cause disease and to use this knowledge to develop treatments (Kampmann, 2020).

Epigenetic Modifications in Neuronal Dysfunction

Neuronal function and differentiation are tightly regulated by both the genome and epigenome. Epigenetic modifications, such as methylation of DNA, chromatin remodeling, and changes in microRNAs (regulatory RNAs), significantly influence memory and neurodegenerative diseases (Ghosh and Saadat, 2023). DNA methylation, histone modifications, chromatin remodeling, non-coding RNAs, and additional gene control mechanisms outside of DNA sequences are the main topics of epigenetics (Liu et al., 2023a). While aberrant epigenetic regulation leads to various forms of CNS (central nervous system) disorders and diseases, epigenetic machinery governs brain growth and activities (Wang, 2020). Disorders that primarily impact complex brain functions, including memory, motor function, and cognitive function, such as autoimmune disorders, tumors, and neurodegenerative diseases, are intimately linked to epigenetic alterations (Wang et al., 2021).

Molecular Pathways in Neurodegenerative Processes

The basic region leucine-zipper transcription factor Nrf2 is essential for the coordinated production of detoxifying and antioxidant genes, which help cells survive in unfavorable environmental or metabolic circumstances. Age-related and hereditary neurodegenerative diseases can be prevented or delayed by targeting Nrf2/ARE signaling molecular intervention and the enhancement of specific gene transcriptional activity, as oxidative stress is one of the hallmarks of neurodegenerative pathologies, along with neuroinflammation and mitochondrial dysfunction (Zgorzynska et al., 2021).

Neurodegenerative disorders like Alzheimer's disease are among the most prevalent ones. Dysfunctional insulin signaling may lead to its main pathological hallmarks: amyloid-β plaques and neurofibrillary tangles (NFTs). The primary molecules implicated in compromised insulin signaling include GSK-3β, PI3K, Akt, and IRS (Akhtar and Sah, 2020). In addition to carrying out a wide range of crucial tasks that define the nervous system's proper operation, astrocytes actively participate in the etiology of neurodegenerative diseases like Alzheimer's (Preman et al., 2021). Numerous studies show that the Ras family's small GTPases are implicated in the pathways that lead to neurodegeneration (Arrazola Sastre et al., 2020).

The regulation of signal transmission and biological processes, including cell proliferation, apoptosis, metabolism, and angiogenesis, is significantly influenced by the PI3K/AKT/mTOR signaling pathway. The PI3K/AKT/mTOR signaling pathway's components are more intricate than those of other signaling pathways. The PI3K/AKT/mTOR signaling system has biological roles and regulatory mechanisms that are relevant to a wide range of human disorders, such as tumors, neurodegenerative diseases, and ischemic brain injury. Inhibitors of the PI3K/AKT/mTOR signaling pathway can be single or dualcomponent (Xu et al., 2020). The molecular pathways in neurodegenerative processes are demonstrated in the form of a flow chart in Figure 12.2.

PROTEIN AGGREGATION AND MISFOLDING

The process by which proteins fold into their native shape is still not entirely understood, despite advancements in machine learning-based protein structure prediction. The concept that polypeptides fold into their physiologically active forms automatically has been increasingly superseded by the knowledge that cellular protein folding frequently needs context-dependent assistance from molecular chaperones to prevent misfolding (Louros et al., 2023).

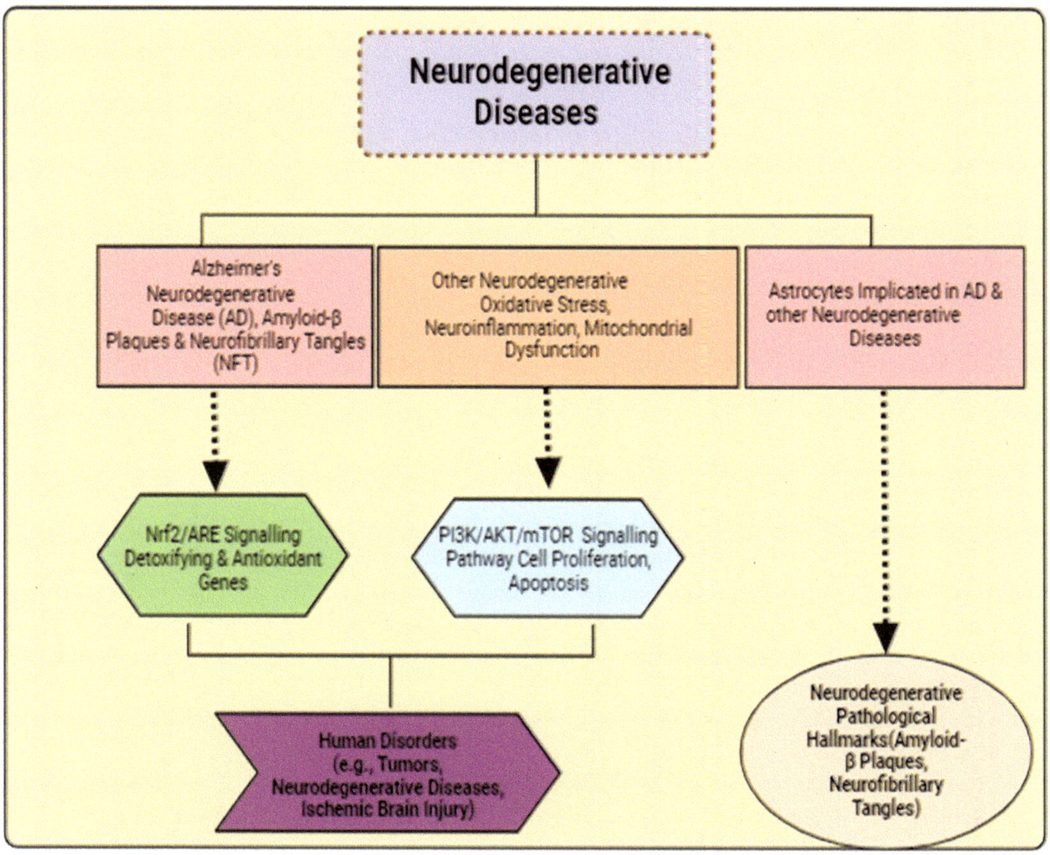

FIGURE 12.2 Molecular pathways in neurodegenerative processes.

The coordinated action of thousands of proteins, each of which must take on a specific structure in order to perform its function, is what allows living things to function as intended. Protein folding is a strictly controlled process that depends on the protein's fundamental sequence. The energetic landscape affects folding because thermodynamic constraints drive a protein to adopt a conformation that reduces the system's free energy, and because chaperone activity aids in the protein's transition to its functional structure (Candelise et al., 2021). Protein aggregates are ultimately formed when partially or misfolded proteins accumulate as a result of disrupted protein folding or diminished protein stability (Housmans et al., 2023).

Basics of Epigenetics

A heritable alteration in gene function that can affect gene expression and subsequently protein expression levels without changing the DNA sequence is known as epigenetics. DNA methylation, histone modifications, non-coding RNA, chromatin structure, and RNA methylation are examples of epigenetic molecular factors. Neurological conditions such as Parkinson's disease, Huntington's disease, and mood disorders (such as anxiety and depression) may be brought on by epigenetic dysregulation (Yu et al., 2021). DNA methylation may be one of the key epigenetic markers that have been studied the most extensively among other epigenetic components. DNA methylation is a common post-replication modification that results in the addition of a methyl group from S-adenosyl methionine to a cytosine in the CpG dinucleotide sequence (Liu et al., 2018). DNA structure is changed by methyl groups when DNA is symmetrically methylated. The primary effect of methylation is that different transcription factors are unable to recognize the DNA, which results in transcription inhibition (Rasmi et al., 2022).

DNA Methylation

One of the main mechanisms of epigenetic variation is DNA methylation (Akhter et al., 2021). One important biological mechanism controlling DNA activity is DNA methylation. It entails introducing a methyl group to either the number-6 nitrogen of the purine ring of adenine or the fifth position of the pyrimidine ring of cytosine, producing 5-methylcytosine (5-MeC). Typically, methylation takes place at the promoter of a gene and prevents gene transcription. This alteration is linked to the etiology of numerous disorders and can be inherited through cell differentiation. Three mechanisms make up DNA methylation: active and passive demethylation, 5-meC recognition, and de novo methylation (Lee et al., 2020).

Histone Modifications

One of the most significant and critical mechanisms of epigenetic regulation is histone modification. Histone proteins that have undergone acetylation, ubiquitination, phosphorylation, or methylation are among the most significant post-translational modifications. The two most prevalent modifications to histone proteins are acetylation and methylation. By modifying non-covalent interactions inside and between nucleosomes, these alterations can modulate the structure of the chromatin. Additionally, they act as docking sites for certain proteins that have distinct domains and may identify these alterations. Histone acetyltransferase (HAT), histone deacetylase (HDAC), histone methyltransferase (HMT), histone demethylase (HDMT), kinases, and E3-ubiquitin are among the enzymes that regulate post-translational histone modification (Neganova et al., 2022).

Numerous biological processes, including chromosomal packing, mitosis, meiosis, apoptosis, transcriptional activation/inactivation, and DNA damage/repair, are impacted by histone alterations (Ramazi et al., 2020). These changes are caused by a variety of histone-modifying enzymes, which can also undo them. As a result, they serve as key mechanisms for cellular adaptation to changing environmental conditions (Cavalieri, 2021).

Non-Coding RNAs

Untranslated RNA molecules known as non-coding RNAs (ncRNAs) control the expression of some genes. ncRNAs comprise piwi-interacting RNAs (piRNAs), transfer RNAs (tRNAs), ribosomal RNAs (rRNAs), small nuclear RNAs (snRNAs), small nucleolar RNAs (snoRNAs), and circular RNAs (cRNAs) (Kumar et al., 2020). The canonical function of miRNAs is to bind to the 3′ untranslated region (UTR) as recognition sites to reduce the stability of messenger RNA (mRNA) for decreasing gene expression. The biogenesis of miRNAs is followed by different steps in the cytoplasm and nucleus (Zandieh et al., 2023).

Nutrient Sensing and Signaling

The primary elements that keep life going are amino acids, fats, and carbohydrates, or macronutrients. Cells must be able to detect changes in these nutrients and react accordingly in order to survive. Thus, nutrient-sensing pathways are established to control metabolic homeostasis, cellular energy, and a variety of biological functions. As such, disruptions in various routes of sensing are linked to a multitude of illnesses, particularly those related to metabolism (Sung et al., 2023). The deep connection between epigenetic control and cellular metabolism has come to light in recent years. But our understanding of how nutrition and diet affect human health through epigenetic mechanisms, and how metabolism contributes to a variety of diseases through epigenetic gene regulation and heredity, is still developing (Bartke and Schneider, 2020). For cells to identify stressors and, ultimately, to survive, they depend on nutrition sensing. Cellular senescence and an imbalance in metabolic homeostasis complement each other, because nutrient availability and perception are critical building blocks for sustaining cell growth and appropriate function (Zhang et al., 2023).

EPIGENETIC MODULATION BY DIETARY COMPONENTS

The effects of nutritional phenolic compounds on human health have garnered a lot of attention during the past 30 years. Even with more evidence at hand, different conclusions are still reached, necessitating more research. These days, research on phenolic chemicals found in food is also conducted with an eye toward how they might affect epigenetic pathways (Číž et al., 2020). Curcumin, resveratrol, catechins, quercetin, genistein, sulforaphane, epigallocatechin-3-gallate, alkaloids, vitamins, and peptides are the principal bioactive substances found in food. Additionally, the host epigenome is significantly impacted by bioactive chemicals that are generated during fermentation by the action of bacteria (Sedley, 2021; Kumari et al., 2022). The makeup and function of the gut microbiota are significantly influenced by the consumption of particular dietary components, including dietary fat, fiber, glucosinolates, and polyphenols. These food ingredients' microbial metabolism controls crucial epigenetic processes, which in turn affect host health (Shock et al., 2021). It is undeniable that a healthy diet with a relative richness of nutrients has an impact on brain function. A growing body of research indicates that nutritional supplementation can help prevent and treat neurologic symptoms in a variety of pathological disorders (Mao et al., 2021).

Ginkgo biloba

Ginkgo biloba L., a member of the Ginkgoaceae family with no near living relatives, is one of the most unusual plants to have ever grown on the globe. The most significant mechanisms underlying these actions include modifying the production of ROS at elevated levels, blocking the activation of NADPH oxidases, modifying the expression of antioxidant enzymes, downregulating AP-1 and MAPKs (p38 MAPK, ERK, and JNK), raising cAMP, deactivating Stat5, and triggering the AMP-activated protein kinase (AMPK) signaling pathway (Mohammadi Zonouz et al., 2024). Human neurodegenerative diseases like Parkinson's and Alzheimer's are primarily caused by the disruption of cerebral redox equilibrium. 1MeTIQ has an advantage over the other chemicals since it is an endogenous amine that is produced in the brains of humans and animals (Wąsik and Antkiewicz-Michaluk, 2017). Inflammatory and/or oxidative insults are frequently involved in the neurodegeneration seen in these diseases (Sunkaria et al., 2017).

Curcumin

Curcumin has drawn attention as a potential treatment for AD since it is an essential part of traditional medicine in many cultures across the world. Because of curcumin's strong anti-inflammatory and antioxidant characteristics, recent research suggests that it may have therapeutic promise in neurodegenerative diseases. It has been shown that curcumin and its derivatives can alter cellular pathways through epigenetic processes (Abdul-Rahman et al., 2024). Polyphenols have the ability to counteract the effects of oxidative stress, inflammation, and apoptosis by modifying several cellular activities. As a result, they play a significant role in neuroprotection. It's interesting to note that neurodegeneration is impacted by all of these pathways (Di Meo et al., 2020). Curcumin's ability to chelate and reduce iron may be the basis for its antiapoptotic actions, according to research by Xi-Xun Du and colleagues. The aging process and chronic neurodegenerative disorders like Alzheimer's and Parkinson's disease both cause neuronal death due to the cytotoxic effects of ROS. But ROS also regulate neurotrophic factors, which in turn serve as signaling molecules that underlie physiological functions like cell migration, proliferation, and survival (Berry et al., 2021).

Bacopa monnieri

The herb *Bacopa monnieri* (BM) has been shown to provide potential benefits for brain health. These benefits include active components such as bacosides A and B, betulinic acid, loliolide, asiatic acid, and quercetin. Few studies have been done on the potential therapeutic uses of BM in neurodegenerative diseases. Projecting BM's helpful effect in brain illnesses is the goal of this systematic review. Antiapoptotic and antioxidant in nature, BM can enhance neuroprotection, improve nerve transmission, activate kinase activity, heal injured neurons, and restore synaptic function (Valotto Neto et al., 2024).

Purusothaman et al. (2021) provide brief information about *B. monnieri*, along with a discussion of the medications and nutraceuticals that are frequently used to improve cognition and their methods of action. Afterward, *B. monnieri's* impact on a number of signal transduction-related parameters, including neurotransmitters, receptors, the second messenger system, gene expression, and neuronal structural components, is examined.

Withania somnifera

Withania somnifera (*W. somnifera*) affects hyperthermia, memory loss, neurotoxicity, and neuroinflammation brought on by MDMA. Mice were given a vehicle or *W. somnifera* extract (WSE) pretreatment for three days. Following that, four groups of mice—one each for saline, WSE, MDMA alone, and WSE plus MDMA—were randomly assigned to receive either vehicle or WSE treatment (Costa et al., 2023). *W. somnifera* is a naturally occurring plant whose phytoconstituents, which include antioxidant and anti-inflammatory qualities, have been shown to have a variety of medical uses, including the treatment of AD. Root extract from *W. somnifera* has been proposed as a potential chemotherapeutic treatment against the host oncoprotein gankyrin, whose expression was changed by *Helicobacter pylori* and the changing cellular milieu associated with Epstein–Barr virus (Kashyap et al., 2023).

Resveratrol

In a variety of organisms, the dietary polyphenol resveratrol produces neuroprotection through its antioxidant and cytoprotective properties. ROS buildup, which has been suggested as the reason for aging, leads to oxidative stress (OS). DNA damage is one of the most detrimental impacts of ROS in the cell. However, there is also data showing that OS can result in various biological alterations, including apoptosis, inflammation, mitochondrial malfunction, and epigenetic abnormalities. It is noteworthy that resveratrol exhibits strong antioxidant properties and pleiotropic effects, operating via many molecular mechanisms (Griñán-Ferré et al., 2021).

A natural phytoalexin found in peanuts, grapes, and their derivatives, resveratrol (RS (3,5,4′-trihydroxy-trans-stilbene) has many advantageous properties, including the ability to induce neural plasticity and reduce inflammation, as well as alter epigenetic modifications. Additionally, resveratrol provides neuroprotection in a number of neuropathological circumstances. Resveratrol improves cognitive decline in the senescence-accelerated mouse prone 8 (SAMP8) and its progeny, as previous research from our group has shown, by modulating important pathways for proper neural function. Though there is a lack of full explanation for the pleiotropic mechanism of action of resveratrol, mounting data from animal models highlight the generation-to-generation health benefits of polyphenols like resveratrol (Izquierdo et al., 2021). Resveratrol is easily absorbed orally and is excreted in the urine and feces after metabolism. Many studies have shown that Resveratrol has antioxidative, anticancer, and cardiovascular protective effects (Huang et al., 2021).

Omega-3 Fatty Acids

It has been discovered that ω-3 polyunsaturated fatty acids (PUFAs) have numerous benefits, including neuroprotective ones (Ceccarini et al., 2022). An important factor in the pathophysiology of age-related brain disorders is inflammation. Therefore, it makes sense to focus on inflammation in order to prevent disease. Because of their anti-inflammatory properties, long-chain omega-3 polyunsaturated fatty acids (LCPUFAs), such as eicosapentaenoic acid (EPA) and docosahexaenoic acid (DHA), are important for maintaining human health and providing protection. Increased glutamatergic and dopaminergic synaptic activity and improved signal transduction are caused by higher DHA incorporation into the synaptic membrane (Boga and Basak, 2023).

Omega-3 PUFAs are essential for the fluidity and permeability of cell membranes. The majority of the phospholipid in the cell membrane is esterified to omega-3 PUFAs. Following neurotransmitter receptor-mediated activation of particular phospholipase A2 (PLA2) enzymes, omega-3 PUFAs are released from the membrane and can function as secondary messengers to modulate signal transmission, either directly or indirectly through their bioactive derivatives (Zhou et al., 2022).

Coenzyme Q10

A molecule that resembles a vitamin, coenzyme Q10 (CoQ10) is produced in the majority of bodily tissues. CoQ10 is a potent lipid-soluble antioxidant that guards cell membranes against oxidative damage caused by free radicals. It also plays a crucial role in the process of cellular energy supply via oxidative phosphorylation within mitochondria, transporting electrons from complexes I and II to complex III of the mitochondrial respiratory chain (Mantle et al., 2021).

As the third most popular dietary supplement, CoQ10 may one day be used to treat a number of noncommunicable disorders (Arenas-Jal et al., 2020). The majority of ROS are formed in the electron transport chain (ETC), which has CoQ10 as a key cofactor. CoQ10 is essential for moving electrons between complexes I, II, and III inside the ETC. It is acknowledged to possess antioxidant properties as well. Therefore, the placement of CoQ10 positions it as a critical antioxidant to prevent oxidative damage when a mitochondrial shock disrupts ETC function. On the other hand, persistently high levels of oxidative stress might cause CoQ10 depletion and hinder proper ROS elimination (Zimmerman et al., 2021).

Alpha-Lipoic Acid

There are two forms of alpha-lipoic acid (α-LA): reduced (di-thiol: dihydro-lipoic acid, DHLA) and oxidized (disulfide, LA). Lipoic acid (C8H14O2S2) is also known as thioctic acid. With its strong antioxidant properties, α-LA exhibits great promise in the treatment of neurodegenerative illnesses (Khan et al., 2022). Examples of its antioxidant influence include recycling, metal chelation, direct radical scavenging, the regeneration of endogenous antioxidants, and the modulation of the expression of several antioxidant and anti-inflammatory genes. The intramolecular disulfide's five-membered ring's stretched conformation gives α-LA radioprotective and antioxidant properties, which accounts for its comparatively high scavenging activity (Abdelkader et al., 2022). Furthermore, LA may have a function as a metal chelator. Additionally, it replenishes intracellular glutathione and ascorbic acid through nonenzymatic regeneration in the event that they are depleted due to senescence, environmental contaminants, or toxicants (Andreeva-Gateva et al., 2020).

Ginseng

The most active component of ginseng, ginsenosides, is also a secondary metabolite of triterpene saponins and steroid glycosides. Studies have shown that ginsenosides have several protective benefits for the CNS. Ginsenosides may have neuroprotective effects via the Akt/CREB/BDNF pathway in the central nervous system. In the current investigation, ginsenosides and related nanoformulations target the Akt/CREB/BDNF pathway in hopes of achieving possible neuroprotective benefits (Zarneshan et al., 2022).

Dubbed "the king of herbs," ginseng, which is the root of *Panax ginseng* Meyer C.A., Araliaceae (sometimes called Asian ginseng), is a perennial plant that is extensively used in traditional Chinese medicine. It is mostly found in China, Korea, and Russia (Feng et al., 2022). Protopanaxadiol (PPD) and protopanaxatriol (PPT) are the two subtypes of ginsenosides (Smith et al., 2014). PPD compounds are not readily absorbed by the body, yet they have a variety of biological characteristics and neuroprotective actions. After ginsenoside is metabolized by the intestinal flora, ginsenoside compound K, the primary metabolite of ginsenosides in vivo, is produced (Li et al., 2023).

Astaxanthin

One of the marine carotenoids, astaxanthin, was first extracted from a lobster by Kuhn and Sorensen. While astaxanthin is found throughout nature, it is most noticeable in a few marine creatures, such as shrimp, crayfish, and salmonids, where it appears as a reddish-orange pigment. Even though bacteria, plants, and microalgae can also manufacture astaxanthin, it is known that *Haematococcus pluvialis* chlorophyte algae have the greatest capacity to accumulate astaxanthin (Bahbah et al., 2021). Mitogen-activated protein kinase (MAPK)/extracellular signal-regulated protein kinase (ERK) and phosphoinositide 3-kinase (PI3K)/protein kinase B (Akt) pathways are both triggered by astaxanthin. Nuclear factor erythroid 2-related factor 2 (Nrf2) and Kelch-like ECH-associated protein 1 can be dissociated more easily because of these two routes. When Nrf2 enters the nucleus, the Nrf2 antioxidant response element (ARE) signaling pathway is triggered (Si and Zhu, 2022). Because of its strong cell-reinforcing abilities, astaxanthin has become a prominent research topic. Astaxanthin influences ROS levels, which in turn inhibits oxidative stress-induced cell damage and death, lessens DNA damage and aggravation, and sustains and increases mitochondrial activity in neurons. Furthermore, a select few remarkable substances, such as astaxanthin, have the ability to pass through the blood-brain barrier and shield neurons from harm and death (Liu et al., 2023b).

Huperzine A

Huperzine A (HupA) has been isolated from *Lycopodium serratum*, a herb that is commonly utilized in traditional Chinese medicine. Huperzine A has been demonstrated to reduce oxidative stress and enhance cognitive performance in older people. Because oxidative stress disrupts cholinergic neurotransmission, it may hasten cognitive aging (Yang et al., 2020). As such, huperzines A and B are commonly used as natural moieties in the synthesis of stronger AChE inhibitors (AChEIs). Furthermore, novel AChEIs based on the 4-aminoquinoline substructures of tacrine and huperzine A with multiple substituents have been proposed as potential AChEIs. Additionally, a number of heterodimers including huperzine A, pyridone, and donepezil dimethoxyindanone linked by an alternative methylene linker have been identified as AChEIs with potential utility in the treatment of AD (Islam et al., 2022).

Melatonin

The receptors, MT1 (Mel1a, MTNR1A) and MT2 (Mel1b, MTNR1B), are activated by melatonin. Human chromosomes 4 and 11 contain genes that encode the G protein-coupled MT1 and MT2 receptors. In particular, neurodegenerative illnesses like Parkinson's and Alzheimer's disease are protected by melatonin's neuroprotective effects on the CNS. It also guards against ischemic harm to the brain. The primary mechanisms responsible for melatonin's neuroprotective effects are its anti-inflammatory, antiapoptotic, and antioxidant characteristics (Leung et al., 2020; Alamdari et al., 2021).

It is critical to continuously investigate the known and potential effects of melatonin on neuronal and glial transcription factors, as these processes are crucial in determining the pathophysiology, course, and prognosis of neurodegenerative illnesses. These processes include neuronal survival and differentiation (Onaolapo et al., 2020). Melatonin administration is greatly appreciated for maintaining the best possible health in the senior population, most likely against potential neurological conditions (Onaolapo et al., 2020).

L-Theanine

A unique amino acid called L-theanine, or LTN, is present in drinks made from the *Camellia sinensis* L. plant, including black and green teas. It has been reported to have positive benefits on health, including activating pathways that are anticancer, anti-inflammatory, and antioxidant; reducing stress disorders; elevating mood; exhibiting neuroprotective effects; and having antidepressant qualities. It lowers blood levels of glucose and insulin by suppressing the expression of genes linked to the small intestine's glucose transporters (GLUTs) and amino acid absorption (Tekmeh et al., 2023). Catechins have gained interest recently in the realm of neurodegenerative illnesses, such as multiple sclerosis, the most common cause of lifelong non-traumatic impairment in young people, and Huntington's disease, with an estimated prevalence of 2.7–5.7/100,000 (Sebastiani et al., 2021). The main catechins in green tea have been shown to be epicatechin, epicatechin gallate, epigallocatechin, and epigallocatechin gallate (Malar et al., 2020).

Quercetin

When compared to the other phytochemicals, quercetin (3,3,4,5,7-pentahydroxyflavone) is more prevalent. Five OH groups can be found in quercetin at positions 3, 3′, 5, 7, and 4′. Since quercetin is an aglycone, its glycoside is created when a glycosyl group is added at position 3, replacing the OH group (Grewal et al., 2021). It has been demonstrated that quercetin possesses neuroprotective and anti-inflammatory qualities. Additionally, it can activate AMPK, a cellular energy sensor that regulates oxidative stress and inflammation. In treating neurodegenerative illnesses and other brain disorders, quercetin shows promise as a safe and efficacious supplementary therapy since it lowers inflammation and guards against neuroinflammatory damage (Chiang et al., 2023). Nrf2, one of quercetin's most important targets, is able to regulate the synthesis of several cytoprotective enzymes with antioxidative, detoxifying, and neuroprotective properties (Zamanian et al., 2023).

MAPK/ERK PATHWAY AND ITS EPIGENETIC REGULATION

Key signaling pathways known as mitogen-activated protein kinase (MAPK) cascades control a wide range of cellular functions, including stress responses, apoptosis, differentiation, and proliferation. Three primary kinases that phosphorylate and activate downstream proteins are part of the MAPK pathway (Guo et al., 2020b). MAPK signaling is an essential signaling system that is involved in the regulation of glucose absorption by malignant cells and carcinogenesis. Mutations in important molecules involved in MAPK/ERK signaling and deregulation of this pathway are highly prevalent events in diverse human malignancies (Asl et al., 2021).

Proteins that make up the MAPK/ERK signaling cascade have been shown to be upregulated and downregulated in response to mutations in the MAPK/ERK signaling machinery. These findings point to an involvement of MAPK/ERK signaling in both the normal developments of the mature brain and CNS. Therefore, any abnormalities in different MAPK/ERK signaling pathway components that cause disruption may lead to changed brain cytoarchitecture (Iroegbu et al., 2021). Pathologies related to neural crest derivatives and mental dysfunctions associated with autism spectrum disorders are caused by genetic abnormalities in any component of the ERK pathway. A crucial component of the neuroinflammatory route that is set off by glial cells in the course of neurodegenerative illnesses like Parkinson's, Alzheimer's, Huntington's, and amyotrophic lateral sclerosis, as well as prion diseases, is the MAPK/ERK pathway (Albert-Gascó et al., 2020).

PI3K/AKT Signaling and Epigenetic Control

One of the most significant signaling pathways engaged in regular cellular functions is the PI3K/AKT/mTOR pathway (Fattahi et al., 2020). A family of lipid kinases known as phosphatidylinositol 3-kinases (PI3Ks) is involved in intracellular signaling and regulates a broad range of cellular functions in almost all normal tissues. PI3K family members are classified into three unique classes (class I–III) according to their substrate selectivity, expression, structure, and regulation. PI3Kα, PI3Kβ, PI3Kγ, and PI3Kδ are the four well-known PI3Ks of the class I PI3K family. These PI3Ks are involved in a remarkable variety of pathways (Ranganna et al., 2020).

An essential signal transduction link connecting extracellular signals and the signal transduction of cellular processes is the PI3K/Akt signaling pathway. It is essential for numerous cellular processes, such as adhesion, migration, invasion, metabolism, and survival. It achieves this in response to various signals, such as G protein-coupled receptor signaling and growth factor receptor tyrosine kinases (RTKs) (Deng and Zhou, 2023).

NRF2/KEAP1 Pathway: Antioxidant Response and Epigenetic Influences

Nucleus erythroid 2–related factor 2/Kelch-like ECH-associated protein 1 (human NRF2/KEAP1 and murine Nrf2/Keap1) signaling has been better understood due to epigenetics (Bhattacharjee and Dashwood, 2020). A key player in mediating the adaptive cellular stress response to oxidative and electrophilic substances is the KEAP1/NRF2/ARE signaling pathway. Over the past 20 years, this canonical pathway has been thoroughly examined and reviewed, but it has rarely been examined from a quantitative signaling standpoint. The robust activation of antioxidant genes to sufficient levels that can effectively counteract the stressors requires ultrasensitivity, or signal amplification (Liu et al., 2022). The 66 kDa protein Nrf2 is made up of seven functionally conserved areas that are referred to as Nrf2-ECH homology domains (Neh). The fundamental leucine zipper motif found in the Neh1 domain allows it to heterodimerize with its transcriptional partner Maf TF and attach to DNA (Ulasov et al., 2022). As an oxidative stress sensor, KEAP1 controls the body's or cell's NRF2 turnover, and NRF2 activation results in the synthesis of compounds that reduce OS by acting as an antioxidant and detoxifier (Polverino et al., 2021).

NUTRITIONAL EPIGENETICS IN NEURONAL PLASTICITY AND COGNITIVE FUNCTION

Synaptic Plasticity and Learning

Nutrition is important because the gut–brain axis, which connects the two organ systems, is very close to the body and starts in utero, as the embryonic germ layer is where both the CNS and the enteric nervous system (ENS) develop (Pizzorusso and Tognini, 2020). Dynamic alterations of epigenetic marks allow extracellular signals to be incorporated at the chromatin level, fine-tuning gene transcription in the process (Saw et al., 2020). The primary immunological defense mechanism in the central nervous system is provided by microglia. PI3K is expressed by microglia and has been demonstrated to be involved in inflammation and synaptic plasticity in neurons (Zhou and Ryan, 2023).

This plasticity of the epigenome enables optimal adaptation and more favorable responses to certain environmental conditions through modifications of gene activity. The best understood and most stable epigenetic mechanism is DNA methylation, which commonly involves the addition of a methyl group at the CpG dinucleotide in the regulatory promoter regions of genes (Creighton et al., 2020).

Epigenetic Regulation of Memory Formation

Current research indicates that age-related memory deterioration may be influenced by changes in the function of the aging brain's epigenome, supporting the acknowledged role of epigenetic control in memory. The aging process is generally acknowledged to modify several epigenetic marks, but the specifics of these modifications and their relationship to memory impairment remain highly disputed (Zovkic et al., 2013).

Epigenetic indicators such as post-translational changes of histone tails and DNA methylation have become prominent memory regulators in the past ten years. The consolidation of long-term memory is supported by their capacity to dynamically control gene transcription in response to neural stimulation (Yuan et al., 2023). Changes in epigenetic modifications lead to aberrant neuroendocrine responses, neurotransmission, impaired neuroplasticity, and dysfunctional neuroglia, all of which are implicated in the pathophysiology of major depressive disorder (MDD).

ROLE OF HERBS AND NUTRIENTS IN COGNITIVE HEALTH

Growing older impairs cognitive function, increasing the risk of dementia and other brain illnesses. The course of cognitive aging and its transition to pathological states is determined by both genetic predisposition and environmental exposures (Howes et al., 2020).

Aside from age-related brain illnesses like dementia and mild cognitive impairment, including Alzheimer's disease, cognitive decline can also occur with normal aging and can be exacerbated by prescription medication therapy. The evaluation takes into account pharmacological (in vitro and in vivo), clinical, and epidemiological data regarding commonly consumed plants and their dietary forms, such as tea, coffee, chocolate, red wine, grapes, citrus fruits, and other fruits, as well as plants that are used less frequently in specific diets (Gutierrez et al., 2021). The effectiveness of dietary supplements was found to be varied; there was little support for PUFAs, vitamin D, some proteins, amino acids, and other forms of supplements, but there was substantial evidence for the advantages of polyphenols and nutritional combinations (Rasheed et al., 2021).

CLINICAL INSIGHTS: MOLECULAR AND EPIGENETIC OUTCOMES

Human Studies on Epigenetic Changes in Neurodegenerative Disorders

Recent advances in the modulation of neuroinflammatory variables in Parkinson's disease have been largely attributed to epigenetic mechanisms, which offer a promising avenue to break this cycle. The primary epigenetic pathways underlying the glial cell-mediated neuroinflammation in Parkinson's disease and its phenotypic change were reviewed in this research (Saul and Kosinsky, 2021). The accumulation of histone variations, adjustments to chromatin accessibility, loss of histones and heterochromatin, aberrant histone modifications, and dysregulated expression/activity of miRNAs are among the epigenetic changes that directly contribute to aging and aging-related disorders (Gupta et al., 2020).

Biomarkers of Epigenetic Modifications in Response to Herbal Interventions

Dietary herbs are recognized to target these epigenetic gene regulation pathways and are rich in a variety of phytochemicals that offer additional nutrition and health benefits. Because of their beneficial effects on health, they have been used since the dawn of human civilization (Han et al., 2023). Dietary flavonoids are naturally occurring epigenetic modulators that aid in the identification of biomarkers for the prevention of diabetes and the creation of complementary therapies. Diabetes and its complications are caused, maintained, and progressed by DNA methylation, histone changes, and non-coding RNAs (Mead et al., 2023).

Challenges and Opportunities in Translating Molecular Findings to Clinical Practice

The main neurodegenerative diseases have made the development of disease-modifying treatments difficult for a number of reasons (Palermo et al., 2020). Degeneration of motor neurons is the cause of the debilitating disease known as ALS. For a variety of reasons, the development of disease-modifying medicines has proven difficult, as it does with all major neurodegenerative conditions. It can make it possible for therapeutic medicines to be successfully translated from preclinical research to clinical benefit for ALS patients (Bai et al., 2023). The validation and standardization of pathophysiological biomarkers for various contexts of use, such as early detection, diagnosis, prognosis, and treatment response prediction, is one of the biggest challenges facing the clinical research of neurodegenerative disorders (Gravandi et al., 2023).

FUTURE TRENDS IN RESEARCH

Unexplored Pathways and Targets

A traditional nuclear transcription factor, nuclear factor erythroid 2-related factor 2 (Nrf2), controls the body's response to oxidative stress. The system's ability to combat oxidative stress is enhanced by Nrf2 activation, which also causes the production of antioxidant proteins (Sharifi-Rad et al., 2020). One emerging tactic for treating neurodegenerative disorders such as Parkinson's disease, Alzheimer's disease, stroke, aging, and other neurodegenerative disorders seems to be targeting the Ras/Raf/MAPKs pathway. Therefore, secondary metabolites found in plants have demonstrated encouraging potential for concurrently modulating the Ras/Raf/MAPKs pathway and are crucial in the development of neurodegenerative disorders (Moradi et al., 2020).

Impact of Omics in Precision-Based Neuroprotective Drugs

The methods for high-throughput biological molecule characterization and quantification have changed in the last few years. Omics is the study of genomes, epigenomics, transcriptomics, proteomics, and metabolomics. These platforms offer opportunities to examine molecular pathway dynamics, function, and changes in biological samples from both healthy and sick states (Steg et al., 2022). The simultaneous examination of hundreds of genes, RNAs, proteins, and metabolites is made possible by the integration of numerous omics data, which reveals networks of relationships between various molecular levels. Integrated multi-omics study will be beneficial in gaining an understanding of etiology, therapeutic target identification, and biomarker development (Silva and Pogacnik, 2020). The scientific fields known as "omics" include a variety of molecular disciplines with the shared goal of characterizing global sets of biological molecules, including proteins, metabolites, DNAs, and RNAs (Khan et al., 2020; Pathak et al., 2022).

Potential Side Effects

Natural micronutrients called polyphenols can be found in a wide variety of food sources. Not only are these compounds strong antioxidants, but they might also have anti-inflammatory qualities. Numerous research endeavors have emphasized their possible function in the mitigation and management of an array of pathological ailments associated with oxidative stress and inflammation, such as cancer, cardiovascular diseases, and neurodegenerative disorders (Bono et al., 2021). The primary concern pertains to the natural chemicals' bioavailability and digestibility within the human body. Another obstacle to using phytochemicals for nervous system-related issues is the blood–brain barrier. Additionally, it takes a lot of time for the therapy to effectively combat the illness (Barkat et al., 2021).

Regulatory Status of Herbal Products

Because individuals with a variety of health issues can tolerate herbal medications that might have high therapeutic values, herbal medications are useful in treating a wide range of illnesses. In herbal medicine, the whole plant, a portion of it, or specific isolated phytoconstituents may be used. The scientific era's quest for novel medications rekindled interest in the discovery of herbal medications derived from various natural resources. Fifty percent of the medications used in the current healthcare system are natural in origin (Thakkar et al., 2020).

The information from traditional medical practices may serve as the foundation for the use of numerous dietary supplements and herbal remedies. Because naturally derived compounds are used widely and are becoming more so globally, herbal therapies require a thorough and rigorous evaluation of their pharmacological properties and safety concerns. The current state of affairs indicates that extensive and thorough study on the safety of herbal medicines is still lacking internationally (Gregory et al., 2021). Herbal and medicinal plant cures are becoming more popular as complementary and alternative therapies, and they can be a great source of potential new drugs. Preclinical and clinical research has assessed a broad range of natural compounds from various sources for their neuroprotective properties in averting and reducing multifactorial disorders (Chen et al., 2021).

Natural substances that have the ability to scavenge free radicals, shield cells from oxidative damage, and have neuroprotective potential have drawn a lot of attention lately (Rahman et al., 2021a). Although there have been significant advances in our understanding of neurodegenerative diseases, there has not been much success in developing effective treatments. However, several clinical concerns have surfaced about their usage, mostly due to the absence of scientific backing or proof of their effectiveness and patient safety. Natural products may offer enormous potential opportunities in the prevention and management of neurodegenerative diseases. Targeting several modes of action and neuroprotective strategies that incorporate averting cell death and reinstating the function of injured neurons have to be utilized, as neurodegeneration is linked to a multitude of pathological processes.

Natural products and their bioactive chemicals have been the subject of extensive research and exploration in recent years due to their biological activity, nutritional values, and possible health and therapeutic advantages. Numerous studies conducted in recent decades have documented the preventive properties of natural products and their bioactive components against a range of illnesses, including cancer, diabetes, heart disease, and neurological disorders. Potential neuroprotective drugs for the treatment of neurodegenerative illnesses have been identified in natural products (Dubey et al., 2022).

CONCLUSION

Neurodegenerative disorders are long-term, incurable illnesses that mostly cause neurons to die off and are marked by high levels of oxidative damage and swelling. The multiplicity of neurodegenerative diseases, each with a unique mechanism and set of symptoms, highlights how difficult it is to create treatments that work for everyone. Herbal extracts and other natural substances can vary greatly in composition, which can have a substantial impact on their pharmacological activity and therapeutic potential. For this reason, quality control and formulation optimization must be carefully considered. Concerns about possible herb–drug interactions, side effects, and contraindications are also raised by the interaction of herbs, natural products, and traditional pharmacotherapies.

Moreover, the regulatory environment pertaining to natural products and herbal supplements continues to be diverse, with differing levels of supervision. In summary, although natural products and herbs have potential as supplemental treatments for neurodegenerative diseases, their broad use depends on resolving important clinical, scientific, and regulatory issues. The chapter primarily summarizes a few significant medicinal plants and their formulations that are efficacious in treating Parkinson's disease and Alzheimer's disease pharmacologically. It is clear that these herbs both lessen the symptoms and slow the spread of the illness. To fully understand the mechanisms of many herbal medications and their formulations, more research is needed.

REFERENCES

Abdelkader NF, El-Batal AI, Amin YM, Hawas AM, Hassan SHM, Eid NI. Neuroprotective effect of gold nanoparticles and alpha-lipoic acid mixture against radiation-induced brain damage in rats. *Int J Mol Sci* (2022) 23(17):9640.

Abdul-Rahman T, Awuah WA, Mikhailova T, Kalmanovich J, Mehta A, Ng JC, Coghlan MA, Zivcevska M, Tedeschi AJ, de Oliveira EC, Kumar A, Cantu-Herrera E, Lyndin M, Sikora K, Alexiou A, Bilgrami AL, Al-Ghamdi KM, Perveen A, Papadakis M, Ashraf GM. Antioxidant, anti-inflammatory and epigenetic potential of curcumin in Alzheimer's disease. *Biofactors* (2024) 50(4):693–708. https://doi.org/10.1002/biof.2039.

Akhtar A, Sah SP. Insulin signaling pathway and related molecules: role in neurodegeneration and Alzheimer's disease. *Neurochem Int* (2020) 135:104707.

Akhter Z, Bi Z, Ali K, Sun C, Fiaz S, Haider FU, Bai J. In response to abiotic stress, DNA methylation confers EpiGenetic changes in plants. *Plants* (2021) 10:1096.

Alcalá-Vida R, Garcia-Forn M, Castany-Pladevall C, Creus-Muncunill J, Ito Y, Blanco E, Golbano A, Crespí-Vázquez K, Parry A, Slater G, Samarajiwa S. Neuron type-specific increase in lamin B1 contributes to nuclear dysfunction in Huntington's disease. *EMBO Mol Medicine* (2021) 13(2):e12105.

Alamdari AF, Rahnemayan S, Rajabi H, Vahed N, Kashani HR, Rezabakhsh A, Sanaie S. Melatonin as a promising modulator of aging related neurodegenerative disorders: role of microRNAs. *Pharmacol Res* (2021) 173:105839.

Albert-Gascó H, Ros-Bernal F, Castillo-Gómez E, Olucha-Bordonau FE. MAP/ERK Signaling in developing cognitive and emotional function and its effect on pathological and neurodegenerative processes. *Int J Mol Sci* (2020) 21:4471.

Andreeva-Gateva P, Traikov L, Sabit Z, Bakalov D, Tafradjiiska-Hadjiolova R. Antioxidant effect of Alpha-Lipoic acid in 6-Hydroxydopamine unilateral intrastriatal injected rats. *Antioxid (Basel)* (2020) 9(2):122.

Arenas-Jal M, Suñé-Negre JM, García-Montoya E. Coenzyme Q10 supplementation: efficacy, safety, and formulation challenges. *Compr Rev Food Sci Food Saf* (2020) 19(2):574–594.

Arrazola Sastre A, Luque Montoro M, Gálvez-Martín P, Lacerda HM, Lucia AM, Llavero F, Zugaza JL. Small GTPases of the ras and rho families switch on/off signaling pathways in neurodegenerative diseases. *Int J Mol Sci* (2020) 21(17):6312.

Asl ER, Amini M, Najafi S, Mansoori B, Mokhtarzadeh A, Mohammadi A, Lotfinejad P, Bagheri M, Shirjang S, Lotfi Z, Rasmi Y, Baradaran B. Interplay between MAPK/ERK signaling pathway and MicroRNAs: a crucial mechanism regulating cancer cell metabolism and tumor progression. *Life Sci* (2021) 278:119499.

Azam S, Haque ME, Balakrishnan R, Kim IS, Choi DK. The ageing brain: molecular and cellular basis of neurodegeneration. *Front Cell Dev Biol* (2021) 9:683459.

Bahbah EI, Ghozy S, Attia MS, Negida A, Emran TB, Mitra S, Albadrani GM, Abdel-Daim MM, Uddin MS, Simal-Gandara J. Molecular mechanisms of Astaxanthin as a potential neurotherapeutic agent. *Mar Drugs* (2021) 19(4):201.

Bai X, Bian Z, Zhang M. Targeting the Nrf2 signaling pathway using phytochemical ingredients: A novel therapeutic road map to combat neurodegenerative diseases. *Phytomed* (2023) 109:154582.

Barkat MA, Goyal A, Barkat HA, Salauddin M, Pottoo FH, Anwer ET. Herbal medicine: clinical perspective and regulatory status. *Comb Chem High Throughput Screen* (2021) 24(10):1573–1582.

Bartke T, Schneider R. You are what you eat–How nutrition and metabolism shape the genome through epigenetics. *Mol Metab* (2020) 38:100987.

Berry A, Collacchi B, Masella R, Varì R, Cirulli F. Curcuma Longa, the "Golden Spice" to counteract neuroinflammaging and cognitive decline—what have we learned and what needs to be done. *Nutrients* (2021) 13(5):1519.

Bertogliat MJ, Morris-Blanco KC, Vemuganti R. Epigenetic mechanisms of neurodegenerative diseases and acute brain injury. *Neurochem Int* (2020) 133:104642.

Bhat BA, Almilaibary A, Mir RA, Aljarallah BM, Mir WR, Ahmad F, Mir MA. Natural therapeutics in aid of treating alzheimer's disease: a green gateway toward ending quest for treating neurological disorders. *Front Neurosci* (2022) 16:884345.

Bhattacharjee S, Dashwood RH. Epigenetic regulation of NRF2/KEAP1 by phytochemicals. *Antioxid* (2020) 9(9):865.

Bianchi VE, Herrera PF, Laura R. Effect of nutrition on neurodegenerative diseases. A systematic review. *Nutr Neurosci* (2021) 24(10):810–834.

Birajdar SV, Mulchandani M, Mazahir F, Yadav AK. Dementia and neurodegenerative disorder: an introduction. In *Nanomedicine-based approaches for the treatment of dementia*, ed U Gupta, P Kesharwani (2023) Jan 1 (pp. 1–36). Academic Press.

Boga NS, Basak S. Omega-3 fatty acids and ageing brain. In Evidence-based functional foods for prevention of age-related diseases, ed S Pathak, A Banerjee, AK Duttaroy (2023) Jul 7 (pp. 101–128). Springer Nature Singapore.

Bono S, Feligioni M, Corbo M. Impaired antioxidant KEAP1-NRF2 system in amyotrophic lateral sclerosis: NRF2 activation as a potential therapeutic strategy. *Mol Neurodegen* (2021) 16:1–26.

Bordoni M, Scarian E, Rey F, Gagliardi S, Carelli S, Pansarasa O, Cereda C. Biomaterials in neurodegenerative disorders: a promising therapeutic approach. *Int J Mol Sci* (2020) 21:3243.

Candelise N, Scaricamazza S, Salvatori I, Ferri A, Valle C, Manganelli V, Garofalo T, Sorice M, Misasi R. Protein aggregation landscape in neurodegenerative diseases: clinical relevance and future applications. *Int J Mol Sci* (2021) 22:6016.

Cao Q, Liu J, Dong P, Han H. Different ratios of *Schisandra chinensis* and *Evodia rutaecarpa* combination for treating Alzheimer's disease. *Exp Biol Med* (Maywood) (2023) 248(20):1877–1886.

Cavalieri V. The expanding constellation of histone post-translational modifications in the epigenetic landscape. *Genes* (2021) 12(10):1596.

Ceccarini MR, Ceccarelli V, Codini M, Fettucciari K, Calvitti M, Cataldi S, Albi E, Vecchini A, Beccari T. The -polyunsaturated fatty acid EPA, but not DHA, enhances neurotrophic factor expression through epigenetic mechanisms and protects against Parkinsonian neuronal cell death. *Int J Mol Sci* (2022) 23:16176.

Chaki J, Marcin Woźniak, Deep learning for neurodegenerative disorder (2016 to 2022): a systematic review. *Biomed Signal Process Control* (2023) 80(1):104223.

Chen X, Drew J, Berney W, Lei W. Neuroprotective Natural Products for Alzheimer's Disease. Cells (2021) 10(6):1309.

Chiang M-C, Tsai T-Y, Wang C-J. The potential benefits of Quercetin for brain health: a review of anti-inflammatory and neuroprotective mechanisms. *Int J Mol Sci* (2023) 24:6328.

Číž M, Dvořáková A, Skočková V, Kubala L. The role of dietary Phenolic compounds in epigenetic modulation involved in inflammatory processes. *Antioxid* (2020) 9:691.

Costa G, Serra M, Maccioni R, Casu MA, Kasture SB, Acquas E, Morelli M. Withania somnifera influences MDMA-induced hyperthermic, cognitive, neurotoxic and neuroinflammatory effects in mice. *Biomed Pharmacother* (2023) 161:114475.

Creighton SD, Stefanelli G, Reda A, Zovkic IB. Epigenetic mechanisms of learning and memory: implications for aging. *Int J Mol Sci* (2020) 21:6918.

Deneubourg C, Ramm M, Smith LJ, Baron O, Singh K, Byrne SC, Duchen MR, Gautel M, Eskelinen EL, Fanto M, Jungbluth H. The spectrum of neurodevelopmental, neuromuscular and neurodegenerative disorders due to defective autophagy. *Autophagy* (2022) 18(3):496–517.

Deng RM, Zhou J. The role of PI3K/AKT signaling pathway in myocardial ischemia-reperfusion injury. *Int Immunopharmacol* (2023) 123:110714.

Di Meo F, Valentino A, Petillo O, Peluso G, Filosa S, Crispi S. Bioactive Polyphenols and neuromodulation: molecular mechanisms in neurodegeneration. *Int J Mol Sci* (2020) 21:2564.

Dubey A, Ghosh NS, Agnihotri N, Kumar A, Pandey M, Nishad S. Herbs derived bioactive compounds and their potential for the treatment of neurological disorders. *Clin Schizophr Relat Psychoses* (2022) 16(2):1–11 Doi: 10.3371/CSRP.DANG.081922.

Dumurgier J, Tzourio C. Epidemiology of neurological diseases in older adults. *Revue Neurologique* (2020) 176(9):642–648.

Fattahi S, Amjadi-Moheb F, Tabaripour R, Ashrafi GH, Akhavan-Niaki H. PI3K/AKT/mTOR signaling in gastric cancer: epigenetics and beyond. *Life Sci* (2020) 262:118513.

Feng H, Xue M, Deng H, Cheng S, Hu Y, Zhou C. Ginsenoside and its therapeutic potential for cognitive impairment. *Biomolecules* (2022) 12(9):1310.

Ghosh P, Saadat A. Neurodegeneration and epigenetics: a review. *Neurología* (English Edition) (2023) 38(6):e62–e68.

Gravandi MM, Abdian S, Tahvilian M, Iranpanah A, Moradi SZ, Fakhri S, Echeverría J. Therapeutic targeting of Ras/Raf/MAPK pathway by natural products: a systematic and mechanistic approach for neurodegeneration. *Phytomed* (2023) 115:154821.

Gregory J, Vengalasetti YV, Bredesen DE, Rao RV. Neuroprotective herbs for the management of Alzheimer's disease. *Biomolecules* (2021) 11(4):543.

Grewal AK, Singh TG, Sharma D, Sharma V, Singh M, Rahman MH, Najda A, Walasek-Janusz M, Kamel M, Albadrani GM, Akhtar MF, Saleem A, Abdel-Daim MM. Mechanistic insights and perspectives involved in neuroprotective action of quercetin. *Biomed Pharmacother* (2021) 140:111729.

Grinán-Ferré C, Bellver-Sanchis A, Izquierdo V, Corpas R, Roig-Soriano J, Chillón M, Andres-Lacueva C, Somogyvári M, Sőti C, Sanfeliu C, Pallas M. The pleiotropic neuroprotective effects of resveratrol in cognitive decline and Alzheimer's disease pathology: from antioxidant to epigenetic therapy. *Ageing Res Rev* (2021) 67:101271.

Guo T, Zhang D, Zeng Y, Huang TY, Xu H, Zhao Y. Molecular and cellular mechanisms underlying the pathogenesis of Alzheimer's disease. *Mol Neurodegener* (2020a) 15:1–37.

Guo YJ, Pan WW, Liu SB, Shen ZF, Xu Y, Hu LL. ERK/MAPK signalling pathway and tumorigenesis. *Exp Ther Med* (2020b) 19(3):1997–2007.

Gupta J, Sharma S, Sharma NR, Kabra D. Phytochemicals enriched in spices: a source of natural epigenetic therapy. *Arch Pharm Res* (2020) 43(2):171–186.

Gutierrez L, Folch A, Rojas M, Cantero JL, Atienza M, Folch J, Camins A, Ruiz A, Papandreou C, Bulló M. Effects of nutrition on cognitive function in adults with or without cognitive impairment: a systematic review of randomized controlled clinical trials. *Nutrients* (2021) 13:3728.

Guzman-Martinez L, Maccioni RB, Andrade V, Navarrete LP, Pastor MG, Ramos-Escobar N. Neuroinflammation as a common feature of neurodegenerative disorders. *Front Pharmacol* (2019) 10:1008.

Hajizadeh Tekmeh H, Vanizor Kural B, Kör S, Arıkan Malkoç M, Yuluğ E, Kutlu A, Abidin İ, Orem A. How does l-theanine treatment affect the levels of serum and hippocampal BDNF, insulin and adipocytokines in diabetic rats? *Biochem Biophys Res Commun* (2023) 667:95–103.

Han S, Luo Y, Liu B, Guo T, Qin D, Luo F. Dietary flavonoids prevent diabetes through epigenetic regulation: advance and challenge. *Crit Rev Food Sci Nutr* (2023) 63(33):11925–11941.

Houldsworth A. Role of oxidative stress in neurodegenerative disorders: a review of reactive oxygen species and prevention by antioxidants. *Brain Commun* (2024) 6(1):fcad356.

Housmans JA, Wu G, Schymkowitz J, Rousseau F. A guide to studying protein aggregation. *FEBS J* (2023) 290(3):554–583.

Howes MR, Perry NSL, Vásquez-Londoño C, Perry EK. Role of phytochemicals as nutraceuticals for cognitive functions affected in ageing. *Br J Pharmacol* (2020) 177(6):1294–1315.

Huang J, Huang N, Xu S, Luo Y, Li Y, Jin H, Yu C, Shi J, Jin F. Signaling mechanisms underlying inhibition of neuroinflammation by resveratrol in neurodegenerative diseases. *J Nutr Biochem* (2021) 88:108552.

Iroegbu JD, Ijomone OK, Femi-Akinlosotu OM, Ijomone OM. ERK/MAPK signalling in the developing brain: Perturbations and consequences. *Neurosci Biobehav Rev* (2021) 131:792–805.

Islam F, Khadija JF, Harun-Or-Rashid M, Rahaman MS, Nafady MH, Islam MR, Akter A, Emran TB, Wilairatana P, Mubarak MS. Bioactive compounds and their derivatives: an insight into prospective phytotherapeutic approach against Alzheimer's disease. *Oxid Med Cell Longev* (2022) 2022:5100904.

Izquierdo V, Palomera-Ávalos V, Pallàs M, Griñán-Ferré C. Resveratrol supplementation attenuates cognitive and molecular alterations under maternal high-fat diet intake: epigenetic inheritance over generations. *Int J Mol Sci* (2021) 22(3):1453.

Kabir MT, Rahman MH, Shah M, Jamiruddin MR, Basak D, Al-Harrasi A, Bhatia S, Ashraf GM, Najda A, El-Kott AF, Mohamed HRH, Al-Malky HS, Germoush MO, Altyar AE, Alwafai EB, Ghaboura N, Abdel-Daim MM. Therapeutic promise of carotenoids as antioxidants and anti-inflammatory agents in neurodegenerative disorders. *Biomed Pharmacother* (2022) 146:112610.

Kampmann M. CRISPR-based functional genomics for neurological disease. *Nat Rev Neurol* (2020) 16:465–480.

Kashyap D, Roy R, Varshney N, Baral B, Bagde PH, Kandpal M, Kumar S, Kar P, Jha HC. *Withania somnifera* extract reduces gastric cancerous properties through inhibition of gankyrin in cellular milieu produced by *Helicobacter pylori* and Epstein Barr virus. *J Biomol Struct Dyn* (2023) 1–17.

Khan A, Jahan S, Imtiyaz Z, Alshahrani S, Antar Makeen H, Mohammed Alshehri B, Kumar A, Arafah A, Rehman MU. Neuroprotection: targeting multiple pathways by naturally occurring phytochemicals. *Biomedicines* (2020) 8(8):284.

Khan H, Singh TG, Dahiya RS, Abdel-Daim MM. α-Lipoic Acid, an Organosulfur Biomolecule a novel therapeutic agent for neurodegenerative disorders: an mechanistic perspective. *Neurochem Res* (2022) 47(7):1853–1864.

Kovacs GG. Concepts and classification of neurodegenerative diseases. In *Handbook of clinical neurology, ed* Gabor G. and KovacsIrina Alafuzoff (2018) 145: (pp.301–307). Elsevier.

Kumar P, Banik SP, Goel A, Chakraborty S, Bagchi M, Bagchi D. Revisiting the multifaceted therapeutic potential of Withaferin A (WA), a novel steroidal Lactone, W-ferinAmax Ashwagandha, from *Withania Somnifera* (L) Dunal. *J Am Nutr Assoc* (2024) 43(2):115–130.

Kumar S, Gonzalez EA, Rameshwar P, Etchegaray JP. Non-coding RNAs as mediators of epigenetic changes in malignancies. *Cancers (Basel)* (2020) 12(12):3657.

Kumari A, Bhawal S, Kapila S, Yadav H, Kapila R. Health-promoting role of dietary bioactive compounds through epigenetic modulations: a novel prophylactic and therapeutic approach. *Crit Rev Food Sci Nutr* (2022) 62(3):619–639.

Lam S, Bayraktar A, Zhang C, Turkez H, Nielsen J, Boren J, Shoaie S, Uhlen M, Mardinoglu A. A systems biology approach for studying neurodegenerative diseases. *Drug Discov Today* (2020) 25(7):1146–1159.

Lamptey RNL, Chaulagain B, Trivedi R, Gothwal A, Layek B, Singh J. A review of the common deurodegenerative disorders: current therapeutic approaches and the potential role of nanotherapeutics. *Int J Mol Sci* (2022) 23(3):1851.

Lee HT, Oh S, Ro DH, Yoo H, Kwon YW. The key role of DNA Methylation and Histone Acetylation in epigenetics of Atherosclerosis. *J Lipid Atheroscler* (2020) 9(3):419–434.

Leuci R, Brunetti L, Poliseno V, Laghezza A, Loiodice F, Tortorella P, Piemontese L. Natural compounds for the prevention and treatment of cardiovascular and neurodegenerative diseases. *Foods* (2020) 10(1):29.

Leung JW, Cheung KK, Ngai SP, Tsang HW, Lau BW. Protective effects of Melatonin on neurogenesis impairment in neurological disorders and its relevant molecular mechanisms. *Int J Mol Sci* (2020) 21(16):5645.

Levenson RW, Sturm VE, Haase CM. Emotional and behavioral symptoms in neurodegenerative disease: a model for studying the neural bases of psychopathology. *Annu Rev Clin Psychol* (2014) 10:58--606.

Li N, Pang Q, Zhang Y, Lin J, Li H, Li Z, Liu Y, Fang X, An Y, Bai H, Li D, Cao Z, Liu J, Yang Q, Hu S. Ginsenoside compound K reduces neuronal damage and improves neuronal synaptic dysfunction by targeting Aβ. *Front Pharmacol* (2023) 14:1103012.

Liu DD, Zhang CY, Zhang JT, Gu LM, Xu GT, Zhang JF. Epigenetic modifications and metabolic memory in diabetic retinopathy: beyond the surface. *Neural Regen Res* (2023a) 18(7):1441–1449.

Liu J, Zhang W, Wu Z, Dai L, Koji T. Changes in DNA Methylation of Oocytes and Granulosa Cells Assessed by HELMET during Folliculogenesis in Mouse Ovary. *Acta Histochem Cytochem* (2018) 51(2):93–100.

Liu N, Lyu X, Zhang X, Zhang F, Chen Y, Li G. Astaxanthin attenuates cognitive deficits in Alzheimer's disease models by reducing oxidative stress via the SIRT1/PGC-1α signaling pathway. *Cell Biosci* (2023b) 13(1):173.

Liu S, Pi J, Zhang Q. Signal amplification in the KEAP1-NRF2-ARE antioxidant response pathway. *Redox Biol* (2022) 54:102389.

Louros N, Schymkowitz J, Rousseau F. Mechanisms and pathology of protein misfolding and aggregation. *Nat Rev Mol Cell Biol* (2023) 24(12):912–933.

Luo F, Sandhu AF, Rungratanawanich W, Williams GE, Akbar M, Zhou S, Song BJ, Wang X. Melatonin and autophagy in aging-related neurodegenerative diseases. *Int J Mol Sci* (2020) 21(19):7174.

Luthra R, Roy A. Role of medicinal plants against neurodegenerative diseases. *Curr Pharm Biotechnol* (2022) 23(1):123–139.

Makkar R, Behl T, Bungau S, Zengin G, Mehta V, Kumar A, Uddin MS, Ashraf GM, Abdel-Daim MM, Arora S, Oancea R. Nutraceuticals in neurological disorders. *Int J Mol Sci* (2020) 21(12):4424.

Malar DS, Prasanth MI, Brimson JM, Sharika R, Sivamaruthi BS, Chaiyasut C, Tencomnao T. Neuroprotective properties of green tea (*Camellia sinensis*) in Parkinson's disease: a review. *Molecules* (2020) 25(17):3926.

Mantle D, Heaton RA, Hargreaves IP. Coenzyme Q10, Ageing and the nervous system: an overview. *Antioxid (Basel)* (2021) 11(1):2.

Mao XY, Yin XX, Guan QW, Xia QX, Yang N, Zhou HH, Liu ZQ, Jin WL. Dietary nutrition for neurological disease therapy: Current status and future directions. *Pharmacol Ther* (2021) 226:107861.

Martínez-Iglesias O, Naidoo V, Cacabelos N, Cacabelos R. Epigenetic biomarkers as diagnostic tools for neurodegenerative disorders. *Int J Mol Sci* (2021) 23(1):13.

Mead RJ, Shan N, Reiser HJ, Marshall F, Shaw PJ. Amyotrophic lateral sclerosis: a neurodegenerative disorder poised for successful therapeutic translation. *Nat Rev Drug Discov* (2023) 22(3):185–212.

Mohammadi Zonouz A, Ghasemzadeh Rahbardar M, Hosseinzadeh H. The molecular mechanisms of ginkgo (Ginkgo biloba) activity in signaling pathways: a comprehensive review. *Phytomed* (2024) 126:155352.

Mohd Murshid N, Aminullah Lubis F, Makpol S. Epigenetic changes and its intervention in age-related neurodegenerative diseases. *Cell Mol Neurobiol* (2022) 42(3):577–595.

Mor A, Tankiewicz-Kwedlo A, Krupa A, Pawlak D. Role of Kynurenine pathway in oxidative stress during neurodegenerative disorders. *Cells* (2021) 10(7):1603.

Moradi SZ, Momtaz S, Bayrami Z, Farzaei MH, Abdollahi M. Nanoformulations of herbal extracts in treatment of neurodegenerative disorders. *Front Bioeng Biotechnol* (2020) 8:238.

Neganova ME, Klochkov SG, Aleksandrova YR, Aliev G. Histone modifications in epigenetic regulation of cancer: perspectives and achieved progress. *Semin Cancer Biol* (2022) 83:452–471.

Olajide OJ, Suvanto ME, Chapman CA. Molecular mechanisms of neurodegeneration in the entorhinal cortex that underlie its selective vulnerability during the pathogenesis of Alzheimer's disease. *Biol Open* (2021) 10(1):bio056796.

Onaolapo OJ, Onaolapo AY, Olowe OA, Udoh MO, Udoh DO, Nathaniel TI. Melatonin and melatonergic influence on neuronal transcription factors: implications for the development of novel therapies for neurodegenerative disorders. *Curr Neuropharmacol* (2020) 18(7):563–577.

Palermo G, Mazzucchi S, Della Vecchia A, Siciliano G, Bonuccelli U, Azuar C, Ceravolo R, Lista S, Hampel H, Baldacci F. Different clinical contexts of use of blood neurofilament light chain protein in the spectrum of neurodegenerative diseases. *Mol Neurobiol* (2020) 57(11):4667–4691.

Pathak N, Vimal SK, Tandon I, Agrawal L, Hongyi C, Bhattacharyya S. Neurodegenerative disorders of Alzheimer, Parkinsonism, Amyotrophic Lateral Sclerosis and Multiple Sclerosis: an early diagnostic approach for precision treatment. *Metab Brain Dis* (2022) 37(1):67–104.

Pirooznia SK, Rosenthal LS, Dawson VL, Dawson TM. Parkinson disease: translating insights from molecular mechanisms to neuroprotection. *Pharmacol Rev* (2021) 73(4):33–97.

Pizzorusso T, Tognini P. Interplay between metabolism, nutrition and epigenetics in shaping brain DNA methylation, neural function and behavior. *Genes (Basel)* (2020) 11(7):742.

Polverino A, Sorrentino P, Pesoli M, Mandolesi L. Nutrition and cognition across the lifetime: an overview on epigenetic mechanisms. *AIMS Neurosci* (2021) 8(4):448–476.

Preman P, Alfonso-Triguero M, Alberdi E, Verkhratsky A, Arranz AM. Astrocytes in Alzheimer's disease: pathological significance and molecular pathways. *Cells* (2021) 10(3):540.

Purusothaman D, Chalichem NSS, Bethapudi B, Murugan S, Mundkinajeddu D. Bacopa monnieri for cognitive health—a review of molecular mechanisms of action. In *Nutraceuticals in brain health and beyond*, ed D Ghosh (2021) (pp. 15–30). Elsevier.

Quan Y, Ma A, Yang B. Preventive and therapeutic effect of Ganoderma (Lingzhi) on brain injury. *Adv Exp Med Biol* (2019) 1182:159–180.

Rahman MH, Akter R, Kamal MA. Prospective function of different antioxidant containing natural products in the treatment of neurodegenerative diseases. *CNS Neurol Disord Drug Targets* (2021a) 20(8):694–703.

Rahman MH, Bajgai J, Fadriquela A, Sharma S, Trinh TT, Akter R, Jeong YJ, Goh SH, Kim CS, Lee KJ. Therapeutic potential of natural products in treating neurodegenerative disorders and their future prospects and challenges. *Molecules* (2021b) 26(17):5327.

Ramazi S, Allahverdi A, Zahiri J. Evaluation of post-translational modifications in histone proteins: a review on histone modification defects in developmental and neurological disorders. *J Biosci* (2020) 45:135.

Ranganna K, Selvam C, Shivachar A, Yousefipour Z. Histone Deacetylase inhibitors as multitarget-directed Epi-drugs in blocking PI3K oncogenic signaling: a polypharmacology approach. *Int J Mol Sci* (2020) 21(21):8198.

Rasheed M, Liang J, Wang C, Deng Y, Chen Z. Epigenetic regulation of neuroinflammation in Parkinson's disease. *Int J Mol Sci* (2021) 22(9):4956.

Rasmi Y, Shokati A, Hassan A, Aziz SG, Bastani S, Jalali L, Moradi F, Alipour S. The role of DNA methylation in progression of neurological disorders and neurodegenerative diseases as well as the prospect of using DNA methylation inhibitors as therapeutic agents for such disorders. *IBRO Neurosci Rep* (2022) 14:28–37.

Rekatsina M, Paladini A, Piroli A, Zis P, Pergolizzi JV, Varrassi G. Pathophysiology and therapeutic perspectives of oxidative stress and neurodegenerative diseases: a narrative review. *Adv Ther* (2020) 37(1):113–139.

Saul, D, Kosinsky, RL. Epigenetics of aging and aging-associated diseases. *Int J Mol Sci* (2021) 22:401.

Saw G, Krishna K, Gupta N, Soong TW, Mallilankaraman K, Sajikumar S, Dheen ST. Epigenetic regulation of microglial phosphatidylinositol 3-kinase pathway involved in long-term potentiation and synaptic plasticity in rats. *Glia* (2020) 68(3):656–669.

Sebastiani G, Almeida-Toledano L, Serra-Delgado M, Navarro-Tapia E, Sailer S, Valverde O, Garcia-Algar O, Andreu-Fernández V. Therapeutic effects of Catechins in less common neurological and neurodegenerative disorders. *Nutrients* (2021) 13(7):2232.

Sedley L. Advances in nutritional epigenetics- a fresh perspective for an old idea. Lessons learned, limitations, and future directions. *Epigenet Insights* (2021) 113:2516865720981024.

Seritan AL. Advances in the diagnosis and management of psychotic symptoms in neurodegenerative diseases: a narrative review. *J Geriatr Psychiatry Neurol* (2023) 36(6):435–460.

Sharifi-Rad M, Lankatillake C, Dias DA, Docea AO, Mahomoodally MF, Lobine D, Chazot PL, Kurt B, Tumer TB, Moreira AC, Sharopov F, Martorell M, Martins N, Cho WC, Calina D, Sharifi-Rad J. Impact of natural compounds on neurodegenerative disorders: from preclinical to pharmacotherapeutics. *J Clin Med* (2020) 9(4):1061.

Sharma R, Garg N, Verma D, Rathi P, Sharma V, Kuca K, Prajapati PK. Indian medicinal plants as drug leads in neurodegenerative disorders. In *Nutraceuticals in brain health and beyond*, ed D Ghosh (2021) 31–45. Academic Press.

Shock T, Badang L, Ferguson B, Martinez-Guryn K. The interplay between diet, gut microbes, and host epigenetics in health and disease. *J Nutr Biochem* (2021) 95:108631.

Si P, Zhu C. Biological and neurological activities of astaxanthin. *Mol Med Rep* (2022) 26(4):1–2.

Silva RFM, Pogačnik L. Polyphenols from food and natural products: neuroprotection and safety. *Antioxid* (2020) 9(1):61.

Singh P, Sivanandam TM, Konar A, Thakur MK. Role of nutraceuticals in cognition during aging and related disorders. *Neurochem Int* (2021) 143:104928.

Smith I, Williamson EM, Putnam S, Farrimond J, Whalley BJ. Effects and mechanisms of ginseng and ginsenosides on cognition. *Nutrition Rev* (2014) 72(5):319–333.

Steg A, Oczkowicz M, Smołucha G. Omics as a tool to help determine the effectiveness of supplements. *Nutrients* (2022) 14(24):5305.

Sung Y, Yu YC, Han JM. Nutrient sensors and their crosstalk. *Exp Mol Med* (2023) 55:1076–1089.

Sunkaria A, Yadav A, Sharma SK, Sandhir R. Molecular mechanisms involved in the neuroprotective action of phytochemicals. In Neuroprotective Effects of Phytochemicals in Neurological Disorders, ed T Farooqui, AA Farooqui (2017) (pp. 515–559). John Wiley & Sons.

Teleanu DM, Niculescu A-G, Lungu II, Radu CI, Vladâcenco O, Roza E, Costăchescu B, Grumezescu AM, Teleanu RI. An overview of oxidative stress, neuroinflammation, and neurodegenerative diseases. *Int J Mol Sci* (2022) 23:5938.

Thakkar S, Anklam E, Xu A, Ulberth F, Li J, Li B, Hugas M, Sarma N, Crerar S, Swift S, Hakamatsuka T, Curtui V, Yan W, Geng X, Slikker W, Tong W. Regulatory landscape of dietary supplements and herbal medicines from a global perspective. *Regul Toxicol Pharmacol* (2020) 114:104647.

Tomoshige S, Ishikawa M. PROTACs and other chemical protein degradation technologies for the treatment of neurodegenerative disorders. *Angew Chem Int Ed Engl* (2021) 60(7):3346–3354.

Ulasov AV, Rosenkranz AA, Georgiev GP, Sobolev AS. Nrf2/Keapl/ARE signaling: towards specific regulation. *Life Sci* (2022) 291:120111.

Valori CF, Possenti A, Brambilla L, Rossi D. Challenges and opportunities of targeting Astrocytes to halt neurodegenerative disorders. *Cells* (2021) 10:2019.

Valotto Neto LJ, Reverete de Araujo M, Moretti Junior RC, Mendes Machado N, Joshi RK, Dos Santos Buglio D, Barbalho Lamas C, Direito R, Fornari Laurindo L, Tanaka M, Barbalho SM. Investigating the neuroprotective and cognitive-enhancing effects of *Bacopa monnieri*: a systematic review focused on inflammation, oxidative stress, mitochondrial dysfunction, and apoptosis. *Antioxid (Basel)* (2024) 13(4):393.

Wang L, Yu CC, Liu XY, Deng XN, Tian Q, Du YJ. Epigenetic Modulation of Microglia Function and Phenotypes in Neurodegenerative Diseases. *Neural Plast* (2021) 29:9912686.

Wang T, Zhang J, Xu Y. Epigenetic basis of lead-induced neurological disorders. *Int J Environ Res Public Health* (2020) 17:4878.

Wąsik A, Antkiewicz-Michaluk L. The mechanism of neuroprotective action of natural compounds. *Pharmacol Rep* (2017) 69(5):851–860.

Xu F, Na L, Li Y, Chen L. Roles of the PI3K/AKT/mTOR signalling pathways in neurodegenerative diseases and tumours. *Cell Biosci* (2020) 10(1):54.

Yan Y, Omisore OM, Xue YC, Li HH, Liu QH, Nie ZD, Fan J, Wang AL. Classification of neurodegenerative diseases via topological motion analysis—a comparison study for multiple gait fluctuations. *IEEE Access* (2020) 8:96363–96377.

Yang X, Wei HM, Hu GY, Zhao J, Long LN, Li CJ, Zhao ZJ, Zeng HK, Nie H. Combining antioxidant astaxantin and cholinesterase inhibitor huperzine A boosts neuroprotection. *Mol Med Re*p (2020) 21(3):1043–1050.

Yeerlan J, He B, Hu X, Zhang L. Global research trends and hotspots for Ferroptosis, Necroptosis, and Pyroptosis in Alzheimer's disease from the past to 2023: a combined bibliometric review. *J Alzheimers Dis Rep* (2024) 8(1):129–142.

Yu G, Su Q, Chen Y, Wu L, Wu S, Li H. Epigenetics in neurodegenerative disorders induced by pesticides. *Genes and Environ* (2021) 43:55 (2021).

Yuan M, Yang B, Rothschild G, Mann JJ, Sanford LD, Tang X, Huang C, Wang C, Zhang W. Epigenetic regulation in major depression and other stress-related disorders: molecular mechanisms, clinical relevance and therapeutic potential. *Sig Transduct Target Ther* (2023) 8:309.

Zamanian MY, Soltani A, Khodarahmi Z, et al. Targeting Nrf2 signaling pathway by quercetin in the prevention and treatment of neurological disorders: an overview and update on new developments. *Fund Clin Pharmacol* (2023) 37(6):1050–1064.

Zandieh MA, Farahani MH, Rajabi R, Avval ST, Karimi K, Rahmanian P, Razzazan M, Javanshir S, Mirzaei S, Paskeh MD, Salimimoghadam S. Epigenetic regulation of autophagy by non-coding RNAs in gastrointestinal tumors: Biological functions and therapeutic perspectives. *Pharmacol Res* (2023) 187:106582.

Zarneshan SN, Fakhri S, Khan H. Targeting Akt/CREB/BDNF signaling pathway by ginsenosides in neurodegenerative diseases: a mechanistic approach. *Pharmacol Res* (2022) 177:106099.

Zgorzynska E, Dziedzic B, Walczewska A. An overview of the Nrf2/ARE pathway and its role in neurodegenerative diseases. *Int J Mol Sci* (2021) 22:9592.

Zhang J, Wang S, Liu B. New insights into the genetics and epigenetics of aging plasticity. *Genes* (2023) 14:329.

Zhang Q, Yan Y. The role of natural flavonoids on neuroinflammation as a therapeutic target for Alzheimer's disease: a narrative review. *Neural Regen Res* (2023) 18(12):2582–2591.

Zhang W, Xiao D, Mao Q, Xia H. Role of neuroinflammation in neurodegeneration development. *Signal Transduct Target Ther* (2023) 8(1):267.

Zhao J, Han Z, Ding L, Wang P, He X, Lin L. The molecular mechanism of aging and the role in neurodegenerative diseases. *Heliyon* (2024) 10(2):e24751.

Zhou A, Ryan J. Biological embedding of early-life adversity and a scoping review of the evidence for intergenerational epigenetic transmission of stress and trauma in humans. *Genes* (2023) 14:1639.

Zhou L, Xiong JY, Chai YQ, Huang L, Tang ZY, Zhang XF, Liu B, Zhang JT. Possible antidepressant mechanisms of omega-3 polyunsaturated fatty acids acting on the central nervous system. *Front Psychiatry* (2022) 13:933704.

Zimmerman MA, Hall M, Qi Q, Mehta SL, Chen G, Li PA. Ubisol Coenzyme Q10 promotes mitochondrial biogenesis in HT22 cells challenged by glutamate. *Exp Ther Med* (2021) 22(5):1–7.

Zovkic IB, Guzman-Karlsson MC, Sweatt JD. Epigenetic regulation of memory formation and maintenance. *Learn Mem* (2013) 20(2):61–74.

13 Role of Nanotechnology in Multiple Sclerosis

Anjana Sharma[1] *and Nitin Sharma*[2*]

[1]Lloyd Institute of Management and Technology (Pharmacy), Greater Noida, Uttar Pradesh, India

[2]Department of Pharmaceutics, Amity Institute of Pharmacy, Amity University, Noida, Uttar Pradesh, India

*Corresponding author

MULTIPLE SCLEROSIS

One of the most challenging, prevalent, chronic inflammatory demyelinating, and neurodegenerative diseases of the central nervous system, multiple sclerosis (MS) primarily affects young adults (Wingerchuk et al., 2001; Yamout and Alroughani, 2018). MS is an unpredictable disorder that causes harm to people all around the world. Estimates place MS sufferers worldwide at around 2.8 million. Since 2013, the occurrence of MS has increased internationally by 30%. The incidence rate for the 75 reporting nations is 2.1 per 100,000 people annually. The mean age at which MS is diagnosed is 32 years old, and women are twice as likely to have the disease as men. The number of MS patients in India is 145,800, or 1 in every 9500 individuals (Salmon et al., 2014; Filippi et al., 2018; Dobson and Giovannoni, 2019).

Myelin is charged by the immune system that cushions the nerve fibers in MS patients, which slows or stops neuronal communications. Immune cells from the body target the nervous system, resulting in inflammation and nerve damage. Localized primary demyelination plaques and various forms of neurodegeneration are brought on by MS in the brain, spinal cord, and gray and white matter (Mahad et al., 2015; Walton et al., 2020).

PATHOPHYSIOLOGY

Several pathogenic processes, including demyelination and axonal degeneration, contribute to the development of multiple sclerosis (Ciccarelli et al., 2014). The primary targets for inflammation are the white and gray matter components of the central nervous system (CNS) in MS patients, which are triggered by localized immune cell infiltration and the cytokines associated with it (Ghasemi et al., 2017). Genetic and environmental factors trigger T-cell activation peripherally. Their penetration in the blood–brain barrier (BBB) is facilitated by the expression of integrins on the cell surface of inflammatory cells such as VL-4 that mediate their interaction with vascular cell adhesion molecule (VCAM-1) (Gold and Wolinsky, 2010). These interactions lead to the degradation of the BBB, which is driven by the secretion of matrix metalloproteases. Upon reaching the CNS, these T-cells get reactivated when they encounter antigens by interacting with antigen-presenting cells like macrophages, microglia, or B-cells. This interaction results in the generation of cytokines like interleukin (IL)-12, IL-23, and IL-4 (Ghasemi et al., 2017). Additionally, it has been discovered that a recently identified T-lymphocyte subtype that secretes interleukin-17, while being regulated by interleukin-23, plays a significant role in inflammation (Langrish et al., 2005; Amedie et al., 2016). These two interleukins allow Th17 cells to enter the brain and destroy neurons by rupturing the BBB (De Jonga et al., 2006;

DOI: 10.1201/9781003506317-13

Tzartos et al., 2008). Because lymphotoxin and tumor necrosis factor (TNF) are produced by B cells and their cytokines, MS progresses as a result (Ghasemi et al., 2017) (Figure 13.1).

The causes of MS are numerous, much like those of other diseases. Given the multiple etiologies of MS, it has been determined that both environmental and genetic variables significantly contribute to the development of the condition (Ascherio et al., 2016). Environmental factors like vitamin deficiency, nutrition, exposure to UV radiation, smoking, and exposure to microbes like the Epstein–Barr virus (EBV) may initiate the progression of the disease (Ascherio, 2013; Speer, 2013). A published study found that 20% to 25% of MS patients had blood antibodies that bind tightly to both EBNA1, an Epstein–Barr virus protein, and GlialCAM (a protein discharged from the brain and spinal cord), which is also known as a glial cell adhesion molecule (Gandhi et al., 2010; Bar-Or et al., 2020). In this case, a part of the EBV protein mimics GlialCAM, which is found in the insulating sheath on neurons. This suggests that when our immune system assaults EBV to get rid of the virus, it also makes GlialCAM, which is found in the myelin, a target (Pender, 2011). There has been speculation that MS may have a genetic propensity in addition to environmental factors (Patsopoulos et al., 2018). According to how much genetic information they share, family members may be more likely to get MS (Ebers et al., 1995). Clinical concordance rates were observed to be about 25%–30% between identical and monozygotic twins as compared to dizygotic or fraternal twins, which is about 3%–7%. According to research, 15%–20% of MS patients have the same condition as MS patients who are genetically similar (Compston et al., 2002).

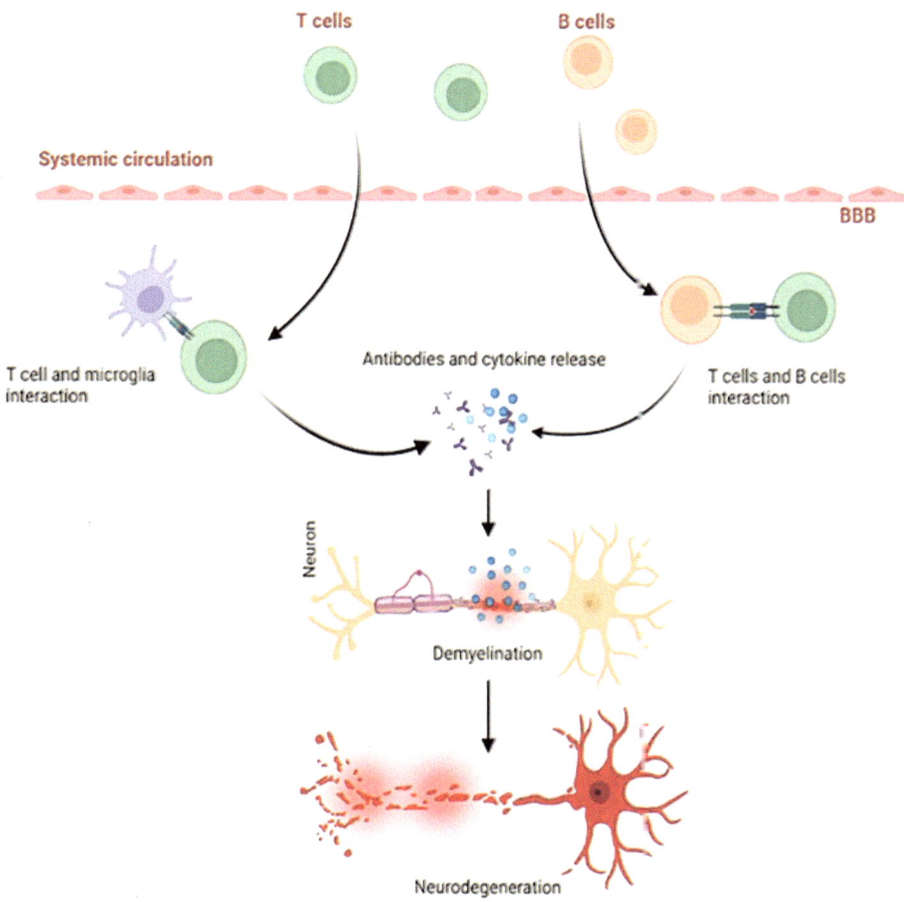

FIGURE 13.1 Pathophysiology of multiple sclerosis.

Clinical Symptoms and Treatment

The clinical manifestations of MS can vary and include fatigue, spasticity, Lhermitte's sign, urine retention, Uhthoff's phenomenon, and cognitive loss. The clinical symptoms depend upon the region of the brain or spinal cord that is affected. The symptoms may subside once MS is treated, but a relapse can have long-lasting effects or prominent effects on functional abilities such as vision problems, cognitive and emotional impairments, and depression (Ghasemi et al., 2017).

Currently, no medications are available to cure MS, although several medications have been shown to lessen the impact of MS episodes and may even slow the disease's development. These treatments are referred to as disease-modifying therapy. The specific mechanism is still unknown for the first disease-modifying medicine, interferon beta-1b, or IFN, approved by the US Food and Drug Administration (FDA) in 1993 for MS (Rossi et al., 2014). These treatments frequently try to lessen neuroinflammation, which may subtly decrease the development of neurodegeneration. MS patients who relapse are often treated by decreasing inflammation with intravenous corticosteroids or injections of adrenocorticotropic hormone (Diebold et al., 2016). These therapies lack long-term neuroprotective benefits but may hasten recovery and shorten the length of relapses by preventing the generation of proinflammatory cytokines. These therapies also call for routine parenteral administration, which may be troublesome to the patient's way of life. This is where oral therapies come into play. Due to the development of innovative drugs that have become standard therapies for MS, MS treatment has evolved dramatically in recent years. In terms of efficiency, these include glatiramer acetate and interferons (Faissner et al., 2019). The FDA has granted market authorization to more than ten medications since 2010 for use as oral MS medications. Dimethyl fumarate, teriflunomide, fingolimod, cladribine, and dalfampridine are a few of these drugs; more details are provided in Table 13.1.

NANOTECHNOLOGY IN TREATMENT OF MULTIPLE SCLEROSIS

Since many drugs lose their effectiveness due to the barriers between the blood and brain, which prevents their entry into the CNS, novel drug delivery systems play a pivotal role in drug transport to the target tissues by improving their pharmacokinetic profiles. The effectiveness and side effects of treating MS with medications that are delivered directly to the CNS are therefore expected to increase (Ortiz et al., 2014; Kabanov et al., 2007). Because they are easily absorbed by cells due to their nano size, nanomaterials become effective transporters for medicinal drugs, immunomodulatory agents, or antigens. Nanoparticles thereby address two of the main drawbacks of neurotherapeutics, namely, their limited ability to penetrate the brain or overcome barriers in order to enter brain cells and their limited ability to access immune cells there. It has been demonstrated that adding specific ligands or polymers to the nanoparticle's surfaces can effectively increase their penetrating power. Here, many kinds of nanocarriers that are used to treat MS are discussed (Table 13.2) (Nagpal et al., 2013).

Nanoparticles

Over the past 20 years, numerous nanoparticle compositions loaded with therapeutic agents have been considered for the effective treatment of multiple sclerosis. It is possible to prepare polymer nanoparticles utilizing a variety of processed and natural polymers and various preparation techniques. Additionally, their facets are modified for brain targeting. Endocytosis or transcytosis through endothelial cells are the means by which the polymeric nanoparticles enter the brain and release drugs into the CNS. A lot of work has been done to create a nano drug delivery system that will transport therapeutic substances to the demyelinated tissues of the CNS with high bioavailability and improved therapeutic effectiveness (Demetzos and Pippa, 2014; Ojha and Kumar, 2018). Additionally, remyelination is a captivating, cutting-edge approach to MS treatment, where

TABLE 13.1
Therapeutics Used in Treatment of Multiple Sclerosis

Drug	Chemical Constituents	Mechanism of Action
Glatiramer acetate	Composed of synthetic polypeptides having four amino acids mimicking myelin basic protein	It penetrates BBB by inducing T helper 2 (Th2) suppressor cells and produces anti-inflammatory effects.
Mitoxantrone	Dihydroxyanthraquinone	It interferes with antigen presentation and release of pro-inflammatory cytokines. Also, it increases T-cell suppressor function and inhibits B-cell function, which prevents myelin degradation.
Cladribine	Deoxyadenosine analogue	Cladribine-phosphates prevent DNA synthesis through DNA polymerase and ribonucleotide reductase inhibition.
Fingolimod	Structural analogue of sphingosine	Fingolimod-phosphate primarily triggers lymphocyte $S1P_1$ but later on brings about $S1P_1$ downregulation that forbids lymphocyte exit from lymphoid tissues, thereby preventing lymphocyte infiltration into the CNS (Chun and Hartung, 2010).
Teriflunomide	Enamide derivative	Teriflunomide prevents de novo pyrimidine synthesis pathway via inhibiting dihydro-orotate dehydrogenase, leading to the inhibition of proliferating T and B lymphocytes.
Dimethyl fumarate (DMF)	Methyl ester of fumaric acid	DMF prevents neuroinflammation by two pathways: Nrf2-dependent and Nrf2-independent pathway. The NrF2-dependent pathway promotes neuroprotection and oligodendrocyte protection while deactivating pro-inflammatory activation of astrocytes in the Nrf2-independent pathway (Yadav et al., 2019).
Alemtuzumab	Human monoclonal antibody	Alemtuzumab targets CD52 receptors of T and B cells specifically and depletes them via antibody-dependent cytolysis and complement-dependent cytolysis. After a week, the T and B lymphocyte depletion is followed by the regrowth of these cells, which normally takes place by 6 months post-treatment (Havrdova et al., 2015)
Natalizumab	Human monoclonal antibody with an IgG4 structurally	It targets to CD49, preventing the binding between the endothelial cell and the immune cell, thus hampering the movement of leukocytes into the central nervous system.

nanoparticles can also assist. Through the targeted distribution of remyelinating agents to particular cells, this approach will improve the therapeutic efficacy of the cells (Chountoulesi et al., 2020).

Al-Ghobashy et al. developed poly(caprolactone)-based nanoparticles for the delivery of recombinant human myelin basic protein in the treatment of multiple sclerosis. They demonstrated that the prepared nanoparticles showed better therapeutic efficacy and better protection compared to the free recombinant human myelin basic protein when administered subcutaneously in the experimental autoimmune encephalomyelitis animal model. The prepared nanoparticles provide an enhanced drug delivery method for recombinant human myelin basic protein, allowing it to pass across the BBB and enter the brains of mice (Al-Ghobashy et al., 2017). Similarly, Lunin et al. (2019) prepared poly(butylcyanoacrylate)-based nanoparticles for the delivery of thymulin in the treatment of multiple sclerosis. The prepared formulation was evaluated in the relapsing–remitting experimental

TABLE 13.2
Details of Nanocarriers, Drugs, and Study Outcomes

Nanocarrier	Drug	Observation	Reference
Nanoparticles	Recombinant human myelin basic protein	Improved therapeutic value as compared to the free protein	Al-Ghobashy et al., 2017
	Thymulin	Enhanced circulation time of thymulin	Lunin et al., 2019
	Interferon-β	Prolonged released and thus improved bioavailability of interferon-β as compared to subcutaneous commercial formulation (Rebif®)	Kondiah et al., 2013
	LINGO-1-directed small interfering RNA (siRNA)	Better recovery from the intranasal drug loaded nanoparticles	Youssef et al., 2019
Liposome	Prednisolone	Availability of a higher amount of prednisolone in brain	Schmidt et al., 2003
	Methylprednisolone	Enhancement of CNS delivery	Gaillard et al., 2012
	Myelin basic protein	Improved therapeutic efficacy as compared to naked peptides	Alexey et al., 2013
Dendrimers	Amino bis (methylene phosphonate)	Better prevention of the disease	Hayder et al., 2015
	2-(Phosphonomethyl)-pentanedioic acid	Selective targeting of therapeutic substances to the microglia cell	Hollinger et al., 2022
Nanogel	CaMK4 inhibitors	Prevention of Th17 cell	Otomo et al., 2015

autoimmune encephalomyelitis in mice. It was observed that nanoparticles ensure the availability of thymulin for a prolonged period of time in the biological media of mice as compared to free thymulin (Lunin et al., 2019).

Liposomes

Liposomes are one of the most explored formulations for the delivery of therapeutic substances in multiple sclerosis. The lipid membranes in liposomes, which can be bilayer or unilayer, may originate from a synthetic or natural source. The liposomes range in size from 20 nm to 1000 nm, where an aqueous core region is surrounded by a spherical hydrophobic phospholipid bilayer structure. The pharmaceutical ingredient that is enclosed is guarded from environmental deterioration throughout the circulatory system thanks to the non-toxic molecule's ability to encapsulate both hydrophilic and hydrophobic therapeutic substances (Kumari et al., 2016). They have the capacity to penetrate the brain's capillary endothelial cells and cross the BBB, lessening their side effects on surrounding tissues. Additionally, by appropriately decorating their surfaces, liposomes can be created to engage with specific cell receptors found in the BBB, even delivering therapeutics that ordinarily cannot cross the BBB.

Schmidt et al. prepared a PEG-coated dipalmitoyl phosphatidylcholine-based liposomal formulation for the delivery of prednisolone in the treatment of MS. They reported a higher concentration of prednisolone in the brain compared to the free drug when administered to rats exhibiting experimental autoimmune encephalomyelitis (Schmidt et al., 2003). In similar work, Gaillard et al. (2012) developed hydrosoy phosphatidylcholine lipid-based liposomes for the delivery of methylprednisolone. The prepared liposomes were actively targeted to the brain by the ligation of glutathione. It was observed that the prepared delivery system possessed prolonged plasma circulation, increased brain uptake, and showed better protection from multiple sclerosis during the study conducted in animal models (Nagpal et al., 2013).

Dendrimers

Dendrimers are macromolecules with a substantially defined system that combines a large number of functional groups with a small molecular size. These molecules are nanoscale in size, symmetric, homogeneous, and monodisperse morphology with arms or branches that resemble a tree (Abbasi et al., 2014). Dendrimers have additional advantages over the others nanocarriers as they can easily permeate the BBB due to their low molecular weight. Additionally, with functionalization with brain cell receptors, dendrimers can be actively targeted to the brain cells. The low molecular weight may also cause less damage to brain cells and thus result in lower toxicity as compared to other nanocarriers (Rodrigues et al., 2018).

Hayder et al. demonstrate the accelerated uptake of amino-bis(methylene phosphonate)-capped ABP dendrimer by the brain in an experimental autoimmune encephalomyelitis mouse model. They observed that the prepared dendrimers were able to prevent the development of multiple sclerosis and thus inhibit the disease score (Hayder et al., 2015). The hydroxyl dendrimers are ~4 nm to increase the brain delivery of the therapeutic substance. 2-(Phosphonomethyl)-pentanedioic acid was attached to hydroxyl poly(amidoamine) dendrimers for the selective targeting of microglia cells (Hollinger et al., 2022).

Nanogel

Nanogels are small, swollen, mechanically or chemically cross-linked aggregates formed either by water-soluble or amphiphilic (anionic or cationic) polymer networks. By forming biomaterial interactions such as hydrophobic or ionic bonds, nanogels can easily absorb active therapeutic substances physiologically, thus improving the bioavailability. By optimizing the chemical nature and morphology, nanogels can readily enclose a variety of classes of biomolecules and therapeutic substances, ensuring controlled release. By allowing the spontaneous association between therapeutics and polymers, various biological or drug components can be incorporated into the nanogel form. This helps to provide highly distributed hydrophilic particles (Rahman et al., 2017), and formulations comprised of CaMK4 inhibitors have been developed (Otomo et al., 2015).

Nanoneuroprotection

Neurodegeneration, assumed to be the primary cause of progressive multiple sclerosis, is believed to be a major contributor toward the neurological deficit seen in MS. The disease's increasing variants are frequently more difficult to cure than its relapsing–remitting counterparts with today's therapies, which focus on immunological control. The destruction of neural cells over time in particular areas of the brain is known as neurodegeneration, and it is a fundamental pathologic sign of many disorders, especially MS. As a consequence of such decline, cognitive abilities including memory slowly deteriorate. Neurodegeneration is a common trait of several illnesses classified as "neurodegenerative disorders" (Dhib-Jalbut et al., 2006; Przedborski et al., 2003). Numerous neuroprotective treatments have already been utilized to safeguard the nervous system. MS was previously largely managed by disease-modifying drugs like glatiramer acetate and interferon; however, these drugs just substantially relieve the symptoms and fail to stop the neurodegeneration. Nanotechnology has been proven to protect the nervous system. When nanomaterials created and used for nanotechnology have functional organization at the nanoscale, it is known as nanoneuroprotection. These nanomaterials interact biologically (Singh et al., 2012). The regeneration of the neurological system is facilitated by neuroprotection. Additionally, the advancement and application of nanotechnological approaches may assist in addressing and perhaps removing challenges related to neurotherapeutics. A range of nanomaterials with significant antioxidant capacity may be used to reduce free radicals in the brain. In a study, the nanoparticle form of ascorbic acid prevents the cell from oxidative damage in micromolar concentrations (Chakraborty et al., 2017). Since they displayed reactive oxygen species (ROS) mitigation in *in vitro* investigations using hippocampal neural cells, platinum and nanoceria nanoparticles in particular have drawn interest as a class of novel nanomaterials for neuroprotection (Orive et al., 2009).

CONCLUSION

Currently, multiple sclerosis is being treated symptomatically only, and the major reason for this is low accessibility of therapeutic substances to the brain. It has been proved that reversal of the disease is possible by ensuring the therapeutic substance reaches the brain. For this purpose, various nano drug delivery systems have proven their potential. Due to their size, they can easily cross the BBB. Apart from this, several other approaches such as surface engineering, coating, and ligation make it a suitable carrier system for brain targeting. Nanocarrier-mediated brain delivery systems are gaining attraction with improved therapeutic efficacy, and as a result, many such products are ready for commercialization. However, the nanotoxicity concern should be properly addressed for better disease management.

REFERENCES

Abbasi E, Aval SF, Akbarzadeh A, Milani M, Nasrabadi HT, Joo SW, Hanifehpour Y, Nejati-Koshki K, Pashaei-Asl R. Dendrimers: synthesis, applications, and properties. *Nanoscale Res Lett* (2014) 9:247.

Alexey A, Belogurov Jr, Alexey V, Stepanov IV, Smirnov DM, Bacon A, Mamedov AE, Boitsov, VM, Sashchenko LP, Ponomarenko NA, Sharanova SN, Boyko AN, Dubina MV, Friboulet A, Genkin DD, Gabibov AG. Liposome-encapsulated peptides protect against experimental allergic encephalitis. *FASEB J* (2013) 27:222–231.

Al-Ghobashy MA, El Meshad AN, Abdelsalam RM, Nooh MM, Al-Shorbagy M, Laible G. Development and pre-clinical evaluation of recombinant human myelin basic protein nano therapeutic vaccine in experimental autoimmune encephalomyelitis mice animal model. *Sci Rep* (2017) 7:46468.

Ascherio. Environmental factors in multiple sclerosis. *Expert Rev Neurother* (2013) 13(12 Suppl):3–9.

Ascherio A, Munger KL. Epidemiology of multiple sclerosis: From risk factors to prevention—an update. *Semin Neurol* (2016) 36:103–114.

Amedie A, Prisco D, D'Elios MM. Multiple sclerosis: the role of cytokines in pathogenesis and in therapies. *Int J Mol Sci* (2016) 13(10):13438–13460.

Bar-Or A, Pender MP, Khanna R, Steinman L, Hartung HP, Maniar T, et al. Epstein-barr virus in multiple sclerosis: theory and emerging immunotherapies. *Trends Mol Med* (2020) 26:296–310.

Chakraborty A, Jana NR. Vitamin C conjugated nanoparticle protects cells from oxidative stress at low dose but induces oxidative stress and cell death at high dose. *ACS Appl Mater Interfaces* (2017) 9:48.

Chountoulesi M, Demetzos C. Promising nanotechnology approaches in treatment of autoimmune diseases of central nervous system. *Brain Sci* (2020) 10(6):338.

Chun J, Hartung HP. Mechanism of action of oral fingolimod (FTY720) in multiple sclerosis. *Clin Neuropharmacol* (2010) 33(2):91-101.

Ciccarelli O, Barkhof F, Bodini B, De Stefano N, Golay X, Nicolay K, et al. Pathogenesis of multiple sclerosis: Insights from molecular and metabolic imaging. *Lancet Neurol* (2014) 13(8):807–822.

Compston A, Coles A. Multiple sclerosis. *Lancet* (2002) 359:1221–1231.

De Jonga BA, Huizingab TWJ, Bollenc ELEM, Uitdehaagd BMJ, Bosmae GPTh, van Bucheme MA, et al. Production of IL-1β and IL-1Ra as risk factors for susceptibility and progression of relapse-onset multiple sclerosis. *J Neuroimmunol* (2006) 126:172–179.

Demetzos C, Pippa N. Advanced drug delivery nanosystems (addnss): a mini-review. *Drug Deliv* (2014) 21(4):250–257.

Diebold M, Derfuss T. Immunological treatment of multiple sclerosis. *Semin Hematol* (2016) 53:S54–57.

Dhib-Jalbut S, Arnold DL, Cleveland DW, Fisher M, Friedlander RM, Mouradian MM, et al. Neurodegeneration and neuroprotection in multiple sclerosis and other neurodegenerative diseases. *J Neuroimmunol* (2006) 176:198–215.

Dobson R, Giovannoni G. Multiple sclerosis: a review. *Eur J Neurol* (2019) 26(1):27–40.

Ebers GC, Sadovnick AD, Risch NJ. A genetic basis for familial aggregation in multiple sclerosis. *Nature* (1995) 377:150–151.

Faissner S, Gold R. Oral therapies for multiple sclerosis. *Cold Spring Harb Perspect Med* (2019) 9(1):a03201.

Filippi M, Bar-Or A, Piehl F, Preziosa P, Solari A, Vukusic S, et al. Multiple sclerosis. *Nat Rev Dis Primers* (2018) 4:1–27.

Gaillard PJ, Appeldoorn CC, Rip J, Dorland R, van der Pol SM, Kooij G, de Vries HE, Reijerkerk A. Enhanced brain delivery of liposomal methylprednisolone improved therapeutic efficacy in a model of neuroinflammation. *J Control Release* (2012) 164:364–369.

Gandhi R, Laroni A, Weiner HL. Role of the innate immune system in the pathogenesis of multiple sclerosis. *J Neuroimmunol* (2010) 221:7–14.

Ghasemi N, Razavi S, Nikzad E. Multiple sclerosis: pathogenesis, symptoms, diagnoses and cell-based therapy. Cell J (2017) 19(1):1–10.

Gold R, Wolinsky JS. Pathophysiology of multiple sclerosis and the place of teriflunomide. *Acta Neurol Scand* (2011) 124(2):75-84.

Hayder M, Varilh M, Turrin CO, Saoudi A, Caminade AM, Poupot R, Liblau RS. Phosphorus-based dendrimer ABP treats neuroinflammation by promoting IL-10-producing CD4(+) T cells. *Biomacromolecules* (2015) 16(11):3425–3433.

Havrdova E, Horakova D, Kovarova I. Alemtuzumab in the treatment of multiple sclerosis: key clinical trial results and considerations for use. *Ther Adv Neurol Disord* (2015) 8(1):31–45.

Hollinger KR, Sharma A, Tallon C, Lovell L, Thomas AG, Zhu X, Wiseman R, Wu Y, Kambhampati SP, Liaw K, Sharma R, Rojas C, Rais R, Kannan S, Kannan RM, Slusher BS. Dendrimer- 2PMPA selectively blocks upregulated microglial GCPII activity and improves cognition in a mouse model of multiple sclerosis. *Nanotheranost* (2022) 6(2):126–142.

Kabanov AV, Gendelman HE. Nanomedicine in the diagnosis and therapy of neurodegenerative disorders. *Prog Poly Sci* (2007) 32:1054–1082.

Kondiah PP, Tomar LK, Tyagi C, Choonara YE, Modi G, du Toit LC, Kumar P, Pillay V. A novel pH-sensitive interferon-β (INF-β) oral delivery system for application in multiple sclerosis. *Int J Pharm* (2013) 456:459–472.

Kumari P, Ghosh B, Biswas S. Nanocarriers for cancer-targeted drug delivery. *J Drug Target* (2016) 24:179–191.

Langrish CL, Chen Y, Blumenschein WM, Mattson J, Basham B, Sedgwick JD et al. Il-23 drives a pathogenic t cell population that induces autoimmune inflammation. *J Exp Med* (2005) 201:233–240.

Lunin SM, Khrenov MO, Glushkova OV, Parfenyuk SB, Novoselova TV, Novoselova EG. Protective effect of PBCA nanoparticles loaded with Thymulin against the relapsing- remitting form of experimental autoimmune encephalomyelitis in mice. *Int J Mol Sci* (2019) 20(21):5374.

Mahad DH, Trapp BD, Lassmann H. Pathological mechanisms in progressive multiple sclerosis. *Lancet Neurol* (2015) 183–193.

Nagpal K, Singh SK, Mishra DN. Drug targeting to brain: a systematic approach to study the factors, parameters and approaches for prediction of permeability of drugs across BBB. *Expert Opin Drug Deliv* (2013) 10:927–955.

Ojha S, Kumar B. A review on nanotechnology-based innovations in diagnosis and treatment of multiple sclerosis. *J Cell Immunother* (2018) 4(2):56–64.

Orive G, Anitua E, Pedraz JL, Emerich DF. Biomaterials for promoting brain protection, repair and regeneration. *Nat Rev Neurosc* (2009) 10:682–692.

Ortiz GG, Pacheco-Moisés FP, Macías-Islas M, Flores-Alvarado LJ, Mireles-Ramírez MA, González-Renovato ED et al. Role of the blood-brain barrier in multiple sclerosis. *Arch Med Res* (2014) 45:687–697.

Otomo K, Koga T, Mizui M, Yoshida N, Kriegel C, Bickerton S, Fahmy TM, Tsokos GC. Cutting edge: nanogel-based delivery of an inhibitor of CaMK4 to CD4+ T cells suppresses experimental autoimmune encephalomyelitis and lupus-like disease in mice. *J Immunol* (2015) 195(12):5533–5537.

Patsopoulos NA. Genetics of multiple sclerosis: an overview and new directions. *Cold Spring Harb Perspect Med* (2018) 8(7):a028951..

Pender MP. The essential role of Epstein-Barr virus in the pathogenesis of multiple sclerosis. *Neuroscientist* (2011) 17(4):351–367.

Przedborski S, Vila M, Jackson-Lewis V. Series introduction: neurodegeneration: what is it and where are we? *J Clin Investig* (2003) 111:3–10.

Rahman FA, Magbool F, Elnima EI, Shayoub ME, Elhassan AM, Hussein SEO. Nanogel as a pharmaceutical carrier – review article. *Sch J App Med Sci* (2017) 5(11F):4730–4736.

Rodrigues DB, Oliveira JM, Santos TC, Reis RL. Dendrimers: breaking the paradigm of current musculoskeletal autoimmune therapies. *J Tissue Eng Regen Med* (2018) 12(3):e1796–e1812.

Rossi S, Motta C. et al. Interleukin-1β causes excitotoxic neurodegeneration and multiple sclerosis disease progression by activating the apoptotic protein p53. *Mol Neurodegen* (2014) 9:56.

Salmon J, Desai M, Douglas GW. *Multiple sclerosis in India: a qualitative study of access, adherence and other issues in Southern India*. Proceedings of the Business and Health Administration Association Annual Conference, Chicago, IL, USA (2014).

Schmidt J, Metselaar JM, Wauben MHM, Toyka KV, Storm G, Gold R. Drug targeting by long-circulating liposomal glucocorticosteroids increases therapeutic efficacy in a model of multiple sclerosis. *Brain* (2003) 126:1895–1904.

Singh AV, Khare M, Gade W, Zamboni P. Theranostic implications of nanotechnology in multiple sclerosis: a future perspective. *Autoimmune Dis* (2012) 2012:160830.

Speer G. Impact of vitamin d in neurological diseases and neurorehabilitation: from dementia to multiple sclerosis. Part I: The role of vitamin D in the prevention and treatment of multiple sclerosis. *Ideggyogy Szle* (2013) 66:293–303.

Tzartos JS, Friese MA, Craner MJ, Palace J, Newcombe J, Esiri MM, et al. Interleukin-17 production in central nervous system-infiltrating T cells and glial cells is associated with active disease in multiple sclerosis. *Am J Pathol* (2008) 172:146–155.

Walton C, King R, Rechtman L, Kaye W, Leray E, Marrie RA, et al. Rising prevalence of multiple sclerosis worldwide: insights from the atlas of MS, third edition. *Mult Scler J.* (2020) 26(14):1816–1821.

Wingerchuk DM, Lucchinetti CF, Noseworthy JH. Multiple sclerosis: current pathophysiological concepts. *Lab Investig* (2001) 263–281.

Yadav SK, Soin D, Ito K, Dhib-Jalbut S. Insight into the mechanism of action of dimethyl fumarate in multiple sclerosis. *J Mol Med (Berl)* (2019) 97(4):463–472.

Yamout BI, Alroughani R. Multiple sclerosis. *Semin Neurol* (2018) 38(2):212–225.

Youssef AEH, Dief AE, El Azhary NM, Abdelmonsif DA, El-Fetiany OS. LINGO-1 siRNA nanoparticles promote central remyelination in ethidium bromide-induced demyelination in rats. *J Physiol Biochem* (2019) 75(1):89–99.

14 Therapeutic Advancement in the Treatment of Neurodegenerative Diseases

Praveen Kumar Gaur[1], Rosaline Mishra[1], Raju Singh[1], Moumita Barman[2], and Kanak Lata[1]*

[1]Metro College of Health Science and Research, Greater Noida, Uttar Pradesh, India

[2]I. T. S. College of Pharmacy, Ghaziabad, Uttar Pradesh, India

*Corresponding author

INTRODUCTION

Degenerative nerve diseases affect many body processes, including balance, movement, talking, breathing, and cardiac function. Many of these diseases are inherited. A disease such as alcoholism, a tumor, or a stroke can sometimes be the source of the issue. Toxins, chemicals, and pathogens are all possible reasons. There are instances when there is no recognized cause. Neurodegenerative disorders impact millions of individuals globally. Alzheimer's disease (AD) and Parkinson's disease (PD) are the two most common neurological diseases (Hippius and Neundorfer, 2003; Querfurth and LaFerla, 2010). According to the Alzheimer's Disease Association, 6.2 million individuals in the United States may have AD by 2022 (Luca et al., 2019). According to the Parkinson's Foundation, approximately 1 million individuals in the USA have the disease (Hirsch et al., 2016; Gillis et al., 2019; Tom et al., 2020). A neurodegenerative disease occurs when the nerve cells in the brain or the peripheral nervous system continue to lose efficacy over time and ultimately perish. Although certain medicines can relieve some neurodegenerative disease signs, there is currently no cure and only a slowing of the disease's development (Checkoway et al., 2011; Lamptey et al., 2022; Walia et al., 2022).

TYPES OF NEURODEGENERATIVE DISEASES

The following are the types of neurodegenerative diseases:

- Alzheimer's disease
- Amyotrophic lateral sclerosis
- Friedreich ataxia
- Huntington's disease
- Lewy body disease
- Parkinson's disease
- Spinal muscular atrophy (Dugger et al., 2017)

In the following, we shed light on two major neurodegenerative disorders and their clinical manifestations.

DOI: 10.1201/9781003506317-14

ALZHEIMER'S DISEASE

Alzheimer's disease is the most prevalent neurodegenerative disease, contributing to nearly half of all dementia occurrences. AD affects approximately 30% of people age 85 and up. Incidence rates increase significantly 0.5% per year for people age 65 to 75, and between 6% and 8% per year for persons age 85 and above (Lunn et al., 2011; Merriam et al., 1988). Except in instances of early-onset AD, which accounts for about 5–10% of cases, AD development prior to the age of 50 is rare. The primary clinical symptom of AD is memory loss and a rapid decline in cognitive function beyond that produced by regular aging (Kang et al., 2017). Dementia frequently starts with mood and attitude changes, then progresses to memory loss, confusion, and aphasia. The brain cortex and hippocampus are particularly susceptible to Alzheimer's disease. Neurofibrillary tangles (NFTs) and senile or neuritic plaques are the two pathological abnormalities that differentiate afflicted tissues. Amyloid peptide clusters constitute the bulk of neuropathic plaques identified in the hippocampus's blood vessels and neurons (Miguelez et al., 2020; Montazersaheb et al., 2021). AD is a brain condition that gradually ruins memory and reasoning skills, as well as the ability to perform the most basic activities. Symptoms of AD usually show later in life in the majority of individuals. Estimates differ, but experts believe that over six million Americans, the majority of whom are over the age of 65, may have AD-related dementia (Dugger et al., 2017).

Alzheimer's Disease and Effects on the Brain

Scientists are still working to understand the complicated brain alterations associated with AD, and this is illustrated in Figure 14.1. Brain changes can begin over a decade or more prior to symptoms being displayed. Toxic changes occur in the brain during this early period of Alzheimer's, including abnormal protein buildups that create plaques of amyloid and tau tangles. Previously healthy neurons cease working, lose connections to other neurons, and eventually perish. Many other complicated brain changes are believed to contribute to AD (Jain and Chen-Plotkin, 2018; Silva et al., 2019).

Clinical Manifestations

Memory issues are frequently one of the first symptoms of Alzheimer's-related cognitive impairment. Mild cognitive impairment (MCI) is a condition that affects some individuals who have memory difficulties (Giri et al., 2016). People with MCI have more memory issues than is typical for their age, but the signs have no impact on their daily activities. MCI has also been related to issues with movement and the perception of smell. Older individuals with MCI are more likely to develop Alzheimer's disease, but not all of them do. Some people may even regain normal memory. The first signs of Alzheimer's differ from individual to individual. Non-memory elements of mental processes, such as word-finding, vision/spatial issues, and impaired reasoning or judgment, may indicate the very early phases of the illness for many people. Biomarkers (biological indicators of illness discovered in brain images, cerebrospinal fluid, and blood) are being studied by researchers to identify early changes in the brains of people with MCI and mentally normal people who may be at higher risk for Alzheimer's (Pickett et al., 2019). More study is required before these methods can be used widely and regularly to diagnose Alzheimer's disease in the office of a healthcare provider (Merriam et al., 1988; Liu et al., 2017).

PARKINSON'S DISEASE

Parkinson's disease is a degenerative neurological disorder characterized by motor symptoms (slow movement, trembling, stiffness, walking, and imbalance) as well as a broad range of non-motor consequences (cognitive impairment, mental health disorders, sleep disorders, pain, and other sensory

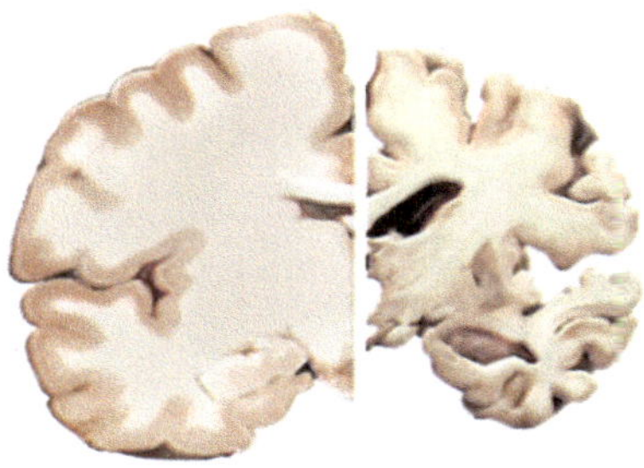

FIGURE 14.1 Normal brain morphology versus affected brain morphology.

disturbances). Motor impairments, such as dyskinesias (involuntary motions) and dystonia (painful involuntary muscle contractions), add to speaking and mobility limitations, as well as limits in many areas of life (Yu et al., 2018). The progression of these symptoms leads to a high incidence of impairment and care needs. Many individuals with Parkinson's disease acquire dementia as the disease progresses. While Parkinson's disease is the most prevalent movement condition, there are others, including multiple system atrophy, progressive supranuclear palsy, chorea, ataxia, and dystonia. Tremor, slow movement, and rigidity are signs of some movement conditions that are comparable to Parkinson's disease. All movement disorders face the same diagnostic and therapy gaps as Parkinson's disease, as well as limited drug availability, especially in nations with low or middle incomes (Bhidayasiri et al., 2023; Bloem and Okun, 2021).

Globally, disability and death due to PD are increasing faster than for any other neurological disorder.

Clinical diagnosis of PD by trained non-specialized healthcare workers and simplified treatment guidelines offer better management in primary care settings. Clinical diagnostic accuracy has been studied in early and late stages of PD (Adler et al., 2014).

Levodopa/carbidopa, the most effective medicine for improving symptoms, functioning, and quality of life, is not accessible, available, or affordable everywhere, particularly in low- and middle-income countries.

Rehabilitation can help improve functioning and quality of life for people with PD (Dickson, 2018).

Clinical Manifestations

Parkinson's disease indications and symptoms can vary from person to person. Early symptoms may be subtle and go undetected. Symptoms typically begin on one side of the body and worsen on that side, even after symptoms begin to impact extremities on both sides (Vázquez-Vélez and Zoghbi 2021), as illustrated in Figure 14.2.

Parkinson's symptoms and indications may include:

Tremors, or rhythmic trembling, that typically start in a limb, most commonly the hand or digits. Individuals may press their thumbs and fingers together. This is referred to as a pill-rolling spasm. When a person's hand is at rest, it may quiver. When subjects are completing activities, the shaking may lessen.

Parkinson's disease may slow movement (bradykinesia) over time, making basic chores challenging and time-consuming. When the individual walks, his or her movements may

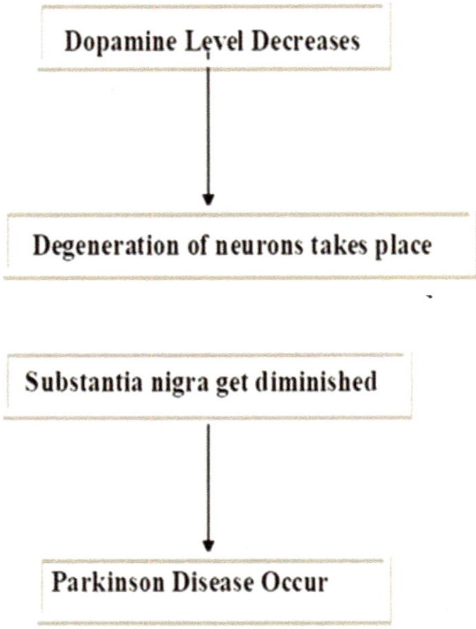

FIGURE 14.2 Flowchart of disease progression.

become shorter. Getting off a settee may be challenging. When attempting to move, an individual's feet may drag or stumble.

Muscle rigidity can appear anywhere in the body. Muscle stiffness can be uncomfortable and restrict the range of motion.

The posture may become stooped, or the person may fall or have balance problems.

Loss of automatic movements. The person may have a decreased ability to perform unconscious movements, including blinking, smiling, or swinging the arms when they walk.

The individual may speak softly, quickly, slur, or hesitate before talking. A person's speech may be more monotone rather than having the usual speech patterns.

It may become hard to write, and the writing may appear small (Yemula et al., 2021; Miguelez et al., 2020).

CURRENT CHALLENGES ASSOCIATED WITH CNS-TARGETED GENE DELIVERY

Brain Structure Complexity

Despite the intricacy and heterogeneity of the brain's structure, neurons remain the most researched central nervous system (CNS) cell group. However, recent biological breakthroughs have revealed that distinct cell subpopulations in the brain play key roles in the pathophysiology of a variety of neurological diseases. Microglia are resident immune cells of the CNS that perform routine brain surveillance and account for 10% to 15% of all brain cells. Oligodendrocytes myelinate and chemically insulate axons in the CNS, allowing saltatory transmission and rapid nerve propagation. Astrocytes, the most common cell group in the brain, are the main glial cells. Given the unique etiology of neurological diseases, gene therapy delivery techniques that can target specific cell subpopulations are required. These specific genes have been delivered using a variety of techniques, including parenchymal injections. The primary problem that needs to be addressed is obtaining universal gene expression in the brain (Giri et al., 2016; Ahmadlou et al., 2017; Kimura et al., 2020).

Low Blood–Brain Barrier Permeability

Neurological diseases are difficult to cure due to limited access to brain regions protected by physical barriers such as the blood–brain barrier (BBB). The BBB acts as a defense system, keeping harmful substances out of the brain and maintaining CNS equilibrium (Ford et al., 2020). The basal lamina covers the barrier's main cellular components, which include a monoendothelial layer of cells connected by tight junctions, pericytes, and perivascular glial processes. Endothelial cells in the BBB are differentiated by their sparse fenestrations and pinocytotic vesicles, which further block molecules from passing through and infiltrating the CNS. If molecules are tiny enough or soluble in lipids, they can passively cross the BBB. They could do so through active receptor uptake, transporters, or carriers that are expressed by the cells of the BBB. Despite decades of research, there are still no effective therapies for neurodegenerative illnesses, which place a heavy burden on society (Hippius and Neundorfer, 2003; Hirsch et al., 2016).

THERAPEUTIC ADVANCEMENT IN THE TREATMENT OF NEURODEGENERATIVE DISEASES

Major advances in tailored gene therapy have opened up new avenues for treating serious neurological diseases with a variety of flexible techniques ranging from exogenous gene regulation to endogenous gene suppression. Recent technical advances have focused on virus vector designs, plasmid transfection, nanoparticles, polymer-mediated gene transport, tailored microRNA, and *in vivo* CRISPR-based therapies to improve gene sequence delivery (Huang and Zhang, 2018; Hussain et al., 2020). These innovative techniques have significant applications in the treatment of rare diseases such as amyotrophic lateral sclerosis (ALS), spinal muscular atrophy, lysosomal storage diseases, X-linked adrenoleukodystrophy, and oncological conditions, as well as highly prevalent neurological diseases and neurodevelopmental disorders such as Parkinson's disease, Alzheimer's disease, and autism spectrum disorder. Barker et al. (2012), Scheper et al. (2019), Armstrong and Okun (2020) and Bhattacharya and colleagues (2022) investigated the challenges and possible paths for gene therapy in the management of neurological diseases using targeted gene delivery.

Stem Cell Technology

The capacity of stem cells to proliferate and differentiate into a variety of cellular lines is noteworthy (Dantuma et al., 2017). The various categories of stem cells show the diversity of cell types that they can produce and the ways by which they are derived. Mesenchymal stem cells (MSCs), precursor cells, embryonic stem (ES) cells, and induced pluripotent stem (iPS) cells are among them (Staff et al., 2019; Sivandzade and Cucullo, 2021). Understanding the characteristics of the many available stem cells, as well as the potential effects of cellular treatments on disease pathways, is critical to comprehending the potential applications of stem cell technology to neurodegenerative diseases (Dantuma et al., 2010; Lunn et al., 2011; Doi et al., 2020; Ford et al., 2020; Wang et al., 2020).

Each variety of stem cells has unique characteristics and benefits, and the justification for using each type is determined by the desired goals and outcomes. Progenitor cells derived from more developed fetal or adult tissues are more multipotent than ES cells, allowing them to give birth to a narrower spectrum of lineages. These possible lines are frequently determined by the germ layer of origin. Neural stem cells (NSCs) and neural progenitor cells (NPCs), for example, can develop into a variety of cell types that comprise the neural lineage. NPCs can be generated directly from fetal or adult neural tissue or through cell culture manipulation to regulate the fate of ES cells. Although this ability may be retained in culture, transplantation in vivo leads cells to cease proliferating and begin to differentiate (Il-Kwon Kim et al., 2021).

Cellular Therapy Strategies and Applications

In cellular treatments, cell or tissue transplants are used to repair illnesses or wounds. The primary objectives of stem cell treatment are usually cellular replacement or environmental enhancement (Figure 14.3). Cellular replacement therapy for neurodegenerative diseases involves obtaining specific neuronal subtypes lost to the disease and then transplanting them into affected areas of the nervous system. The newly transplanted neurons may then integrate, establish synaptic connections, and reconstruct a disease-damaged neuronal network. Stem cells, on the other hand, may form supplementary neuronal networks around affected regions, make neurotrophic factors, or scavenge toxic components to provide ambient enrichment to aid host neurons. Glial-derived neurotrophic factor (GDNF), neurotrophic factor derived from the brain (BDNF), insulin-like growth factor-1 (IGF-I), and vascular endothelial growth factor (VEGF) are examples of growth hormones that provide in situ support at the main foci of illness. These variables are protective in neurological disease models. Each neurodegenerative disease's appropriate cellular treatment objective must be identified based on its distinct neuronal pathophysiology. Cellular replacement treatment may be effective in diseases such as PD, where a particular neuronal subpopulation is lost. Cellular therapies that enhance the local spinal cord environment to support the surviving motor neurons, on the other hand, are more likely to be helpful for ALS (Fortier, 2005; Carradori et al., 2017).

Considerations include the ease with which grafted neurons integrate and move through the host tissue, as well as the length to which axons must travel to reach their objectives.

Cellular Therapy for Neurodegenerative Diseases

The first stage in creating and delivering cellular therapies to patients is choosing the right stem cell type and comprehending the desired mechanism of support. Although each disease and cellular therapy is unique, the path from bench to bedside is lengthy and difficult, and certain universal issues must be taken into account for a secure transition to patient therapies (Malgrange et al., 2011; Emerich et al., 2014; Drouin-Ouellet and Barker, 2014; Gorabi et al., 2019).

Alzheimer's Disease

Memory loss and cognitive decline are hallmarks of Alzheimer's disease, the most prevalent form of dementia. As the illness progresses, the cortex, hippocampus, amygdala, and basal forebrain

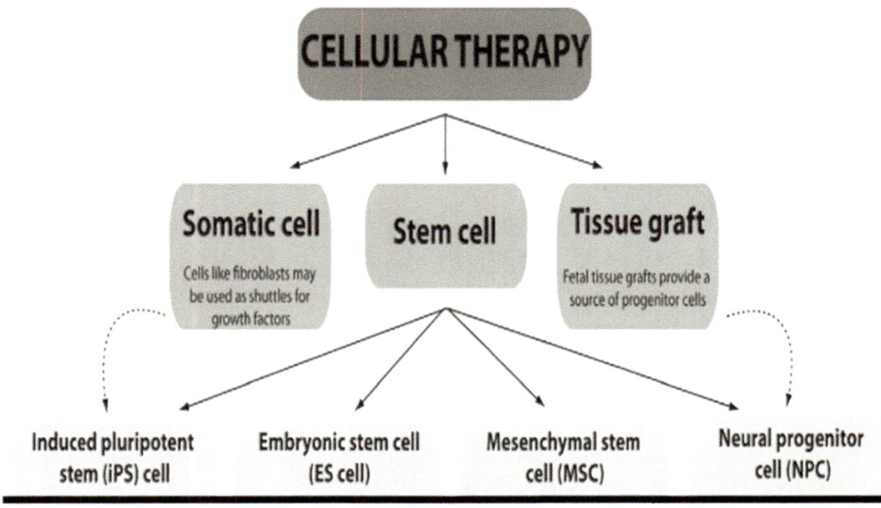

FIGURE 14.3 Cell therapy classification.

all suffer extensive loss of neurons and synaptic connections. Although the exact mechanism of Alzheimer's disease is unclear, Ab plaques and neurofibrillary tangles, etc. are pathological markers (Luca et al., 2019). Age increases the chance of developing Alzheimer's disease, and most cases of the disease appear after the age of 65. The prevalence of Alzheimer's disease is expected to rise as the populace matures (Kalra et al., 2014; Kang et al., 2017). The focus of current AD therapies is on controlling neurotransmitter activity. The present emphasis of Alzheimer's disease treatments is on controlling neurotransmitter function. Cholinergic function can be enhanced to ameliorate the behavioral and cognitive symptoms of Alzheimer's disease. Environmental supplementation could be accomplished through the use of stem cell treatments that target the acetylcholine system. Because neurogenesis in the hippocampus diminishes with age and is exacerbated in AD, cellular therapies that promote neurogenesis or replace damaged neurons may help delay the progression of the disease (Kapasi et al., 2017; Kim et al., 2021). Raising BDNF levels, which are reduced with age and in Alzheimer's disease, promotes regeneration and protects neural function. The production of nerve growth factor (NGF) is another way of increasing the efficacy of cellular treatments. In a phase I study for Alzheimer's disease, patient fibroblasts that have been genetically modified to generate NGF are presently being tested (Knopman et al., 2021).

Parkinson's Disease

The decline of dopaminergic (DA) neurons in the substantia nigra over time produces Parkinson's disease. Patients have significant motor impairments, which manifest as tremors, muscular stiffness, and instability in their movement and posture (Kimura et al., 2020). Deep brain stimulation is presently available as a treatment choice, as are therapies that use L-dopa, dopamine agonists, or dopamine precursors to increase dopamine levels. These treatments are helpful in reducing symptoms early in the course of the disease, but their long-term efficacy is unclear because they do not correct the deficit, have long-term adverse effects, and become increasingly less effective as PD advances. Cellular therapies for Parkinson's disease, on the other hand, focus on the restoration of destroyed DA neurons. Early cellular treatments for Parkinson's disease used DA neurons taken from fetal ventral midbrain tissue (Ilie et al., 2020; Lill et al., 2016). Clinical studies have backed cellular treatments for a possible cognitive advantage in Parkinson's disease, despite varying degrees of efficacy. Yet there could be constraints when employing fetal cells, such as ethical concerns and a lack of adequate tissue for therapy. As an alternative, ES cells allow for the bulk production of neurons with the midbrain DA phenotype. Grafting DA neurons from both ES and MSC into rodent PD models results in functional repair. Recent studies using iPS cells have demonstrated that it is feasible to generate DA neurons that are unique to each patient. These cells were transferred into a mouse Parkinson's disease model, which decreased functional deficits and demonstrated cell implantation in the host tissue. These results are among the first to demonstrate how iPS cells can be used to treat a neurodegenerative disease (Balestrino et al., 2019).

POTENTIAL NANOSTRUCTURES EMPLOYED FOR THE TREATMENT OF NEURODEGENERATIVE DISORDERS

Polymeric nanoparticles are the most common type of drug delivery system used in nanotechnology to treat neurodegenerative disorders (Modi et al., 2010; Rahman et al., 2020; Montazersaheb et al., 2021). As they can traverse the BBB, pass through tight cell junctions, have a high drug-loading capacity, and be targeted at the mutagenic proteins in AD and PD, polymeric nanoparticles are promising for the treatment of AD and PD. Table 14.1 lists drugs and formulation advancement details, and Figure 14.4 shows various types of nanotechnology-based drug delivery systems (Farokhzad and Langer, 2009; Saeedi et al., 2019; Naqvi et al., 2020; Hussain et al., 2020; Sim et al., 2020; Zhu et al., 2021).

TABLE 14.1
Drugs and Advanced Formulation Systems Targeting Neurodegenerative Disorders

Drug	Restrictions	Formulations	Advancement	Reference
Selegiline	Low bioavailability	Nanoemulsion	Nanoemulsion reduces dopamine loss, avoiding the free radical damage, followed by curing the PD.	Kumar et al., 2018
Deferoxamine mesylate	Provides several side effects with high doses and stumpy patient compliance as the drug half-life is low.	Intranasal saline solutions	Memory retention has been enhanced.	Farr et al., 2020
Rivastigmine	The hydrophilic nature of the drug showed reduced CNS bioavailability when it undergoes first-pass metabolism.	Liposomes	Quick onset of action. Better bioavailability. Amplified half-life. Diminished clearance.	Kumar et al., 2020; Rompicherla et al., 2021
Piperine	Low bioavailability.	Gold nanoparticles	In both *in vitro* and *in vivo* models of PD piperine gold nanoparticles successfully repelled paraquat's toxic effects.	Srivastav et al., 2020
Silibilin	Low solubility and reduced bioavailability.	Nanoparticles	Silibinin–albumin nanoparticles improved the neuroprotective and antioxidant activity of silibilin.	Pan et al., 2021
Baicalein	Impaired BBB can destroy the mitochondrial structure and the composition.	Pluronic P85/ F68 Micelles	The stable nanoparticles result in enhanced BBB crossing ability, progressed cell concentration, and sustained release dosage form also can be prepared.	Chen et al., 2017
Carbamazepine	There was no tolerance for acid and alkali. The concentration was higher and lessened the competence for drug sustainability.	Dendrimers	Dendrimers have improved the product's strength and quality, reduced the toxicity in cells, and enhanced the bioavailability.	Igartúa et al., 2018
Memantine	The toxicity in blood is the major issue.	Dendrimers	The formulation results in excellent encapsulation capacity, enhanced efficiency to deliver the drug, elevated entrance in BBB, and can target the brain properly. The drug can be formulated in the form of a sustained release dosage form.	Gothwal et al., 2019

These nanosystems have promising qualities in targeted CNS medication delivery, including:

1. Chemical stability characteristics can be easily changed to allow selective medication delivery to certain organs, tissues, or cells.

2. By hiding the physicochemical properties, they boost the bioavailability and effectiveness of included medications and, as a result, the transfer of drugs across the BBB.
3. They safeguard medicines integrated against enzymatic breakdown.
4. They have fewer negative side effects.

Polymeric Nanoparticles, Nanocapsules, and Nanospheres

The size range of polymeric nanoparticles and nanocapsules is 10 to 1000 nm. They have high drug-loading capabilities and the ability to prevent the incorporated drug load from degrading, which increases the likelihood that the drug will reach the brain. Because their surface characteristics can be altered to prevent recognition by macrophages of the reticuloendothelial system, they are stable and can target the delivery of medicines to the CNS (Moradi and Dashti, 2022). Doxorubicin and other medications have been delivered to the CNS using polymeric nanoparticles. Nanospheres are dense polymeric matrices created by microemulsion polymerization in which drugs are disseminated. Nanospheres and nanocapsules are two different types of nanoparticle systems that are made up of a solid core and a dense polymeric matrix, respectively (Figure 14.5).

Polymeric Nanogels and Nanosuspensions

Nanogels are cross-linked networks of polymers that are often composed of both ionic and nonionic polymeric chains (Figure 14.5). They have been made utilizing an emulsification solvent evaporation method. Nanogels can include molecules such as oligonucleotides, siRNA, DNA, proteins, and low-molecular-mass medicines and swell in water. The drug-loading capacity is up to 40–60%. For transport over the BBB, Vinogradov and colleagues have encased oligonucleotides inside a

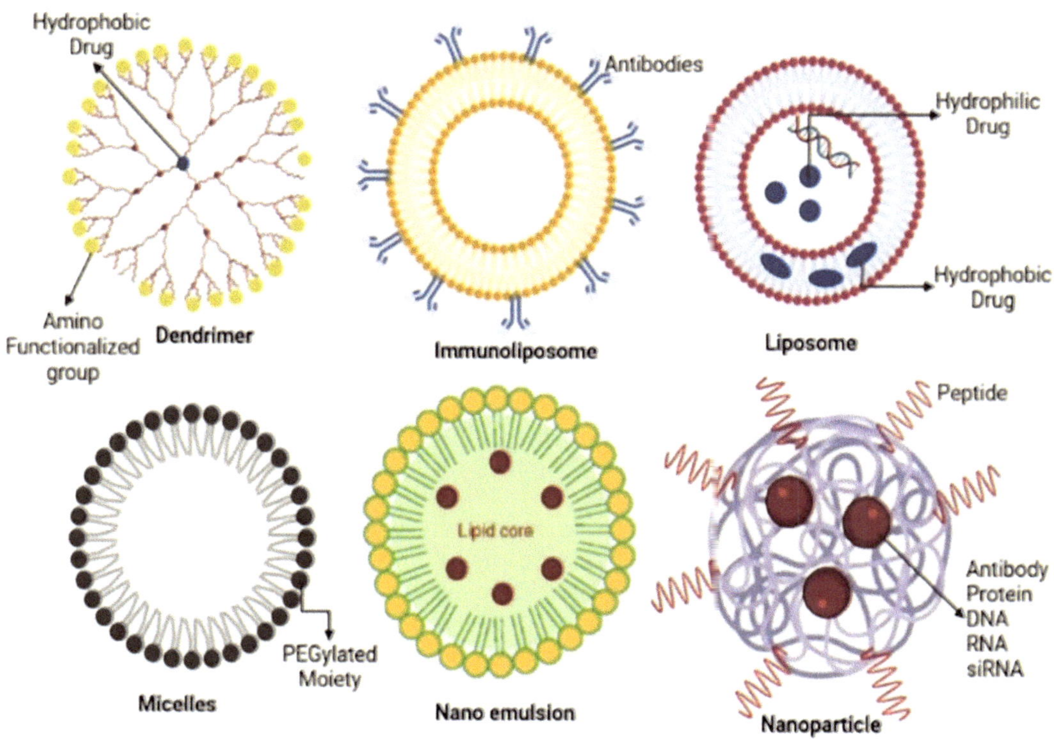

FIGURE 14.4 CNS delivery system based on nanotechnology.

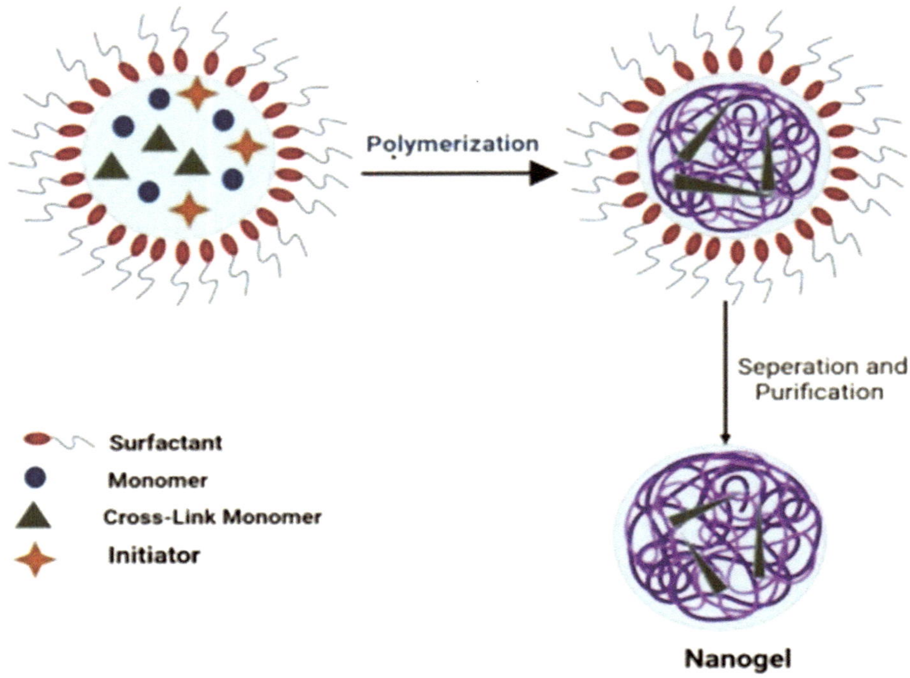

FIGURE 14.5 Constitution of nanogel.

cross-linked nanogel. According to in vivo experiments, the nanogel boosted oligonucleotide uptake in the brain while decreasing it in the liver and spleen. Crystalline drug particles stabilized by nonionic surfactants or lipid mixtures are known as drug-loaded nanosuspensions. The ease of use, large drug-loading capacity, and application to numerous medicines for CNS delivery are key benefits of nanosuspensions (Wang et al., 2020).

Carbon Nanotubes and Nanofibers

Carbon nanotubes are being explored to improve chronic CNS electrical stimulation. Functional electrical stimulation implants are becoming more popular in clinical settings for the treatment of PD. The failure of recording or stimulating chronic CNS electrodes due to the fibrotic response mediated by glial and immunological cells is a serious obstacle. Compressed carbon nanofiber-based electric arrays for CNS stimulation might be developed, and they could be injected into areas of degeneration to offer a physical substrate and the molecular signals required to stimulate and maintain tissue healing in the treatment of neurodegenerative diseases. As shown in Figure 14.6, polymeric nanomicelles have a hydrophobic core and a shell made of blocks of hydrophilic polymer that define the core–shell architecture of polymeric nanomicelles. Up to 20–30% weight-per-weight of hydrophobic medicines can be incorporated into the core, preventing early drug release and degradation. Nanomicelles are stabilized by the shell, which also protects the medication from interactions with serum proteins and unintended cells. Once the medicine has reached the target cells, it is released via diffusion. Although there hasn't yet been any effective research on the transport of DNA molecules to the central nervous system, polymeric nanomicelles are adaptable and have been demonstrated to efficiently transfer DNA molecules *in vitro* and *in vivo*. Carbon nanotubes prepared by chemical vapor deposition have characteristics like electrical conductive capacity and strong mechanical properties. They can be cost-effectively prepared, but these carbon nanotubes indicate toxicity to the cells in their biological background, which can be minimized by providing a coat to

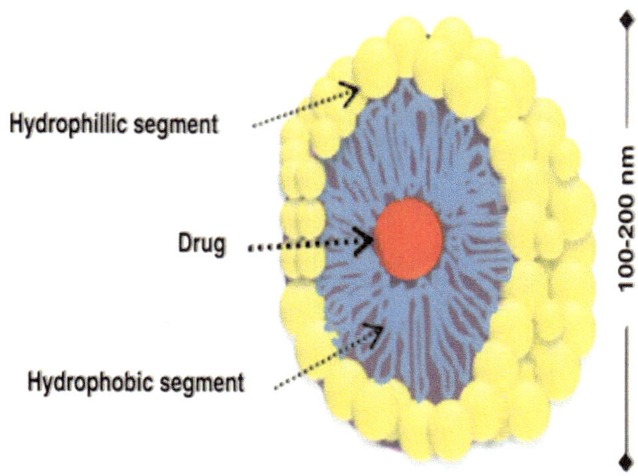

FIGURE 14.6 Structure of carbon nanotubes/nanofibers.

the carbon nanotubes. This coated nanotube diminishes the correlation among the cells and carbon nanotubes (Abdul Waris et al., 2022).

Polymeric Nanoliposomes

Although no effective study on the delivery of DNA molecules to the CNS has been reported to date, nanotechnology and neurodegenerative disorders are flexible and have been proven to efficiently deliver DNA molecules *in vitro* and *in vivo*. Polymeric nanoliposomes are intracellular aqueous compartments surrounded by uni- or multilamellar lipid bilayers. Drugs can be incorporated in relatively significant amounts into lipid bilayers or nanoliposome aqueous compartments. Nanoliposomes with modified surfaces that diminish opsonization in plasma and decrease their detection and removal by the liver and spleen can be used to extend systemic circulation periods. Nanoliposome evaluation for targeted CNS medication delivery has been researched for a variety of applications (Liu et al., 2021).

Quantum Dots

Quantum dots (QDs) are nanoparticles that have a broad array of purposes in treating several neurodegenerative diseases (Gorabi et al., 2019). The properties of QDs, such as sensitivity and selectivity, are effective when formulated in nanosized particles. QDs can cross BBB and can circulate in the systemic circulation after entering the brain due to their very small particle size. QDs have exceptional optical properties when formulated in the form of nanocrystals, which are used to envisage the structure of the brain and mechanisms of brain functions. QDs can also be used for drug delivery systems. Additionally, QDs can also be a distinct agent used in tissue imaging (Utkin et al., 2018).

NANOSYSTEMS EXPLORED FOR ADVANCED EXPERIMENTAL TREATMENT OF AD

Poly(butyl)cyanoacrylate Nanoparticles

With some degree of effectiveness, poly(butyl)cyanoacrylate (PBCA) nanoparticles have been employed to carry medications to the CNS. These particles typically have a diameter of 250 nm and are drug-loaded either by incorporating the drug into the initial polymerization process of the particle or by adsorption onto the surface of the manufactured particle.

Thioflavin-T Nanocapsules for B-Amyloid Detection

Thioflavine T (ThT) interaction with aggregated fibrils of the synthetic beta/A4-derived peptides (beta(1-28) and beta(1-40)) and susequent changes in absorption has been found useful toward diagnosis of AD. ThT and insulin fibrils show intense fluorescence while various polyhydroxy/polyanionic/polycationic materials fails to interact (LeVine, 1993).

D-Penicillamine Nanoparticles for the Treatment of AD

With advancing age, the concentration of metal ions in the brain rises, having fatal effects on the AD brain. An increase in the amount of copper ions (Cu^{2+}) causes oxidative stress, which produces harmful hydroxyl radicals that damage proteins and lipids and alter DNA. Compared to a healthy brain, amyloid plaques have higher concentrations of Cu^{2+} and Zn^{2+}. These discoveries have served as the foundation for innovative therapy strategies for AD that use iron-chelating substances with low neurotoxicity.

Macrophage Delivery of Nanozymes for PD

Macrophages loaded with nanozymes enhance tyrosine hydroxylase related to dopaminergic neurons when compared to untreated mice. This results in an increase in therapeutic activity of macrophage-loaded nanozymes, as it illustrates the slow release of the enzyme in plasma.

CONCLUSION

Neurodegenerative diseases place a hefty burden on society due to their devastating consequences, high expenses, and scarcity of viable treatments. Cellular therapies are appealing alternatives, and stem cell research is increasingly being used to treat neurological illnesses. This chapter starts with a short overview of the different stem cell techniques available before diving into the current state of neurodegenerative disease stem cell research. In this chapter, with the help of factual representations, we covered the primary therapies to ameliorate neurodegenerative disorders, namely, stem cells, cellular therapies, and nanostructures. Stem cells provide cellular replacement and/or environmental supplementation to slow neurodegeneration. Cellular therapies offer considerable potential for treating a variety of diseases. In disorders characterized by specific cell subpopulations or significant neuronal loss, cellular replacement may reproduce or stabilize neural networks. Environmental nourishment may also sustain surviving cells by providing neurotrophic factors or preventing the creation or accumulation of toxic chemicals that harm neurons. Cellular medicines frequently produce good results through both methods (Rahman et al., 2022). We have also discussed the reasoning for, and ongoing challenges in, moving stem cell therapy from the bench to the bedside; namely, both these methods are not very cost-effective, and other alternatives are being explored.

Formulation of nanostructures using already known drugs is one of them. These structures are targeted to cross the BBB and reach the site of action easily and promptly control release of drugs. These structures also enhance the absorption and bioavailability profiles. Because they are targeted, fewer side effects are seen. The drawback that we could find was reduced long-term stability leading to the accumulation of drug particles, thus increasing particle size. The future challenge thus remains in solving the solubility issue, which might be achieved by adjusting the surfactant concentration and replacing excipients to enhance stability.

REFERENCES

Adler CH, Beach TG, Hentz JG, Shill HA, Caviness JN, et al. Low clinical diagnostic accuracy of early vs. advanced Parkinson disease: clinicopathologic study. *Neurol.* (2014) 83: 406–412.

Ahmadlou M, Adeli H. Complexity of weighted graph: a new technique to investigate structural complexity of brain activities with applications to aging and autism. *Neurosci Lett.* (2017) 650: 103–108.

Armstrong MJ, Okun MS. Diagnosis and treatment of Parkinson disease: a review. *JAMA.* (2020) 323(6): 548–560.

Balestrino R. Parkinson disease. *Eur J Neurol.* (2019) 27(1): 27–42.

Barker R, Kehoe PG, Love S. Activators and inhibitors of the plasminogen system in Alzheimer's disease. *J Cell Mol Med.* (2012) 16: 865–876.

Bhattacharya T, Soares GABE, Chopra H, Rahman MM, Hasan Z et al. Applications of Phyto-Nanotechnology for the treatment of neurodegenerative disorders. *Materials. (Basel)* (2022) 15(3): 804.

Bhidayasiri R, Kalia LV, Bloem BR. Tackling Parkinson's disease as a global challenge. *J Parkinsons Dis.* (2023) 13(8): 1277–1280.

Bloem BR, Okun MS, Klein C. Parkinson's disease. *Lancet.* (2021) 397(10291): 2284–2303.

Carradori D, Eyer J, Saulnier P, Préat V, Des Rieux, A. The therapeutic contribution of nanomedicine to treat neurodegenerative diseases via neural stem cell differentiation. *Biomaterials.* (2017) 123: 77–91.

Checkoway H, Lundin JI, Kelada SN. Neurodegenerative diseases. *IARC Sci Pub.* (2011) 163: 407–419.

Chen T, Li Y, Li C, Yi X, Wang R, et al. Pluronic P85/F68 micelles of baicalein could interfere with mitochondria to overcome MRP2-mediated efflux and offer improved anti-Parkinsonian activity. *Mol Pharm.* (2017) 14(10): 3331–3342.

Dantuma E, Merchant S, Sugaya K. Stem cells for the treatment of neurodegenerative diseases. *Stem Cell Res Ther.* (2010) 1: 37.

Dickson DW. Neuropathology of Parkinson disease. *Parkinsonism Relat Disord.* (2018) 46(Suppl 1): S30–S33.

Doi D, Magotani H, Kikuchi T, Ikeda M, Hiramatsu S, et al. Pre-clinical study of induced pluripotent stem cell-derived dopaminergic progenitor cells for Parkinson's disease. *Nat Commun.* (2020) 11: 3369.

Drouin-Ouellet J, Barker RA. Stem cell therapies for Parkinson's disease: are trials just around the corner? *Regen Med.* (2014) 9: 553–555.

Dugger BN, Dickson DW. Pathology of neurodegenerative diseases *Cold Spring Harb Perspect Biol.* (2017) 9(7): a028035.

Emerich DF, Orive G, Thanos C, Tornoe J, Wahlberg LU. Encapsulated cell therapy for neurodegenerative diseases: from promise to product. *Adv Drug Deliv Rev.* (2014) 67–68: 131–141.

Farokhzad OC, Langer R. Impact of nanotechnology on drug delivery. *ACS Nano.* (2009) 3(1): 16–20.

Farr AC, Xiong MP. Challenges and opportunities of deferoxamine delivery for treatment of Alzheimer's disease, Parkinson's disease, and intracerebral hemorrhage. *Mol Pharm.* (2021) 18(2): 593–609.

Ford E, Pearlman J, Ruan T, Manion J, Waller M, Neely GG, Caron L. Human Pluripotent Stem Cells-based therapies for neurodegenerative diseases: current status and challenges. *Cells.* (2020) 9(11): 2517.

Fortier LA, Stem cells: classifications, controversies, and clinical applications. *Vet Surg.* (2005) 34(5): 415–423.

Gillis C, Mirzaei F, Potashman M, Ikram MA, Maserejian N. The incidence of mild cognitive impairment: a systematic review and data synthesis. *Alzheimers Dement.* (2019) 11: 248–256.

Giri M, Zhang M, Lu Y. Genes associated with Alzheimer's disease: an overview and current status. *Clin Interv Aging.* (2016) 11: 665–681.

Gorabi AM, Kiaie N, Barreto GE, Read MI, Tafti HA, Sahebkar A. The therapeutic potential of Mesenchymal stem cell-derived exosomes in treatment of neurodegenerative diseases. *Mol Neurobiol.* (2019) 56: 8157–8167.

Gothwal A, Kumar H, Nakhate KT, Ajazuddin A, Dutta A, et al. Lactoferrin coupled lower generation PAMAM dendrimers for brain targeted delivery of memantine in aluminum-chloride-induced Alzheimer's disease in mice. *Bioconjug Chem.* (2019) 30(10): 2573–2583.

Hippius H. Neundorfer G. The discovery of Alzheimers disease. *Dialogues Clin Neurosci.* (2003) 5(1): 101–108.

Hirsch L, Jette N, Frolkis A, Steeves T, Pringsheim T. The incidence of Parkinson's disease: a systematic review and meta-analysis. *Neuroepidemiol.* (2016) 46: 292–300.

Huang W, Zhang C. Tuning the Size of Poly(lactic-co-glycolic Acid) (PLGA) Nanoparticles Fabricated by Nanoprecipitation. *Biotechnol J.* (2018) 13: 1700203.

Hussain Z, Thu HE, Elsayed I, Abourehab MA, Khan S, et al. Nano-scaled materials may induce severe neurotoxicity upon chronic exposure to brain tissues: a critical appraisal and recent updates on predisposing factors, underlying mechanism, and future prospects. *J Control Release* (2020) 328: 873–894.

Igartúa DE, Martinez CS, Temprana CF, Alonso SDV, Prieto MJ. PAMAM dendrimers as a carbamazepine delivery system for neurodegenerative diseases: a biophysical and nanotoxicological characterization. *Int J Pharm.* (2018) 544(1): 191–202.

Jain N, Chen-Plotkin AS. Genetic modifiers in neurodegeneration. *Curr Genet Med Rep.* (2018) 6: 11–19.

Kalra K, Tomar PC. Stem cell: basics, classification and applications. *Am J Phytomed Clin Therap.* (2014) 2(7): 919–930.

Kang S, Lee YH, Lee JE. Metabolism-centric overview of the pathogenesis of Alzheimer's disease. *Yonsei Med J.* (2017) 58(3): 479–488.

Kapasi A, DeCarli C, Schneider JA. Impact of multiple pathologies on the threshold for clinically overt dementia. *Acta Neuropathol.* (2017) 134: 171–186.

Kim IK, Park JH, Kim B, Hwang KC, Song BW. Recent advances in stem cell therapy for neurodegenerative disease: three-dimensional tracing and its emerging use. *World J Stem Cells.* (2021) 13(9): 1215–1230.

Kimura S, Harashima H. Current status and challenges associated with CNS-targeted gene delivery across the BBB. *Pharmaceutics* (2020) 12(12): 1216.

Knopman DS, Amieva H, Petersen RC, Chételat G, Holtzman DM, et al. Alzheimer disease. *Nat Rev Dis Primers.* (2021) 7(1): 33.

Kumar S, Dang S, Nigam K, Ali J, Baboota S. Selegiline nanoformulation in attenuation of oxidative stress and upregulation of dopamine in the brain for the treatment of Parkinson's disease. *Rejuvenation Res.* (2018) 21(5): 464-476.

Lamptey RNL, Chaulagain B, Trivedi R, Gothwal A, Layek B, Singh J. A review of the common neurodegenerative disorders: current therapeutic approaches and the potential role of nanotherapeutics. *Int J Mol Sci* (2022) 23(3): 1851.

LeVine H 3rd. Thioflavine T interaction with synthetic Alzheimer's disease beta-amyloid peptides: detection of amyloid aggregation in solution. *Protein Sci* (1993) 2(3): 404–410.

Ilie OD, Ciobica A, McKenna J, Doroftei B, Mavroudis I. Minireview on the relations between gut microflora and Parkinson's disease: further biochemical (oxidative stress), inflammatory, and neurological particularities. *Oxid Med Cell Longev* (2020) 2020: 4518023.

Lill CM, Mashychev A, Hartmann C, Lohmann K, Marras C, Lang AE, Klein C, Bertram L. Launching the movement disorders society genetic mutation database (MDSGene). *Mov Disord* (2016) 31(5): 607-609.

Liu Y, Castro Bravo KM, Liu J. Targeted liposomal drug delivery: a nanoscience and biophysical perspective. *Nanoscale Horiz.* (2021) 6: 78–94.

Liu Z, Zhou T, Ziegler AC, Dimitrion P, Zuo L. Oxidative stress in neurodegenerative diseases: from molecular mechanisms to clinical applications. *Oxid Med Cell Longev.* (2017) 2017: 2525967.

Luca M, Di Mauro M, Di Mauro M, Luca A. Gut microbiota in Alzheimer's disease, depression, and type 2 diabetes mellitus: the role of oxidative stress. *Oxid Med Cell Longev* (2019) 2019: 4730539.

Lunn JS, Sakowski SA, Hur J, Feldman EL. Stem cell technology for neurodegenerative diseases. *Ann Neurol.* (2011) 70: 353–361.

Malgrange B, Borgs L, Grobarczyk B, Purnelle A, Ernst P, Moonen G, Nguyen L. Using human pluripotent stem cells to untangle neurodegenerative disease mechanisms. *Cell Mol Life Sci.* (2011) 68: 635–649.

Merriam AE, Aronson MK, Gaston P, Wey SL, Katz I. The psychiatric symptoms of Alzheimer's disease. *J Am Geriatric Soc.* (1988) 36(1): 7–22.

Miguelez C, De Deurwaerdere P, Sgambato V. Editorial: non-dopaminergic systems in Parkinson's disease. *Front Pharmacol.* (2020) 11: 593822.

Modi G, Pillay V, Choonara YE. Advances in the treatment of neurodegenerative disorders employing nanotechnology. *Ann N Y Acad Sci.* (2010) 1184: 154–172.

Montazersaheb S, Ahmadian E, Maleki Dizaj S, Jahanbani Y, Davaran S, Huseynova I, Zhdanov R, Keskin C, Khalilov R, Eftekhari A. Emerging nanotherapeutic strategies in Alzheimer's disease. In *Frontiers in clinical drug research-dementia. Vol. 2.,* ed Atta-ur-Rahman. Bentham Science Publishers Pte. Ltd (2021) pp. 173–203.

Moradi F, Dashti N. Targeting neuroinflammation by intranasal delivery of nanoparticles in neurological diseases: a comprehensive review. *Naunyn Schmiedebergs Arch Pharmacol.* (2022) 395(2): 133–148.

Naqvi S, Panghal A, Flora SJS, Nanotechnology: a promising approach for delivery of neuroprotective drugs. *Front Neurosci.* (2020) 14: 494.

Pan Q, Ban Y, Lijun Xu L. Silibinin-Albumin nanoparticles: characterization and biological evaluation against oxidative stress-stimulated neurotoxicity associated with Alzheimer's disease. *J Biomed Nanotechnol.* (2021) 17(6):1123–1130.

Pickett EK, Herrmann AG, McQueen J, Abt K, Dando O, Tulloch J, Jain P, Dunnett S, Sohrabi S, Fjeldstad MP. Amyloid beta and tau cooperate to cause reversible behavioral and transcriptional deficits in a model of Alzheimer's disease. *Cell Rep.* (2019) 29: 3592–3604.

Querfurth HW, LaFerla FM. Alzheimer's disease. *N Engl J Med* (2010) 362(4): 329–344.

Rahman MM, Ferdous KS, Ahmed M. Emerging promise of nanoparticle-based treatment for Parkinson's disease. *Biointerface Res Appl Chem.* (2020) 10: 7135–7151.

Rahman MM, Islam MR, Shohag S, Hossain ME, Rahaman MS, et al. The multifunctional role of herbal products in the management of diabetes and obesity: a comprehensive review. *Molecules.* (2022) 27(5): 1713.

Rompicherla SKL, Arumugan K, Bojja SL, Kumar N, Rao CM. Pharmacokinetic and pharmacodynamic evaluation of nasal liposome and nanoparticle based rivastigmine formulations in acute and chronic models of Alzheimer's disease. *Naunyn Schmiedebergs Arch Pharmacol.* (2021) 394(8): 1737–1755.

Saeedi M, Eslamifar, M, Khezri K, Dizaj SM. Applications of nanotechnology in drug delivery to the central nervous system. *Biomed Pharmacother.* (2019) 111: 666–675.

Scheper V, Schwieger J, Hamm A, Lenarz T, Hoffmann A. BDNF-overexpressing human mesenchymal stem cells mediate increased neuronal protection in vitro. *J Neurosci Res.* (2019) 97:1414–1429.

Silva MVF, Loures CMG, Alves LCV, de Souza LC, Borges KBG, Carvalho MG. Alzheimer's disease: risk factors and potentially protective measures. *J Biomedical Sci.* (2019) 26: 33.

Sim TM, Tarini D, Dheen ST, Bay BH, Srinivasan DK. Nanoparticle-based technology approaches to the management of neurological disorders. *Int J Mol Sci.* (2020) 21: 6070

Sivandzade F, Cucullo L. Regenerative stem cell therapy for neurodegenerative diseases: an overview. *Int J Mol Sci.* (2021) 22: 2153.

Srivastav S, Anand BG, Fatima M, Prajapati KP, Yadav SS et al. Piperine-coated gold nanoparticles alleviate Paraquat-induced neurotoxicity in Drosophila melanogaster. *ACS Chem Neurosci.* (2020) 11(22): 3772–3785.

Staff NP, Jones DT, Singer W. Mesenchymal stromal cell therapies for neurodegenerative diseases. *Mayo Clin Proc.* (2019) 94: 892–905.

Tom SE, Phadke M, Hubbard RA, Crane PK, Stern Y, Larson EB. Trends in incident dementia and early life socioeconomic status by birth cohort in the adult changes in thought study. *JAMA Open.* (2020) 3: e2011094.

Utkin YN. Brain and quantum dots: benefits of nanotechnology for healthy and diseased brain. *Cent Nerv Syst Agents Med Chem (Former Curr Med Chem - Cent Nerv Syst Agents).* (2018) 18: 193–205.

Vázquez-Vélez GE, Zoghbi HY. Parkinson's disease genetics and pathophysiology. *Ann Rev Neurosci.* (2021) 44: 87–108.

Walia V, Kaushik D, Mittal V, Kumar K, Verma R et al. Delineation of neuroprotective effects and possible benefits of antioxidants therapy for the treatment of Alzheimer's diseases by targeting mitochondrial-derived reactive oxygen species: bench to bedside. *Mol Neurobiol.* (2022) 59(1): 657–680.

Wang J, Hu WW, Jiang Z, Feng MJ. Advances in treatment of neurodegenerative diseases: perspectives for combination of stem cells with neurotrophic factors. *World J Stem Cells.* (2020) 12(5): 323–338.

Wang K, Zhu X, Yu E, Desai P, Wang H, et al. Therapeutic nanomaterials for neurological diseases and cancer therapy. *J Nanomaterials.* 2020 (2020): 2047379.

Waris A, AliAtta A, Khan U, Asim M, Zamel D, et al. Applications of various types of nanomaterials for the treatment of neurological disorders. *Nanomaterials.* (2022) 12(13): 2140.

Yemula N, Dietrich C, Dostal V, Hornberger M. Parkinson's disease and the gut: symptoms, nutrition, and microbiota. *J Parkinsons Dis.* (2021) 11(4): 1491–1505.

Yoo J, Kim HS, Hwang DY. Stem cells as promising therapeutic options for neurological disorders. *J Cell Biochem.* (2013) 114: 743–753.

Yu QJ, Yu SY, Zuo LJ, Lian TH, Hu Y, et al. Parkinson disease with constipation: clinical features and relevant factors. . (2018) 8: 567.

Zhu FD, Hu YJ, Yu L, Zhou XG, Wu JM, et al. Nanoparticles: a hope for the treatment of inflammation in CNS. Front Pharmacol. (2021) 12: 683935.

15 Recent Developments in Nanotechnology-Based Therapeutics for the Treatment of Alzheimer's Disease

Harsh Yadav[1], Subash P[1], Satish Dubey[2], Anand Maurya[3], Ramu Singh[1], and Sabyasachi Maiti[1]*

[1]Department of Pharmacy, Indira Gandhi National Tribal University, Amarkantak, Madhya Pradesh, India
[2]Laboratory of Molecular Taxonomy & Medicinal Plant Biology, Department of Botany, Guru Ghasidas Vishwavidyalaya, Bilaspur Chhattisgarh, India
[3]Department of Medicinal chemistry, Faculty of Ayurveda, Institute of Medical Sciences, Banaras Hindu University, Varanasi, U.P., India
*Corresponding author

INTRODUCTION

Alzheimer's disease (AD) seems to be a degenerative neurological condition in which neuronal cell death promotes memory loss, changes in cognitive capacity, personality, and habits. AD affects about 5% of humans aged 65 to 74 years of age and 50% of humans over 85 (Alzheimer's Association, 2009). AD kills neurons and their connections in the brain's memory-related areas, such as the entorhinal cortex and hippocampus, initially. Thus, it affects the regions of the cerebral cortex that control expression, reasoning, and social behavior. More than 90% of cases occur in those older than 60 years. Mild cognitive impairment (MCI) is when one's cognitive abilities, such as memory and thinking skills, deteriorate gradually but noticeably. A person with MCI is more prone to acquire Alzheimer's disease or another dementia. Neuropsychological tests, including the Mini-Mental State Examination (MMSE), are used to screen cognitive function. A low MMSE score requires further evaluation, such as brain imaging techniques (Ding et al., 2015). By 2020, more than 50 million people will be living with dementia globally. The population will almost double every 20 years, reaching 82 million in 2030 and 152 million in 2050 (Prince et al., 2009). The condition has no cure and is worsening. The disease progresses and eventually leads to death. However, at present, autopsies are the only technique to confirm an AD diagnosis with a high degree of certainty because there are currently no proven biomarkers for AD. Patients with AD are identified for clinical assessment of brain scans showing signs of memory loss signs. Biomarkers are needed to find and track the development of neurological disorders, whether they are caused by genetic traits, biochemical changes, or changes in the way the brain looks or works. Biomarkers are also used to assess the effectiveness of new medicines and therapeutic strategies as they are developed (Ward, 2010). According to current studies, the scientific factors underlying AD etiology need to be investigated further, with specific attention paid to the fact that there are twice as many female Alzheimer's sufferers as men, according to current studies (Alzheimer's Association,

DOI: 10.1201/9781003506317-15

2017). Unfortunately, there hasn't been a detailed investigation into this sex-based variance in AD incidence. AD has a number of neuropsychiatric symptoms, such as apathy, loss of appetite, anxiety, signs of depression, and trouble sleeping (Ishii et al., 2015). Figure 15.1 shows five of the most important therapeutic targets for Alzheimer's disease (Ahmed et al., 2021).

VARIOUS PATHOLOGICAL CAUSES OF ALZHEIMER'S DISEASE

The two hallmark pathological causes of AD pathogenesis are the plaques of β-amyloid (Aβ) peptide accumulation caused by abnormal amyloid precursor protein (APP) synthesis and cholinergic defects. These cause significant loss of neurons and neuronal alterations in memory-related brain areas, such as the hippocampus (Bhushan et al., 2018). Figures 15.2 and 15.3 depict the major hypotheses involved in AD pathogenesis.

B-Amyloid Hypothesis

The Aβ hypothesis was first put forward in 1991. It says that the disease is caused by the buildup of Aβ peptide plaques in brain tissue and the clumping together of Aβ into fibrillar plaques (Hardy et al., 1991; Carter et al., 2010; Hardy, 2009). The 42-residue peptide Aβ (1–42) has the most remarkable ability to cluster among the most frequent forms of Aβ peptides, Aβ (1–40). Furthermore, the exact mechanisms that cause neurotoxicity are unknown (Benilova et al., 2011). Another hypothesis is that oligomers of various sizes bind to different receptors, activating neurotoxic mechanisms exclusive to them. In this regard, it was recently discovered that oligomers with high and low molecular weights (HMW and LMW) on neuronal surfaces do not compete for the same receptor. Also, HMW is much more damaging than LMW, and reducing HMW through therapy can stop AD from worsening by reducing direct damage to neuronal membranes (Yasumoto et al., 2019). Therefore, better AD models should be the focus of significant effort. They incorporated familial AD-linked mutations into mice, resulting in abnormally elevated Aβ levels and depositions resembling AD pathophysiology, the main focus of the most recent in vivo models. Even though these transgenic models have shown essential things about how AD develops, they can only be used to model familial (early-onset) AD, which makes up less than 5% of the complications of the disease

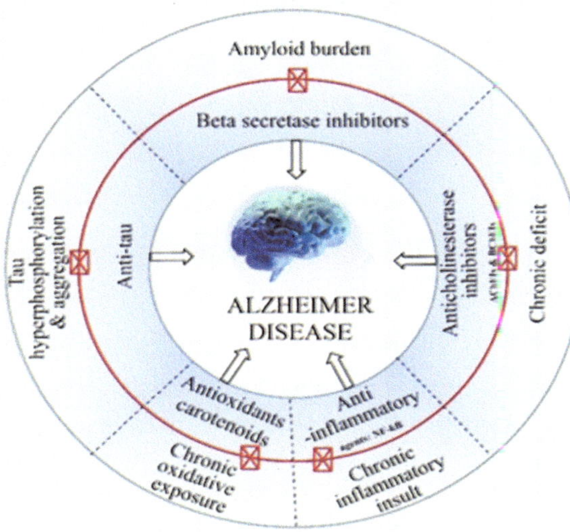

FIGURE 15.1 Five of the most important treatment targets in Alzheimer's disease.

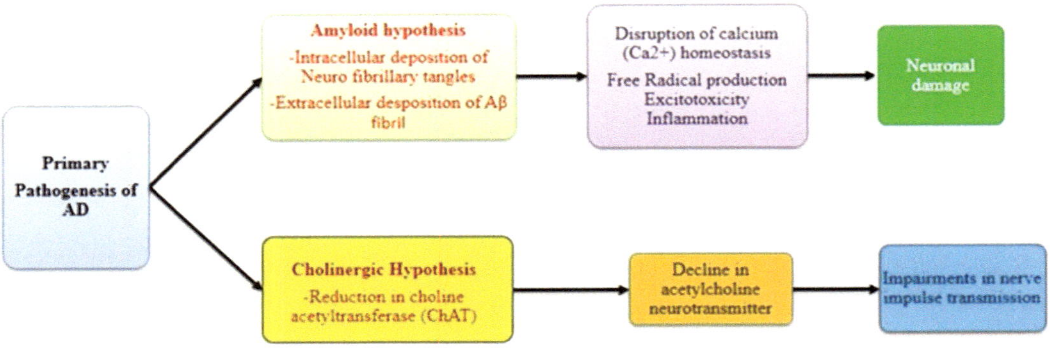

FIGURE 15.2 Major hypothesis involved in AD pathogenesis.

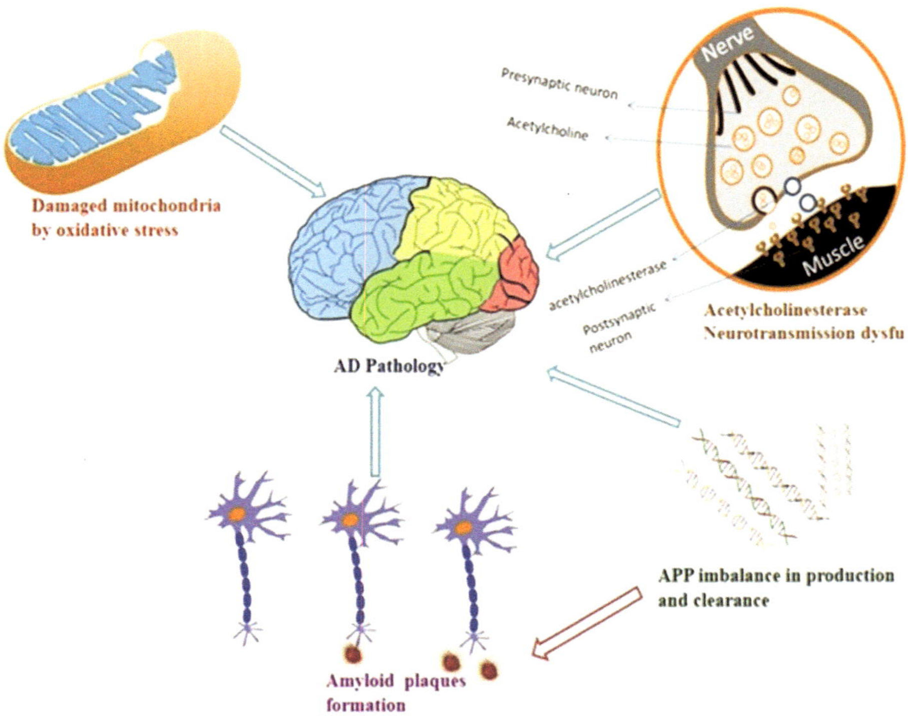

FIGURE 15.3 Pathophysiology of AD.

(LaFerla et al., 2005). The Aβ aggregation causing cognitive impairment in the AD brain is shown in Figure 15.4.

Cholinergic Hypothesis

The cholinergic paradigm of Alzheimer's disease is the earliest idea, dating from the mid-1970s. Cholinergic neurotransmission loss is linked to the degeneration of the cholinergic system and neuronal degeneration (Meyer et al., 2012; Francis et al., 1999; Bartus et al., 1982; Struble et al., 1982; Cummings et al., 1998). Furthermore, drugs that act on both presynaptic and postsynaptic levels can increase acetylcholine (ACh) (Cooper, 1996). In response to nerve impulses, ACh is secreted from nerve fibers into muscle cells. By hydrolyzing ACh, AChE, a cholinergic enzyme

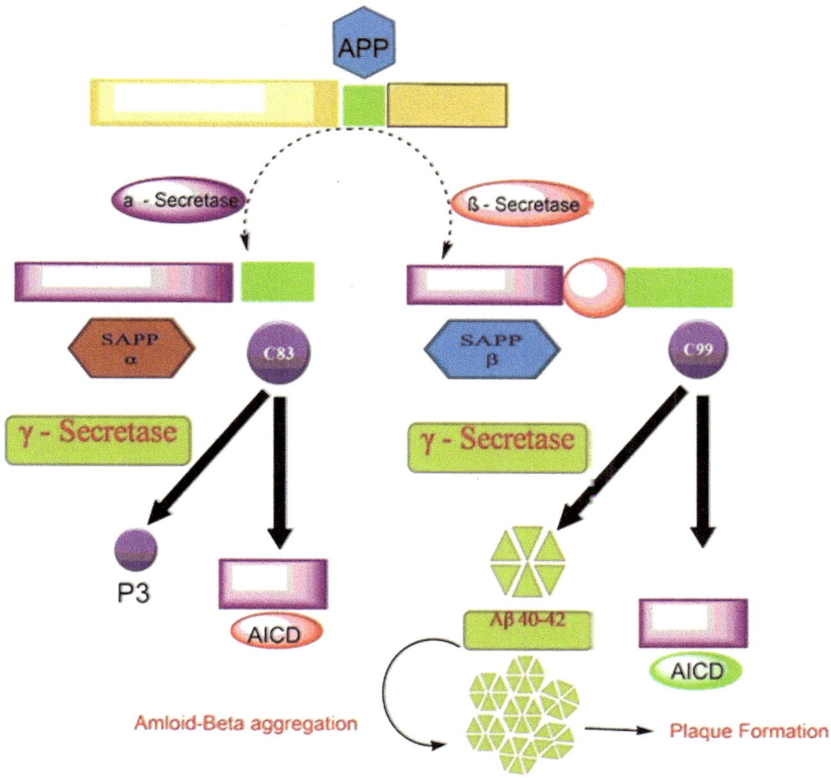

FIGURE 15.4 Mechanism of β-amyloid aggregation causing cognitive impairment in the AD brain.

found largely at the postsynaptic membrane, stops signal transmission. Acetylcholine is a parasympathetic nervous system neurotransmitter that can be found in a variety of bodily regions, but it is most commonly connected with the neuromuscular junction. In order for muscles in the body to be activated, motor neurons in the ventral spinal column need to be connected to the neuromuscular junction. Acetylcholine is a neurotransmitter and the autonomic nervous system's ultimate releasing mediator from parasympathetic postganglionic neurons (Waxenbaum et al., 2019). In terms of recognizing AD and developing drugs, the cholinergic theory is the oldest and most promising. AChE inhibitors prevent ACh from being degraded, allowing more ACh to reach the brain and preventing ACh depletion caused by cholinergic cell death (Purves et al., 2008). Figure 15.5 depicts the method through which AChE causes cognitive impairment in the AD brain.

Alternative Hypothesis

Another prevalent theory for oxidative stress is mitochondrial damage in Alzheimer's disease (Markesbery, 1997). An imbalance in metabolic processes that results in reactive oxygen species is referred to as oxidative stress (Markesbery, 1997; Harman et al., 1981). This idea is linked to AD pathogenesis, based on the free radical theory of aging (Nunomura et al., 1999; Sayre et al., 1999; Nunomura et al., 2001; Praticò, 2005; Sayre et al., 2005). Oxidative stress is a condition that occurs when free radicals overcome the body's ability to control them. One remedy to this problem is antioxidants, which neutralize the negative effects of reactive chemicals. Unfortunately, anti-medications for Alzheimer's disease that are now available on the market have minimal effect on these

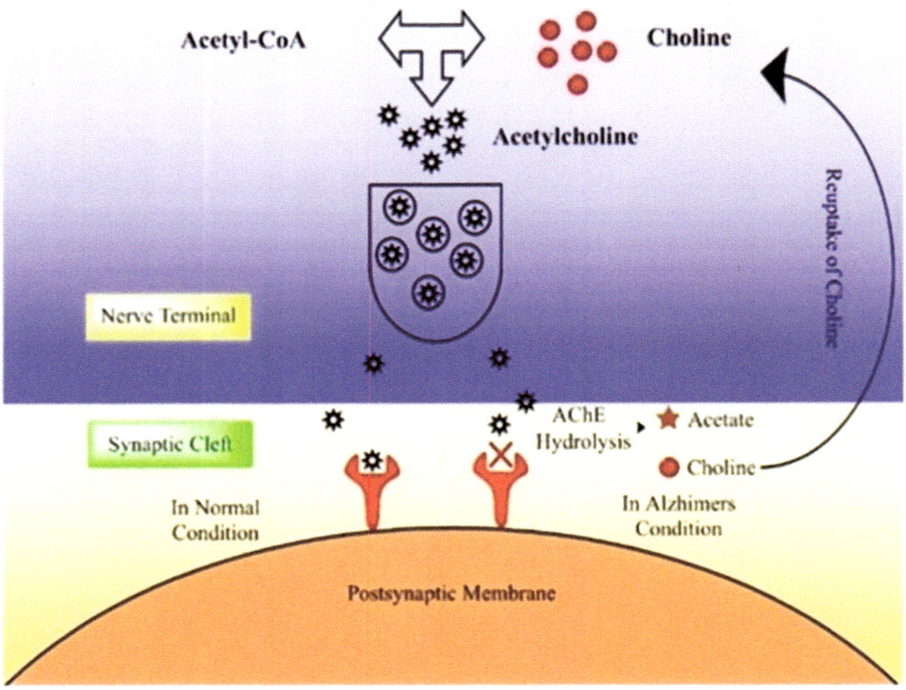

FIGURE 15.5 Mechanism of acetylcholinesterase causing cognitive impairment in the AD brain.

structures. Natural products are required in the development of novel pharmaceuticals (Wojtunik-Kulesza et al., 2016).

ETIOLOGY OF ALZHEIMER'S DISEASE

Aβ is generated by the proteases β-secretase (BACE1) and β-secretase, which process the β-amyloid precursor protein (APP). Although Aβ42 appears to be more aggregation-friendly than Aβ40, both species have been found in senile plaques. Several studies have found a correlation between a patient's degree of dementia in AD and their total brain Aβ levels (Näslund et al., 2000). About 90% of the Aβ in normal cells (Selkoe, 1999) is the less aggregative Aβ40, which is far more prevalent in healthy cells. For those who believe the amyloid theory, Alzheimer's disease is caused by an abnormal buildup of Aβ in brain cells. Aβ40 and Aβ42 are considered the most important Aβ isoforms out of all of them. Despite the fact that these two forms of Aβ differ by just two amino acid residues, a current study found that their metabolism, physiological activities, toxicities, and aggregation mechanisms are all the same (Qiu et al., 2015). According to a study by Ohno et al. (2007), Aβ is neurotoxic and causes neuron death. In contrast to neuritic Aβ plaques, diffuse Aβ plaques, which lack the compact structure of neuritic plaques, have been found.

Diffuse plaques are amorphous and granular, almost entirely composed of Aβ42 and have few amyloidogenic filaments and fibers present in neuritic plaques (Selkoe, 1994, 1999). This diffuse plaque in the same brain regions of healthy patients has no role in the development of AD. Advances in these fields have resulted in targeted drug therapy to reduce Aβ plaque accumulation in the brain. It's formed by the choroid plexus and protects the CNS from shocks and transports waste. The study of biomarkers in the cerebrospinal fluid (CSF) has grown in popularity because AD is a neurodegenerative disease. Several biomarkers in CSF have been identified for AD disorder's main pathological components, such as decreased Aβ, increased total tau, and phosphorylated tau protein. However, these two primary pathological mediators of illness and several other facets of this complex disease include oxidative stress, calcium-mediated toxicity, and neuroinflammation (Hampel et al., 2003; Sunderland et al., 2003; Blennow, 2004).

SYMPTOMS OF ALZHEIMER'S DISEASE

Alzheimer's disease and dementia are growing increasingly prevalent, and this is becoming a global health issue. In dementia patients, non-cognitive symptoms and habits known as "behavioral and social signs of dementia" (BPSD) or "neuropsychiatric symptoms" might develop. No matter what subtype of dementia you have, BPSD is a big part of it. Almost as essential in clinical terms are the severity of physical and cognitive impairments, which are linked to it. Symptoms of bipolar disorder include agitation, apathy, anxiety, delusions, sadness, irritability aberrant motor behavior, tiredness, apathy, delusions, hallucinations, sleep issues, and hunger abnormalities. It is believed that BPSD affects up to 90% of patients with dementia during their illness and is associated with several adverse effects, including anxiety of the patient and the caregiver, long-term hospitalization, and drug abuse. While these symptoms can develop on their own, numerous psychopathological features are more likely to co-occur in the same patient (Ricci, 2019). As a result, grouping BPSDs into clusters based on their typical course, prognosis, and therapy response could be useful in clinical practice. Although the specific causation of BPSD is unknown, it is most likely caused by a complex combination of psychological, social, and biological variables. Neurochemical, neuropathological, and genetic variables have all been implicated in the clinical symptoms of BPSD in recent studies. A high level of professional expertise is required to accurately recognize and treat the neuropsychiatric symptoms of a dementia patient. The recommended treatment for BPSD is a combination of non-pharmacological and pharmacological medications used consistently. Alternative pharmaceutical aims and non-pharmacological techniques are urgently needed to improve the unfavorable outcomes associated with BPSD, given the low success rate of current therapies. In experimental animal models, changes in brain blood arteries, including BBB breakdown, have been identified (Cerejeira et al., 2012; Takeda et al., 2014).

MANAGEMENT OF ALZHEIMER'S DISEASE

The primary goals of therapy seem to be to increase a patient's medical ability to function normally in daily life, to enhance their well-being, to delay the onset of symptoms, and to control symptoms of depression or disruptive behavior. It is important to identify and cease or substitute appropriate medications for certain substances that can exacerbate Alzheimer's disease, including liquor, tranquilizers, and even antidepressants (Selkoe, 1999). Furthermore, because present treatments only provide symptomatic alleviation and do not stop neuronal degeneration, innovative therapeutic approaches that can prevent neuronal failure are frequently required. Since Mother Nature has created numerous important nanoparticles (NPs) from plants, there has been a lot of interest in natural medicinal agents over the last two to three decades.

One example is galantamine, an AChE inhibitor isolated from *Galanthus nivalis* (Amaryllidaceae) and licensed for the treatment of AD symptoms. Clinical and preclinical research on huperzine A, EGb761 from *Ginkgo biloba* extract, essential oil of *Salvia lavandulifolia*, and other lead chemicals and plant extracts have shown promise (Natarajan et al., 2013; Hussain et al., 2018). A wide range of herbs have been prescribed by many traditional medical systems to treat a wide range of ailments. Several herbs classified as "Medhya Rasayana" in Ayurveda, an Indian herbal medicine system, help with neurodegenerative disorders (brain tonic that improves overall body functioning). Rasayana is a tonic or preparation that promotes endurance by improving physical, immunological, and cognitive functioning in the body (Kulkarni et al., 2012).

ACETYLCHOLINESTERASE INHIBITORS

Treatment with a cholinesterase inhibitor (ChEI) has been associated with improved memory in Alzheimer's patients, while untreated patients had worse cognitive abilities (Burns et al., 1999; Rösler et al., 1999; Tariot et al., 2000). The majority of clinical trials lasted three to six months, with open-label extensions demonstrating longer-term improvements in global control and cognition (Rogers et al., 2000; Raskind et al., 2000; Farlow et al., 2000). Despite the fact that the symptoms of AD might last for years, phase III studies with cholinesterase inhibitors, the current basis of symptomatic medication for the current scenario, usually last only three to six months (Hogan, 2014). Tacrine, donepezil, rivastigmine, galantamine, and memantine are examples of approved AD inhibitors (Figure 15.6) (Wilkinson et al., 2004).

TACRINE

In 1993, the Food and Drug Administration (FDA) authorized tacrine (9-amino-1,2,3,4-tetrahydroacridine) as another drug to treat Alzheimer's disease cognitive symptoms. Tacrine is just a transitory AChE antagonist; even so this carries a modest risk of hepatotoxicity, and hence its use has fallen dramatically. This can bind to both nicotinic and muscarinic receptors in the central nervous system. For N-receptors, it exhibits a 100-fold greater affinity. Tacrine also stimulates M1-receptors, which increases ACh release and inhibits both monoamine oxidase isoforms (Perry et al., 1988; Adem et al., 1989). Tacrine further inhibits dopamine and serotonin's re-uptake in the nerve endings, contributing to its antidepressant effect (Stenström et al., 1985).

DONEPEZIL

Donepezil was first approved in 1997 for the management of mild to moderate Alzheimer's disease as a nonclassical, reversible, and noncompetitive AChEI. It is regarded as one of the most effective and safest treatments for Alzheimer's disease. However, like other AChE antagonists, donepezil can only delay the development of AD to a certain degree and cannot cure it (Sabbagh et al., 2006; Sabbagh et al., 2008).

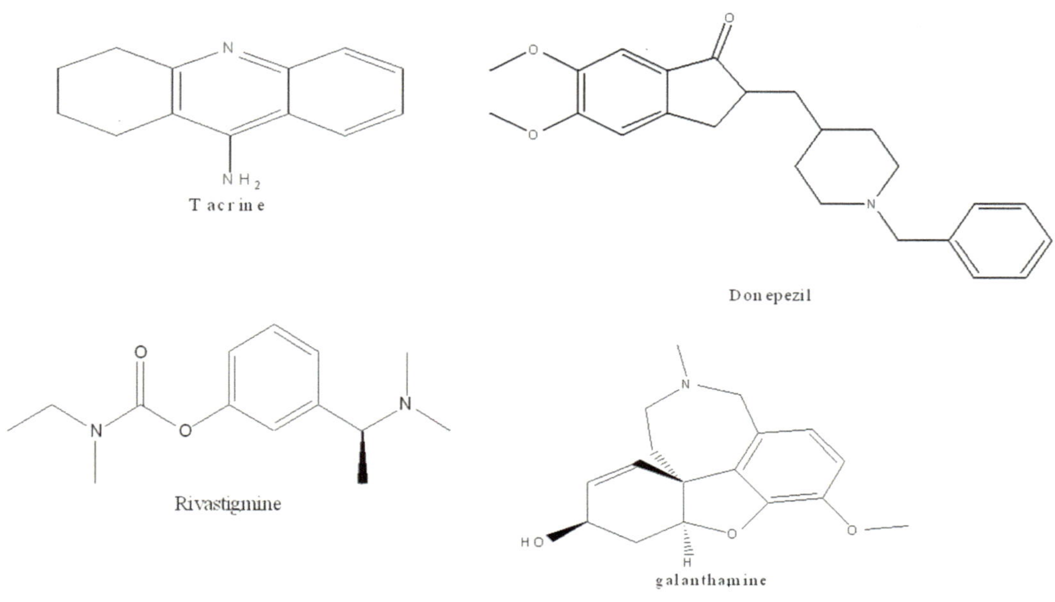

FIGURE 15.6 Chemical structures of some approved AD inhibitors.

Rivastigmine

Rivastigmine works by carbamylating the serine in the active site of AChE. In the year 2000, it was recommended for routine use. (*S*)-3-[1-(dimethylamino)ethyl]phenyl N-ethyl methylcarbamate is another name for the carbamate rivastigmine, which causes carbamoylation whether taken orally or intravenously. Gastrointestinal issues, such as vomiting, nausea, diarrhea, and anorexia, are the most common adverse effects. A daily dose schedule, on the other hand, can help to ease these issues (Molinuevo et al., 2012; Birks et al., 2015).

Galantamine

Galantamine is a plant alkaloid that inhibits AChE by competing with it. It's an inhibitory tertiary amine. AChE affects nicotinic acetylcholine receptors as well. Its chemical structure is ((4aS,6R,8 aS)-5,6,9,10,11,12-hexahydro-3-methoxy-11-methyl-4aH-[1]-benzofuro-[3a,3,2ef] [2] benzazepin-6-ol, [3] benzazepin-6-ol, [4] benzaze. Galantamine binds to the receptor on the opposing side of ACh and causes a conformational alteration in the receptor. In terms of Alzheimer's disease treatment, this is the most frequently recommended medication (Maelicke et al., 2000). By potentiating N-receptors, the entire process improves the postsynaptic response. Presynaptic N-receptors are required for the release of ACh as well as the regulation of other transmitters such as aminobutyric acid, glutamic acid, and serotonin-norepinephrine. These neuromodulators influence mood and emotions and are critical for memory (Levin et al., 1998; Alkondon et al., 2000).

Anti-Amyloidogenic Therapy

Reduced Aβ generation, stimulation of Aβ degrading catabolic routes, degrading catabolic pathways, immunotherapy for Aβ, and suppression of Aβ aggregation are all techniques used in anti-amyloidogenic therapy (Hamaguchi et al., 2006).

Memantine

Memantine causes headaches, sleeplessness, constipation, lethargy, auditory hallucinations, dizziness, and mild allergies in persons having mild to serious Alzheimer's dementia (Bullock, 2006). Curcumin was created as an early diagnostic agent with a high affinity for the amino acid Aβ, with the goal of preventing a buildup in the brain. BACE1 degrades the APP, which in turn degrades important targets for avoiding Aβ buildup (Thompson et al., 2005; Yang et al., 2005; Ghosh et al., 2008; Vassar et al., 2011). The term "memantine" refers to (3,5-dimethyladamantan-1-amine). This non-competitive inhibitor has a moderate specificity for the NMDA (N-methyl-D-aspartate) receptor, which is linked to the glutamate receptor, and has a medium affinity for NMDA. At normal glutamate concentrations, the NMDA receptor is only slightly blocked. However, increased glutamate concentrations considerably increase memantine efficacy, resulting in a severe inhibition of the NMDA receptor. Parkinson's illness, epilepsy, CNS trauma, and amyotrophic lateral sclerosis can all be treated with memantine (Koch et al., 2005; Sonkusare et al., 2005; Herrmann et al., 2011).

Curcumin

Curcumin was extracted from *Curcuma longa*, also known as turmeric, and has been used for centuries to treat wounds and inflammation due to its antibacterial and anti-inflammatory properties (Marathe et al., 2011). Curcumin has been demonstrated in clinical studies to benefit patients with AD (Hamaguchi et al., 2006; Salehi et al., 2019). Curcumin is non-toxic even at higher doses; however, its poor permeability into the BBB is a significant disadvantage of this model (Zhang et al., 2018). There is no effective medication that can stop the disease from progressing.

The therapeutic benefits of numerous medicinal plants utilized in various traditional medical systems, as well as the correlation of folkloric medicinal plant use with scientific data, have become increasingly recognized. Therefore, at present, there is a great need to develop new anti-AD drugs. Medicinal plants are used to cure a wide range of diseases, and as a result, they have become more important in the drug development process for identifying the active molecules that provide therapeutic benefits. For this reason, we aimed to evaluate the anti-Alzheimer's activity of three medicinal plants selected based on their ethno-medicinal use and an *in silico* approach to identify potential lead compounds from medicinal plants for preventing and treating AD. The production of plant-based drugs in bulk is one of the essential criteria for the pharmaceutical industry.

MERITS OF NANOMATERIALS AND THEIR APPLICATION TO DIAGNOSIS OF AD

Nanomaterials have unique physicochemical properties due to unusual effects such as macro quantum, surface, and quantum dimension effects. These nanomaterials benefit from a large separate cavity for loading large amounts of various drugs and auspices as a result of enzymatic decay, improving drug consistency and quality. To increase target specificity and reduce off-target and adverse reactions, drugs incorporating nanoparticles could be processed and deposited into erythema cells and tissue as well as aberrant portions of tissue and cell units (Figure 15.7) (del Burgo et al., 2014; Garbayo et al., 2014; Jembrek et al., 2018; Polanco et al., 2018; Nakamura et al., 2018). As AD progresses, neuronal loss, NFTs, and amyloid deposition become more prevalent. AD biomarkers include overall tau protein (T-Tau), Aβ42, and abnormal tau protein phosphorylation (P-Tau) (Blennow et al., 2018; Jagust, 2018; Song et al., 2018). A biomarker that is sensitive, precise, and noninvasive would be tremendously helpful in the early detection of Alzheimer's disease symptoms. When used in conjunction with certain molecular probes, brain MRIs and PETs (Kincses et al., 2015; Yilmaz, 2015; Li et al., 2019; Zhang et al., 2017; Jack, 2012; McKhann et al., 2011) provide a powerful diagnostic tool for monitoring tau protein and Aβ42 (Table 15.1).

In biomedical applications, biopolymers have shown promise in fields such as medicine, pharmaceutics, and tissue design. Chitosan (Ch) and poly(lactic-co-glycolic) acid (PLGA) are polymer-based nanocomposites (Bharadwaz et al., 2020). PLGA has long been used in medical and engineering applications (Ignjatović et al., 2016). PLGA is an eco-friendly polymer. It is easy to process, has established syntheses, and is 100% biodegradable and biocompatible. PLGA-encapsulated

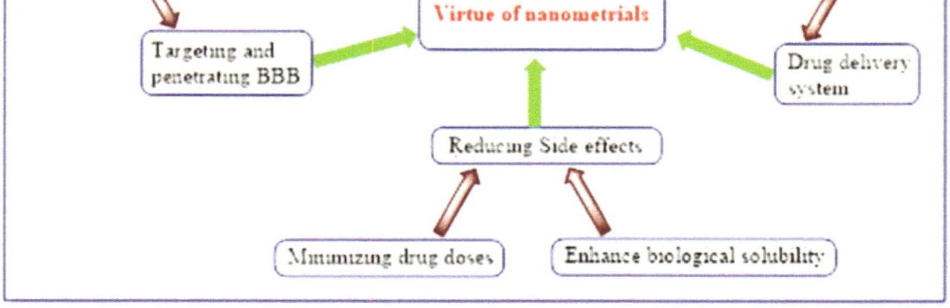

FIGURE 15.7 Merits of nanomaterials.

bevacizumab increased availability while decreasing toxicity in ocular angiogenesis (Zhang et al., 2018). As a result of its biocompatibility and interesting features, Ch can be used in a number of biomedical applications (Yin et al., 2009; Yien et al., 2012; Ararcibia et al., 2013). Bioglass scaffolds can also be utilized instead of collagen, preventing allergic responses and pathogen transmission (Vukajlovic et al., 2019). Ch and bioglass also form chemical bonds that increase mechanical strength (Vukajlovic et al., 2019). As shown in the following experiment, hydroxyapatite/Ch (HAp/Ch) amalgamated particles are antimicrobial, but they also promote inflammation in the developing tissue around the defect (Ignjatović et al., 2016). A polymeric mixture of Ch and PLGA covering HAp may minimize its reactivity and antibacterial activity while improving the quality of freshly produced bone tissue. Changes in PLGA solubility and surface charges can lead to effective cancer drug transporters. Epirubicin-PLGA-Ch NPs outperform free Epi in MCF-7 breast cancer cells. A 5TR1 DNA aptamer targeting the MUC1 receptor (Epi-PLGA-Ch-Apt) also suppressed cancer progression in mice xenograft experiments, suggesting that Epi-PLGA-Ch NPs may diffuse from the cancer site to release Epi into the extracellular space, lowering drug concentration in the target tissue (Taghavi et al., 2017). PLGA-Ch NPs show promise for *in vivo* cancer growth treatment and

TABLE 15.1
Uses and Biological Mechanisms of Nanomaterials in AD Analysis

Nanomaterials	Biological Effect and Mechanism	Reference
Antibodies specific to the Aβ oligomer-linked MNPs	*In vitro* and *in vivo* screenings of the Aβ oligomers that are most likely to cause neurotoxicity may be performed using magnetic resonance imaging.	(Viola et al., 2015; Pansieri et al., 2018)
DNA nanomachine	Lysosomal pH and chloride may be measured simultaneously and quantitatively using a combination of DNA-based reporters, which might be used to research lysosomal disease progression, manage disease progression, and assess the treatment value in AD, among other uses.	(Leung et al., 2019)
AGulX-NPs	Penetrating the blood–brain barrier with minimal toxicity, target amyloid agglutination in the brain, and allow clear visualization of amyloid plaques without side effects.	(Pansieri et al., 2018)
HRP and gelsolin in combination with Au nanoparticles	Very sensitive for determining Aβ concentrations in the normal and AD rat brains.	(Yu et al., 2015)
NP@SiO2@F-SLOH	Compiled as an effective and active multimodal contrast agent for the analysis and visualization of Aβ contents in various age groups of recombinant adult mice. Showed apparent antagonistic effect on Aβ fibrillation and neuroprotective effects toward Aβ-induced toxicity.	(Wang et al., 2020)
A biosensor based on boron-dipyrromethene	Analyzing Aβ aggregation in AD may be possible using this method, which monitors the identity of diphenylalanine.	(Quan et al., 2019)
CuInS2/ZnS quantum dots complexed with DA	Using a linear concentration range of 10 pM to 200 nM, the emission intensity of the immunoassay demonstrated satisfactory accuracy in recognizing tau protein.	(Chen et al., 2019)
Nanobiosensors based on poly-L-lysine-mediated chemical method	A chemical reagent that activates the enhanced amino groups to correct the Aβ analysis's compassion.	(Kang et al., 2019)

tissue engineering. It can promote mental openness by modifying BBB penetrability. When paired with nanoparticles, the most common penetrant, mannitol, can help nanoconveyance frameworks enter and cross the BBB (Rousseau et al., 1997; Son et al., 2011; Doolittle et al., 2014). NPs can also enter cells via receptor-mediated BBB piercing. It is possible to penetrate the BBB by improving the specificity of ligand-binding to receptors on the surface of brain endothelial cells (Jnaidi et al., 2020). Surface angiopep-2-modified NPs could improve drug delivery to glioblastoma multiforme cells by binding to lipoprotein receptor-related protein 1 on the BBB (Jnaidi et al., 2020). According to another study, DOX-SLNs-CBSA had the maximum transcytosis and successfully delivered DOX to brain tissues with the lowest immunogenicity. On the other hand, free DOX-SLNs or unconjugated SLNs were more cytotoxic in HNGC-1 cell lines (Agarwal et al., 2011). Transporter-intervened components can also transport substances to the brain via the BBB (Pardridge, 2005). One has used degradable oxidized permeable silicon (PSiO2) as a transporter to deliver NGF into the brain in order to treat Alzheimer's (Zilony-Hanin et al., 2019). MEM-PEG-PLGA nanoparticles loaded with memantine could target the BBB after oral administration, minimizing drug organization recurrence *in vivo* by delivering memantine gradually. For treating AD, MEM-PEG-PLGA NPs outperformed open medicine.

POTENTIAL NANOTECHNOLOGY THERAPY FOR ALZHEIMER'S DISEASE

Lipid-Based Carrier System

When it comes to improving the bioavailability of poorly soluble medications, using a lipid-based carrier system holds a lot of potential. Even though the ways to make a drug more bioavailable are not the same as those used in other carrier systems (Figure 15.8), these systems do use a unique characterization procedure to make the drug look like it was made while it was traveling through the gastrointestinal tract. Despite the development of an *in vitro* dynamics lipolysis model, progress

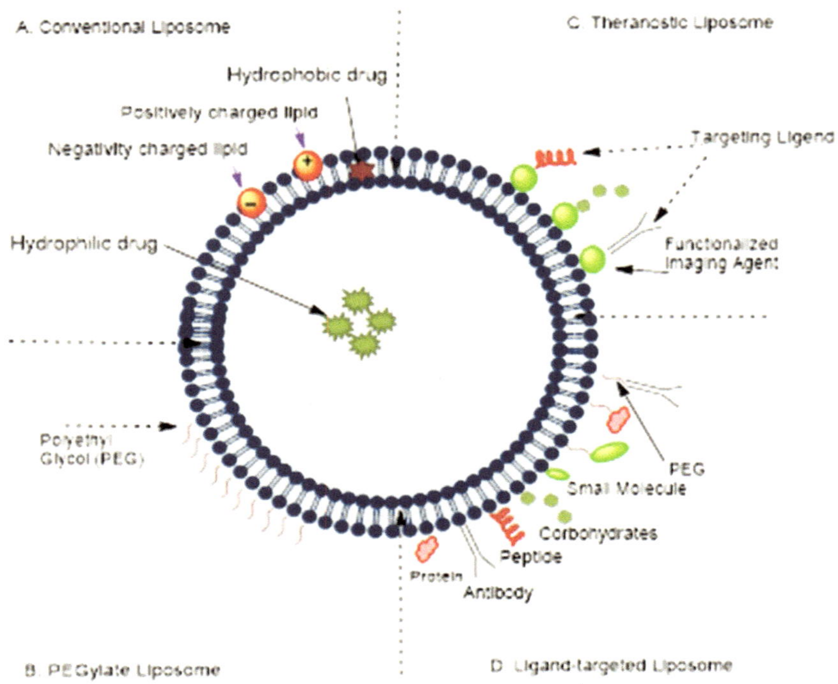

FIGURE 15.8 Various types of liposomal drug delivery systems.

in developing an *in vitro–in vivo* correlation has been made. Scientists are still working to develop robust and simple methodologies for forecasting these systems' *in vitro* fate. The basic issue is that, because of their liquid nature, lipid-based carrier systems have traditionally been linked to poor stability and challenging formulation management. Different approaches for solidification of liquid formulations have been used to successfully overcome these difficulties. Spray-drying, freeze-drying, granulation, and spray-coating inert pellets with these industrially practicable procedures permitted the large-scale manufacture of these products, which should result in an increase in their number on the market (Krstić et al., 2018).

Liposome

Liposomes are well-developed nanometric lipid vesicles that may be useful for administering topical medications. Liposomes are the first generation of nanoparticulate drug delivery devices, and they are made up of one or more vesicular bilayers composed of amphiphilic lipids that form a barrier around an interior aqueous compartment (Masserini, 2013) Although liposomes have a lipophilic membrane on the surface, certain cellular obstacles, like the BBB, prevent the permeation of nanoparticulate drug delivery devices, and they are made up of one or more vesicular bilayers composed of amphiphilic lipids that form a barrier around an interior aqueous compartment (Masserini, 2013). Although liposomes have a lipophilic membrane on the surface, certain cellular obstacles, like BBB, stay impermeable. The outermost membrane's charges affect liposome distribution and stability. According to research, negatively charged lipid bilayers are eliminated from the circulatory system faster than neutral or oppositely charged liposomes.

Because of this, researchers have been experimenting with ways to make liposomes more permeable so that they can reach their intended organ destinations via biological membranes. After regular stealth liposomes failed to stop the active uptake of chemicals by sensitive normal cells or nonspecific targets, a more specific technique based on liposomes was developed. The ability to store a large number of drug molecules in a single compartment is the fundamental advantage of these architectures. In contrast, the liposome structure offers a number of issues in administration. The stomach's low pH and the presence of bile salts destroy the liposome complex, making oral delivery challenging.

Liposomes are excellent nanocarriers because they have a high capacity for medication loading. Particles have been used as a medication delivery method to the brain for a long time because they can entrap the chemicals and prevent their quick removal or degradation, as well as improve BBB penetration (Arora et al., 2020). AD is the most common kind of neurodegenerative illness among the elderly. It's still unclear how this disease manifests itself. Depressive symptoms include a loss of short-term memory, language difficulties, and a lack of time awareness. As the disease progresses, behavioral and psychological problems begin to emerge, and motor abilities begin to deteriorate. Alzheimer's disease may be caused by the accumulation of Aβ; however, there appears to be no medicine that may entirely cure the disease. Aβ peptide accumulation and removal are the focus of the most advanced therapies. Rivastigmine, a cholinesterase inhibitor, is FDA-approved specifically for the treatment of AD (Agrawal et al., 2017). The use of a current pharmaceutical that has been developed using a unique nanotechnology-based formulation was the subject of a prior study, which we address here. Liposome technology was used to create rivastigmine liposomes for intranasal delivery into the brain. According to this study, the use of liposomes to deliver the medicine boosted the drug's exposure and concentration in the brain. Recently, researchers produced a new liposome formulation based on rivastigmine cholesterol. When compared to drug delivery without liposomes, this investigation found that intraperitoneal and oral administration increased drug exposure to the brain considerably (Ordóñez-Gutiérrez et al., 2020).

Another example of a medication that benefits when administered in liposomal form is quercetin. Although oral treatment with quercetin helped Alzheimer's patients' learning and memory, the fundamental challenge with AD is poor absorption and difficulty bridging the blood–brain barrier. Nebulized quercetin liposomes administered intravenously in an animal model of AD prevented hippocampal cell death and cholinergic neurons in the cortex. Amyloid peptide immunization has shown promise in the treatment and prevention of Alzheimer's disease throughout the past decade. It has been demonstrated that vaccinations can prevent or even remove amyloid plaques in transgenic rat models of Alzheimer's disease (Spuch et al., 2011).

When it comes to dealing with neurodegenerative diseases, NPs are an intriguing biochemical instrument that has the potential to treat a variety of issues. On a therapeutic level, they allow for more than one function. For example, one ligand may let a second ligand, the drug, pass through the blood–brain barrier. They can also been used in therapy, with one ligand making it easier to attach to amyloid plaques and the other making molecular imaging techniques easier to use (Patra et al., 2018).

Nanoemulsion

Nanoemulsions are systems of two liquids that don't mix and are spread out evenly in all directions to form tiny droplets or oily phases. Usually, an oily phase is spread out in an aqueous medium, or an aqueous phase is spread out in an oily medium. In some situations, nanoemulsions can hold on to more active molecules that are lipophilic than microemulsions can (Simonazzi et al., 2018). Unfortunately, none of the treatments that are currently available can completely prevent AD. This is because of a number of factors, such as how hard it is to cross the blood–brain barrier. This makes drug concentrations rise in the periphery compared to the brain, which causes more side effects and requires more frequent dosing. To overcome these obstacles, a novel drug delivery technique that improves not only cognitive function but also targets specific transport areas in the AD brain is required. More and more people are using nasal medicine to treat brain diseases like neurodegenerative and psychiatric diseases, epilepsy, and so on. When a drug is given through the nose, it can be absorbed into the CNS without going through the blood–brain barrier. This allows for therapeutic targeting of the CNS with minimal systemic exposure and quick drug levels in the target tissue.

In the past few years, a lot of research has been done on nanocarriers to improve the bioavailability and effectiveness of some drugs used to treat AD. Nanotechnology has shown that it could be used to deliver drugs directly to the site where they are needed. As part of this study, a nanoemulsion was chosen to make new nanoparticle formulations, which were then tested by being breathed in through the nose. Following the discovery of the antagonistic capabilities of nanoemulsion and NMDA receptors, a one-of-a-kind medication known as a non-competitive NMDA binding site agonist was manufactured. The medicine was developed to assist those who are suffering from mild-to-moderate memory decline. Additional clinical studies are required before the European Medicines Agency authorizes nanoemulsion for the management of moderate-to-severe neurodevelopmental disorders. The medication was authorized by the Food and Drug Administration in 2003 for moderate-to-severe AD. It is typically given twice a day by mouth and has a half-life of around 24 hours (Mukherjee et al., 2009). It has been found that the oral administration of nanoparticles for memory enhancement seems to be more efficient than free medicine in certain cases.

Solid Lipid Nanoparticles (SLNs)

Traditional formulations have a number of limitations, including biocompatibility, safety, and toxicity, which led to the development of a lipid-based drug delivery technique to address these difficulties. These formulations have advantages in terms of drug chemical degradation prevention and flexibility in altering drug release when compared to liposomes. High drug loading, drug protection from the environment, an enlarged surface area, and increased bioavailability are all advantages of solid lipid nanoparticles (SLNs). SLNs can deliver site-specific drug delivery, controlled drug delivery, and targeted therapy (Gautam et al., 2022). The use of solid lipid nanoparticles in the management of AD has recently been demonstrated, circumventing previously encountered difficulties in the treatment of dementia. After being administered orally or intravenously, SLNs containing bioactive components demonstrated high bioavailability in the central nervous system, possibly due to their capability to cross the blood–brain barrier, thereby preventing inflammation and future AD progression in the study participants. The effectiveness of a quercetin-loaded SLN with a particle size of around 200 nm in the AD model was examined, and it revealed efficient quercetin dispersion as well as improved antioxidative activity in brain cells. When bioactive chemicals are transported to the brain by endocytosis of brain capillaries, they are able to pass all the way across the blood–brain barrier (Ravi et al., 2017).

The results of the study show that SLNs coated with piperine are more bioavailable to brain cells. This means that they may be able to stop the progression of AD. Studies show that piperine may boost acetylcholinesterase activity, which minimizes the production of cranial plaques and boosts mental performance. Ferulic acid-loaded SLNs have been shown to protect brain cells from oxidation without inflicting any damage, according to new research on neurotoxicity. *In vitro* and *in vivo* studies have shown outstanding anti-Alzheimer's benefits for curcumin, another possible anti-Alzheimer's chemical, but because of its low brain content and poor oral bioavailability, it has not been able to produce a substantial impact. Curcumin bioavailability has been improved in recent research by using SLNs for oral transport of curcumin to the brain (Ganesan et al., 2017).

Nanostructured Lipid Carriers (NLCs)

Nanostructured lipid carriers (NLCs) are a type of medicinal formulation that combines physiologically and biocompatible surfactants and co-surfactants with lipids. Nanoparticles of the first generation have been replaced by second-generation lipid nanocarriers. As a second generation of lipid nanoparticles, NLCs have been developed to overcome the shortcomings of first-generation SLNs. NLCs are created using biodegradable and appropriate lipids and emulsifiers. Structure flaws are created in the crystal structure when liquid oil is added, which limits drug leakage while still providing sufficient drug loading. SLNs, polymeric microparticles, nanoparticle liposomes, emulsions, and other nanoparticles have all been researched as alternatives to NLCs in the past. In addition to hydrophilic and lipophilic medicines, these nanoparticles may distribute them. There are several ways in which NLCs might be employed for medication administration, including oral and transdermal. In the food and cosmetic industries, as well as the nutraceutical industry, NLCs are now being used for a variety of applications (Chauhan et al., 2020). It is difficult for Alzheimer's disease patients to carry out their everyday duties because of memory loss and other cognitive difficulties. With a life expectancy of three to nine years, each patient's illness advances differently. Studies on Alzheimer's disease treatment using lipid nanoparticles are the subject of this section. An antidementia medicine known as huperzine A (HupA) has been shown to improve cognitive function in people with AD. SLN-encapsulated HupA has been the subject of two studies in the past seven years.

There are several possible therapeutic options for AD, but one of the most promising is to target the neurotoxic Aβ peptide that forms fibrils, oligomers, and plaque in the brain. One of the studies that employed SLNs to construct a tailored medicine delivery system for AD targeted these (Aβ) amyloid clumps (Tapeinos et al., 2017).

Inorganic Carrier Systems

Quantum Dots

Quantum dots (QDs) are nanoparticles that act as theranostic agents in the treatment of neurological illnesses like AD. QDs have useful fluorescent and electronic properties, which have a significant impact on quantum physics. According to Azria et al. (2017), the primary advantages of QDs lie in their innovative and unique optical properties, making them suitable for use as nanomedicine due to their innovative and unique optical properties. They are used for the measurement of high sensitivity, selectivity, and stability at the nanoscale range. The use of QDs and elucidation is one of the promising applications for diagnosis and therapeutic activities in AD. Patients with AD may use QDs to detect any toxicity across the BBB. Due to its tiny size, they are able to pass through the bloodstream and into the brain. QDs have made a positive impact on AD therapy by binding to a protein called amyloid, which is essential for protein clumping in AD therapy. In recent times, hazards have increasingly become a major concern. Not only AD but cancer is also a major issue that has been growing exponentially in recent decades. Millions of people die of cancer each year, and it cannot be easily cured by doctors. On a nanometer scale, quantum dots have the ability to do molecular imaging by acting as a fluorescent probe, which is also a pivotal matter of concern in this study. The formation of toxic proteins can also be blocked by quantum dots, which are intermingled with the treatment of brain disease.

Drug-resistant strains are also an important aspect that needs to be intermingled with the pathogens of periodontitis. As per the view of Liang et al. (2020), antibacterial and antibiofilm treatments can be effectively driven by carbon quantum dots. Hydrothermal methods can be frequently used for preparing carbon quantum dots. Two quantum dot methods have been used in this scenario, namely, tinidazole carbon quantum dots (TCDs) and metronidazole carbon quantum dots (MCDs). On the other hand, according to the views of Sikorska et al. (2020), microglial cells are beneficial for clearing the pathogen in the brain. The inclusion of Aβ in AD is beneficial for reducing hazards in brain cells. There are some risk factors for AD that can be well understood by the expression of some genes. The genes are Msrl, CD36, and AGER. Quantum dots are beneficial for tracking all of the lethal hazards in human microglial cells and mitigating health-related issues. The effects of silver nanoparticles and cadmium telluride quantum dots are beneficial for tracing these gene expressions. These two nanoparticles are not only beneficial for tracking the hazards in microglial cells; they can also investigate the dysregulation of the P53 gene.

Magnetic Nanoparticles

Magnetic nanoparticles (MNPs) are used for biomedical applications (Figure 15.9), such as magnetic hyperthermia; they are enabled to create magnetic resonance imaging (MRI) data; and they help deliver drugs in complex positions. Diagnostic techniques such as positron-emission tomography (PET), MRI, and computed tomography (CT) are employed for brain imaging in AD. According to Fernández et al. (2018), different kinds of small particles, such as curcumin, Congo red, and methoxy-XO_4, can be used as fluorescent tracers with the help of a hydrophilic functional group. A higher concentration of magnetic nanoparticle doses is necessary for a significant impact on AD (Hammond et al., 2021). The potential cytotoxicity can be used as the nanocarriers to the brain (Morales-Zavala et al., 2017). Recently, it has been observed that nanoprobes, with the help of L-poly(glycolic acid) (L-PGS), are conjugated to Au nanocages for the detection of Aβ in the brain. The use of various formulations of iron oxide nanoparticles (IONPs) and peptide cages for measuring beta amyloid in the brain during atomic MRI is equally useful. Amyloid plaques are the main cause of this disease, and beta-amyloid 42 is thought to be the most harmful in this case. In this illness, every naturally occurring protein clumps in an inappropriate amount between neurons. All of the plaques accumulate between the end brushes of the neurons; thus, this disease can be enhanced by disrupted cell function. The function of disrupted cells can be traced by magnetic nanoparticles. All of the consequences enhance some neurodegenerative diseases, which can be managed by these particles. Methoxy-XO4 and Congo red are derivatives of each other. It is also a derivative of chrysamine G that contains no acid groups.

These two nanoparticles are also used in brain imaging, which is the first and foremost importance of the nanoparticles. According to Sánchez-López et al. (2018), understanding the aggregation and accumulation of Aβ requires *in vitro* and *in vivo* profiling of the nanoparticle. Vitamin

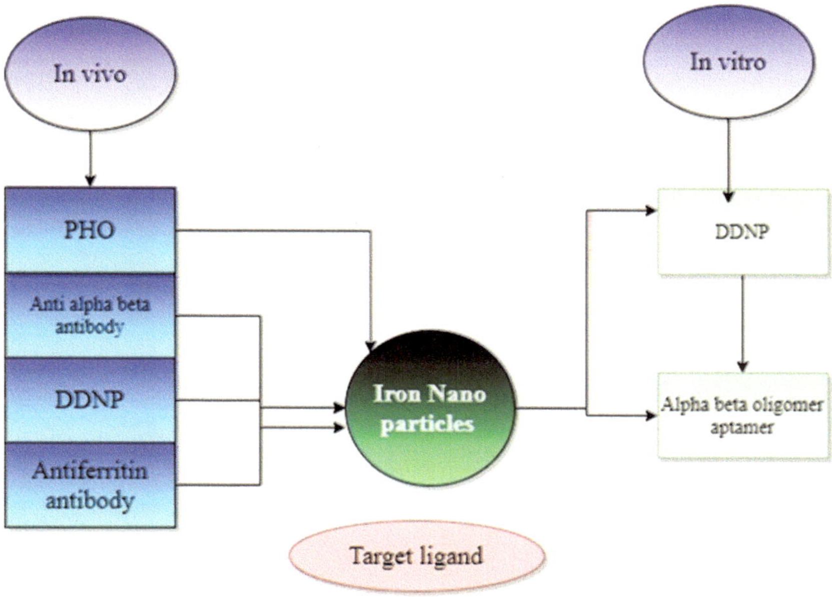

FIGURE 15.9 Diagnosis of a neurodegenerative disease.

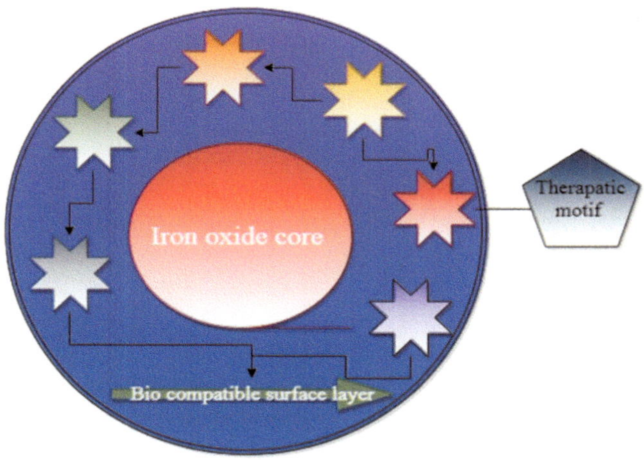

FIGURE 15.10 Magnetic nanoparticles as a key *in vivo* tracer of Alzheimer's disease.

D binding protein and Aβ show different patterns of regulation in this scenario. Conversely, as mentioned by Jeon et al. (2019), PLGA nanoparticles are always beneficial for suppressing protein-coding genes of AD. These two points of view show that aggregation and the accumulation of Aβ gene suppression are beneficial for regulating the disease. Figure 15.10 shows the schematic for adjustable magnetic tracers used as diagnostic tools for Alzheimer's disease.

The tunable functionalization of the iron oxide core has biocompatible surface layers that are the main theme of the characteristics of magnetic nanoparticles. The extracellular space of the brain has pores larger than 100 nm, which is the first and foremost important aspect of calculation in the development of AD probes (Sharma et al., 2020). These nanoparticles are also useful for

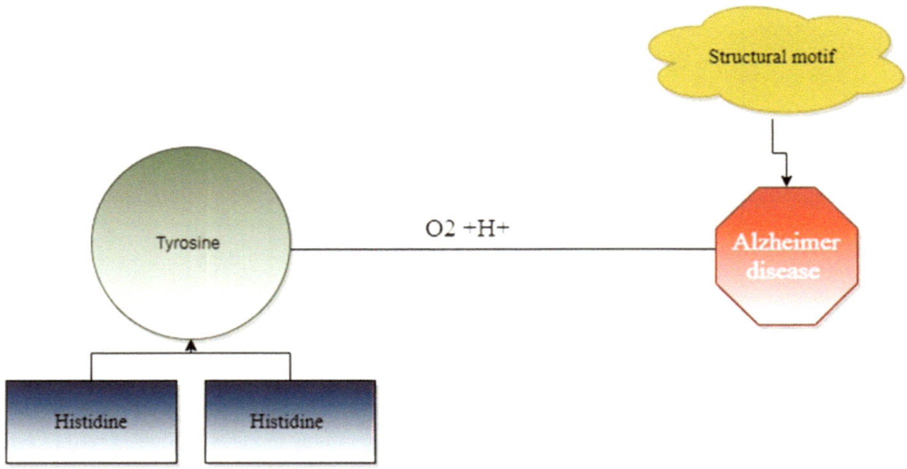

FIGURE 15.11 Rescue from memory deficit by chiral gold nanoparticles.

MRI in detecting some brain-related problems. Magnetic nanoparticles can also detect deposits of cholesterol with the aid of amyloid plaques in a broad array. A NANOCHOAD consists of four important components: an iron oxide magnetic core, PEG+ streptavidin, anti-CHO antibody, and PEG+ transferrin.

Gold Nanoparticles

Drug delivery via the blood–brain barrier in neurological diseases like AD is extremely difficult in the present era. According to Amin et al. (2017), gold nanoparticles (AuNPs) are used to provide authentic treatment and deliver drugs in the BBB (Figure 15.11). Furthermore, they are is crucial for the sustainability of neuronal activity in AD. In Alzheimer's disease management, AuNPs contain antioxidant and anti-inflammatory characteristics. The presence of okadaic acid (OA) and OA-AuNP helps to increase the amount of interleukin (IL)-1β in the cortex and hippocampus. AuNP is enabled to reduce oxidative stress. In AD, patients suffer from spatial memory loss; AuNP treatment is enabled to prevent deficiencies in this disease. AuNP is enabled to maintain normal brain function for the maintenance of different activities in the brain. Long-term AuNP treatment can help to reduce neuroinflammation and modulate mitochondrial activity. Improvement of cognitive deficits is essential for the eradication of this neuron disease. Nontoxic control can be provided by gold nanoparticles. Inflammation of mitochondria can also be controlled by these particles, which is the foremost crucial power of them. In brain microglial cells, high levels of iron can be easily replaced by AuNPs, which is the main beneficial aspect of these particles. In addition, AuNPs can be effectively used for resonance scattering dark field microscopy. Reinforced Self-Dispersing Formulation Matrix (RSDFM) is always beneficial in this scenario for the detection of microbial metabolites.

In the opinion of Hou et al. (2020), chiral gold nanoparticles are always beneficial for rescuing memory deficit enantioselectivity. This experiment was carried out on a mouse model to ensure the quality of the particles. It has been established from the present study that human brain cells are also applicable to this experiment. Memory deficits in the human body can also be rescued by the application of gold nanoparticles (dos Santos et al., 2020). Compared to other nanoparticles, gold nanoparticles have lower toxicity. They work as an Aβ inhibitor, which is also beneficial in the treatment of the disease.

OTHER NANOCARRIER SYSTEMS

Hydrogel

Hydrogel is known as a cross-linked hydrophilic polymer; it is insoluble in water. Due to its high absorbency, it maintains an appropriate structure. There are numerous uses of hydrogel in the biomedical sector. According to Martin-Rapun et al. (2017), the physicochemical structure of hydrogel is used in 3D culture. This 3D culture is used for cell morphology in AD (Martin-Rapun et al., 2017). Hydrogel is enabled to produce type 1 collagen; it has low melting properties that make a significant impact on brain tissue. Hydrogel is useful in addressing issues caused by a lack of progenitor cells. Delivering medicines as a local transport system is important for providing sufficient nutrients, protection, and oxygen for new cell formation in the brain. Moreover, it helps in extracellular matrix formation. The primary role of hydrogel is to control the host neural tissues in the brain. There are different types of chemical properties in hydrogel, such as chitosan, methacrylates, polyethylene glycol, and hyaluronic acid to facilitate brain functions. Although hydrogel has a higher concentration, it requires high stability for the modification of neurodegenerative disease. Moreover, it has some self-healing properties, which are required for effective treatment in AD. Hydrogel formation is used for releasing more than one growth factor to improve a patient's health status. As hydrogel is insoluble in water, it is beneficial in producing type 1 collagen. This collagen is crucial for assembling fibers, which forms the structural and mechanical scaffold of bone. Not only can the bone but also the connective tissue in the brain be reinforced by the application of hydrogel. In the brain, collagen XVII is always found widely, which is the first and foremost matter of concern in the case of AD.

An abnormal buildup of normal protein can easily develop in the brain (Choi et al., 2018). Not only can the brain, but the brain outer layer, which is known as the meninges, can also be affected by this abnormal accumulation of proteins. The reduction of amyloid plaques can also be aided by hydrogel. In the brain, there are various lobes that have different functions. The fifth lobe of the brain is known as the insula or "Island of Reil" and is mostly affected by the accumulation of uncontrolled proteins. These uncontrolled proteins can also cause damage to the temporal lobe of the brain. The application of hydrogel is also beneficial for curing this disease with the help of cross-linking polymers.

Biodegradable Scaffold

The biodegradable scaffold is used to consider the indispensable elements in living tissue. These are made up of biodegradable polymers, which are used for tissue regeneration in AD. According to Maher, it enables delivering a nanofibrous matrix, a 3D porous matrix, a microsphere, and a thermosensitive hydrogel in the brain. The natural extracellular matrix has special biological and mechanical qualities that are applied to the care for neurological illnesses such as AD (Maher, 2019). The biodegradable scaffold is used for the maintenance of cell delivery strategies for the improvement of health conditions. This material is mainly used for the stimulation and interaction of tissues and cells to produce effective physiological responses in AD. The use of nanotechnology provides an authentic route for delivering therapeutic agents in human cells. Biodegradable scaffolds can be prepared from peptide RADA16, which is made up of a sequence of arginine–alanine–aspartic acid alanine. The uses of the biodegradable scaffold provides protection to the cells for the prevention of AD. Cell delivery is one of the innovative ways for the prevention of risk factors in AD. Tissue regeneration is possible in this scenario with the help of biodegradable polymers. Nowadays, biodegradable scaffolds are widely used for physiological responses to some extent, such as atrophy. It is one of the problematic aspects that affects brain cells to a wide extent. Shrinkage of brain cells has been observed in this scenario. Peptide RADA16 disintegrates harmful proteins in the brain cells (Gelain et al., 2020). This has some adverse effects related to the development of brain cancer as well. As per the view of Sankar et al. (2021), D and R residue of amino acids can be effectively used for self-assembling. The delivery of therapeutic agents in human cells can be efficiently traced by biodegradable scaffolds.

CONCLUSION

Since degenerative diseases of the central nervous system were discovered, it has become a major medical task to diagnose and treat them. Across the globe, AD is the most common reason for mortality. AD's societal and economic costs are rising at an ever-increasing rate. Alzheimer's disease has no cure, but early detection is essential to preventing the development of degenerative symptoms and preparing caregivers and loved ones for the inevitable changes that will accompany it. Efforts to administer drugs directly to the brain have seen remarkable success in the field of nano-based medication delivery. A lot more has to be learned in the field of nanomedicine for brain delivery, even though nanotechnology has opened up many new possibilities. There are still considerable questions concerning which formulations of nanomedicine are best. AD is the most frequent form of dementia, and it's associated with a variety of unusual risk factors and illnesses. AD researchers have made tremendous strides in their understanding of the disease and developing new treatments. There must be a thorough analysis of AD diagnostic pathways that differ among persons with varied causatives in order to gain a better grasp of the underlying processes.

In this chapter, we look at the latest research on AD. We also make some guesses about how the disease works, what causes it, and what its symptoms are. Diseases like Alzheimer's induce memory loss and cognitive decline, as well as personality changes and behavioral abnormalities due to the death of brain neurons. The management of AD is influenced by a plethora of other factors. By utilizing the nanomaterials approach, it is possible to monitor the signs of AD in great detail. It was also claimed that nanotechnology therapy for Alzheimer's disease might include lipid-based carrier systems, inorganic carrier systems, hydrogels, and biodegradable scaffold nanocarriers. When we have more advanced nanotechnology-based drug delivery frameworks, we hope to use them for the treatment of drug-resistant diseases or invasive neurological illnesses.

LIST OF ABBREVIATIONS

ACh: acetylcholine
AChE: acetylcholine esterase
AChEI: acetylcholine esterase inhibitor
AD: Alzheimer's disease
ANS: autonomic nervous systems
APP: amyloid precursor protein
AuNP: gold nanoparticle
Aβ: β-amyloid
BACE1: β-secretase
BBB: blood–brain barrier
BPSD: behavioral and social signs of dementia
Ch: chitosan
ChEI: cholinesterase inhibitor
CNS: central nervous system
CSF: cerebrospinal fluid
CT: computed tomography
FDA: Food and Drug Administration
HupA: huperzine A
IONP: iron oxide nanoparticle
MCDs: metronidazole carbon quantum dots

MNPs: magnetic nanoparticles
MRI: magnetic resonance imaging
NLC: nanostructured lipid carrier
NMDA: N-Methyl-D-aspartic acid
NPs: nanoparticles
OA: okadaic acid
PET: positron-emission tomography
PLGA: poly(lactic-co-glycolic) acid
P-Tau: tau protein phosphorylation
QDs: quantum dots
SLNs: solid lipid nanoparticles
TCDs: tinidazole carbon quantum dots

ACKNOWLEDGMENT

The authors would like to express their gratitude to the Department of Pharmacy at Indira Gandhi National Tribal University, Madhya Pradesh, India, for providing the resources required to complete this work successfully.

REFERENCES

Adem A, Jossan SS, Oreland L. Tetrahydroaminoacridine inhibits human and rat brain monoamine oxidase. *Neurosci lett* (1989) 107(1–3):313–317.

Agarwal A, Agrawal H, Tiwari S, Jain S, Agrawal GP. Cationic ligand appended nanoconstructs: a prospective strategy for brain targeting. *Int J Pharm* (2011) 421(1):189–201

Agrawal M, Tripathi DK, Saraf S, Saraf S, Antimisiaris SG, Mourtas S, Hammarlund-Udenaes M, Alexander A. Recent advancements in liposomes targeting strategies to cross blood-brain barrier (BBB) for the treatment of Alzheimer's disease. *J Control Release* (2017) 260:61–77.

Ahmed S, Khan ST, Zargaham MK, Khan AU, Khan S, Hussain A, Uddin J, Khan A, Al-Harrasi A. Potential therapeutic natural products against Alzheimer's disease with Reference of Acetylcholinesterase. *Biomed Pharmacother* (2021) 139:111609.

Alkondon M, Pereira EF, Eisenberg HM, Albuquerque EX. Nicotinic receptor activation in human cerebral cortical interneurons: a mechanism for inhibition and disinhibition of neuronal networks. *J Neurosci* (2000) 20(1):66–75.

Alzheimer's Association. 2017 Alzheimer's disease facts and figures. *Alzheimer Dement* (2017) 13(4): 325–373.

Alzheimer's Association. Alzheimer's disease facts and figures. (2009) Available at: www.alz.org/alzheimers_disease_facts_figures.asp;report_alzfactsfigures; www.alz.org/news_and_events_2009_facts_figures.asp Accessed February 8, 2010.

Amin FU, Hoshiar AK, Do TD, Noh Y, Shah SA, Khan MS, Yoon J, Kim MO. Osmotin-loaded magnetic nanoparticles with electromagnetic guidance for the treatment of Alzheimer's disease. *Nanoscale* (2017) 9(30):10619–10632.

Arancibia R, Maturana C, Silva D, Tobar N, Tapia C, Salazar JC, Martínez J, Smith PC. Effects of chitosan particles in periodontal pathogens and gingival fibroblasts. *J Dent Res* (2013) 92(8):740–745.

Arora S, Layek B, Singh J. Design and validation of liposomal ApoE2 gene delivery system to evade blood–brain barrier for effective treatment of Alzheimer's disease. *Mol Pharm* (2020) 18(2):714–725.

Azria D, Blanquer S, Verdier JM, Belamie E. Nanoparticles as contrast agents for brain nuclear magnetic resonance imaging in Alzheimer's disease diagnosis. *J Mater Chem B* (2017) 5(35):7216–7237.

Bartus RT, Dean R3, Beer B, Lippa AS. The cholinergic hypothesis of geriatric memory dysfunction. *Science* (1982) 217(4558):408–414.

Benilova I, De Strooper B. An overlooked neurotoxic species in Alzheimer's disease. *Nat Neurosci* (2011) 14(8):949–950.

Bharadwaz A, Jayasuriya AC. Recent trends in the application of widely used natural and synthetic polymer nanocomposites in bone tissue regeneration. *Mater Sci Eng C* (2020) 110:110698.

Bhushan I, Kour M, Kour G, Gupta S, Sharma S, Yadav A. Alzheimer's disease: Causes & treatment–a review. *Ann Biotechnol* (2018) 1(1):1002.

Birks JS, Evans JG. Rivastigmine for Alzheimer's disease. *Cochrane Database Syst Rev* (2015) (4):CD001191.

Blennow K. CSF biomarkers for mild cognitive impairment. *J Intern Med* (2004) 256(3):224–234.

Blennow K, Zetterberg H. Biomarkers for Alzheimer's disease: current status and prospects for the future. *J Int Med* (2018) 284(6):643–663.

Bullock R. Efficacy and safety of memantine in moderate-to-severe Alzheimer disease: the evidence to date. *Alzheimer Dis Assoc Discord* (2006) 20(1):23–29.

Burns A, Rossor M, Hecker J, Gauthier S, Petit H, Möller HJ, Rogers SL, Friedhoff LT. The effects of Donepezil in Alzheimer's disease–results from a multinational triall. *Dement Geriatr Cogn Disord* (1999) 10(3):237–244.

Carter MD, Simms GA, Weaver DF. The development of new therapeutics for Alzheimer's disease. *Clin Pharm Therap* (2010) 88(4):475–486.

Cerejeira J, Lagarto L, Mukaetova-Ladinska E. Behavioral and psychological symptoms of dementia. *Front Neurol* (2012) 3:73.

Chauhan I, Yasir M, Verma M, Singh AP. Nanostructured lipid carriers: a groundbreaking approach for transdermal drug delivery. *Adv Pharm Bull* (2020) 10(2):150.

Chen L, Lin J, Yi J, Weng Q, Zhou Y, Han Z, Li C, Chen J, Zhang Q. A tyrosinase-induced fluorescence immunoassay for detection of tau protein using dopamine-functionalized CuInS 2/ZnS quantum dots. *Anal Bioanal Chem* (2019) 411(20):5277–5285.

Choi W, Yeom SY, Kim J, Jung S, Jung S, Shim TS, Kim SK, Kang JY, Lee SH, Cho IJ, Choi J. Hydrogel micropost-based qPCR for multiplex detection of miRNAs associated with Alzheimer's disease. *Biosens Bioelectron* (2018) 101:235–244.

Cooper JR, Bloom FE, Roth RH. *Amino acid transmitters: The biochemical basis of neuropharmacology*, Oxford University Press (1991): 133–189.

Cummings JL, Back C. The cholinergic hypothesis of neuropsychiatric symptoms in Alzheimer's disease. *Am J Geriatr Psychiatry* (1998) 6(2):64–78.

del Burgo LS, Hernández RM, Orive G, Pedraz JL. Nanotherapeutic approaches for brain cancer management. *Nanomed Nanotechnol Biol Med* (2014) 10(5):905–919.

Ding Y, Zhang C, Lan T, Qin Z, Zhang X, Wang W. Classification of Alzheimer's disease based on the combination of morphometric feature and texture feature. In 2015 *IEEE International Conference on Bioinformatics and Biomedicine (BIBM)* (2015) 409–412. IEEE.

Doolittle ND, Muldoon LL, Culp AY, Neuwelt EA. Delivery of chemotherapeutics across the blood–brain barrier: challenges and advances. *Adv Pharmacol* (2014) 71:203–243.

dos Santos Tramontin N, da Silva S, Arruda R, Ugioni KS, Canteiro PB, de Bem Silveira G, Mendes C, Silveira PC, Muller AP. Gold nanoparticles treatment reverses brain damage in Alzheimer's disease model. *Mol Neurobiol* (2020) 57(2):926–936.

Farlow M, Anand R, Messina Jr J, Hartman R, Veach J. A 52-week study of the efficacy of rivastigmine in patients with mild to moderately severe Alzheimer's disease. *Eur Neurol* (2000) 44(4):236–241.

Fernández T, Martínez-Serrano A, Cussó L, Desco M, Ramos-Gómez M. Functionalization and characterization of magnetic nanoparticles for the detection of ferritin accumulation in Alzheimer's disease. *ACS Chem Neurosci* (2018) 9(5):912–924.

Francis PT, Palmer AM, Snape M, Wilcock GK. The cholinergic hypothesis of Alzheimer's disease: a review of progress. *J Neurol Neurosurg Psychiatry* (1999) 66(2):137–147.

Ganesan P, Narayanasamy D. Lipid nanoparticles: different preparation techniques, characterization, hurdles, and strategies for the production of solid lipid nanoparticles and nanostructured lipid carriers for oral drug delivery. *Sustain Chem Pharm* (2017) 6:37–56.

Garbayo E, Estella-Hermoso de Mendoza A, Blanco-Prieto M. Diagnostic and therapeutic uses of nanomaterials in the brain. *Curr Med Chem* (2014) 21(36):4100–4131.

Gautam L, Shrivastava P, Yadav B, Jain A, Sharma R, Vyas S, Vyas SP. Multicompartment systems: a putative carrier for combined drug delivery and targeting. *Drug Discov* (2022) 27(4):1184–1195.

Gelain F, Luo Z, Zhang S. Self-assembling peptide EAK16 and RADA16 nanofiber scaffold hydrogel. *Chem Rev* (2020) 120(24):13434–13460.

Ghosh AK, Gemma S, Tang J. β-Secretase as a therapeutic target for Alzheimer's disease. *Neurotherap* (2008) 5(3):399–408.

Hamaguchi T, Ono K, Yamada M. Anti-amyloidogenic therapies: strategies for prevention and treatment of Alzheimer's disease. *Cell Mol Life Sci* (2006) 63(13):1538–1552.

Hammond J, Maher BA, Ahmed IA, Allsop D. Variation in the concentration and regional distribution of magnetic nanoparticles in human brains, with and without Alzheimer's disease, from the UK. *Sci Rep* (2021) 11(1):1–3.

Hampel H, Goernitz A, Buerger K. Advances in the development of biomarkers for Alzheimer's disease: from CSF total tau and Aβ1–42 proteins to phosphorylated tau protein. *Brain Res Bulletin* (2003) 61(3):243–253.

Hardy J. The amyloid hypothesis for Alzheimer's disease: a critical reappraisal. *J Neurochem* (2009) 110(4):1129–1134.

Hardy J, Allsop D. Amyloid deposition as the central event in the aetiology of Alzheimer's disease. *Trends Pharmacol Sci* (1991) 12:383–388.

Harman D. The aging process. *Proc Nat Acad Sci* (1981) 78(11):7124–7128.

Herrmann N, Chau SA, Kircanski I, Lanctot KL. Current and emerging drug treatment options for Alzheimer's disease. *Drugs* (2011) 71(15):2031–2065.

Hogan DB. Long-term efficacy and toxicity of cholinesterase inhibitors in the treatment of Alzheimer disease. *The Can J Psychiatry* (2014) 59(12):618–623.

Hou K, Zhao J, Wang H, Li B, Li K, Shi X, Wan K, Ai J, Lv J, Wang D, Huang Q. Chiral gold nanoparticles enantioselectively rescue memory deficits in a mouse model of Alzheimer's disease. *Nat Commun* (2020) 11(1):1–1.

Hussain G, Rasul A, Anwar H, Aziz N, Razzaq A, Wei W, Ali M, Li J, Li X. Role of plant derived alkaloids and their mechanism in neurodegenerative disorders. *Int J Biol Sci* (2018) 14(3):341.

Ignjatović N, Wu V, Ajduković Z, Mihajilov-Krstev T, Uskoković V, Uskoković D. Chitosan-PLGA polymer blends as coatings for hydroxyapatite nanoparticles and their effect on antimicrobial properties, osteoconductivity and regeneration of osseous tissues. *Mater Sci Eng C* (2016) 60:357–364.

Ishii M, Iadecola C. Metabolic and non-cognitive manifestations of Alzheimer's disease: the hypothalamus as both culprit and target of pathology. *Cell Metabolism* (2015) 22(5):761–776.

Jack Jr CR. Alzheimer disease: new concepts on its neurobiology and the clinical role imaging will play. *Radiol* (2012) 263(2):344–361.

Jagust W. Imaging the evolution and pathophysiology of Alzheimer disease. *Nat Rev Neurosci* (2018) 19(11):687–700.

Jembrek MJ, Slade N, Hof PR, Šimić G. The interactions of p53 with tau and Aß as potential therapeutic targets for Alzheimer's disease. *Prog Neurobiol* (2018) 168:104–127.

Jeon SG, Cha MY, Kim JI, Hwang TW, Kim KA, Kim TH, Song KC, Kim JJ, Moon M. Vitamin D-binding protein-loaded PLGA nanoparticles suppress Alzheimer's disease-related pathology in 5XFAD mice. *Nanomed Nanotechnol Biol Med* (2019) 17:297–307.

Jnaidi R, Almeida AJ, Gonçalves LM. Solid lipid nanoparticles and nanostructured lipid carriers as smart drug delivery systems in the treatment of Glioblastoma Multiforme. *Pharmaceut* (2020) 12(9):860.

Kang MS, Son KH, Lee TH, Chang SM, Han SW, Shin HK. Development of chemically signal amplified nano-biosensor mediated by poly-l-lysine. *J Nanosci Nanotechnol* (2019) 19(3):1786–1789.

Kincses ZT, Kiraly A, Veréb D, Vécsei L. Structural magnetic resonance imaging markers of Alzheimer's disease and its retranslation to rodent models. *J Alzheimer Dis* (2015) 47(2):277–290.

Koch HJ, Uyanik G, Fischer-Barnicol D. Memantine: a therapeutic approach in treating Alzheimer's and vascular dementia. *CNS Neurol Disord* (2005) 4(5):499–506.

Krstić M, Medarević Đ, Đuriš J, Ibrić S. Self-nanoemulsifying drug delivery systems (SNEDDS) and self-microemulsifying drug delivery systems (SMEDDS) as lipid nanocarriers for improving dissolution rate and bioavailability of poorly soluble drugs. In Alexandru MG and William A (ed) *Lipid nanocarriers for drug targeting* (2018) 1:473–508. William Andrew Publishing.

Kulkarni R, Girish KJ, Kumar A. Nootropic herbs (Medhya Rasayana) in Ayurveda: an update. *Pharmacogn Rev* (2012) 6(12):147.

LaFerla FM, Oddo S. Alzheimer's disease: Aβ, tau and synaptic dysfunction. *Trends Mol Med* (2005) 11(4):170–176.

Leung K, Chakraborty K, Saminathan A, Krishnan Y. A DNA nanomachine chemically resolves lysosomes in live cells. *Nat Nanotechnol* (2019) 14(2):176–183.

Levin ED, Simon BB. Nicotinic acetylcholine involvement in cognitive function in animals. *Psychopharmacol* (1998) 138(3):217–230.

Li F, Takechi H, Saito R, Ayaki T, Kokuryu A, Kuzuya A, Takahashi R. A comparative study: visual rating scores and the voxel-based specific regional analysis system for Alzheimer's disease on magnetic resonance imaging among subjects with Alzheimer's disease, mild cognitive impairment, and normal cognition. *Psychogeriatrics* (2019) 19(2):95–104.

Liang G, Shi H, Qi Y, Li J, Jing A, Liu Q, Feng W, Li G, Gao S. Specific anti-biofilm activity of Carbon Quantum Dots by destroying P. gingivalis biofilm related genes. *Int J Nanomed* (2020) 15:5473.

Maelicke A, Albuquerque EX. Allosteric modulation of nicotinic acetylcholine receptors as a treatment strategy for Alzheimer's disease. *Eur J Pharmacol* (2000) 393(1–3):165–170.

Maher BA. Airborne magnetite-and iron-rich pollution nanoparticles: Potential neurotoxicants and environmental risk factors for neurodegenerative disease, including Alzheimer's disease. *J Alzheimers Dis* (2019) 71(2):361–375.

Marathe SA, Dasgupta I, Gnanadhas DP, Chakravortty D. Multifaceted roles of curcumin: two sides of a coin. *Expert Opin Biol Ther* (2011) 11(11):1485–1499.

Markesbery WR. Oxidative stress hypothesis in Alzheimer's disease. *Free Radic Biol Med* (1997) 23(1):134–147.

Martin-Rapun R, De Matteis L, Ambrosone A, Garcia-Embid S, Gutierrez L, M de la Fuente J. Targeted nanoparticles for the treatment of Alzheimer's disease. *Curr Pharm Des* (2017) 23(13):1927–1952.

Masserini M. Nanoparticles for brain drug delivery. *Int Sch Res Notices* (2013) 2013:238428.

McKhann GM, Knopman DS, Chertkow H, Hyman BT, Jack Jr CR, Kawas CH, Klunk WE, Koroshetz WJ, Manly JJ, Mayeux R, Mohs RC. The diagnosis of dementia due to Alzheimer's disease: recommendations from the National Institute on Aging-Alzheimer's Association workgroups on diagnostic guidelines for Alzheimer's disease. *Alzheimer Dement* (2011) 7(3):263–269.

Meyer EM, Simpkins JW, Yamamoto J, Crews FT, eds. *Treatment of Dementias: a new generation of progress*. Springer Science + Business Media (2012).

Molinuevo JL, Arranz FJ. Impact of transdermal drug delivery on treatment adherence in patients with Alzheimer's disease. *Expert Rev Neurother* (2012) 12(1):31–37.

Morales-Zavala F, Arriagada H, Hassan N, Velasco C, Riveros A, Álvarez AR, Minniti AN, Rojas-Silva X, Muñoz LL, Vasquez R, Rodriguez K. Peptide multifunctionalized gold nanorods decrease toxicity of β-amyloid peptide in a Caenorhabditis elegans model of Alzheimer's disease. *Nanomed Nanotech Biol Med* (2017) 13(7):2341–2350.

Mukherjee S, Ray S, Thakur RS. Solid lipid nanoparticles: a modern formulation approach in drug delivery system. *Ind J Pharm Sci* (2009) 71(4):349.

Nakamura A, Kaneko N, Villemagne VL, Kato T, Doecke J, Doré V, Fowler C, Li QX, Martins R, Rowe C, Tomita T. High performance plasma amyloid-β biomarkers for Alzheimer's disease. *Nature* (2018) 554(7691):249–254.

Näslund J, Haroutunian V, Mohs R, Davis KL, Davies P, Greengard P, Buxbaum JD. Correlation between elevated levels of amyloid β-peptide in the brain and cognitive decline. *JAMA* (2000) 283(12):1571–1577.

Natarajan S, Shunmugiah KP, Kasi PD. Plants traditionally used in age-related brain disorders (dementia): an ethanopharmacological survey. *Pharm Biol* (2013) 51(4):492–523.

Nunomura A, Perry G, Aliev G, Hirai K, Takeda A, Balraj EK, Jones PK, Ghanbari H, Wataya T, Shimohama S, Chiba S. Oxidative damage is the earliest event in Alzheimer disease. *J Neuropathol Exp Neurol* (2001) 60(8):759–767.

Nunomura A, Perry G, Pappolla MA, Wade R, Hirai K, Chiba S, Smith MA. RNA oxidation is a prominent feature of vulnerable neurons in Alzheimer's disease. *J Neurosci* (1999) 19(6):1959–1964.

Ohno M, Cole SL, Yasvoina M, Zhao J, Citron M, Berry R, Disterhoft JF, Vassar R. BACE1 gene deletion prevents neuron loss and memory deficits in 5XFAD APP/PS1 transgenic mice. *Neurobiol Dis* (2007) 26(1):134–145.

Ordóñez-Gutiérrez L, Wandosell F. Nanoliposomes as a therapeutic tool for Alzheimer's disease. *Front Synaptic Neurosci* (2020) 12:20.

Pansieri J, Gerstenmayer M, Lux F, Mériaux S, Tillement O, Forge V, Larrat B, Marquette C. Magnetic nanoparticles applications for amyloidosis study and detection: a review. *Nanomaterials* (2018) 8(9):740.

Pardridge WM. Molecular biology of the blood-brain barrier. *Mol Biotech* (2005) 30(1):57–69.

Patra JK, Das G, Fraceto LF, Campos EV, del Pilar Rodriguez-Torres M, Acosta-Torres LS, Diaz-Torres LA, Grillo R, Swamy MK, Sharma S, Habtemariam S. Nano based drug delivery systems: recent developments and future prospects. *J Nanobiotech* (2018) 16(1):1–33.

Perry EK, Smith CJ, Court JA, Bonham JR, Rodway M, Atack JR. Interaction of 9-amino-1, 2, 3, 4-tetrahydroaminoacridine (THA) with human cortical nicotinic and muscarinic receptor binding in vitro. *Neurosci Lett* (1988) 91(2):211–216.

Polanco JC, Li C, Bodea LG, Martinez-Marmol R, Meunier FA, Götz J. Amyloid-β and tau complexity—towards improved biomarkers and targeted therapies. *Nat Rev Neurol* (2018) 14(1):22–39.

Praticò D. Peripheral biomarkers of oxidative damage in Alzheimer's disease: the road ahead. *Neurobiol Aging* (2005) 26(5):581–583.

Prince M, Ali GC, Guerchet M, Yu-Tzu W, Prina M. The global prevalence of dementia. *World Alzheimer Rep* (2009) 25–45.

Purves D, Augustine GJ, Fitzpatrick D, Hall WC, LaMantia AS, McNamara JO, White L (2014) De Boeck (Ed.), *Neuroscience*, Sinauer 2008 (pp. 15–16).

Qiu T, Liu Q, Chen YX, Zhao YF, Li YM. Aβ42 and Aβ40: similarities and differences. *J Pept Sci* (2015) 21(7):522–529.

Quan L, Gu J, Lin W, Wei Y, Lin Y, Liu L, Ding H, Pan C, Xie Z, Wu T. A BODIPY biosensor to detect and drive self-assembly of diphenylalanine. *Chem Commun* (2019) 55(59):8564–8566.

Raskind MA, Peskind ER, Wessel T, Yuan W, Galantamine USA-Study Group. Galantamine in AD: a 6-month randomized, placebo-controlled trial with a 6-month extension. *Neurol* (2000) 54(12):2261–2268.

Ravi G, Gupta NV. Development of Solid lipid Nanoparticles of Rivastigmine Tartrate by using full factorial design for the treatment of Alzheimer's disease. *J Pharm Sci Res* (2017) 9(12):2447–2452.

Ricci G. Social aspects of dementia prevention from a worldwide to national perspective: a review on the international situation and the example of Italy. *Behav Neurol* (2019) 8:2019.

Rogers SL, Doody RS, Pratt RD, Ieni JR. Long-term efficacy and safety of donepezil in the treatment of Alzheimer's disease: final analysis of a US multicentre open-label study. *Eur Neuropsychopharmacol* (2000) 10(3):195–203.

Rösler M, Bayer T, Anand R, Cicin-Sain A, Gauthier S, Agid Y, Dal-Bianco P, Stähelin HB, Hartman R, Gharabawi M. Efficacy and safety of rivastigmine in patients with Alzheimer's disease: international randomised controlled trial Commentary: Another piece of the Alzheimer's jigsaw. *BMJ* (1999) 318(7184):633–640.

Rousseau V, Denizot B, Pouliquen D, Jallet P, Le Jeune JJ. Investigation of blood-brain barrier permeability to magnetite-dextran nanoparticles (MD3) after osmotic disruption in rats. *Magn Reson Mater Phys Biol Med* (1997) 5(3):213–222.

Sabbagh MN, Farlow MR, Relkin N, Beach TG. Do cholinergic therapies have disease-modifying effects in Alzheimer's disease? *Alzheimer's Dement* (2006) 2(2):118–125.

Sabbagh MN, Richardson S, Relkin N. Disease-modifying approaches to Alzheimer's disease: challenges and opportunities—Lessons from donepezil therapy. *Alzheimer's Dement* (2008) 4(1):109–118.

Salehi B, Stojanović-Radić Z, Matejić J, Sharifi-Rad M, Kumar NV, Martins N, Sharifi-Rad J. The therapeutic potential of curcumin: a review of clinical trials. *Eur J Med Chem* (2019) 163:527–545.

Sánchez-López E, Ettcheto M, Egea MA, Espina M, Cano A, Calpena AC, Camins A, Carmona N, Silva AM, Souto EB, García ML. Memantine loaded PLGA PEGylated nanoparticles for Alzheimer's disease: in vitro and in vivo characterization. *J Nanobiotech* (2018) 16(1):1–6.

Sankar S, O'Neill K, Bagot D'Arc M, Rebeca F, Buffier M, Aleksi E, Fan M, Matsuda N, Gil ES, Spirio L. Clinical use of the self-assembling peptide RADA16: a review of current and future trends in biomedicine. *Front Bioeng Biotechnol* (2021) 9:465.

Sayre LM, Moreira PI, Smith MA, Perry G. Metal ions and oxidative protein modification in neurological disease. *Ann Ist Super Sanita* (2005) 41(2):143–164.

Sayre LM, Perry G, Smith MA. In situ methods for detection and localization of markers of oxidative stress: application in neurodegenerative disorders. *Meth Enzymol* (1999) 309:133–152.

Selkoe DJ. Cell biology of the amyloid beta-protein precursor and the mechanism of Alzheimer's disease. *Annu Rev Cell Biol* (1994) 10(1): 373–403.

Selkoe DJ. Translating cell biology into therapeutic advances in Alzheimer's disease. *Nature* (1999) 399(6738): 23–31.

Sharma B, Pervushin K. Magnetic nanoparticles as in vivo tracers for Alzheimer's disease. *Magnetochemist* (2020) 6(1):13.

Sikorska K, Grądzka I, Wasyk I, Brzóska K, Stępkowski TM, Czerwińska M, Kruszewski MK. The impact of ag nanoparticles and cdte quantum dots on expression and function of receptors involved in amyloid-β uptake by bv-2 microglial cells. *Materials* (2020) 13(14):3227.

Simonazzi A, Cid AG, Villegas M, Romero AI, Palma SD, Bermúdez JM. Nanotechnology applications in drug controlled release. In Alexandru MG (ed.) *Drug targeting and stimuli sensitive drug delivery systems.* William Andrew Publishing (2018) 81–116.

Son S, Jang J, Youn H, Lee S, Lee D, Lee YS, Jeong JM, Kim WJ, Lee DS. A brain-targeted rabies virus glycoprotein-disulfide linked PEI nanocarrier for delivery of neurogenic microRNA. *Biomaterials* (2011) 32(21):4968–4975.

Song C, Deng P, Que L. Rapid multiplexed detection of beta-amyloid and total-tau as biomarkers for Alzheimer's disease in cerebrospinal fluid. *Nanomed Nanotechnol Biol Med* (2018) 14(6):1845–1852.

Sonkusare SK, Kaul CL, Ramarao P. Dementia of Alzheimer's disease and other neurodegenerative disorders—memantine, a new hope. *Pharmacol Res* (2005) 51(1):1–7.

Spuch C, Navarro C. Liposomes for targeted delivery of active agents against neurodegenerative diseases (Alzheimer's disease and Parkinson's disease). *J Drug Deliv* (2011) 2011:469679.

Stenström A, Oreland L, Hardy J, Wester P, Winblad B. The uptake of serotonin and dopamine by homogenates of frozen rat and human brain tissue. *Neurochem Res* (1985) 10(5):591–599.

Struble RG, Cork LC, Whitehouse PJ, Price DL. Cholinergic innervation in neuritic plaques. *Science* (1982) 216(4544):413–435.

Sunderland T, Linker G, Mirza N, Putnam KT, Friedman DL, Kimmel LH, Bergeson J, Manetti GJ, Zimmermann M, Tang B, Bartko JJ. Decreased β-amyloid1-42 and increased tau levels in cerebrospinal fluid of patients with Alzheimer disease. *JAMA* (2003) 289(16):2094–2103.

Taghavi S, Ramezani M, Alibolandi M, Abnous K, Taghdisi SM. Chitosan-modified PLGA nanoparticles tagged with 5TR1 aptamer for in vivo tumor-targeted drug delivery. *Cancer Lett* (2017) 400:1–8.

Takeda S, Sato N, Morishita R. Systemic inflammation, blood-brain barrier vulnerability and cognitive/non-cognitive symptoms in Alzheimer disease: relevance to pathogenesis and therapy. *Front Aging Neurosci* (2014) 6:171.

Tapeinos C, Battaglini M, Ciofani G. Advances in the design of solid lipid nanoparticles and nanostructured lipid carriers for targeting brain diseases. *J Control Release* (2017) 264:306–332.

Tariot PN, Solomon PR, Morris JC, Kershaw P, Lilienfeld S, Ding C, Galantamine USA-Study Group. A 5-month, randomized, placebo-controlled trial of galantamine in AD. *Neurol* (2000) 54(12):2269–2276.

Thompson LA, Bronson JJ, Zusi FC. Progress in the discovery of BACE inhibitors. *Curr Pharm Des* (2005) 11(26):3383–3404.

Vassar R, Kandalepas PC. The β-secretase enzyme BACE1 as a therapeutic target for Alzheimer's disease. *Alzheimer's Res Ther* (2011) 3(3):1–6.

Viola KL, Sbarboro J, Sureka R, De M, Bicca MA, Wang J, Vasavada S, Satpathy S, Wu S, Joshi H, Velasco PT. Towards non-invasive diagnostic imaging of early-stage Alzheimer's disease. *Nat Nanotech* (2015) 10(1):91–98.

Vukajlovic D, Parker J, Bretcanu O, Novakovic K. Chitosan based polymer/bioglass composites for tissue engineering applications. *Mater Sci Eng C* (2019) 96:955–967.

Wang C, Wang X, Chan HN, Liu G, Wang Z, Li HW, Wong MS. Amyloid- β oligomer-targeted gadolinium-based NIR/MR dual-modal theranostic nanoprobe for Alzheimer's disease. *Adv Funct Mater* (2020) 30(16):1909529.

Ward M. Biomarkers for brain disorders. *Clin Pract* (2010) 7(4):3.

Waxenbaum JA, Reddy V, Bordoni B. *Anatomy, head and neck, cervical nerves.* Eu PMC. StatPearls Publishing (2019).

Wilkinson DG, Francis PT, Schwam E, Payne-Parrish J. Cholinesterase inhibitors used in the treatment of Alzheimer's disease. *Drugs Aging* (2004) 21(7):453–478.

Wojtunik-Kulesza KA, Oniszczuk A, Oniszczuk T, Waksmundzka-Hajnos M. The influence of common free radicals and antioxidants on development of Alzheimer's Disease. *Biomed Pharmacother* (2016) 78:39–49.

Yang F, Lim GP, Begum AN, Ubeda OJ, Simmons MR, Ambegaokar SS, Chen PP, Kayed R, Glabe CG, Frautschy SA, Cole GM. Curcumin inhibits formation of amyloid β oligomers and fibrils, binds plaques, and reduces amyloid in vivo. *J Biol Chem* (2005) 280(7):5892–5901.

Yasumoto T, Takamura Y, Tsuji M, Watanabe-Nakayama T, Imamura K, Inoue H, Nakamura S, Inoue T, Kimura A, Yano S, Nishijo H. High molecular weight amyloid β1-42 oligomers induce neurotoxicity via plasma membrane damage. *FASEB J* (2019) 33(8):9220–9234.

Yien L, Zin NM, Sarwar A, Katas H. Antifungal activity of chitosan nanoparticles and correlation with their physical properties. *Inter J Biomater* (2012) 2012:632698.

Yilmaz U. Alzheimer-Demenz. *Radiologe* (2015) 55(5):386–388.

Yin H, Du Y, Zhang J. Low molecular weight and oligomeric chitosans and their bioactivities. *Curr Top Med Chem* (2009) 9(16):1546–1559.

Yu Y, Sun X, Tang D, Li C, Zhang L, Nie D, Yin X, Shi G. Gelsolin bound β-amyloid peptides (1–40/1–42): electrochemical evaluation of levels of soluble peptide associated with Alzheimer's disease. *Biosens Bioelectron* (2015) 68:115–121.

Zhang N, Yan F, Liang X, Wu M, Shen Y, Chen M, Xu Y, Zou G, Jiang P, Tang C, Zheng H. Localized delivery of curcumin into brain with polysorbate 80-modified cerasomes by ultrasound-targeted microbubble destruction for improved Parkinson's disease therapy. *Theranostics* (2018) 8(8):2264.

Zhang XP, Sun JG, Yao J, Shan K, Liu BH, Yao MD, Ge HM, Jiang Q, Zhao C, Yan B. Effect of nanoencapsulation using poly (lactide-co-glycolide)(PLGA) on anti-angiogenic activity of bevacizumab for ocular angiogenesis therapy. *Biomed Pharmacother* (2018) 107:1056–1063.

Zhang XY, Yang ZL, Lu GM, Yang GF, Zhang LJ. PET/MR imaging: new frontier in Alzheimer's disease and other dementias. *Front Mol Neurosci* (2017) 10:343.

Zilony-Hanin N, Rosenberg M, Richman M, Yehuda R, Schori H, Motiei M, Rahimipour S, Groisman A, Segal E, Shefi O. Neuroprotective effect of nerve growth factor loaded in porous silicon nanostructures in an Alzheimer's disease model and potential delivery to the brain. *Small* (2019) 15(45):1904203.

Index